DIFFUSION IN SOLIDS
Recent Developments

MATERIALS SCIENCE AND TECHNOLOGY

EDITORS

ALLEN M. ALPER
*GTE Sylvania Inc.
Precision Materials Group
Chemical & Metallurgical
Division
Towanda, Pennsylvania*

JOHN L. MARGRAVE
*Department of Chemistry
Rice University
Houston, Texas*

A. S. NOWICK
*Henry Krumb School
of Mines
Columbia University
New York, New York*

A. S. Nowick and B. S. Berry, ANELASTIC RELAXATION IN CRYSTALLINE SOLIDS, 1972

E. A. Nesbitt and J. H. Wernick, RARE EARTH PERMANENT MAGNETS, 1973

W. E. Wallace, RARE EARTH INTERMETALLICS, 1973

J. C. Phillips, BONDS AND BANDS IN SEMICONDUCTORS, 1973

H. Schmalzried, SOLID STATE REACTIONS, 1974

J. H. Richardson and R. V. Peterson (editors), SYSTEMATIC MATERIALS ANALYSIS, VOLUMES I, II, AND III, 1974

A. J. Freeman and J. B. Darby, Jr. (editors), THE ACTINIDES: ELECTRONIC STRUCTURE AND RELATED PROPERTIES, VOLUMES I AND II, 1974

A. S. Nowick and J. J. Burton (editors), DIFFUSION IN SOLIDS: RECENT DEVELOPMENTS, 1975

J. W. Matthews (editor), EPITAXIAL GROWTH, PARTS A AND B, 1975

In preparation

J. M. Blakely (editor), SURFACE PHYSICS OF MATERIALS, VOLUMES I AND II

DIFFUSION IN SOLIDS

RECENT DEVELOPMENTS

Edited by A. S. NOWICK

Henry Krumb School of Mines
Columbia University
New York, New York

J. J. BURTON

ESSO Research Center
Linden, New Jersey

ACADEMIC PRESS   New York San Francisco London 1975
A Subsidiary of Harcourt Brace Jovanovich, Publishers

COPYRIGHT © 1975, BY ACADEMIC PRESS, INC.
ALL RIGHTS RESERVED.
NO PART OF THIS PUBLICATION MAY BE REPRODUCED OR
TRANSMITTED IN ANY FORM OR BY ANY MEANS, ELECTRONIC
OR MECHANICAL, INCLUDING PHOTOCOPY, RECORDING, OR ANY
INFORMATION STORAGE AND RETRIEVAL SYSTEM, WITHOUT
PERMISSION IN WRITING FROM THE PUBLISHER.

ACADEMIC PRESS, INC.
111 Fifth Avenue, New York, New York 10003

United Kingdom Edition published by
ACADEMIC PRESS, INC. (LONDON) LTD.
24/28 Oval Road, London NW1

Library of Congress Cataloging in Publication Data

Nowick, A   S
    Diffusion in solids.

    (Materials science and technology series)
    Includes bibliographies.
    1. Diffusion.   2. Solids.   I. Burton, James
Joseph, Date   joint author.   II. Title.
QC176.8.D5N68   531'.3   74-5704
ISBN 0-12-522660-8

PRINTED IN THE UNITED STATES OF AMERICA

Contents

LIST OF CONTRIBUTORS ix
PREFACE xi

1. Classical and Quantum Theory of Diffusion in Solids
Wilbur M. Franklin

I.	Introduction	1
II.	Jump Frequency	5
III.	Lattice Vibration Theory	14
IV.	Defect Modes	26
V.	Temperature Dependence—Classical Case	50
VI.	Mass Dependence—Classical Case	56
VII.	Quantum and Anharmonic Effects	58
VIII.	Lattice Vibration Theory and Diffusion Experiments	69
	References	70

2. Exact Defect Calculations in Model Substances
Charles H. Bennett

I.	Introduction	74
II.	The Molecular Dynamics, Monte Carlo, and Molecular Statics Methods	76
III.	Discussion of Molecular Dynamics and Monte Carlo Results on Point Defects at Thermal Equilibrium	91
Appendix A.	Monte Carlo Acceptance Ratio Method for Free Energy Differences	104
Appendix B.	Lennard-Jones Vacancy Jump Calculations	105
Appendix C.	Molecular Dynamics Calculation of the Isotope Effect	107
	References	112

3. Isotope Effects in Diffusion
N. L. Peterson

I.	Preface	116
II.	Introduction	117

III.	Self-Diffusion in Pure Metals	124
IV.	Diffusion in Dilute Alloys	138
V.	Diffusion in Concentrated Alloys	146
VI.	Diffusion in Alkali and Silver Halide Crystals	149
VII.	Diffusion in Transition Metal Oxides	157
VIII.	Correlation Effects in Grain Boundary Diffusion	163
	References	167

4. Fast Diffusion in Metals

W. K. Warburton and D. Turnbull

I.	Introduction	172
II.	Experience on Fast Diffusion	174
III.	Corroboration of Existence of Interstitial-Type Defects	196
IV.	Fast Diffusion Mechanisms	206
V.	Interpretation of Fast Diffusion Behavior of Particular Systems	221
	References	226

5. Hydrogen Diffusion in Metals

J. Völkl and G. Alefeld

I.	Introduction	232
II.	Site Location, Phase Diagrams, and Solubility	233
III.	Experimental Methods	240
IV.	Values for the Diffusion Coefficients at Small Concentrations (α Phases)	246
V.	High Hydrogen Concentrations	258
VI.	Isotope Dependence	260
VII.	Deviations of the Diffusion Coefficient from the Arrhenius Relation	264
VIII.	Dependence on Alloying	266
IX.	Influence of Traps	267
X.	Influence of Structure	268
XI.	Conclusions	269
	References	295

6. Electromigration in Metals

H. B. Huntington

I.	Introduction	303
II.	Formal Background	306
III.	Techniques for Measurement	314
IV.	The Nature of the Driving Force	321
V.	Interstitial Electromigration	325
VI.	Monovalent Metals and Their Alloys	328
VII.	Divalent Metals—Anisotropy in Single Crystals	336
VIII.	Electromigration in Trivalent Metals	338
IX.	Electromigration in Metals of More Complex Electronic Structure	340
X.	Electromigration in Thin Films: Problem for Integrated Circuitry	345
	References	349

7. Atom Currents Generated by Vacancy Winds

T. R. Anthony

	List of Symbols	353
I.	Introduction	355
II.	Theory	355
III.	Measurement of the Vacancy Wind and the Wind-Generated Solute Current	369
IV.	Solute Segregation around Vacancy Sinks	374
	References	378

8. Diffusion in Alkali Halides

W. J. Fredericks

I.	Introduction	381
II.	Defects and Their Interactions	382
III.	Theory of Diffusion in Ionic Crystals	388
IV.	Experimental Methods	398
V.	The Experimental Situation and Numerical Results	408
VI.	Conclusion	438
	References	439

9. Very Rapid Ionic Transport in Solids

Robert A. Huggins

I.	Introduction	445
II.	Special Characteristics of Fast Ionic Conductors	446
III.	Materials Which Exhibit Fast Ionic Motion	452
IV.	Theoretical Approaches to Fast Ion Conduction	475
V.	Outlook for the Future	482
	References	483

INDEX   487

List of Contributors

Numbers in parentheses indicate the pages on which the authors' contributions begin.

G. ALEFELD (231), Physik-Department der Technischen Universität München, 8046 Garching, West Germany

T. R. ANTHONY (353), Metallurgy and Ceramics Laboratory, Research and Development Center, General Electric Company, Schnectady, New York

CHARLES H. BENNETT (73), IBM Thomas J. Watson Research Center, Yorktown Heights, New York

WILBUR M. FRANKLIN (1), Physics Department, Kent State University, Kent, Ohio

W. J. FREDERICKS (381), Chemistry Department, Oregon State University, Corvallis, Oregon

ROBERT A. HUGGINS (445), Center for Materials Research, Stanford University, Stanford, California

H. B. HUNTINGTON (303), Department of Physics, Rensselaer Polytechnic Institute, Troy, New York

N. L. PETERSON (115), Materials Science Division, Argonne National Laboratory, Argonne, Illinois

D. TURNBULL (171), Division of Engineering and Applied Physics, Harvard University, Cambridge, Massachusetts

J. VÖLKL (231), Physik-Department der Technischen Universität München, 8046 Garching, West Germany

W. K. WARBURTON (171), Division of Engineering and Applied Physics, Harvard University, Cambridge, Massachusetts

Preface

The study of diffusion in crystalline solids has become an increasingly active field since 1945 when a wide variety of radioactive isotopes of the elements began to be available. Because of this increased activity, our concepts of diffusion mechanisms have developed rapidly in the last 25 years or so; simultaneously, our detailed knowledge of the point defects which are so often responsible for the elementary diffusion step increased. There have been many fine review articles and several textbooks on diffusion, which provide an adequate systematic introduction to the subject. Such is not the objective of the present volume. Rather, it was desired to bring up to date several of the most active areas of study of the past decade by presenting reviews written by foremost authorities in these areas. Often the present status of the subject will be found to be still in a state of flux, but it is hoped that each of these critical reviews will clarify the field for the reader and better help him to understand the rapidly evolving literature.

The first two chapters deal with aspects of the theory of diffusion. The elementary diffusion jump has continued to fascinate the theoretician as one of the simplest examples of a chemical rate process, and yet it has turned out to be an elusive one. In Chapter 1, the problem is attacked from several different viewpoints, for example, harmonic and anharmonic theories, and the use of quantum mechanical concepts. In Chapter 2, a relatively new approach is presented; detailed and exact information about hypothetical materials is calculated using high-speed computers.

Almost all the experimental chapters deal with new tools or approaches which supplement conventional diffusion measurements. Examples are isotope effects, vacancy wind effects, ionic conductivity, and electromigration. Some of the topics have become especially active in the last few years because of their great technological potential, for example, electromigration and very rapid ionic transport. Others, such as fast diffusion in metals, and the unique aspects of diffusion of the lightest element, hydrogen, are topics of enormous intrinsic interest, although not without technological importance too. Finally, Chapter 8 shows how diffusion in alkali halides, long thought to be the simplest of systems, is really extremely complex, to an extent that is only just beginning to be understood.

While these topics are not the only active ones in the field of diffusion, in order to limit the size of this book, it was necessary to restrict our coverage. Nevertheless, we do feel that those areas presented here represent a set of topics of prime importance at this stage in the development of the field of diffusion.

DIFFUSION IN SOLIDS
Recent Developments

Classical and Quantum Theory of Diffusion in Solids

WILBUR M. FRANKLIN

PHYSICS DEPARTMENT
KENT STATE UNIVERSITY
KENT, OHIO

I. Introduction	1
II. Jump Frequency	5
A. Introduction	5
B. Classical Harmonic Approximation	6
C. Model with Anharmonicity and Deformed Force Constants	10
III. Lattice Vibration Theory	14
A. Introduction	14
B. Harmonic Theory	17
C. Anharmonic Theory	24
IV. Defect Modes	26
A. A descriptive Model for Diffusion in Terms of Localized, Resonance, and Band Modes	27
B. Inhomogeneity Matrix	33
C. Linear Chain with Impurity Adjacent to a Vacancy	36
D. Green's Function in Terms of Defect Lattice Eigenfunctions	40
E. Displacement and Momentum Correlation Functions	41
F. Internal Degrees of Freedom	42
G. Defect Calculations Related to Diffusion	44
V. Temperature Dependence—Classical Case	50
VI. Mass Dependence—Classical Case	56
VII. Quantum and Anharmonic Effects	58
VIII. Lattice Vibration Theory and Diffusion Experiments	69
References	70

I. Introduction

The seemingly simple process of diffusion in solids is, when analyzed in detail, one of the most complex theoretical problems in solid state physics and materials science. As a simple example, consider the Arrhenius plot for the diffusivity, $D = D_0 \exp(-\Delta H/kT)$, which shows an exponential temperature dependence and an activation enthalpy ΔH and a preexponential diffusion constant D_0, which are constant with respect to the temperature. This empirical result is surprising in view of the anharmonic terms, including many-body forces, that come into play with the very large saddle-point strains inherent in the migration event. In the classical work of Wert (1950) on the diffusion of carbon in α iron (which is reproduced in Fig. 1), he said, after finding that ΔH was constant over a large temperature range, "This fact is perhaps the most surprising result of the analysis ...."[‡]

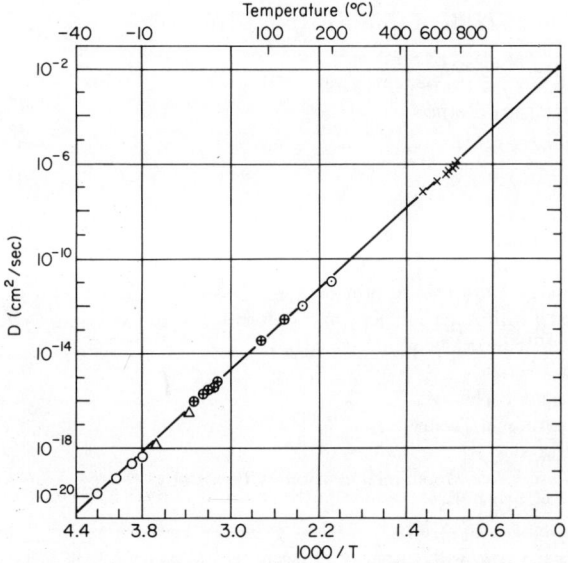

Fig. 1. The diffusion coefficient for C in α iron, taken from Wert (1950). This plot shows that Arrhenius-type behavior is followed by D vs T^{-1} in this system for more than 14 orders of magnitude.

There are, in addition to the linearity of the Arrhenius plot, many other features of diffusion, both overt and obstruse, which challenge theoretical understanding or explanation. The contributions to the understanding of migration events by theorists such as Wigner (1933), Hirschfelder and Wigner (1939), Bardeen and Herring (1952), Eyring (1935), Montroll and Potts (1955), and Bak and Prigogine (1959) indicate the complexity to the problems involved as well as the intrigue of attempting to formulate viable

---

[‡] While more recent data for the diffusion of C in Fe show a definite departure from the Arrhenius relation, other systems (e.g. N in Fe) do obey the relation very well (Lord and Beshers, 1966).

theoretical paradigms and constructs for a difficult problem. However, despite the work of many capable theorists, a good attempt has not yet been made to find the localized and resonance modes of the equilibrium configuration‡ (EC) or the saddle-point configuration (SPC) of even the most commonly studied diffusion defect—an impurity adjacent to a vacancy—in a three-dimensional model. There are numerous other examples of complex problems in atomic migration theory which have not yet been solved or are only partially solved. Some of these will be stated or inferred in this chapter which deals, principally, with the theory of the basic migration event in the solid state.

The thermodynamic description of rate processes was developed initially by Eyring (1935) and was followed by Wert (1950), who corrected the description of the activated state to include the proper number of degrees of freedom. Wert and Zener (1949) and Zener (1951) have given a theoretical basis for the preexponential part of the diffusivity, D_0. Basic background material and contact with experimental aspects of diffusion can be found in the articles by LeClaire (1949, 1953) and Lazarus (1960) and in the Pocono Manor Symposium (1952). The phonon theory of point defects, which considers the $3N$ frequencies of the N-body system, was formulated initially by Huntington *et al.* (1955) who stated that the theory could be extended to diffusion. Application to the migration process followed in the work by Vineyard (1957) and, in a different approach, by Rice (1958), Manley and Rice (1960), and Lawson *et al.* (1960). Vineyard, in addition to applying the many-body theory to the jump frequency, introduced the mass dependence, noting that the simple $m^{-1/2}$ mass dependence expected for free particle motion or for the motion of an atom in a simple harmonic lattice does not hold exactly in a solid. Later, the essential equivalence of the harmonic theories of Vineyard (1957) and Rice (1958) was shown by Franklin (1967) and by Glyde (1967). Rice and Frisch (1960) developed an anharmonic model of the migration process in one dimension. An alternative development of the theory of migration has been provided by Flynn (1968, 1972) and by Feit (1971, 1972).

As the elementary jump process was placed on a reasonably firm theoretical basis utilizing reaction rate theory, more attention was paid to fundamental and complex questions such as the role played by correlations, anharmonicity, many-body effects in the sense of more than two-body interactions, quantum effects, and refinements of the isotope effect. The first work on anharmonic effects in rate processes, which was done by Slater

‡ The abbreviations EC (equilibrium configuration) and SPC (saddle-point configuration) are used throughout this article. The SPC is often considered to be an equilibrium state of the system but we used SPC, nontheless, simply to denote the different atomic arrangement from that of the EC.

(1959), was followed by that of Rice and Frisch (1960), Franklin (1967, 1969), and Flynn (1971b). The thermalization process of an activated species, which involves anharmonic effects, was studied by Geszti (1967) based on the original work on anharmonicity in solids by Bak and Prigogine (1959). Wigner (1933) and Hirschfelder and Wigner (1939) studied the quantum effects of tunneling on small mass diffusion. Quantum statistics were studied by LeClaire (1966) and Franklin (1969) and tunneling effects were included, in addition, by Weiner and Partom (1969, 1970), Flynn (1971a) and Ebisuzaki *et al.* (1967a,b). The zero-point energy, another observable quantum effect which is a part of quantum statistics, was discussed by Flynn and Stoneham (1970). The zero-point energy in quantum crystals was discussed by Varma (1971). The many-body terms of the generalized anharmonic terms (three- and higher body interactions) were developed by Burton (1969). The effects of correlations, studied initially by Bardeen and Herring (1952) and Compaan and Haven (1956, 1958), were developed further (Manning, 1968). The effects of di-, tri-, and multivacancy diffusion mechanisms on the Arrhenius plot were discussed by Seeger and Mehrer (1968, 1969a,b, 1970), Burton (1970), Burton and Jura (1967), Burton and Lazarus (1970), and Rothman and Peterson (1969). The effects of local modes were studied first by Montroll and Potts (1955) and then by Schottky (1965), with reference to the Soret effect.

Most of the topics cited above have become specialized areas of study so that significant contributions to the theory of diffusion now often require a detailed theoretical knowledge of contemporary statistical or quantum mechanics, lattice vibration theory with defects, pseudopotential theory, etc. Consequently, it is too imposing a task to treat all the significant areas of diffusion theory adequately and critically in a short article. Therefore, certain areas, which are outlined below, have been chosen for emphasis since the remaining areas cited have been treated elsewhere or, on the other hand, do not fit comfortably within the framework of the topics selected here, which are couched in the theory of lattice vibrations.

In Section II, the Vineyard theory is developed for the atomic jump frequency, utilizing the harmonic approximation for the associated lattice vibrations. Then, anharmonic terms and expansions of deformed saddle-point force constants are used in one-dimensional models to obtain the saddle-point frequencies and anharmonic corrections using known quantities. Section III begins with a brief introduction giving some of the limitations and domain of applicability of phonon theory. Then, the harmonic theory is presented, followed by the addition of anharmonic terms. Lattice vibration theory with the presence of atomic defects is utilized from the outset; the development is given in second quantized form and the impingement on diffusion theory is assessed from a basic standpoint throughout. The most extensive presentation is given in Section IV. It includes an introductory descriptive model of diffusion in terms of localized, resonance, and band modes, followed by a mathematical development of the inhomogeneity matrix, the localized modes of a linear chain, and Green's function in terms of defect lattice eigenfunctions. A discussion and presentation of displacement and momentum correlation functions and internal degrees of freedom is given and related to diffusion. The section concludes with the results of a few calculations of defect modes taken from the literature and related to atomic migration. In Section V, the classical theory is presented for the high temperature regime. The mass dependence in the classical case is presented in Section VI and quantum effects in Section VII. In Section VII, the anharmonic quantum development of the statistical dynamical theory of diffusion is given, and both the classical and quantum cases are considered with emphasis on the latter. A simple classical derivation of the jump frequency is given in Section V, whereas the complete quantum derivation is outlined in Section VII. Experimental applications are discussed in Section VIII.

II. Jump Frequency

A. Introduction

The jump frequency in a many-body system can be derived in a very simple fashion under the assumption of equilibrium states in both the EC and the SPC. The idea is that the entropy factor together with the frequency in the thermodynamic approach can be expressed in terms of the phonon frequencies of the system in its EC ($3N$-dimensional) and its constrained SPC [($3N - 1$)-dimensional]. Consequently, the theory rests on the assumption that phonon frequencies can be defined which are valid representations of the state of the system in these two configurations. In this regard, no problem exists for the EC since lattice deformations and anharmonic effects are usually small for this configuration and the phonon lifetime is long enough for an elementary excitation to be defined. However, distortions and anharmonicity are normally large in the SPC so that the theory of defect modes must be invoked to obtain the lattice vibration frequencies. Anharmonic effects are an additional complication in the SPC. With strong anharmonicity, the phonon lifetimes may be short. Consequently, the theory may be difficult to describe in terms of simple elementary excitations. In this case, anharmonic corrections to the theory are significant and can provide the necessary modifications for small perturbations.

If anharmonic theory should prove inadequate, then some other means of treating large perturbations, such as a variational technique, must be sought.

B. CLASSICAL HARMONIC APPROXIMATION

Vineyard (1957) has defined the jump rate Γ in terms of a number per unit time of saddle-point crossing events R divided by the configurational partition function Q, so that

$$\Gamma = \tau^{-1} = R/Q \tag{1}$$

Our interest in describing the diffusion event lies in many-body configuration space. Figure 2 shows, in schematic fashion, a $3N$-dimensional

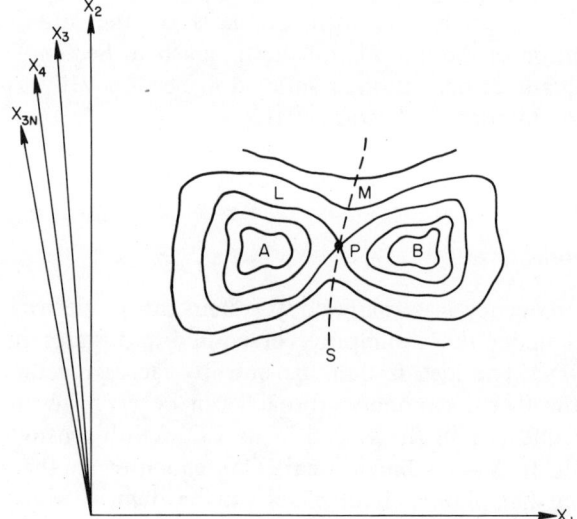

Fig. 2. Configuration space of $3N$ dimensions showing, schematically, hypersurfaces of constant potential energy (—) and an imaginary hypersurface S ($\cdots$). P and M denote the SPC and another point on S, respectively. From Franklin (1967), used with permission.

configuration space with q_1 chosen as the generalized coordinate which represents the migration event in which only one configuration of atoms is displaced—in the q_1 generalized direction. Since q_1 is a generalized coordinate, more than one atom can be involved in permanent displacement. The two equilibrium configurations A and B are separated by a $(3N - 1)$-dimensional hypersurface which we call the saddle surface. This saddle surface passes through the saddle point and is perpendicular to contours of constant potential energy everywhere else. A diffusing atom must pass through this saddle surface at some point along its migration trajectory.

1. CLASSICAL AND QUANTUM THEORY OF DIFFUSION

The normal coordinates of the system, $q_i = m_i^{1/2} x_i$, where x_i is the direction of a generalized coordinate, can be obtained by a linear transformation at the outset of our derivation which makes a quadratic form out of the potential energy. One of the generalized coordinates is chosen to represent the displacements of the atoms in the N-body system in passing from the EC toward the SPC. Quantitatively, only the neighbors near the migrating atom or atoms would make a significant contribution to this generalized coordinate. The system as a whole has $3N$ degrees of freedom, where N is the number of atoms, even though it may have defects with resulting defect modes. The configurational partition function is given by

$$Q = \rho_0 \int_V e^{-\Phi/kT} \, dq \tag{2}$$

where $dq = dq_1 \, dq_2 \cdots dq_{3N}$, ρ_0 is the normalizing constant, Φ is the potential energy, and the integration is over the $3N$-dimensional volume of the system. The differential dR is given by

$$dR = d\mathbf{S} \cdot \int \dot{\mathbf{q}} \rho(\mathbf{q}, \dot{\mathbf{q}}) \, d\dot{\mathbf{q}} \tag{3}$$

where $d\mathbf{S}$ is the normal to an element of area of the saddle surface, $\dot{\mathbf{q}}$ is the velocity, and $\rho(\mathbf{q}, \dot{\mathbf{q}})$ is given by

$$\rho(\mathbf{q}, \dot{\mathbf{q}}) = \rho_1 e^{-H_2/kT} \tag{4}$$

where $\rho_1 = \rho_0 (2\pi kT)^{-3N/2}$ is the normalizing constant and H_2 is the harmonic Hamiltonian. In the SPC, which dR represents, dR is given by

$$dR = \rho_1 e^{-\Phi/kT} \, dS \int_0^\infty \dot{q}_1 \exp(-\dot{q}_1^2/2kT) \, d\dot{q} \prod_{j=2}^{3N} \int_{-\infty}^\infty \exp(-\dot{q}_j^2/2kT) \, d\dot{q}_j \tag{5}$$

where the assumption has been made that migrating atoms, by definition, do not return through the SPC during the lifetime of the migration event dealt with here. Hence, the first integral in Eq. (5) is in the positive space of $\dot{q}_1$ where "positive" means in the diffusion direction for the atoms involved in the generalized diffusion coordinate. For diffusion, the projection of the system velocity $\dot{\mathbf{q}}$ onto the components of $d\mathbf{S}$ in the positive sense is what is required by Eq. (5). When the integrals in Eq. (5) are evaluated, we obtain

$$dR = \rho_1 e^{-\Phi/kT} kT \, dS \, (2\pi kT)^{(3N-1)/2} \tag{6}$$

Therefore,

$$R = \rho_0 \left(\frac{kT}{2\pi}\right)^{1/2} \int_S e^{-\Phi(S)/kT} \, dS \tag{7}$$

and

$$\Gamma = \frac{R}{Q} = \left(\frac{kT}{2\pi}\right)^{1/2} \frac{\int_S e^{-\Phi(S)/kT} \, dS}{\int_V e^{-\Phi(V)/kT} \, dV} \tag{8}$$

Equation (8) gives the jump rate in units of second^{-1}, since the hypersurface integral over the generalized coordinates, each of which includes (mass)$^{1/2}$, in the numerator has one less dimension than the volume integral in the denominator. Since one generalized coordinate can be chosen to represent the system's motion from the EC to the SPC, Eq. (8) provides the fundamental equation for the classical jump frequency. The interactions between atoms in the system can be limited to harmonic terms in the potential or, in addition, anharmonic terms can be included.

When the harmonic approximation is made for the lattice potential energy in both the EC and SPC with no external forces on the system, we have

$$\Phi(S) = \Phi^0(S) + \tfrac{1}{2} \sum_i^{3N-1} \omega_i'^2 q_i'^2 \tag{9a}$$

$$\Phi(V) = \Phi^0(V) + \tfrac{1}{2} \sum_i^{3N} \omega_i^2 q_i^2 \tag{9b}$$

where $\omega_i = 2\pi v_i$ (in which v_i is the frequency of the system in the EC), a prime denotes the SPC, and $\Phi^0(S) - \Phi^0(V)$ is the activation energy of the static lattice configuration. The combination of Eqs. (8) and (9) yields the Vineyard equation for the jump frequency in diffusion, which is given by

$$\Gamma = \left(\prod_i^{3N} v_i \bigg/ \prod_i^{3N-1} v_i'\right) \exp\{-[\Phi^0(S) - \Phi(V)]/kT\} \tag{10}$$

In Eq. (10), the frequencies of the EC (v_i) must, in most cases of interest to diffusion, include some modes which have resonance or localized mode character. While the defect character of these modes may be weak, since the EC often does not present a case widely different from that of the perfect lattice, this is not the case for the SPC, which is represented by the frequencies v_i'. In the SPC, the strains in the lattice are substantial and the defect mode structure is expected to differ, in some cases considerably, from that of the perfect lattice. In addition, anharmonic effects will be much more significant on v_i' than on v_i, since the displacements in the SPC are much larger than those in the EC.

It is considerably easier to obtain an estimate of v_i' through the utilization of the deformed force constant procedure than it is to solve for the localized and resonance modes of the SPC in a three-dimensional lattice. The latter could, perhaps, be estimated from known theoretical work on simpler defects than those encountered in diffusion in one- and three-dimensional lattices, but there is no formal procedure available for making such an approximation. The facile route involves the deformed lattice methodology of Born and Huang (1956) in which the force constants of the deformed lattice are written as series expansions of those of the undeformed lattice. It is then easy to add the anharmonic terms of the crystal potential in order to obtain a reasonable estimate of the frequency factor, utilizing data which are available for certain materials.

It is important to note in our discussion of localized or resonance modes and of anharmonic effects that Γ contains no limitation on the time required for the system to progress from the EC to the SPC. It represents, quite simply, the ratio of probabilities for the system to be in the SPC to that for the EC [multiplied by $(kT/2\pi)^{1/2}$] as shown by Eq. (8). The system may oscillate such that the potentially diffusing atoms move back and forth building up momentum until the system finally reaches the SPC and migration occurs. In such a view of the mechanism of diffusion, there is sufficient time for the buildup of defect mode structure and of close to an equilibrium situation, when the SPC is finally attained. The alternative view, as stated, in part, by Montroll and Potts (1955), is that the actual jump time in an elementary migration event is so short ($\approx 10^{-13}$ sec in Cu, for example) that the localized or resonant mode structure will not have sufficient time to be developed, so that a near equilibrium state could not be attained in the SPC. It is true that the jump time for migration events, in which the activation energy is substantially greater than kT, is short compared with the inverse frequencies of many lattice vibration modes but the migration jump represents the last step in a process of energy and momentum buildup leading to migration. The points to be made in this argument are (1) that an equilibrium analysis of the SPC may not be far from the truth for many models and (2) that there is time for the formation of SPC defect mode frequencies so that, when the migration jump occurs, saddle-point defect modes make a significant contribution to Γ. In this model, the migration event, which takes place in the very short time of $\approx 10^{-13}$ sec for an activation energy of about 1 eV, is essentially adiabatic in character (only a small percentage of the total activation energy is imparted by the lattice during one pass from the EC to the SPC); nevertheless, the requisite energy for the jump has been supplied to the defect region over a longer prior history of nonadiabatic events, leading to a localized concentration of energy and development of defect mode structure.‡ In this nonadiabatic buildup of both energy and momentum, anharmonic effects provide the mechanism for the transfer

‡ The particulars of the migration event are discussed in greater detail by Bennett in Chapter 2.

of energy and momentum from the lattice to the localized region containing the defect. Umklapp processes provide the increase in localized momentum and the increase in local temperature preceding migration. Following migration, anharmonic processes convey energy and momentum from the defect area to the lattice as a whole. The validity of the fundamental postulate of rate theory, that equilibrium can be assumed for both the EC and the SPC, rests, in the classical case, on the magnitude of the exchange of energy near the SPC during the migration process.

C. Model with Anharmonicity and Deformed Force Constants

In order to obtain an estimate of the frequencies v_i' in the SPC, without resorting to the difficult task of evaluating the localized and resonance mode structure of a three-dimensional lattice with defects, we use the deformed force constant approach of Born and Huang (1956). This can be done for the inhomogeneous deformation around a defect almost as easily as for a homogeneous deformation applied to a perfect lattice. In addition, anharmonic effects can be included in both the EC and SPC potentials.

The classical case will be presented here in a slighty different fashion than done before (Franklin, 1967). The potential of the SPC is expanded about the equilibrium SPC, rather than the EC, utilizing, however, the deformed force constants as before. In addition, the subscript notation for the potential energy terms is inverted in order to conform with the notation used in the quantized approach (Franklin, 1969) and errors are corrected.‡

The aim of the classical derivation presented here is to obtain v_i' utilizing v_i and the deformed force constant expansions and the anharmonic correction factors for both v_i and v_i' in the Born–Oppenheimer adiabatic approximation. Utilizing this procedure, higher order atomic force constants, which can be estimated from higher order elastic constants, when available, or from interatomic potentials, can be used to find v_i' and the anharmonic corrections to v_i and v_i'.

The most convenient point to introduce anharmonic and deformed potential effects into the classical theory is Eq. (8) which contains a ratio of configurational partition functions. The potentials of the EC and SPC, which appear in these partition functions, can be formulated as follows in the adiabatic approximation:

$$\Phi(V) = \Phi^0(V) + \tfrac{1}{2}\phi_2 + \tfrac{1}{6}\phi_3 + \tfrac{1}{24}\phi_4 \qquad (11)$$

$$\Phi(S) = \Phi^0(S) + \tfrac{1}{2}\phi_2{}^S + \tfrac{1}{6}\phi_3{}^S + \tfrac{1}{24}\phi_4{}^S \qquad (12)$$

‡ The lengthy equations (27) and (28) of Franklin (1967) contain some misprints and errors. The corrected classical derivation is presented in Sections II and V.

where, for a crystal with one atom per unit cell,

$$\phi_i = \sum_{\substack{\alpha\beta\cdots\sigma_i \\ ll'\cdots l^{i-1}}} \phi_{\alpha\beta\cdots\sigma_i}(ll'\cdots l^{i-1})x_\alpha(l)x_\beta(l')\cdots x_{\sigma_i}(l^{i-1}) \qquad (13)$$

where $x(l)$ is the displacement of the lth atom from its equilibrium lattice position. When the series is truncated at second order, $\phi_{\alpha\beta}(ll')$ is the atomic force constant for interaction between atoms l and l' and $\alpha, \beta = x, y, z$. The force constants of the SPC are written in terms of those of the EC utilizing the deformed force constant expansion procedure such that, for example, in

$$\phi_3{}^S = \sum_{\substack{\alpha\beta\gamma \\ ll'l''}} \phi_{\alpha\beta\gamma}^{\text{def}}(ll'l'')x_\alpha(l)x_\beta(l')x_\gamma(l'') \qquad (14)$$

we have

$$\phi_{\alpha\beta\gamma}^{\text{def}}(ll'l'') = \phi_{\alpha\beta\gamma}^0(ll'l'') + \phi_{\alpha\beta\gamma\delta}^0(ll'l''l''')X_\delta(l''') \qquad (15)$$

where $X_\delta(l''')$ is the δ component of the critical displacement of the l'''th atom in the SPC. The critical displacement is defined as the static lattice displacement of the SPC with respect to the EC. In both the anharmonic and deformed force constant expansions, we adopt as the upper limit the fourth order and the product of two third-order force constants, since higher order terms violate the adiabatic principle of separation of electron and phonon states (Born and Huang, 1956).

A simplified notation is adopted here for the potential terms in which the first subscript denotes the number of variable displacements and the second subscript the number of critical saddle-point displacements (which appear in the deformed constants). Terms with no second subscript denote an absence of critical displacements in the term. Using these simplifications, we obtain the complete SPC potential energy by combining Eqs. (12)–(15):

$$\Phi(S) = \Phi^0(S) + \phi_{11} + \tfrac{1}{2}\phi_{12} + \tfrac{1}{6}\phi_{13} + \tfrac{1}{2}(\phi_2 + \phi_{21} + \tfrac{1}{2}\phi_{22})$$
$$+ \tfrac{1}{6}(\phi_3 + \phi_{31}) + \tfrac{1}{24}\phi_4 \qquad (16)$$

where the term ϕ_1 is omitted since the force always vanishes in equilibrium. The ϕ_{13} term is included here, since numerical evaluation done in previous work indicates that it could make a contribution as significant as other terms containing ϕ_4 or $\phi_3{}^2$.‡

---

‡ This is a significant change since it adds the $\phi_{11}\phi_{31}$ and $\phi_{11}\phi_{13}$ terms to Eq. (20). In Eq. (22) of Franklin (1967), the ϕ_{13} (ϕ_{31} in the notation used there) term was omitted since it was assumed to be of order λ^3. A better order, in view of numerical calculations, is that ϕ_{13} is of order λ^2. This seems reasonable since ϕ_{13} contains the quartic force constant, and other terms in the quartic constant are of order λ^2.

The evaluation of the integrals which appear in the numerator and denominator of Eq. (8) can now be accomplished utilizing expansions of the small terms in the exponents. Consider, first, the configurational partition function for the EC. This integral

$$\int_V e^{-\Phi(V)/kT} \, dV = \exp\{-\Phi^0(V)/kT\} \int \exp\{-\phi_2/2kT\}$$
$$\times [1 + \phi_3^2/72(kT)^2 - \phi_3/6kT - \phi_4/24kT] \, dV \quad (17)$$

is given by

$$\int_V e^{-\Phi(V)/kT} \, dV = \exp\{-\Phi^0(V)/kT\} \int \exp\{-ax^2/2kT\}[1 + b^2x^6/72(kT)^2$$
$$- bx^3/6kT - cx^4/24kT] \, dx$$
$$= \exp\{-\Phi^0(V)/kT\}(2\pi kT/a)^{1/2}(1 - kTc/8a^2 + 5kTb^2/24a^3)$$
$$(18)$$

where a, b, and c are the harmonic, cubic, and quartic atomic force constants, respectively, for nearest neighbor interactions in the one-dimensional model where the potential energy is given by $V = V_0 + \tfrac{1}{2}ax^2 + bx^3/6 + cx^4/24$. As an order of magnitude example, consider hypothetical values for these force constants given by $a = 3 \times 10^4$ ergs/cm^2, $b = -2 \times 10^{13}$ ergs/cm^3, and $c = 10^{22}$ ergs/cm^4. Then, for $kT = 10^{-13}$ erg ($T = 725°$K), we find that the anharmonic terms in Eq. (18) give rise to the following contributions: $kTc/8a^2 = 0.14$ and $5kTb^2/24a^3 = 0.30$. Consequently, there is, in this example, a 16% increase in the partition function due to anharmonicity at $T = 725°$K. It should be noted that normally the two anharmonic terms in Eq. (18) are of comparable magnitude, that their difference is a small number, and that the sign of the anharmonic contribution depends on the relative magnitude of the cubic and quartic force constants.

If more than nearest neighbor interactions are included, the harmonic and anharmonic contributions from each interaction may be included, utilizing appropriate force constants. Consider a one-dimensional linear chain of N atoms with interactions between all atoms. Then

$$\int \exp\{-(\Phi - \Phi^0)/kT\} \, dx = \prod_i^N \left(\frac{2\pi kT}{a_i}\right)^{1/2} \left[1 - \frac{kT}{8}\left(\frac{c_i}{a_i^2}\right) + \frac{5kT}{24}\left(\frac{b_i^2}{a_i^3}\right)\right]$$
$$(19)$$

where the subscript i denotes the interaction with the ith atom in the chain. Obviously, the more distant of these interactions will be negligible. In many crystals, nearest and next-nearest neighbor interactions are sufficient for a reasonable approximation. In crystals with coulombic or metallic character, more distant interactions are significant but, even in these cases, truncation at a small number of neighbor interactions can be performed without significant error for the partition function.

To proceed with our simple model of anharmonic effects (on diffusion in solids) we consider now the SPC and the evaluation of the terms in the crystal potential energy which include terms for the deformed force constants. Utilizing Eq. (16) in the exponent, and expanding, except for the $\Phi(S)$ and ϕ_2 terms, we get the terms of significance within the framework of the Born–Oppenheimer adiabatic principle. In performing this expansion we omit, at the outset, terms which contain an odd number of variable displacements, since these terms will vanish anyway upon integration. Then we obtain

$$\int_S \exp\{-[\Phi(S) - \Phi^0(S)]/kT\}\,dS = \int_S \exp\{-\phi_2/2kT\}\,dS\{1 - (2kT)^{-1}$$
$$\times (\phi_{21} - \tfrac{1}{2}\phi_{22} - \tfrac{1}{12}\phi_4) + [2(kT)^2]^{-1}$$
$$\times [\phi_{11}^2 + \phi_{11}\phi_{12} + \tfrac{1}{3}\phi_{11}\phi_3 + \tfrac{1}{4}\phi_{12}^2$$
$$+ \tfrac{1}{6}\phi_{12}\phi_3 + \tfrac{1}{36}\phi_3{}^2 + \tfrac{1}{4}\phi_{21}^2 + \tfrac{1}{6}\phi_{11}\phi_{13}$$
$$+ \tfrac{1}{3}\phi_{11}\phi_{31}]\} \qquad (20)$$

This integral is easily evaluated in terms of our nearest neighbor interaction model in one coordinate. In that case, Eq. (20) can be written as

$$= \int \exp(-ax^2/2kT)\,dx\{1 + [(X/2kT)(-b + cX)$$
$$+ [2(kT)^2]^{-1}(a^2X^2 + abX^3 + \{b^2/4 + ac/6\}X^4)]x^2$$
$$+ [\{c/24kT + [2(kT)^2]^{-1}\}(\{ab/3\}X + \{5b^2/12 + ac/3\}X^2)]x^4$$
$$+ [b^2/72(kT)^2]x^6\} \qquad (21)$$

where X is the critical displacement in the SPC. The integration of Eq. (21) is easily accomplished and rearrangement, to show the temperature dependence, gives

$$\int_S \exp\{-(\Phi - \Phi^0)/kT\}\,dx = (2\pi kT/a)^{1/2}(1 + \delta + \varepsilon/T + \mu T) \qquad (22)$$

where the coefficients are given by

$$\delta = (cX^2 - bX)/2a + (3/2a^2)[abX/3 + (5b^2/12 + ac/3)X^2]$$
$$\varepsilon = (2ak)^{-1}[a^2X^2 + abX^3 + (b^2/4 + ac/6)X^4]$$
$$\mu = ck/8a^2 + 5b^2k/24a^3 \qquad (23)$$

In the case in which we have interaction between the migrating atom and all other atoms in a linear chain, the SPC partition function is given by

$$\prod_{i}^{N-1} (2\pi kT/a_i)^{1/2}(1 + \delta_i + \varepsilon_i/T + \mu_i T) \tag{24}$$

where δ_i, ε_i, and μ_i are functions of a_i, b_i, and c_i. Combination of Eqs. (19) and (24) in the jump frequency for the linear chain gives

$$\Gamma = v_N \frac{\prod^{N-1}(1 + \delta_i + \varepsilon_i/T + \mu_i T)}{\prod^{N}[1 - (kT/8)(c_i/a_i^2 - 5b_i^2/6a_i^3)]} \exp\{-[\Phi^0(S) - \Phi^0(V)]/kT\} \tag{25}$$

where v_N is the frequency corresponding to the extra degree of freedom in the EC compared with the SPC. In the representation for the jump frequency given by Eq. (25), all but one of the frequencies cancel out in numerator and denominator because of our choice of expanding the deformed force constants of the SPC in terms of those of the EC which, in turn, were assumed to be those of the perfect lattice.

If it were possible to obtain the set of lattice frequencies in the SPC in a three-dimensional model, then the jump frequency could be written as the Vineyard equation with anharmonic factors for each of the frequencies in numerator and denominator. For our simple one-dimensional model, the equation for this case is given by

$$\Gamma = \frac{\prod^{N} v_i/[1 - (kT/8)(c_i/a_i^2 - 5b_i^2/6a_i^3)]}{\prod^{N-1} v_i'/[1 - (kT/8)(c_i'/a_i'^2 - 5b_i'^2/6a_i'^3)]} e^{-\Delta E/kT} \tag{26}$$

III. Lattice Vibration Theory

A. Introduction

Since diffusion is a thermal process in which nuclei, with their associated electronic shells, migrate from one site to another in the medium, lattice vibration theory provides a superstructure on which to base a theory of diffusion in crystalline solids. In Section II, we saw that the Vineyard theory was based on lattice vibrations. For these reasons, the lattice vibration theory is developed more extensively in this section.

If the phonons in a system have lifetimes long enough to be defined and if their frequencies are amenable to certain forms of perturbation theory, then an anharmonic lattice vibration theory provides a useful basis for the study of migration. An anharmonic theory which accounts for defects in the lattice is complicated but can, in certain cases, give reliable results for defect-type materials. Although solution of the localized or resonance modes, together with anharmonic effects, has not been obtained and presents tremendous complexities, the insights into the basic migration process which are provided by the defect theory of lattice vibrations more than compensate for the difficulties in the formalism. In addition to the diffusion process itself, many calculations which involve the migration saddle point apply equally well to large vibrational levels in which diffusion does not occur. That is, instead of the system being constrained to go through the saddle surface, another hypersurface can be picked at a lesser energy or at a location where a diffusion defect is not present. In this manner, the characteristics of large vibration amplitudes can be studied without the appearance of an imaginary frequency associated with the saddle-point configuration. For instance, pumped vibronic levels as high as a vibrational level of 30 have been reported by Prokhorov (1972). The characteristics of states of high vibration number resemble, in certain ways, those of diffusion, since the displacements are of the same order of magnitude.

Since diffusion involves the motion of nuclei, the only elementary excitations in a crystal which can be involved in the migration process are those which give rise to nuclear motion directly or indirectly. Consequently, phonons, which are quantized lattice vibrations, together with their interactions with other elementary excitations, affect the process of diffusion. The interaction of magnons, polarons, and electron states with phonons affects diffusion only insofar as it modifies or stimulates nuclear displacement. In metals, degenerate semiconductors and other materials with metallic character electronic states, electron density, and fluctuations in charge affect migration, as shown by Huntington (Chapter 6 and articles referenced there).

Because of the interactions of electrons and other elementary excitations, the theory of migration in metals would not be expected to follow lattice vibration theory closely. However, even though the Born–Oppenheimer adiabatic principle does not hold for metals, unmodified lattice vibration theory is still capable of giving reasonable results for self-diffusion in metals. The reason for this lies in the comparative change of the free electron distribution and the core potential during the migration process. DiVincenzo and Girifalco (1971), using Wannier potential theory, have shown that the binding energy of metals can be divided approximately equally between two-body potentials and the uniform, volume dependent electron distribution. The latter is not expected to change by a large amount during a migration event, whereas the interionic force constants, which are determined by the two-body potentials, are changed significantly during displacements as large as those encountered in diffusion. The alteration of force constants in any material by displacement results in a change of the phonon frequencies. Therefore, in the phonon theory of diffusion, the primary effects of nuclear displacements on the system, when an atom diffuses, are accounted for via alterations in the normal mode frequencies of the saddle-point configuration.

Since jump times in the solid state are typically of the order of 10^{-12}–10^{-13} sec (Franklin and Graddick, 1970) for an activation energy of the order of an electron volt, a lattice adiabatic approximation for the migration event is quite good in the sense that after an atom has attained sufficient energy to migrate, the actual jump process occurs more rapidly than the accrual or dissipation of energy from lattice vibrations. (Here, we are talking about adiabatic in the sense of phonon–phonon interactions—not of electron–phonon interactions, as in the last paragraph.) This argument applies to cases in which the activation energy for migration is considerably greater than kT. On the other hand, for cases which are nonadiabatic in behavior, that is, in which phonon–phonon interaction terms contribute a large percentage of the activation energy during a jump, a proper accounting of this energy transfer from the system's phonon reservoir to the migrating atom is difficult to make. When the activation barrier becomes so low with respect to kT that the atom can reside in the saddle-point region for a large percentage of its time, then the simple concepts of jump times and of nonadiabatic jumps cannot be invoked.

Many authors have made notable contributions to the theories of harmonic, anharmonic, and defect lattice vibrations and some of these are listed in the references which follow. Maradudin *et al.* (1971) have written an introduction to the harmonic theory of lattice vibrations which includes a brief discussion of the effects of defects and disorder. Anharmonic effects have been treated by Maradudin *et al.* (1961), Maradudin and Flinn (1961), Flinn and Maradudin (1963), Leibfried and Ludwig (1961), and Cowley (1963). The effects of defects on lattice vibrations were treated initially by Lifschitz (1956) and subsequently by Montroll and Potts (1955), Maradudin (1965, 1966a, and 1966b), Lifschitz and Kosevitch (1968), and by Ludwig (1963, 1966).

Besides the localized mode treatment, another way in which the large strains in the SPC can be treated within the framework of lattice vibration theory is to utilize the strained force constant approach. This approach was first taken by Born and Huang (1956) when they considered the effects of a homogeneous strain on the atomic force constants of the system and then the effects, in turn, on the system's frequencies. An extension of the theory to the case of inhomogeneous strain can be done (Franklin, 1967) and, in this case, the theory applies to the strained SPC. In this manner, the frequencies of the SPC can be estimated if the strains of this configuration are known. Since these strains can be calculated by computer techniques and an assumed potential (Johnson, 1966; Wynblatt, 1969), it is possible to obtain an estimate of the localized mode frequencies of the SPC via this technique. When anharmonic terms are included, the calculation becomes reasonably complete and represents a viable challenge to the rather formidable technique of calculating the localized modes of the SPC from basic defect lattice vibration theory. The latter has not been done but offers the distinct advantage that, when it is done for a particular lattice, the results will be applicable to any material which has that lattice structure. The former technique, on the other hand, requires computer calculation of the saddle-point strains for each system. Consequently, the calculation of the localized modes for the SPC of the common structures such as the bcc, fcc, and hcp structures would give a major contribution to the understanding of the migration process in solids.

B. Harmonic Theory

The simplicity of the homogeneous wave equation and the equation of motion for a simple harmonic oscillator carries over, in part, to the derivation of the frequencies of a many-body system of oscillating nuclei in the harmonic approximation. The necessity for the inclusion of anharmonicity when large excursions are considered introduces considerable complexity. The Hamiltonian for a vibrating system of atoms, ions, or nuclei with their core electrons is comprised of the kinetic and potential energy parts, T and Φ, respectively,

$$H = T + \Phi \tag{27}$$

where

$$T = \tfrac{1}{2} \sum_{\alpha l \kappa} m_{l\kappa} \left[\dot{u}_\alpha \binom{l}{\kappa} \right]^2 \tag{28}$$

and

$$\Phi = \sum_{n=2}^{\infty} \frac{1}{n!} \sum_{\substack{\alpha\beta \cdots \sigma_n \\ ll' \cdots l^{n-1} \\ \kappa\kappa' \cdots \kappa^{n-1}}} \phi_{\alpha\beta \cdots \sigma_n}\binom{ll' \cdots l^{n-1}}{\kappa\kappa' \cdots \kappa^{n-1}} u_\alpha\binom{l}{\kappa} \cdots u_{\sigma_n}\binom{l^{n-1}}{\kappa^{n-1}} \tag{29}$$

In Eq. (28), $m_{l\kappa}$ is the mass of the κ atom in the l unit cell. The index κ covers the atoms in the crystal's basis. For example, in NaCl, each cell has two atoms. The displacement of the κ atom in the l unit cell is $u\binom{l}{\kappa}$ and its velocity is $\dot{u}\binom{l}{\kappa}$. Cartesian components of vectors and tensors are denoted by $\alpha, \beta, \ldots$. The potential energy is comprised of terms in a series which contain products of force constants and displacements. For example, in the harmonic approximation,

$$\Phi_H = \Phi_2 = \tfrac{1}{2} \sum \phi_{\alpha\beta}\binom{ll'}{\kappa\kappa'} u_\alpha\binom{l}{\kappa} u_\beta\binom{l'}{\kappa'} \tag{30}$$

where $\phi_{\alpha\beta}\binom{ll'}{\kappa\kappa'}$ is the second-rank force constant tensor for interaction between the atom κ in cell l with atom κ' in cell l'. Higher order terms in Φ contain force constant tensors of higher rank, as indicated in Eq. (29). The components of $\phi_{\alpha\beta}$ are given by

$$\phi_{\alpha\beta}\binom{ll'}{\kappa\kappa'} = \partial^2 \Phi \bigg/ \left[\partial u_\alpha\binom{l}{\kappa} \partial u_\beta\binom{l'}{\kappa'}\right] \tag{31}$$

which is symmetric upon interchange of the triplets α, l, κ and β, l', κ'. Therefore, for both perfect and defect lattices,

$$\phi_{\alpha\beta}\binom{ll'}{\kappa\kappa'} = \phi_{\beta\alpha}\binom{ll'}{\kappa\kappa'} \tag{32}$$

Similar relations can be obtained for force constants of higher order.

Since diffusion involves lattice defects in some manner, the theory of lattice vibrations must, in order to be generally applicable, be able to treat defects whose state alters during a migration event within its formalism. Consequently, we set up the theory from the standpoint of the imperfect lattice following the work of Lifschitz (1956) and of Maradudin (1965). As far as lattice vibration theory is concerned, the defect contributions may be significant for different atomic mechanisms of diffusion either in the EC or the SPC, and in between, or both. For example, vacancies and interstitials are defects in their equilibrium state and, during the migration process, their force constants and displacements are modified, thus altering the character of the defect in the lattice. In the following, the lattice vibration theory will be developed, therefore, for a lattice with defects noting, in places, the corresponding relationships for a perfect lattice.

The displacement transformation operator $u_\alpha\binom{l}{\kappa}$ in a lattice with defects, which yields the eigenvalues and eigenvectors as well as the proper potential energy for a defect lattice, is given by

$$u_\alpha\binom{l}{\kappa} = (\hbar/2m_{l\kappa})^{1/2} \sum_s \left[B_\alpha^{(s)}\binom{l}{\kappa}\bigg/\omega_s^{1/2}\right][b_s^\dagger\, b_s] \tag{33}$$

In Eq. (33), $B_\alpha^{(s)}\binom{l}{\kappa}$ is the eigenvector for the mode (degree of freedom) s at the position $l\kappa$ and $b_s^\dagger$ and b_s are the creation and annihilation operators for the s mode which has frequency ω_s. If the crystal is perfect, with no defects, and has periodic symmetry, the Bloch condition applies to the

1. CLASSICAL AND QUANTUM THEORY OF DIFFUSION

eigenvectors and Eq. (33) can then be written directly as

$$u_\alpha\binom{l}{\kappa} = (\hbar/2Nm_{l\kappa})^{1/2} \sum_{kj} [e_\alpha(kj)/\omega_{kj}^{1/2}] \exp(i\mathbf{k}\cdot\mathbf{r})[a_{kj}^\dagger + a_{kj}] \quad (34)$$

where $\hat{e}(kj)$ is the crystal polarization vector for the mode having wavevector $\mathbf{k}$ and branch j. The creation and annihilation operators $a_{kj}^\dagger$ and a_{kj} are similar to their counterparts in the defect lattice except that they operate on modes having definable branches and wavevectors and do not depend on the location in the crystal. Since diffusion normally involves a lattice with defects, Eq. (34) is not used in this article. It is shown merely for comparison with the form involving defect lattice operators and eigenvectors, shown in Eq. (33), which will be used here.

There are three branches for a perfect three-dimensional crystal—two transverse and one longitudinal—for each atom per unit cell. Assuming periodic boundary conditions, there are N values of k, where N is the number of unit cells per crystal, for each branch. Therefore, if there are r atoms per unit cell, the total number of possible modes of vibration is $3rN$. In a defect lattice without molecular-type defects, the number of degrees of vibrational freedom of the system is the same as in the perfect crystal—$3rN$—but wavevectors and branches are not definable in any simple manner. Hence, we simply replace kj by s and sum the modes up to 3γ, where, in the perfect crystal, $\gamma = rN$. In molecular-type defects, the number of degrees of freedom, 3γ, is not equal to $3rN$. For example, for a vacancy, the number is $3(rN-1)$ and for a diatomic interstitial it is $3(rN+2)$.

Lifschitz pointed out that small effects of a defect on the displacement operator could be treated by increments in $\mathbf{k}$ and $\mathbf{r}$ such that

$$\mathbf{k}\cdot\mathbf{r}(l\kappa) = (\mathbf{k}_0 + \Delta\mathbf{k})\cdot[\mathbf{r}_0(l\kappa) + \Delta\mathbf{r}(l\kappa)]$$
$$\approx \mathbf{k}_0\cdot\mathbf{r}_0 + \mathbf{r}_0\cdot\Delta\mathbf{k} + \mathbf{k}_0\cdot\Delta\mathbf{r}_0 \quad (35)$$

If $\Delta r(l\kappa)$ is small, as in the case of a substitutional impurity of similar size to the matrix atoms, then $\mathbf{k}\cdot\Delta\mathbf{r}(l\kappa)$ is negligible. However, in diffusion, $\Delta r(l\kappa)$ is certainly large in most systems in the SPC so that this term is not negligible. If one tries, however, to use Eq. (35) in Eq. (34) together with a polarization vector which is dependent on location, the resulting formalism becomes unwieldy. Hence, the simple representation of the displacement transformation given by Eq. (33) is used in which the eigenvectors $B^{(s)}(l\kappa)$ are unit vectors which depend on location in the medium. It is significant to note however, that the formalism portrayed by Eq. (35) implies that if the periodicity which is implicit in the $\mathbf{k}_0\cdot\mathbf{r}_0$ term is significant, then Umklapp processes may still occur in part, even though perfect periodicity of the lattice is destroyed.

The eigenvalue equation is obtained from the homogeneous harmonic equation of motion for the defect lattice which is given for the κ atom in the l cell by

$$m_{l'\kappa'}\ddot{u}_\beta\binom{l'}{\kappa'} = -\sum_{\alpha l\kappa}\phi_{\alpha\beta}\binom{ll'}{\kappa\kappa'}u_\alpha\binom{l}{\kappa} \qquad (36)$$

The time dependence for simple harmonic motion is given by $\exp(-i\omega t)$ and when this is included in Eq. (33) and the resultant time dependent displacement utilized in Eq. (36), we obtain the eigenvalue equation, which is given by

$$\sum_{\beta l'\kappa'} D_{\alpha\beta}\binom{ll'}{\kappa\kappa'} B_\beta^{(s)}(l'\kappa') = \omega_s^2 B_\alpha^{(s)}(l\kappa) \qquad (37)$$

where the dynamical matrix is defined by

$$D_{\alpha\beta}\binom{ll'}{\kappa\kappa'} = \phi_{\alpha\beta}\binom{ll'}{\kappa\kappa'}\bigg/(m_{l\kappa}m_{l'\kappa'})^{1/2} \qquad (38)$$

The force constants and masses are real; hence, $D_{\alpha\beta}$ is a real matrix. It is also symmetric in the index triplets $l\kappa\alpha$ and $l'\kappa'\beta$ because of the definition of the force constants which is given by Eq. (31). If it is real and symmetric, the dynamical matrix is Hermitian and its eigenvalues, the squared frequencies of the system, must be real. If the $\phi_{\alpha\beta}$ are not all real, then complex eigenvalues may arise.

If the reality condition is applied to the displacement operator, then $B_\alpha(-s) = -B_\alpha^*(s)$. In addition, the eigenvectors may be orthogonalized and normalized. Therefore, the orthonormality and closure conditions, which are given by

$$\sum_s B_\alpha^{(s)}(l\kappa)B_\beta^{(s)}(l'\kappa') = \delta_{\alpha\beta}\,\delta_{ll'}\,\delta_{\kappa\kappa'} \qquad (39)$$

$$\sum_{\alpha l\kappa} B_\alpha^{(s)}(l\kappa)B_\alpha^{(s')}(l\kappa) = \delta_{ss'} \qquad (40)$$

respectively, apply. In Eqs. (39) and (40), the δ_{ij} represent the components of the Kronecker delta function. The normality condition ensures that the $B^{(s)}(l\kappa)$ are unit vectors. Their number, as mentioned above, equals the number of degrees of freedom, 3γ, including those of the crystal as a whole, which is justifiable for large γ.

The annihilation and creation operators are defined in the usual manner (Maradudin et al., 1971) and obey the commutation relations given by

$$[b_s, b_s^\dagger] = \delta_{ss'} \qquad (41a)$$

$$[b_s, b_{s'}] = [b_s^\dagger, b_{s'}^\dagger] = 0 \qquad (41b)$$

In the case of perfect crystals, $s = k, j$ and the Kronecker delta in Eq. (41a) is replaced by $\Delta(k' - k)\delta_{jj'}$ where $\Delta(k' - k) = 1$, if $k' - k = 0$ or a reciprocal lattice vector G, and equals zero otherwise. The Bloch condition gives rise to G in lattices with periodic symmetry.

In the phonon number representation, the state function $\psi(n_s)$ is represented, simply, by n_s—the number of phonons in the mode s. The operation of the creation and annihilation operators on the state function yields

$$b_s|n_s\rangle = n_s^{1/2}|n_s - 1\rangle \qquad b_s|0\rangle = 0$$
$$b_s^\dagger|n_s\rangle = (n_s + 1)^{1/2}|n_s + 1\rangle, \qquad b_s^\dagger|0\rangle = |1\rangle \tag{42}$$

The ground state $|0\rangle$ and the thermal state s represent, respectively, the state of lowest phonon occupation number (corresponding to the zero-point energy) and the thermal state with n_s phonons. The temperature dependence of n_s in thermal equilibrium is given by

$$\bar{n}_s = 1/(e^{x_s} - 1) \tag{43}$$

where $x_s = \hbar\omega_s/kT$. Utilizing this representation for n_s and using Eq. (42), we easily see that

$$\langle n_s|b_s^\dagger b_{s'}|n_{s'}\rangle = \delta_{ss'}\bar{n}_s = 1/(e^{x_s} - 1)$$
$$\langle n_s|b_s b_{s'}^\dagger|n_{s'}\rangle = \delta_{ss'}(\bar{n}_s + 1) = 1/(1 - e^{-x_s}) \tag{44}$$
$$\langle n_s|[b_s + b_s^\dagger][b_{s'} + b_{s'}^\dagger]|n_{s'}\rangle = \delta_{ss'}(2\bar{n}_s + 1) = \coth x_s/2$$

since $\langle n_s|n_{s'}\rangle = \delta_{ss'}$. The results given by Eq. (44) are useful in the evaluation of Hamiltonian and partition function components.

Returning now to the vibrational Hamiltonian, which is given by Eqs. (27)–(29), we can write, for the harmonic approximation,

$$H = \tfrac{1}{2}\sum m_{l\kappa} \dot{u}_\alpha^2\binom{l}{\kappa} + \tfrac{1}{2}\sum \phi_{\alpha\beta}\binom{ll'}{\kappa\kappa'}u_\alpha\binom{l}{\kappa}u_\beta\binom{l'}{\kappa'} \tag{45}$$

which can also be written as, using Eq. (33),

$$H = \hbar\sum_s \omega_s(b_s^\dagger b_s + \tfrac{1}{2}) \tag{46}$$

The vibrational energy of the system is given by

$$E = \langle\psi|H|\psi\rangle = \hbar\sum_s \omega_s(\bar{n}_s + \tfrac{1}{2}) \tag{47}$$

where $\bar{n}_s$ is given by Eq. (43). This shows that the vibrational energy is given, simply, by the sum of the phonon energies for all the modes in the system including their degeneracy.

The partition function and thermodynamic quantities follow directly from a knowledge of the energy. In the harmonic approximation, the partition function is defined by

$$Z = \sum_{n_s} e^{-E(n_s)/kT} \tag{48}$$

which becomes, when Eq. (47) is utilized,

$$Z = \prod_s e^{-x_s/2}/(1 - e^{-x_s}) \tag{49}$$

The Helmholtz free energy is then

$$F = -kT \ln Z = kT \sum_s \ln(2 \sinh x_s/2) \tag{50}$$

and the internal energy, specific heat, Gibbs free energy, and entropy are easily obtained using the usual thermodynamic relations for these quantities in terms of Eqs. (49) and (50).

The frequency spectrum $G(\omega^2)$ for the ω^2 eigenvalues of a lattice with defects is defined such that $G(\omega^2) \, d\omega^2$ is the fraction of normal modes for ω^2 in an interval $d\omega^2$. The frequency distribution function $g(\omega)$ is defined in a similar manner such that $g(\omega) \, d\omega$ is the fraction of normal modes for ω in an interval $d\omega$. The frequency distribution function is related to the frequency spectrum by $g(\omega) = 2\omega G(\omega^2)$. The delta function representation of $G(\omega^2)$ is given by

$$G(\omega^2) = (3\gamma)^{-1} \sum_s \delta(\omega^2 - \omega_s^2) \tag{51}$$

and another representation, which utilizes the dynamical matrix, **D**, is given by

$$G(\omega^2) = (3\gamma\pi)^{-1} \operatorname{Im}(d/d\omega^2) \ln|\omega^2 \mathbf{I} - \mathbf{D}| \tag{52}$$

where **I** is the identity tensor. For a perfect crystal, we replace **D** by $\mathbf{D}_0$, the dynamical matrix of the perfect lattice. Then the change of the frequency spectrum of the band modes of a lattice caused by the introduction of defects is given by

$$\Delta G(\omega^2) = (3\gamma\pi)^{-1} \operatorname{Im}(d/d\omega^2) \ln(|\omega^2 \mathbf{I} - \mathbf{D}|/|\omega^2 \mathbf{I} - \mathbf{D}_0|) \tag{53}$$

In a defect system in which atomic migration is occurring, the change in the frequency spectrum of the band modes, which occurs in going from the equilibrium configuration to a strained unequilibrium position, is given by

$$\Delta G(\omega^2) = (3\gamma\pi)^{-1} \operatorname{Im}(d/d\omega^2) \ln(|\omega^2 \mathbf{I} - \mathbf{D}_{\text{SPC}}|/|\omega^2 \mathbf{I} - \mathbf{D}_{\text{EC}}|) \tag{54}$$

where $\mathbf{D}_{\text{EC}}$ and $\mathbf{D}_{\text{SPC}}$ are the dynamical matrices at the equilibrium configuration and the strained position, respectively.

In the case in which defects introduce localized modes which occur outside of the band of lattice frequencies, the dynamical matrix formalism does not apply since there is a decoupling between the matrix and a localized region of the lattice. In this case, a delta function representation of the difference spectrum applies and

$$\Delta G(\omega^2) = (3\gamma)^{-1} \sum_s{}' [\delta(\omega^2 - \omega_s^2) - \delta(\omega^2 - \omega_{ms}^2)], \qquad \omega^2 \neq \Omega_b^2 \quad (55)$$

where the prime on the sum means that the sum extends over the localized modes only, not the band modes, ω_{ms} denotes the maximum band mode frequency, and Ω_b denotes the set of band mode frequencies. The change in the spectrum of localized modes during a migration event is then given by

$$\Delta G(\omega^2) = (3\gamma)^{-1} \sum_s{}' \{\delta[\omega^2 - (\omega_s^S)^2] - \delta(\omega^2 - \omega_s^2)\}, \qquad \omega^2 \neq \Omega_b^2 \quad (56)$$

where ω_s^S are the localized mode frequencies of the saddle-point configuration. One of the frequencies in the SPC is imaginary since the potential energy on each side of the saddle point in the migration path decreases. Close to the SPC, the slope of the potential energy along the migration path is small so that the frequency for the normal mode involving the vibration of the diffusing atom in a direction parallel to the migration

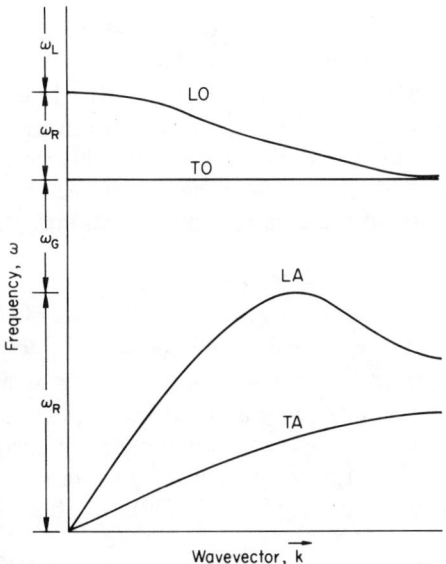

Fig. 3. Typical dispersion relation for a [100] direction in a diatomic lattice showing the ranges for resonance (R), gap (G), and localized (L) modes. The dispersions for acoustic (A) and optical (O) band modes for transverse (T) and longitudinal (L) motions are shown.

trajectory is low and well within Ω_b. Consequently, close to the SPC, this frequency contributes to Eq. (54), not to Eq. (56).

The dispersion of $\omega(k)$ vs k for localized and gap modes is usually insignificant for a small concentration of defects. Resonance mode frequencies lie in or close to the dispersion of the perfect lattice band modes. The ranges of defect-induced modes are portrayed schematically in Fig. 3, which shows the dispersion relations for the band modes in a diatomic crystal together with the possible ranges of values of the localized, gap, and resonance mode frequencies. While there is no dispersion for localized or gap modes for low concentrations it is, nonetheless, significant to note that complex values of the wavevector are necessary, so that $k(\text{loc}) = k' + ik''$ in which $k'' > 0$.

C. Anharmonic Theory

In the anharmonic case, the Hamiltonian is more complex than in the harmonic approximation since the equation is inhomogeneous. However, the anharmonic free energy can be evaluated formally in a fairly direct manner utilizing, largely, the methods applied above in the harmonic case. Anharmonicity enters the diffusion theory in many places including the anharmonic partition functions which provide corrections to the equilibrium and saddle-point configuration frequencies, the interaction of phonons and their lifetimes.

The anharmonic potential energy is given by Eq. (29) without the harmonic, or second-order term. In the Born–Oppenheimer adiabatic approximation (Born and Huang, 1956), which is adopted here, only third- and fourth-order terms—cubic and quartic, respectively—occur since above this order, the electron coordinates are inextricably bound to those of the nuclear motions so that separation of the wave functions and eigenvalues is impossible. Therefore, in the adiabatic approximation, the Hamiltonian is

$$H = H_2 + \lambda \Phi_3 + \lambda^2 \Phi_4 \tag{57}$$

where λ is a perturbation order parameter and H_2, Φ_3, and Φ_4 are the harmonic, cubic, and quartic portions of the Hamiltonian, respectively. H_2 is comprised of the kinetic energy and harmonic potential energy whereas the cubic and quartic terms have no kinetic energy parts and are written, therefore, as potential energy components. Utilizing the normal coordinate transformation, given by Eq. (33), in Eq. (29), we obtain

$$\Phi_3 = \frac{1}{6} \left(\frac{\hbar}{2}\right)^{3/2} \sum \frac{\phi_{\alpha\beta\gamma}\binom{ll'l''}{\kappa\kappa'\kappa''}}{(m_{l\kappa} m_{l'\kappa'} m_{l''\kappa''})^{1/2}} \sum_{ss's''} \frac{B^{(s)}_\alpha\binom{l}{\kappa} B^{(s')}_\beta\binom{l'}{\kappa'} B^{(s'')}_\gamma\binom{l''}{\kappa''}}{(\omega_s \omega_{s'} \omega_{s''})^{1/2}}$$
$$\times [b_s^\dagger + b_s][b_{s'}^\dagger + b_{s'}][b_{s''}^\dagger + b_{s''}] \tag{58}$$

for the cubic term. The quartic term is exactly analogous to Eq. (58) except that it is simply one higher in order in each index which appears in Eq. (29), utilizing Eq. (33).

The partition function in the anharmonic case is given by

$$Z = \text{Tr } e^{-\beta H} = \text{Tr } e^{-\beta H_2(\lambda \Phi_3 + \lambda^2 \Phi_4)} \tag{59}$$

and the operators can be expanded in the interaction picture in the usual manner. The details will not be shown here since they are available in many articles and texts (Cowley, 1963; Maradudin et al., 1961). The expression for the partition function is given by

$$Z = Z_2 + \lambda^2 (Z_4 + Z_3 \overset{*}{*} Z_3) \tag{60}$$

where

$$Z_2 = \text{Tr } e^{-\beta H_2} \tag{61a}$$

$$Z_3 \overset{*}{*} Z_3 = \beta^2 \text{ Tr } e^{-\beta H_2} \int_0^1 dS_1 \int_0^{S_1} dS_2 \, e^{S_1 \beta H_2} H_3 \, e^{-(S_1 - S_2)\beta H_2} H_3 \, e^{-S_2 \beta H_2} \tag{61b}$$

$$Z_4 = -\beta \text{ Tr } e^{-\beta H_2} \int_0^1 dS_1 \, e^{S_1 \beta H_2} H_4 \, e^{-S_1 \beta H_2} \tag{61c}$$

Then the anharmonic free energy can be expanded in the following manner:

$$F = -\beta^{-1} \ln Z = -\beta^{-1} \{\ln Z_2 + \ln[1 + \lambda^2 (Z_4/Z_2 + Z_3 \overset{*}{*} Z_3/Z_2)]\}$$
$$= F_2 - (\lambda^2/\beta)(Z_4/Z_2 + Z_3 \overset{*}{*} Z_3/Z_2) \tag{62}$$

where F_2 is the harmonic free energy and the latter equality holds for order λ^2. Integration of Eqs. (61b) and (61c) is straightforward and the result is

$$Z_3 \overset{*}{*} Z_3 = \beta \sum_{nn'} e^{-\beta E_n} \langle n|H_3|n'\rangle \langle n'|H_3|n\rangle / (E_{n'} - E_n) \tag{63a}$$

$$Z_4 = -\beta \sum_n e^{-\beta E_n} \langle n|H_4|n\rangle \tag{63b}$$

The matrix elements in Eqs. (63) are reasonably easy to evaluate and, consequently, the anharmonic Hamiltonian in the adiabatic approximation can be obtained numerically for materials in which the requisite data are available. The reader is referred to the calculations of Maradudin et al. (1961) for a numerical evaluation of the anharmonic free energy for lead in the cases of both a linear chain and a three-dimensional lattice.

The harmonic, cubic, and quartic partition functions, which are given by Eqs. (61a), (63a), and (63b), respectively, are utilized in the quantum theory of diffusion which is presented in Section V. In that theory, as we shall see, the total partition functions in both the EC and the SPC are required. The former include the components just derived, whereas the latter are comprised of terms which include the effects of the inhomogeneous strain in the SPC, as shown for the simple model in Section II.C. Since the force constants of the SPC are not known, what is done is to expand the saddle-point force constants in terms of those in the EC, which are known. This procedure involves, therefore, a double perturbation expansion —one for each of the force constants in the SPC and one for the potential energy. The anharmonic terms in the EC and the SPC provide for renormalization of the lattice vibration frequencies in the EC and SPC, respectively.

Before concluding this introductory section on anharmonic lattice vibration theory, a brief insight into the domain of experimental and theoretical applications of the formalism is afforded by a consideration of the simple one-dimensional anharmonic oscillator. This model is transparent since it is not dressed with fearsome notations or complexities. Consider the potential

$$V(x) = \tfrac{1}{2}ax^2 + \tfrac{1}{6}bx^3 + \tfrac{1}{24}cx^4 \tag{64}$$

which, with the kinetic energy in the total Hamiltonian, gives rise to the partition function.

$$Z = 2\pi kT(m/a)^{1/2}(1 - ckT/8a^2 + 5b^2kT/24a^3) \tag{65}$$

via the defining relation $Z = \int \exp[-H(pq)/kT] \, dp \, dq$. Energy and specific heat are obtained from Z and are given by

$$E = -\partial \ln Z/\partial(1/kT) = kT(1 - ckT/8a^2 + 5b^2kT/24a^3) \tag{66}$$

$$C_V = \partial E/\partial T = k(1 - ckT/4a^2 + 5b^2kT/12a^3) \tag{67}$$

IV. Defect Modes

In Section III, the formal structure of the theory of lattice vibrations was developed for both the harmonic and anharmonic cases. In this section, specific effects of defect lattice modes are discussed. Sections B, C, D, and F present a theoretical basis for the nonperturbative approach to the solution of the defect mode frequencies which contribute to a diffusion event. This is done through the utilization of Green's function techniques. Sections E and G deal, respectively, with correlation functions and with recent lattice vibration calculations of localized and resonance modes which apply to diffusion. In Section A, we develop a simple physical argument for the changes in defect structure during a migration event.

1. CLASSICAL AND QUANTUM THEORY OF DIFFUSION

A. A Descriptive Model for Diffusion in Terms of Localized, Resonance, and Band Modes

In the theoretical development of the harmonic and anharmonic lattice vibration theory, the eigenvectors and eigenfrequencies of the lattice with defects were utilized. Bloch-type symmetry was not assumed except to show the comparison with perfect lattice theory. The reason that defect lattice vibration theory is utilized is that almost all diffusion mechanisms which have been postulated involve the presence of defects in the lattice. For example, the vacancy, interstitial, relaxed vacancy, crowdion, and interstitialcy mechanisms all depend on the existence of lattice defects. The ring rotation mechanism does not, but this mechanism probably does not make a significant contribution to diffusion in most systems. Therefore, since diffusion models depend on an initial concentration of defects, it is important to utilize defect lattice vibration theory and to note the special effects introduced by these defects. In addition, during a migration event including the processes leading up to the actual jump, the force constants of the system near the migrating atom or atoms are altered significantly so that major changes in things like the resonance nature and root mean square (rms) vibrational amplitude may occur. Because of these changes during the diffusion process, defect lattice vibration theory is applicable to all diffusion mechanisms which have been postulated.

With regard to the wavevector dependence of the lattice vibrations which give rise to migration, opposing motions of nearest neighbor atoms are usually required for most diffusion mechanisms so that one would expect that optical modes—and only optical modes—in compound crystals could give rise to diffusion. In monoatomic crystals, for opposing motions of nearest neighbor atoms, modes having k_{max} would be required on the basis of this argument. For single-mode coherent waves, this argument holds merit so that conditions close to those just mentioned are required for opposing motions of adjacent atoms. However, in a lattice at thermal equilibrium, the temperature determines the number of phonons per mode and their directions are normally random as a result of phonon–phonon and defect interactions. Therefore, the displacement of any given atom in the crystal is made up of the Fourier superposition of all the modes passing through the atom in question. Consequently, modes of any wavelength and type can contribute to diffusion by this argument. Another significant point is that large displacements require weak force constants which correspond to low frequencies. Consequently, low frequencies give rise to large rms displacements. Two very important factors have been left out of the argument, namely, the effects of anharmonicity and defect modes such as localized, gap, and resonance modes. Both are important to diffusion and the latter are the subject of the following discussion and, later, theoretical development.

The optical and acoustic modes of a perfect crystal are usually called band modes since they form bands of allowable lattice vibration frequencies. In addition to the band modes, there are two basic types of defect modes which give rise, in a three-dimensional crystal, to a small discrete set of vibrational frequencies for each point defect. These defect modes are the resonance (and near resonance) modes and the localized (and gap) modes. The former have regions of overlap (or near overlap) with band modes—either acoustic or optical modes. Even if some of the frequencies of a set of modes associated with a point defect lie within the band gap or slightly above the range of band modes, if overlap occurs in some range of frequencies between this set and the band modes, then the defect modes can interact among themselves and can decay via those frequencies which overlap into the frequencies of the band modes, thus displaying resonance character for the set. Since they decay, resonance modes have a width. On the other hand, localized and gap modes do not, by definition, have any of their frequencies which overlap those of the band modes. Consequently, there is little probability for decay into the continuum (the discrete set of band modes goes into a continuum for large N, for all practical purposes) of band modes, so that their lines are very sharp, having little, if any, width. Naturally, as the frequencies of localized or gap modes approach those of band modes, the decay probability (and, therefore, the width) increases.

The displacement amplitudes of localized and gap modes decay exponentially in space upon receding from the location of the defect which gives rise to the mode. However, for resonance modes, the vibration amplitudes of the atoms in the surrounding lattice are affected by the presence of the defect. As a resonance frequency is approached, the vibration amplitudes of both the atoms in the defect and the matrix atoms increase sharply (unless the defect atoms are so massive that they cannot move significantly, in which case the resonance is "seen" by the large amplitudes of neighboring atoms). The effect of resonance frequencies is not localized, as is the case with the localized and gap modes. The magnitude of the vibration in defect modes depends upon the masses and force constants involved. Small masses and high force constants give rise to high frequency modes, so that these conditions lead to localized modes (or gap modes, if they ascend from the acoustic band in a crystal with more than one atom in its basis) provided that a critical mass or force constant difference, which is required by the theory, is attained. On the other hand, weak impurity coupling and large defect masses lead, in general, to low frequencies and, hence, to resonance modes (or to gap modes which descend from an optical band).

1. CLASSICAL AND QUANTUM THEORY OF DIFFUSION

Since diffusion depends on large nuclear displacements, the effects of defect modes on these displacements are of prime significance. In this regard, it is important to note that the squared displacement amplitude for a resonance mode which interacts with the entire crystal is $O(N^{-1})$, whereas that for a localized mode is $O(1)$. Since resonance modes, by definition, interact with existing lattice band modes, the occupation number of these modes must be included at the temperature of the crystal.

It is interesting to note the effect of an external field on the vibrational amplitude of a defect with localized or resonance character. For example, a photon absorbed by a localized optical mode gives rise to a squared displacement with amplitude $O(1)$, whereas absorption at a resonance frequency gives $O(N^{-1})$. Therefore, a single photon can, if conditions are right, have a much larger influence on diffusion at a defect when it is absorbed by a localized (or gap) mode than when it is absorbed by a resonance mode.

Another factor of significance to diffusion is the frequency dependence of resonance amplitudes. Very large amplitudes, which are predicted for certain systems in certain ranges of resonance frequencies, should have a larger effect on diffusion than the nonresonance band modes. In addition, the change of force constants during the atomic migration process leads, in certain systems, to resonances (or localized modes) which develop along the way.

During the migration process in a three-dimensional crystal, the different types of defect modes which are possible can be analyzed qualitatively on the basis of known results of calculations on resonance and localized frequencies and their interaction with the matrix lattice. In lattice dynamics, the assumption is normally made that the forces on the mass points are comprised of two-body interaction potentials and that higher body interactions are neglected, with negligible error for most calculations. In this argument, the harmonic force constants increase rapidly as atoms are brought closer together than their equilibrium distance and decrease, but less rapidly, as they are separated (see Fig. 4). Hence, in the 3D crystal, the force constants in a microscopic view are increased under compression and decreased under tension. As a result of this, we expect a tendency toward resonance mode behavior in regions of tension and toward localized mode behavior in regions of compression.

An example which represents, qualitatively, the arguments presented here is that of the SPC of vacancy diffusion in the fcc structure. In this configuration, there are regions with large tension (or separation between equilibrium positions) both fore and aft of the atom diffusing along its minimum energy trajectory. To the sides of this trajectory, the lattice is under a large compression. As a consequence of this situation, one would expect resonance-type behavior (or a tendency toward resonance behavior)

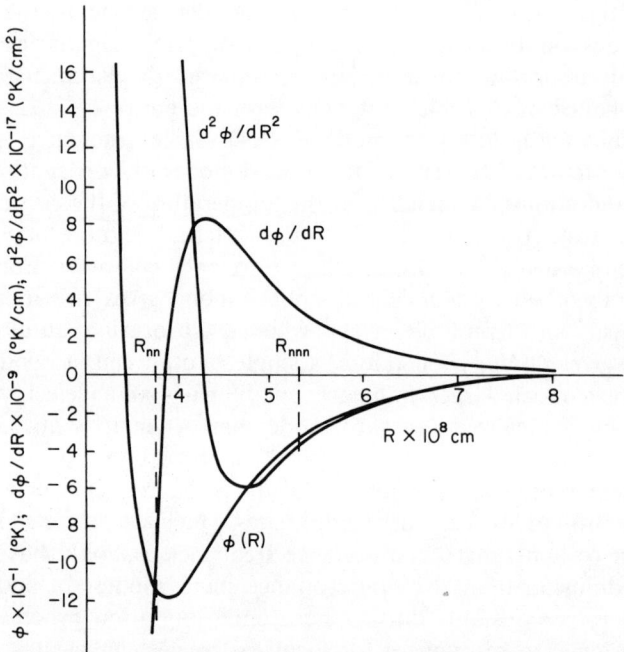

Fig. 4. The 6-12 two-body potential $\phi(r)$ and its first and second derivatives $\phi'(r)$ and $\phi''(r)$, respectively, for Ar at 140°K. The extrema in ϕ, ϕ', and ϕ'' are located at 3.87×10^{-8}, 4.27×10^{-8}, and 4.72×10^{-8} cm, respectively. The nearest and next-nearest neighbor distances, R_{nn} and R_{nnn}, respectively, in solid Ar are shown and are 3.76×10^{-8} and 5.31×10^{-8} cm, respectively.

in the direction of the migration trajectory and a tendency toward localized mode behavior perpendicular to the trajectory and near the defect. In addition, where the force constants are decreased, the velocity of sound and, hence, of phonons is less, and vice versa. Therefore, there is a buildup of phonon numbers (and, hence, energy) in regions of reduced force constants, and vice versa.

Another significant point in the description of the diffusion phenomenon in the fcc vacancy mechanism deals with the relative motions of the migrating atom and the set of neighboring atoms along the migration trajectory which tend toward resonance. In resonance, the vibrations are similar to those in optical modes, so that the migrating atom and the resonance-type neighbors vibrate counter to each other. Consequently, as the migration event progresses from the EC to the SPC, the force constants in and near the migration trajectory decrease and, as a result, the vibration amplitudes and energy density increase along this trajectory. If a resonance situation is attained, the migrating atom vibrates against its neighbors which are in the line of the trajectory and, in so doing, can receive and impart energy to these atoms in a classical billiard ball type of behavior. A similarity could be drawn, in a way, to cars on an air track oscillating against each other. Obviously, the mass of the migrating atom will have a strong effect on the relative vibration amplitudes of the atoms in the resonance system. A very heavy migrating atom will not oscillate with much amplitude but will induce large amplitude oscillations in its lighter neighbors which are in resonance, and vice versa.

The diffusion event might be described in its space–time development in the following manner. An oscillation larger than the thermal average begins as a result of a localized fluctuation of phonons. As the oscillation builds up, the frequency along the trajectory decreases with respect to that perpendicular to the trajectory. Then, a rather slow oscillation develops along the trajectory with faster perpendicular oscillations. During the process, a channeling of energy occurs because of the variation of the speed of sound in the neighborhood of the defect. The phonons slow down in the regions of tension along the trajectory leading to a buildup of energy in those regions, and vice versa, in regions perpendicular to the trajectory. If the stress pattern leads to a substantial focusing of energy along the migration trajectory, then it can have a significant influence on the jump probability. As the vibration amplitude increases, the potentially migrating atom approaches the SPC. If a full resonance condition should occur near the SPC, then the stage of the diffusion event in which the migrating atom surmounts its potential barrier and goes through the SPC might follow the head-on collision of the atom with some of its resonance neighbors. The last oscillation, which results in the atom's jumping its barrier, occurs in approximately 10^{-13} sec for many systems, as pointed out by Franklin and Graddick (1970) for Cu self-diffusing. This seems like a short time but it is still significantly longer than the period of the oscillations perpendicular to the trajectory in such a system. Admittedly, the argument given in these paragraphs is not precise, since the pattern of strain around the migrating atom in the SPC is rather complicated. However, it is a general indication of the pattern of events around the migrating atom in the SPC of the fcc structure.

The discussion in the above paragraphs was couched in terms of the vacancy migration process in the fcc structure. However, other diffusion mechanisms in other systems can be analyzed in a similar manner on a qualitative basis. Defect modes (which we define as including localized, gap, and resonance modes) can, as pointed out above, have a significant influence on the diffusion process. This influence, during the development of a migration event, may replace, or be in addition to, the effects of the superposition and interaction of band modes at the location of the migrating atom. If the migrating atom(s) moves relatively slowly along the diffusion trajectory with respect to oscillations perpendicular to this trajectory, then energy transfer to or from the migrating atom(s) takes place via anharmonic interactions. We define the diffusion event as adiabatic (with respect to phonon–phonon interaction) if the migrating atom interchanges no energy with the surrounding lattice during a single migration movement between the EC and the SPC. Nonadiabatic effects, which are small for many systems, account for the energy exchange during the process of the jump. Consequently, in this model, the matrix lattice imparts most of the energy of migration to the diffusing atom(s) during an unspecified period of time and number of oscillations prior to the actual jump event.

Another argument which points up the relationship between the migrating atom and its hindering neighbors has been given by Girifalco (1962) in a discussion of Soret effect. That discussion applies equally well to diffusion in macroscopic equilibrium, since it is a fluctuation in local phonon density, giving rise to a localized gradient in phonon number, which initiates the diffusion event. If the migrating atom gains sufficient energy during its modified random walk about the EC (modified by the presence of the vacancy) so that it can "push" or "punch" its way through the constraining field of its neighbors on its way to the SPC and reach its new EC site, then the migrating atom is the primary source of the strain imposed on the neighboring atoms on the way through the SPC. On the other hand, if the neighboring atoms which constrain the potential diffusing atom along its trajectory relax, so that the migration event can take place more easily (meaning the motion of the migrating atom), then the restraining neighbors, rather than the migrating atom itself, are the major source of the vibrational energy which provides the saddle-point strain. The defect mode vibrations, which are allowed by the symmetry of the system, and the masses and force constants involved, both before and during the migration event, determine, in large measure, which of these possibilities is the more probable. An argument for the former—the "pushing model"—was given above for the vacancy mechanism in the fcc structure. The latter— saddle-point relaxation model—is expected in systems in which the core–core repulsion is soft, so that the force constants of the compressed system differ little from those of the system in tension.

The above arguments have been very qualitative and are meant only to illustrate possibilities existing for certain systems. The detailed calculations for realistic three-dimensional models are highly mathematical and are often too complex to solve in a reasonable amount of time with any degree of precision using contemporary techniques. However, calculations have been done on defect systems in which results have been obtained which are directly applicable to diffusion. These include calculations of the defect modes at vacancies and interstitials in three-dimensional systems. In the following, some of the necessary introduction to the theory of defect lattice vibrations is given, together with its relevance to the theory of atomic migration. Then the results of realistic calculations for certain systems are summarized and extensions to the complicated problems encountered in diffusion are formulated.

B. Inhomogeneity Matrix

It is not the purpose of the theoretical development presented here to give an exhaustive introduction to lattice vibration theory with defects. Rather, a brief introduction will be given, in summary form, noting the features which are significant to diffusion theory and giving sufficient background for those not familiar with the theory to consult specialized references on the topic for further work (Ludwig, 1966; Maradudin, 1965). First, we will consider Green's function solutions in terms of the eigenvalues and eigenfrequencies of the perfect lattice.

The introduction of a defect into an otherwise perfect lattice can affect the frequencies by the changes of three possible variables—the mass, force constants, and number of degrees of freedom. Examples of these three types of imperfection are an isotopic impurity, a substitutional impurity with different force constants but the same mass, and an interstitial, respectively. We will consider, first, changes in both mass and force constants and defer until later the more complex consideration of changes in number of degrees of freedom.

In a perfect lattice, the masses in each unit cell are identical and the force constants are those of the ideal lattice. Then

$$m_{l'\kappa'} \to m_{\kappa'} \quad \text{and} \quad \phi_{\alpha\beta}\begin{pmatrix} ll' \\ \kappa\kappa' \end{pmatrix} \to \phi^0_{\alpha\beta}\begin{pmatrix} ll' \\ \kappa\kappa' \end{pmatrix} \tag{68}$$

denote the changes in Eq. (36) for the perfect lattice. Then Eq. (36) becomes, when the partial with respect to time is taken,

$$\sum_{\alpha l\kappa} \left[m_\kappa \omega^2 \delta_{\alpha\beta} \delta_{ll'} \delta_{\kappa\kappa'} - \phi^0_{\alpha\beta}\begin{pmatrix} ll' \\ \kappa\kappa' \end{pmatrix} \right] u_\alpha\begin{pmatrix} l \\ \kappa \end{pmatrix} = L_{\alpha\beta}\begin{pmatrix} ll' \\ \kappa\kappa' \end{pmatrix} u_\alpha\begin{pmatrix} l \\ \kappa \end{pmatrix} = 0 \tag{69}$$

or, in matrix form,

$$\mathbf{L}u = 0 \tag{70}$$

thus defining **L**. In a lattice with defects, the eigenvalue equation is given by Eq. (36) since $\ddot{u} = \omega^2 u$. Then an inhomogeneity matrix $\delta\mathbf{L}$ can be defined such that Eq. (36) becomes

$$(\mathbf{L} - \delta\mathbf{L})u = 0 \tag{71}$$

where the components of δ**L** are given by

$$\delta L_{\alpha\beta}\begin{pmatrix}ll'\\\kappa\kappa'\end{pmatrix} = -\omega^2\,\Delta m_{l\kappa}\,\delta_{ll'}\,\delta_{\kappa\kappa'}\,\delta_{\alpha\beta} + \Delta\phi_{\alpha\beta}\begin{pmatrix}ll'\\\kappa\kappa'\end{pmatrix} \tag{72}$$

in which $\Delta m_{l\kappa} = m_{l\kappa} - m_\kappa$ and $\Delta\phi_{\alpha\beta} = \phi_{\alpha\beta} - \phi_{\alpha\beta}^0$.

The solution of the inhomogeneous equation (71) is facilitated by Green's function for the perfect lattice. This formalism applies to eigenvalues which coincide with those of the band modes or, at least, do not differ widely from them. In Green's function theory, a unit influence is represented by a Kronecker delta which replaces δ**L**u in Eq. (71), so that

$$\mathbf{LG} = \mathbf{I} \tag{73}$$

where **I** is the unit dyadic. Therefore, $\mathbf{G} = \mathbf{L}^{-1}$ and the displacement is given by multiplying Eq. (71) from the right by **G** to get

$$u = \delta\mathbf{L}\mathbf{G}u \tag{74}$$

Equation (71) refers to the case in which interaction exists between the defect modes and those of the perfect lattice. If no such interaction exists, then the two systems vibrate independently. In that case **L**u and δ**L**u simply vanish independently. This corresponds to truly localized gap-type defect modes. In the case of interest here, interaction between defect and band modes is possible and Green's function is expandable in terms of the eigenvalues of the perfect lattice so that

$$G_{\alpha\beta}\begin{pmatrix}ll'\\\kappa\kappa'\end{pmatrix},\omega^2\Big) = [N(m_\kappa m_{\kappa'})^{1/2}]^{-1}\sum_{kj}\{w_\alpha(\kappa|kj)w_\beta^*(\kappa'|kj)/[\omega^2 - \omega_j^2(k)]\}$$
$$\times \exp\{i\mathbf{k}\cdot[x(l\kappa) - x(l'\kappa')]\} \tag{75}$$

Since the spatial region in which the effects of a defect on the lattice properties (i.e., the force constants) are usually small compared with the lattice as a whole, the quantities δ**L**, **G**, and *u* can be partitioned into localized, perfect lattice, and interaction regions in the following manner

$$\delta\mathbf{L} = \begin{pmatrix}\delta l & 0\\0 & 0\end{pmatrix}, \quad \mathbf{G} = \begin{pmatrix}g & G_{12}\\G_{21} & G_{22}\end{pmatrix}, \quad u = \begin{pmatrix}u_1\\u_2\end{pmatrix} \tag{76}$$

Then, by matrix multiplication of the quantities in Eq. (74), we obtain two equations for the two different regions of the displacement field

$$u_1 = g\,\delta l u_1 \tag{77a}$$

$$u_2 = \mathbf{G}_{21}\delta l u_2 \tag{77b}$$

These are special solutions for the displacement and the general solutions include, in addition, linear combinations of the displacements of the perfect lattice $u_n{}^0$ so that

$$u_1(\text{gen}) = \sum_n a_n u_n{}^0 + \mathbf{g}\,\delta l u_1$$
$$u_2(\text{gen}) = \sum_n b_n u_n{}^0 + \mathbf{G}_{21}\delta l u_2 \tag{78}$$

where a_n and b_n are linear coefficients, determined by the boundary conditions.

The displacement amplitudes given by the solutions of Eqs. (78) include contributions from only those modes which are perturbed by the introduction of the defect, even in the perfect portion of the lattice. Nontrivial solutions of Eq. (77a) are obtained from the determinantal equation $|\mathbf{I} - \mathbf{g}\,\delta l| = 0$ whose roots are the frequencies of the normal modes which are perturbed by the introduction of the defect. Not all modes are perturbed by the introduction of a defect. For instance, some modes in a lattice with a simple mass defect have nodes at the defect so that the defect is not disturbed. Consequently, the general solution for the displacement must include linear combinations of the perfect lattice displacements with non-vanishing coefficients a_n for those modes which are unaltered by the presence of the defect. For those modes

$$\mathbf{L}u = 0 \quad \text{and} \quad \delta \mathbf{L} u = 0 \tag{79}$$

must be satisfied simultaneously.

The frequencies of the normal modes of the crystal with defects stem from the roots of the determinental equation obtained from

$$|\mathbf{L} - \delta \mathbf{L}| = |\mathbf{L}|\,|\mathbf{I} - \mathbf{G}\,\delta \mathbf{L}| \tag{80}$$

The determinental equation in ω^2 is

$$\begin{aligned}\Delta(\omega^2) &= |\mathbf{I} - \mathbf{G}\,\delta\mathbf{L}| = |\mathbf{L} - \delta\mathbf{L}|/|\mathbf{L}| \\ &= \frac{|\mathbf{M}^{1/2}(\omega^2\mathbf{I} - \mathbf{D})\mathbf{M}^{1/2}|}{|\mathbf{M}_0^{1/2}(\omega^2\mathbf{I} - \mathbf{D}_0)\mathbf{M}_0^{1/2}|} \\ &= \frac{|\mathbf{M}|}{|\mathbf{M}_0|}\prod_s \frac{\omega^2 - \omega_s^2}{\omega^2 - \omega_{0s}^2}\end{aligned} \tag{81}$$

where ω_{0s} and ω_s denote the normal mode frequencies of the unperturbed and perturbed lattices, respectively, and $\mathbf{D}_0 = \mathbf{M}_0^{-1/2}\boldsymbol{\phi}^{(0)}\mathbf{M}_0^{-1/2}$ and $\mathbf{D} = \mathbf{M}^{-1/2}\boldsymbol{\phi}\mathbf{M}^{-1/2}$ are the corresponding dynamical matrices. The solution of the dynamical matrix equation for the frequencies, as shown formally in Eqs. (72) and (81), yields solutions for the band modes, as modified by the presence of the defect. In order to treat more complicated situations in which some of the frequencies emerge from the band to form localized or gap modes which are not amenable to the above treatment, Green's function for the defect lattice must be used. This theory will be developed briefly later. But first, an example of a linear chain problem which is important to diffusion will be given. In this problem, the principles just developed will be applied.

C. Linear Chain with Impurity Adjacent to a Vacancy

Since the complexities of solving the three-dimensional case of an impurity adjacent to a vacancy are imposing, the linear chain case is solved to find the dependence of localized mode frequencies on the defect mass and force constants. The formalism of Montroll and Potts (1955) is utilized here. Consider the simple model shown in Fig. 5 in which γ, γ', and γ'' are

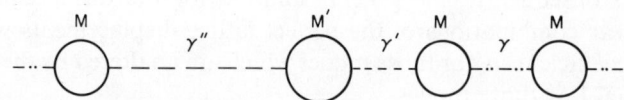

Fig. 5. Linear chain with a vacancy and an adjacent impurity. The matrix atoms have mass M and the impurity, M'. The force constant spanning the vacancy is γ'' and that connecting the impurity to an adjacent matrix atom is γ'.

the nearest neighbor force constants between perfect lattice atoms, between the impurity and a perfect lattice atom, and spanning the vacancy, respectively. The mass of the perfect lattice atoms is m and that of the impurity is m'. The difference equations which represent the disturbance of the vacancy and impurity on the chain are given by

$$Lu(0) = (m - m')\omega^2 u(0) + (\gamma - \gamma'')[u(1) - u(0)] + (\gamma - \gamma')[u(-1) - u(0)]$$
$$Lu(1) = (\gamma'' - \gamma)[u(1) - u(0)] \tag{82}$$
$$Lu(-1) = (\gamma' - \gamma)[u(-1) - u(0)]$$

where $u(0)$, $u(-1)$, and $u(1)$ are the displacements of the impurity, the atom adjacent to the impurity, and the atom across the vacancy from the impurity as shown in Fig. 5. Equations (82) have a general representation, by means of an expansion in terms of the displacements, which is given by

$$Lu(j) = \sum_k w^{(k)}(j+k) u(j+k) \tag{83}$$

1. CLASSICAL AND QUANTUM THEORY OF DIFFUSION

where, for our model,

$$w^{(0)}(0) = (m - m')\omega^2 - 2\gamma + \gamma' + \gamma''$$
$$w^{(1)}(1) = \gamma - \gamma''$$
$$w^{(-1)}(-1) = \gamma - \gamma', \qquad w^{(1)}(0) = \gamma - \gamma'$$
$$w^{(0)}(1) = \gamma'' - \gamma, \qquad w^{(-1)}(0) = \gamma - \gamma''$$
$$w^{(0)}(-1) = \gamma' - \gamma, \qquad w^{(\pm 1)}(\mp 1) = 0$$
(84)

The solution of the problem, which is an inhomogeneous boundary value problem, is accomplished utilizing the Green's function technique. Consequently, the solution for the displacement can be written directly as

$$u(j) = \gamma^{-1} \sum_{kl} g(j + k + l) w^{(k)}(l) u(l) \tag{85}$$

where $g(k)$ is the Green's function for the linear chain with the impurity adjacent to a vacancy. Utilizing Eqs. (84) in Eq. (85), we obtain the complete equation for the displacement.

$$\begin{aligned}u(j) = \gamma^{-1}\{&u(0)[g(j)(m - m')\omega^2 - 2\gamma + \gamma' + \gamma'' + g(j+1)(\gamma + \gamma')\\&+ g(j-1)(\gamma - \gamma'')] + u(1)(\gamma'' - \gamma)[g(j-1) - g(j)]\\&+ u(-1)(\gamma' - \gamma)[g(j+1) - g(j)]\}\end{aligned}$$
(86)

Then we define $P = \gamma'/\gamma$, $T = \gamma''/\gamma$, $Q = m'/m$, and $f = \omega/\omega_L$, where ω_L is the maximum lattice band mode frequency which is given by $\omega_L = (4\gamma/m)^{1/2}$. With these definitions, the displacement can be written

$$\begin{aligned}u(j) = u(0)\{&g(j)[1 - Q]m\omega_L^2 f^2/\gamma - 2 + P + T\\&+ g(j+1)[1 - P] + g(j-1)[1 - T]\}\\&+ u(1)[T - 1][g(j-1) - g(j)]\\&+ u(-1)[P - 1][g(j+1) - g(j)]\end{aligned}$$
(87)

with the resulting inhomogeneity matrix which represents the matrix of interactions between the displacement given by

$u(j)$	-1	0	$+1$
-1	$(P-1)[g(0) - g(1) + \lambda]$	$4f^2 g(1)(1-Q) - 2 + P + T$ $+ g(0)(1-P) + g(2)(1-T)$	$(P'-1)[g(2) - g(1)]$
0	$(P-1)[g(1) - g(0)]$	$4f^2 g(0)(1-Q) - 2 + P + T$ $+ g(1)(1-P) + g(1)(1-T) + \lambda$	$(P'-1)[g(1) - g(0)]$
$+1$	$(P-1)[g(2) - g(1)]$	$4f^2 g(1)(1-Q) - 2 + P + T$ $+ g(2)(1-P) + g(0)(1-T)$	$(P'-1)[g(0) - g(1) + \lambda]$

(88)

The eigenvalue equation can be written, letting $f = \coth(z/2)$, as a cubic equation in λ which is given by

$$\lambda^3 + b\lambda^2 + c\lambda + d = 0 \qquad (89)$$

where

$$\begin{aligned}
b &= [(1-P) + (1-T)][\coth(z/2) - 1] - (1-Q)\coth(z/2) \\
-c &= [(1-P) + (1-T)]e^{-z}(1-Q)\coth(z/2) \\
&\quad + (1-P)(1-T)e^{-2z}[1 - 2e^{-3z/2}/\sinh(z/2)] \\
-d &= (1-T)(1-P)(1-Q)e^{-2z}\coth(z/2)
\end{aligned} \qquad (90)$$

The condition for the existence of localized modes is that each of the λ's must equal 1. Equation (89) is symmetric with respect to the interchange of P and T, as expected. Three special cases, which represent simpler defects, can be recovered from Eq. (89) for certain values or conditions on the parameters. The simplest case is that of the isotope for which $P = T = 1$. In this case, Eq. (89) gives

$$f^2 = 1/Q(2 - Q) \qquad (91)$$

which represents the localized mode of the symmetric mode oscillating about the isotopic impurity. The second case is that of the vacancy, which stems from letting $P = Q = 1$. Then Eq. (89) reduces to

$$f^2 = T^2/(2T - 1) \qquad (92)$$

The third case is the impurity case in which $P = T$ and $Q = 1$. This case gives rise to two modes which are given by

$$\begin{aligned}
f^2 &= P^2/(2P - 1) \\
f^2/(f^2 - 1) &= \{1 + P/[2(P-1)f^2]\}^2
\end{aligned} \qquad (93)$$

From the above special cases, it can be seen that the simplifications do give rise to the expected localized mode frequencies for three well-known cases.

For the case of the impurity with a different mass adjacent to a vacancy, which is represented by Eq. (89) and Fig. 5, numerical calculations were done in order to obtain the localized mode frequencies in terms of the P, Q, and T. The results are portrayed in Figs. 6–8 for $Q = 0.5$, 1.0, and 2.0, respectively. In each figure f is plotted vs P for five different values of T. The results of the numerical analysis reveal the following conclusions. The localized mode frequency increases, as expected, as γ' and γ'' increase and as m' decreases. For large m', however, this increase in f is not significant until the value of P approaches that of T. In addition, for $T \leq 1$, the minimum value of γ' for the existence of a localized mode increases as m' increases.

1. CLASSICAL AND QUANTUM THEORY OF DIFFUSION

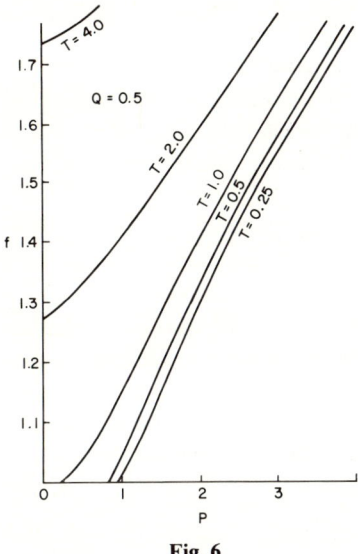

Fig. 6

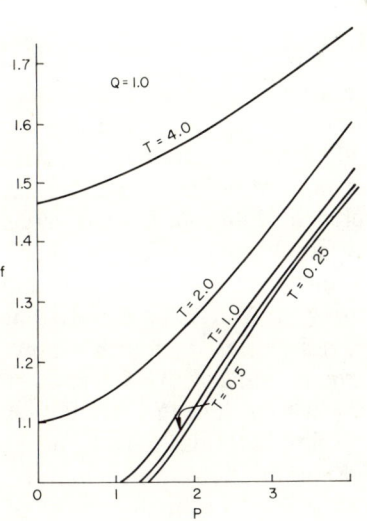

Fig. 7

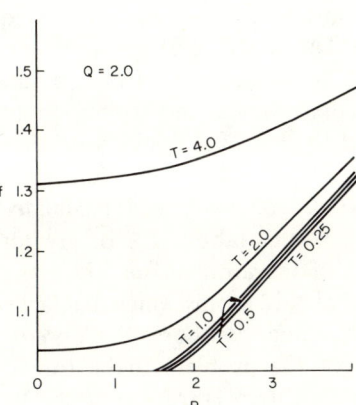

Fig. 8

Fig. 6. Localized mode frequencies for a linear chain with a vacancy and an adjacent impurity atom for $Q = m'/m = 0.5$. The force constant ratios are given by $P = \gamma'/\gamma$ and $T = \gamma''/\gamma$ and the frequency ratio by $f = \omega/\omega_L$ where ω_L is the maximum band mode frequency.

Fig. 7. Same as Fig. 6 except that $Q = 1.0$.

Fig. 8. Same as Fig. 6 except that $Q = 2.0$.

D. GREEN'S FUNCTION IN TERMS OF DEFECT LATTICE EIGENFUNCTIONS

Up to this point, we have used a Green's function defined in terms of the eigenvectors of the perfect lattice whereas, in actuality, we are interested in a lattice with defects. A new Green's function can be defined in terms of the defect lattice by

$$U_{\alpha\beta}\binom{ll'}{\kappa\kappa'},\omega^2\bigg) = [(m_{l\kappa} m_{l'\kappa'})^{1/2}]^{-1} \sum_s [B_\alpha^{(s)}(l\kappa)B_\beta^{(s)}(l'\kappa')/(\omega^2 - \omega_s^2)] \quad (94)$$

where $m_{l\kappa}$ and $B^{(s)}(l\kappa)$ are the mass and eigenvector, respectively, for the κ atom in the l unit cell. The Bloch condition does not apply to a lattice with defects so that no exponent in $k \cdot r$ appears. However, another way of writing U is to utilize a double Fourier series in both kj and $k'j'$. The problem then becomes one of evaluating the Fourier coefficients properly.

The defining relationship for the defect lattice Green's function U is given by the inverse of $L - \delta L$,

$$U = (L - \delta L)^{-1} \quad (95)$$

as opposed to G, which is equal to L^{-1}. Equation (95) can be written, therefore, as $U = (I - G\delta L)^{-1}G$ and can be rearranged to give the alternate form,

$$U = G + G\delta L\, U \quad (96)$$

The matrix U can be partitioned as done above for G and, in certain applications, solution of the secular equation only for the space of δl is required. The space of δl for a detailed calculation of the defect modes caused by an impurity adjacent to a vacancy in a three-dimensional lattice, which is the problem most often encountered in diffusion, is quite large, compared to that of a single point defect at equilibrium, when the effects of the strain during the migration process is included. For instance, in the SPC in Cu, Johnson[‡] has calculated, using a Morse potential, that significant strains (about 2% or greater) extend at least to fourth-nearest neighbors of the atom in its SPC, as shown in Table I. This region of strain includes 22 atoms which is a large number to include in a calculation of the defect modes of the SPC. Consequently, simplifications must be made in order for the problem to be tractable. Simplifications which may be introduced include (1) truncating the strained region at a small number of neighbors, (2) using symmetry of the SPC to reduce calculation time, and (3) letting the mass of the impurity be the same as that of the host (which makes the problem equivalent in difficulty to that of self-diffusion,

---

[‡] R. A. Johnson, unpublished work on Cu similar to that reported for Ni by Johnson (1966b).

1. CLASSICAL AND QUANTUM THEORY OF DIFFUSION

since the force constants would be altered in either case). Utilizing these simplifications, a three-dimensional calculation of the case of an impurity in the SPC might be feasible with contemporary computer capacity and techniques.

E. Displacement and Momentum Correlation Functions

The space–time correlations of displacements and momenta play an important role in diffusion events. The displacement of a potentially migrating atom along its migration trajectory at a certain point in time can be studied via its correlations in space and time with the motions of both neighboring and distant atoms in the system. The displacement of atom κ in the unit cell l in the α direction at time t is denoted by $u_\alpha(l\kappa; t)$ so that the displacement and, in a similar manner, momentum correlation functions, together with their interaction, can be written as

$$\langle u_\alpha(l\kappa; t) u_\beta(l'\kappa'; 0)\rangle = \frac{\hbar}{2m} \sum_s \frac{B_\alpha^{(s)}(l\kappa) B_\beta^{(s)}(l'\kappa')}{\omega_s} [(n_s + 1)e^{-i\omega_s t} + n_s e^{i\omega_s t}] \quad (97a)$$

$$\langle p_\alpha(l\kappa; t) p_\beta(l'\kappa'; 0)\rangle = \frac{\hbar m}{2} \sum_s \omega_s B_\alpha^{(s)}(l\kappa) B_\beta^{(s)}(l'\kappa') [(n_s + 1)e^{-i\omega_s t} + n_s e^{i\omega_s t}] \quad (97b)$$

$$\langle p_\alpha(l\kappa; t) u_\beta(l'\kappa'; 0)\rangle = \frac{\hbar}{2i} \sum_s B_\alpha^{(s)}(l\kappa) B_\beta^{(s)}(l'\kappa') [-(n_s + 1)e^{-i\omega_s t} + n_s e^{i\omega_s t}] \quad (97c)$$

using the transformation given by Eq. (33) together with an exponential time dependence. In addition to Eq. (97c), the displacement–momentum correlation function $\langle u_\alpha(l\kappa; t) p_\beta(l'\kappa'; 0)\rangle$ can be written. We note that the evaluation of these correlation functions for each atom pair requires a knowledge of the eigenvectors for each pair, in addition to the masses and eigenfrequencies of the system. A more useful form for the correlation functions is obtained by Fourier inversion which yields, for displacement correlations,

$$\langle u_\alpha(l\kappa; t) u_\beta(l'\kappa'; 0)\rangle = \frac{\hbar}{\pi} \int_{-\infty}^{\infty} d\omega \frac{\operatorname{sgn} \omega}{1 - \exp(-\beta\hbar\omega)} e^{-i\omega t} \operatorname{Im} U_{\alpha\beta}\!\left(\begin{matrix} ll' \\ \kappa\kappa' \end{matrix}; \omega^2 - i\varepsilon\right) \quad (98)$$

In Eq. (98), the frequency dispersion and the defect lattice Green's function are required for evaluation. Similar relations can be written for the other correlation functions. In the equal-time case, it is easily shown that

$$\langle u_\alpha(l\kappa) u_\beta(l'\kappa')\rangle = -\beta^{-1} \sum U_{\alpha\beta}\!\left(\begin{matrix} ll' \\ \kappa\kappa' \end{matrix}; -\Omega_n^{\,2}\right) \quad (99)$$

where $\Omega_n = 2\pi n/\beta\hbar$. Thus, evaluation of the correlation functions is reduced to evaluation of the elements of the Green's function matrix for the defect crystal, which may be done without deriving the eigenvectors.

An additional application of lattice vibration theory to the theory of diffusion is that of the scattering of phonons by defects. The transfer of energy via lattice vibrations (having the proper wavevector to affect the migration process) to the migrating atom involves the scattering matrix of the defect involved. Utilizing scattering matrix theory, the critical migration displacement can be derived from the scattering of phonons by the defect center. Schottky (1965) has used this type of argument in the theory of thermal diffusion. Even if the system is in thermal equilibrium, diffusion can be analyzed in terms of localized thermal fluctuations so that the scattering of phonons having a net wavevector flux in the migration direction via Umklapp processes, in a region localized around the defect, can give rise to diffusion in a system in equilibrium.

Diffusion in isotropic liquids differs from that in solids since it is essentially a kinetic rather than an activated process (activation energies are very small). An atom with thermal energy $\approx kT$ in a liquid is hindered from translational motion by the liquid's viscosity η rather than potential barriers, which characterize solids, so that $D = kT/\eta$. In kinetic theory, D is given by the time integral of the velocity autocorrelation function and, for an anisotropic liquid such as a nematic liquid crystal,

$$D_{\alpha\beta} = \int_0^\infty dt \langle v_\alpha(t) v_\beta(0) \rangle \tag{100}$$

where $v_\alpha(t)$ is the α component of the molecular velocity at time t. Smectic liquid crystals have translational order in one dimension in which activated diffusion takes place across potential barriers, whereas liquid-like diffusion occurs perpendicular to the direction of translational order. Liquids and liquid crystals have, in addition to translational symmetry, rotational degrees of freedom with rotational diffusion constants given by the time integral of the angular velocity autocorrelation function.

F. Internal Degrees of Freedom

An additional complication is introduced into the theory of lattice vibrations in defect crystals by molecular-type defects which have additional degrees of freedom. In the theory which has been developed up to this point, the atoms or holes in a crystal with defects could be assigned to a particular basis site within a unit cell. However, defects exist which may not be located at atomic sites of the system's basis and which may have extra atoms compared to the perfect lattice. In this case, the effects of the extra degrees of freedom can be taken into account by a theory developed by Wagner (1963, 1964). This type of theory is not necessary for vacancies, even though the introduction of a vacancy reduces the number of degrees of freedom, since the latter can simply be eliminated from the perfect lattice theory. In the case of the introduction of an interstitial into a perfect lattice, on the other hand, three extra degrees of freedom for displacements in Cartesian space are introduced into a system which still has the same number of unit cells and basis atoms per unit cell as the perfect lattice had. Consequently, a method for treating these extra degrees of freedom is needed and is provided by Wagner's theory. The atomic mechanisms of diffusion to which this theory applies include the interstitial, interstitialcy, and crowdion mechanisms, since an extra atom exists in the locale of these defects compared with the perfect lattice.

The addition of atoms to the perfect lattice causes a perturbation in the lattice modes. This is accounted for via $\delta \mathbf{L}$, as in Eq. (71). The interaction of the extra atom(s) with the lattice is accounted for through a coupling matrix $\mathbf{A}$ which couples the "molecular" coordinates ξ to the displacements of the lattice $\mathbf{u}$. Consequently, a pair of coupled equations—one for the lattice coupled to the molecular coordinates and another for the molecular system coupled to the lattice—can be written as

$$(\mathbf{L} - \delta \mathbf{L})\mathbf{u} + \mathbf{A}\xi = 0 \tag{101}$$

$$\tilde{\mathbf{A}}\mathbf{u} + \mathbf{\Lambda}\xi = 0 \tag{102}$$

In Eq. (102), $\mathbf{\Lambda}$ is a matrix which gives rise to the molecular oscillations without any effects from the surrounding lattice. The transpose of the matrix $\mathbf{A}$ describes the interaction of the molecular system with the lattice displacements. If the lattice and the molecular system have $3rN$ and $3s$ degrees of freedom, respectively, then the matrices in Eqs. (101) and (102) have the following dimensions: $\mathbf{L}$ and $\delta \mathbf{L}$ have $3rN \times 3rN$ dimensions while $\mathbf{A}$ and $\tilde{\mathbf{A}}$ have $3rN \times 3s$ and $3s \times 3rN$, respectively, and $\mathbf{\Lambda}$ has $3s \times 3s$ dimensions. The lattice and molecular displacement vectors, $\mathbf{u}$ and ξ, have $3rN$ and $3s$ dimensions, respectively. If no coupling existed between the lattice and molecular systems, then $\mathbf{A} = \tilde{\mathbf{A}} = 0$. This would represent the case of completely localized modes.

A Green's function matrix $\gamma(\omega^2)$ for the molecular system can be defined as

$$\gamma(\omega^2) = \mathbf{\Lambda}^{-1} \tag{103}$$

so that the molecular coordinates ξ can be eliminated from Eqs. (101) and (102) by combining them to give

$$(\mathbf{L} - \delta \mathbf{L} - \mathbf{A}\gamma\tilde{\mathbf{A}})\mathbf{u} = 0 \tag{104}$$

which may also be written, using $\mathbf{L} = \mathbf{G}^{-1}$, as

$$\mathbf{u} = \mathbf{G}(\delta\mathbf{L} + \mathbf{A}\gamma\tilde{\mathbf{A}})\mathbf{u} \tag{105}$$

Another representation of Eq. (104) utilizes the defect lattice Green's function, $\mathbf{U} = (\mathbf{L} - \delta\mathbf{L})^{-1}$, so that

$$\mathbf{u} = \mathbf{U}\mathbf{A}\gamma\tilde{\mathbf{A}}\mathbf{u} \tag{106}$$

Since the interaction of the molecular defect can be limited realistically to a rather small number of atoms, n, of the host lattice, the interaction matrix can be written as

$$\mathbf{A} = \begin{pmatrix} \mathbf{a} \\ 0 \end{pmatrix} \tag{107}$$

where $\mathbf{a}$ is a $3n \times 3s$ matrix. The matrices $\delta\mathbf{L}$, $\mathbf{G}$, and $\mathbf{u}$ can be partitioned in a manner similar to that of Eq. (76) so that $\delta\mathbf{L}$ and $\mathbf{G}$ become $3n \times 3n$ matrices and $\mathbf{u}$ a $3n$-dimensional vector. In Eq. (106), $\mathbf{U}$ would then be a $3n \times 3n$ matrix. The eigenvalue equation obtained from Eqs. (76), (105), and (107) is

$$\Delta(\omega^2) = |\mathbf{I} - \mathbf{g}(\delta\mathit{l} + \mathbf{a}\gamma\tilde{\mathbf{a}})| = 0 \tag{108}$$

Another representation of the eigenvalue equation is obtained utilizing $\mathbf{U}$ in a partitioned form in which $\mathbf{u}_{11}$ is defined in the space of $\delta\mathbf{L}$. It is given by

$$\Delta(\omega^2) = |\mathbf{I} - \mathbf{u}_{11}\mathbf{a}\gamma\tilde{\mathbf{a}}| = 0 \tag{109}$$

Both Eqs. (108) and (109) are $3n \times 3n$ determinants and the difficulty of solution increases rapidly with increasing n. We note that in Eq. (108) the additional term $g a \gamma \tilde{a}$ is added to the determinantal equation for a defect lattice without extra degrees of freedom. It is this term which gives rise to the additional frequencies introduced by the molecular defect. Anharmonic terms can be added rather easily to the formalism by addition of matrices to Wagner's (1963, 1964) theory.

G. Defect Calculations Related to Diffusion

The effects of defect modes on vacancy diffusion were first discussed by Montroll and Potts (1955) for the case of a linear chain with a pulsating localized mode. For that case, it was pointed out that the diffusion jump time was shorter than the period of the slow pulsating mode, so that the structure of the pulsating mode was not altered appreciably by diffusion of an atom through its SPC until some time after the migration event had occurred. The migration process was adiabatic for the linear chain in the sense that the pulsating mode remained essentially unchanged during the diffusion event. In the three-dimensional case, it turns out that nonadiabatic effects are also small during the actual jump but are significant, especially for certain systems, during the time development of the displacement and momentum autocorrelation functions of the migrating atom before migration.

The defect mode calculations which apply directly or indirectly to diffusion include those of Schottky (1965) for phonon scattering in a linear chain with a vacancy, Pegel (1967a,b), who worked with an interstitial in a linear chain (which is topologically equivalent to a substitutional impurity), Ludwig (1966), Land and Goodman (1967), and Mitani and Takeno (1965) who analyzed the effects of vacancies on localized modes in simple cubic, fcc, and NaCl-type crystals, respectively, and of Bellomonte and Pryce (1966) and Brice (1965) who developed the localized and resonance modes, respectively, in Si and of Si and Ge. Some of the results of these calculations which are important to the theory of diffusion, especially to that of three-dimensional crystals, are given here. One important significance of resonance modes to the process of diffusion lies in the transfer of energy between the atoms involved in the migration event and the lattice. If resonance modes exist or develop during an event, then the scattering probability of the defect is enhanced by a resonance factor. Consequently, the resonance frequencies and their temperature dependence may control, in some cases, a large segment of the energy transfer involved in diffusion. In the case of quasi-localized modes, which may be low frequency virtual local modes with frequencies in or adjacent to the band of perfect lattice frequencies, the lower the frequency of the mode the narrower the resonance peak becomes and the more real the virtual mode. True localized modes have no significant interaction with the lattice so that energy transfer between the localized mode and those of the matrix material is negligible. Diffusion in this case is very small unless some external field can be used to excite the localized mode. However, in a strongly resonant system, interaction with the lattice matrix is strong, so that energy transfer to and from the atoms involved in the diffusion process can affect that process if the wavevectors of the resonance modes are directed such that their excitation stimulates diffusion. On the other hand, certain vibrational modes of the defect center may inhibit the migration process if their wavevectors and amplitudes do not contribute to the diffusion mode. Inhibited diffusion would occur when the excitation of an inhibiting mode makes another mode, which aids diffusion, less effective. In the case of a strong resonance between the lattice and a defect mode which aids diffusion, the external excitation of the resonance frequency by some nonthermal means would provide for an external influence on the diffusion process which is frequency selective. This is possible, in certain systems, with external electric or laser light fields.

The resonance and localized mode problem for a vacancy adjacent to an impurity has not been solved. However, each defect has been solved independently. Consequently, the aspects of the independent defects only can be discussed. Since the isolated vacancy problem is applicable to self-diffusion, the results of the solution of the determinantal equation for the frequencies of three-dimensional lattices are useful to diffusion. Ludwig (1966) found that a vacancy in a SC lattice does not show localized vibrations. Land and Goodman (1967) got a similar result for Fe and Na but did find a weak localized mode (frequency slightly above the band modes) in Cu. Mitani and Takeno (1965) studied the NaCl lattice and found that gap modes were produced by a metal ion vacancy only when the mass of the halide is greater than the mass of the metal atom.

Land and Goodman (1967) utilized a molecular method rather than the Green's function technique. They found that the A_{2u} mode in Cu gave rise to a weak localized mode. In the molecular method, the number of movable shells of atoms around the vacancy influenced the results as shown in Fig. 9. However, after the ninth shell for the A_{2u} mode, there was no apparent change in the localized mode frequency. Their work showed that, for the most part, there was a relative softening of the force constants around a vacancy and that only a few of the force constants became stiffer. Consequently, most of the frequencies induced by the presence of a vacancy in a fcc lattice lie within the resonance range. The nature of these resonances is important to diffusion and further work is needed in this area.

The results obtained by Mitani and Takeno (1965) for an F center (Cl ion vacancy) in NaCl are shown in Fig. 10. In this case, $m(\text{Na})/m(\text{Cl}) = 0.65$, which satisfies the required condition on the masses for the appearance of a gap mode. Modes which have s-, p-, and d-like character were obtained. They also showed that gap modes occurred, near the bottom of the optical band, only if the vacancy was on the sublattice of the heavier ion. In addition, results similar to the F center were expected for V centers. From these results, one might induce that if defect modes occurred at all in the case in which the vacancy was on the sublattice of the lighter ion, the frequencies would be in the resonance range. Distortion of the lattice around the vacancy was neglected in this work whereas the work of Land and Goodman (1967) showed that the relaxation around the vacancy was necessary for localized modes to occur. Consequently, the assumption of no distortion may have oversimplified reality in the case of NaCl. Additional assumptions made by Mitani and Takeno (1965) include the harmonic approximation, neglect of long-range forces, and neglect of the electronic polarizability.

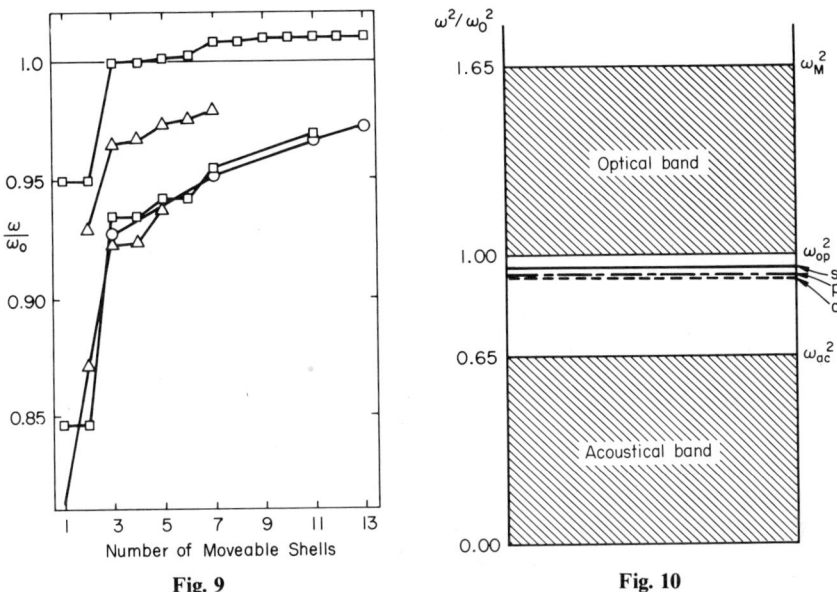

Fig. 9. Frequency versus the number of movable shells for the highest frequency A_{2u} and F_{1u} modes in the molecular derivation of Land and Goodman (1967). The upper and lower curves for the A_{2u} and F_{1u} modes are the frequencies with and without relaxation around the vacancy, respectively. (Used with permission.)

Fig. 10. Gap mode frequencies of an M_1 hole in a diatomic lattice for the case in which $M_2/M_1 = 0.65$. Modes having s-, p-, and d-like character are shown. Taken from Mitani and Takeno (1965).

The localized modes of interstitial ions of Li in Si have been derived by Bellomonte and Pryce (1966). They found that localized modes occurred only if the mass of the interstitial was below a critical value of approximately 12 amu. Their results were close to experimentally observed values for the ^{6}Li and ^{7}Li isotopes. Phonon resonances associated with interstitials in Ge and Si were studied by Brice (1965). The variables were the ratio of interstitial mass to that of Ge and Si and the force constants between the interstitial and the nearest neighbors. Calculations concentrated on resonance frequencies below the transverse acoustic band edge. The results for Ge and Si are shown in Figs. 11 and 12, respectively. These figures display the expected rapid increase of resonance frequency with decreasing mass and with increasing force constant. In both of the studies on interstitials reported here, the distortion of the lattice was not accounted for. Therefore, the results must be corrected for this assumption, if accuracy is desired.

Of the various mechanisms of diffusion, the vacancy and interstitial

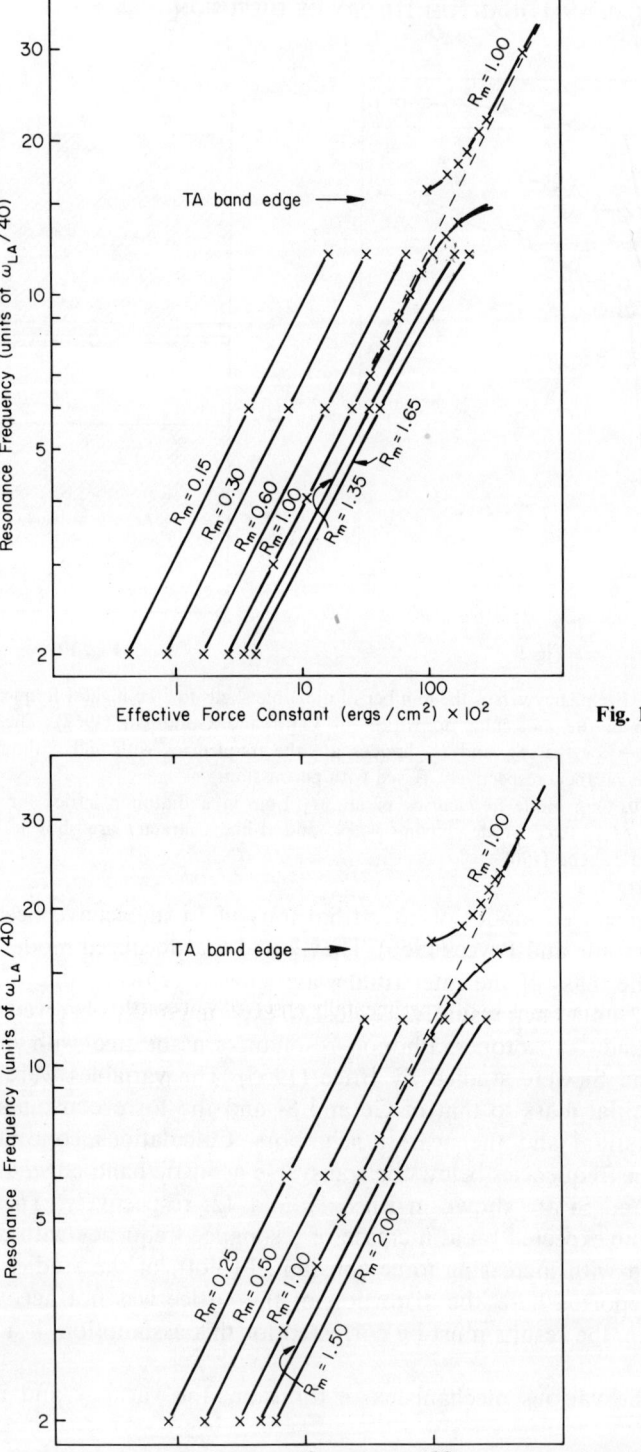

Fig. 11

Fig. 12

mechanisms are the most common. The results of the calculations reported for defect modes for the vacancy and interstitial are helpful to diffusion in that they give the defect modes and other useful information which deals with the EC in the diffusion event. However, no work has yet been done which deals with the changes in the defect mode structure as the diffusion event takes place or as the displacement amplitude in the diffusion mode builds up. The other mechanisms of diffusion, such as the interstitialcy and crowdion mechanisms, demand special consideration. The EC of the interstitialcy mechanism is identical to that of the interstitial mechanism. However, the cooperative and permanent displacement of two atoms in the migration event of the former mechanism is a complex process to analyze. In the crowdion mechanism in the bcc structure, the defect modes of the EC are unknown. Another interesting mechanism in the bcc case which has been postulated is that of the vacancy with metastable SPC (Grimes, 1968). The symmetry of the vacancy diffusion mechanism in the SPC in the bcc structure, and in other structures, makes the SPC more amenable to a derivation of the defect modes than the case in the EC of an impurity adjacent to a vacancy.

It is important to point out the significance of the material presented in this section to the diffusivity which will be developed in the next section. In many-body form, the "preexponential" part of the jump frequency (and, hence, the diffusivity) depends upon the set of system frequencies in the EC and in the SPC. Consequently, any alteration of the system's frequencies by the presence of defects affects the EC and further alterations at the SPC enter the theory. To be explicit, in the harmonic approximation, the "preexponential" part of the jump frequency is given by

$$\Gamma_0 = \prod_i^{3N} v_i \bigg/ \prod_i^{3N-1} v_i' \qquad (110)$$

where v_i and v_i' are the system frequencies in the EC and SPC, respectively. We note that resonance modes have no effect on this term and, hence, should not affect the "preexponential" part of the diffusivity. However, localized modes in the EC and SPC will affect the numerator and denominator of this term, respectively. The effect of resonance modes enters the theory via the temperature in the exponential. Since the temperature in the localized region where the diffusion event is taking place is the quantity that fluctuates, the buildup of energy in the oscillating diffusion

---

Fig. 11. Resonance frequencies versus effective force constants for interstitial atoms in Ge for different values of the ratio R_m of interstitial to Ge atom mass. Taken from Brice (1965).

Fig. 12. Same as Fig. 11 except for Si. Taken from Brice (1965).

mode prior to the actual migration event changes the temperature in this localized region. The temperature is more conveniently described in terms of the equal-time correlation function for displacement. Then the equilibrium temperature of the system is written in terms of the displacement correlation function of the EC of the defect center. In a rather oversimplified manner which is, nonetheless, quite illustrative of the form we are speaking of, the activation energy for migration over kT is given by (Franklin, 1969; Franklin and Sengupta, 1972)

$$\Delta E_m/kT \approx \sum_i u_{ic}^2 \sum_{ij} \langle u_i u_j \rangle \tag{111}$$

where u_{ic} is the ith component of the critical displacement of the system in the SPC and $\langle u_i u_j \rangle$ is the equal-time displacement correlation function in the EC. The latter can be obtained from Eq. (97a), (98), or (99).

V. Temperature Dependence—Classical Case

In this section, the temperature dependence of the diffusivity is developed with the inclusion of a temperature dependent activation energy and pre-exponential part. While it is true that a temperature dependent activation energy implies, thermodynamically, a temperature dependent entropy factor in D, a more basic theoretical construct is needed for quantitative analysis. This is provided by a detailed analysis of the temperature dependent factors in D_0 from a lattice vibration standpoint. The classical case is considered in this section, whereas quantum mechanical effects are considered in Section VII. This section includes a partition function expansion of D_0 which shows that third- and higher nearest neighbor interactions make a negligible contribution to D_0 and that perturbation theory is convergent for the vacancy mechanism in Cu, a proof that the Arrhenius plot is valid for a single-migration mechanism over a large portion of the temperature range, despite a high degree of anharmonicity, and a quantitative calculation of the small deviations from the Arrhenius plot at high temperatures.

In the harmonic classical theory of solid state diffusion, we normally write the diffusivity in the form

$$D = \beta f l^2 \left(\prod_i^{3N} v_i \bigg/ \prod_i^{3N-1} v_i' \right) e^{-\Delta E/kT} \tag{112}$$

where $\beta, f,$ and l are, respectively, the geometrical factor, ‡ correlation factor, and jump distance. Consider the activation energy ΔE_0 as the difference in static lattice energies for the system in the SPC and the EC at zero degrees.

‡ Note that β is also used as $1/k_\beta T$.

Then, if the activation energy had a temperature dependence, the linear term would give rise to a constant multiplicative factor in the diffusivity. For example, if

$$\Delta E = \Delta E_0 + \lambda T \tag{113}$$

where ΔE_0 is the activation energy at zero degrees and λ is the thermal coefficient for a linear temperature dependence, then

$$D = e^{-\lambda/k} D_0 \, e^{-\Delta E_0/kT} \tag{114}$$

where $D_0 = \beta f l^2 (\prod v_i / \prod v_i')$. Thus, the first-order correction accounting for the temperature dependence of ΔE, which is expected to be linear in temperature, gives rise to a constant temperature independent term in D. Consequently, deviations from true Arrhenius behavior with temperature, when they occur because of the fundamental nature of the migration dynamics for a single-migration mechanism, are most likely due to the temperature dependence of the preexponential part of the diffusivity D_0 in first order. The factors which contribute to the temperature dependence of D_0 include the following items: (1) interatomic spacing (thermal expansion effects), (2) the force constants, both harmonic and anharmonic, (3) anharmonic terms which include an inherent temperature dependence, and (4) quantum effects in the expansion of the occupation number for phonons in both harmonic and anharmonic terms. Equation (112) has been written for the harmonic case; anharmonic corrections to the harmonic frequencies will contain terms with a temperature dependence which will contribute to the overall temperature dependence of D. The remaining temperature dependence of D_0 stems from that of the jump distance l and the frequencies which appear in D. When the frequencies of the SPC, v_i', are obtained in terms of those of the EC via force constant expansions for the SPC force constants, then the temperature dependence of D_0 can be analyzed in terms of known quantities. In order to do this, we use the perturbation procedure developed in Section II. We will do this for the quasi-harmonic case, since it is easier to visualize, and then later insert the anharmonic terms which have been calculated (Franklin, 1969). We consider Eq. (8) in which the potential in the EC and the deformed potential in the SPC are given by

$$\Phi(V) = \Phi^0(V) + \tfrac{1}{2} \sum_{ij} \phi^0_{ij} u_i u_j \tag{115}$$

$$\Phi(S) = \Phi^0(S) + \sum_i \phi^0_i u_i + \sum_{ij} \phi^0_{ij} u_i \bar{u}_j + \tfrac{1}{2} \sum_{ij} \phi^0_{ij} u_i u_j \tag{116}$$

where the superscript zero denotes potentials and force constants in the EC and $\bar{u}_i$ is a critical saddle-point displacement. This theory follows that of Born and Huang (1956) for a deformed lattice such that the deformed saddle-point force constants can be written, for small deformations, in terms of those of the EC, which are known or can be estimated realistically. The expansion of $\Phi(S)$ was truncated at terms which are second-order in the force constants. The term in ϕ_i^0 vanishes since $\phi_i^0 = \partial \Phi(V)/\partial u_i = 0$ in equilibrium. When Eqs. (115) and (116) are used in Eq. (8) and the perturbation terms expanded to the fourth power in u_i, we get, utilizing the reduced notation of Section II,

$$\Gamma_0 = \left(\frac{kT}{2\pi}\right)^{1/2} \int_S e^{-\phi_2/2kT}\left(1 + \frac{\phi_{11}^2}{2(kT)^2} + \frac{\phi_{11}^4}{24(kT)^4}\right) dS \bigg/ \int_V e^{-\phi_2/2kT} dV \quad (117)$$

where only the even terms have been displayed since the odd terms vanish anyway on integration. When the matrices are diagonalized, integration of Eq. (117) gives

$$\Gamma_0 = v_{3N} \prod_i^{3N-1} (1 + \phi_i'' \bar{u}_i^2/2kT + (\phi_i'')^2 \bar{u}_i^4/8(kT)^2) \quad (118)$$

where ϕ_i'' is the harmonic force constant for interaction with the ith neighbor. As a numerical example, consider the vacancy diffusion mechanism in Cu in which the SPC has four first-nearest, four second-nearest, and eight third-nearest neighbors. We will evaluate contributions to Γ_0 up to third-nearest neighbors and include the fourth-order term in u_i to show the magnitude of these terms in the product series.

For the numerical evaluation of Eq. (118), we will use the Born–Mayer potential which is good particularly for repulsive conditions. Since the SPC involves mostly repulsive conditions, the Born–Mayer potential provides an adequate approximation for the interaction potential of diffusion in the vacancy mechanism. For the harmonic force constant, the potential is given by

$$\phi'' = (AB^2/r_0^2) e^{-B(r/r_0-1)} \quad (119)$$

where A and B are the constants in the Born–Mayer potential, and, for Cu, they are given by $A = 0.0728$ eV and $B = 12.67$ (Hiki et al., 1967). The nearest neighbor distance is r_0, which is 2.599 Å for Cu at 1000°C. At a temperature of 1000°C, $\phi'' = 2.01 \times 10^4$ ergs/cm² for nearest neighbors. For second- and third-nearest neighbors in the perfect lattice configuration, the values of the force constants obtained from Eq. (119) are 1.08×10^2 and 1.91 ergs/cm², respectively. From Table I, the differences of the first-, second-, and third-nearest neighbors from the migrating atom with respect to the equilibrium positions in the perfect lattice are given by $X(nn) = 0.176$ Å, $X(nnn) = 0.914$ Å, and $X(nnnn) = 0.995$ Å, respectively. Utilization of the

1. CLASSICAL AND QUANTUM THEORY OF DIFFUSION

TABLE I

Saddle-Point Configuration in Copper[a,b]

Neighbor	No. of neighbors	Symmetry	$\|r_{Nom}\|^2$	Relaxed positions			$\|r\|^2$
1	4	1 $\bar{\tfrac{1}{2}}$ $\tfrac{1}{2}$	1.5[c]	1.087	−0.551	0.511	1.746
2	4	0 $\bar{\tfrac{1}{2}}$ $\tfrac{3}{2}$	2.5	0.000	−0.461	1.432	2.263
3	8	1 $\tfrac{1}{2}$ $\tfrac{3}{2}$	3.5	1.002	0.497	1.478	3.435
4	2	0 $\tfrac{3}{2}$ $\tfrac{3}{2}$	4.5	0.000	1.476	1.476	4.357
4	4	2 $\tfrac{1}{2}$ $\tfrac{1}{2}$	4.5	2.033	0.506	0.506	4.645
5	4	1 $\tfrac{3}{2}$ $\tfrac{3}{2}$	5.5	0.998	−1.490	1.490	5.436

[a] R. A. Johnson. (1966a). Department of Materials Science, University of Virginia, private communication.

[b] The saddle-point configuration (SPC) is the configuration in which an atom adjacent to a vacancy is half way between two equilibrium configurations so that the vacancy is split into two equal halves.

[c] Nominal distances squared of neighbors from the center of the SPC, measured in units of $a_0/2$ where a_0 is the fcc unit cell edge.

given values of the displacements and force constants gives, for the pre-exponential jump frequency of migration,

$$\Gamma_0 = v_{3N}(1 + 0.174 + 0.015)^4(1 + 0.025)^4(1 + 0.001)^8 \qquad (120)$$

where the successive exponents denote the numbers of first-, second-, and third-nearest neighbors of the SPC for vacancy diffusion in the fcc lattice. Finally, we obtain $\Gamma_0 = 2.24v_{3N}$. If we let v_{3N} be the Debye frequency which, for Cu, is 7.05×10^{12}/sec, we obtain $\Gamma_0 = 1.58 \times 10^{13}$/sec at 1000°C. This example was meant purely as an exercise since it is essentially one-dimensional in character and does not make use of the usual transformation, Eq. (33), for the displacement. However, it is reasonably close to the better approximation (Franklin, 1969) which, when done in full quantum mechanical form, does not yield an answer which is widely different from that just obtained. The important features to note in the derivation of the result are (1) that the second-nearest neighbors contribute a factor of only about 1.002 and the third-nearest neighbors contribute a negligible amount, (2) the u_i^4 term in the expansion of the exponential contributes a factor of only 1.013/atom for the four nearest neighbors, and (3) that perturbation theory provides a reasonable representation of the SPC since convergence, at least for the case of vacancy diffusion in Cu, is rapid. Even when the $\phi_i'' \bar{u}_i^2 / 2kT$ term in Eq. (118) is slightly greater than 1, convergence is reasonably rapid, so that perturbation theory provides an approach which gives a reasonable approximation for the jump frequency. The temperature dependence of D_0 has been derived in detail in a manner similar to that just completed for the classical case utilizing, from the outset, the quantum mechanical representation (Franklin, 1972). The high temperature regime of this derivation will now be discussed since it represents the classical approximation.

The above derivation of Γ_0 in a simple classical approximation was given to show the general technique for obtaining a reasonably accurate assessment of the jump frequency utilizing perturbation theory. When the theory is put into quantum mechanical form, derived for migration, and inserted into D_{0m}, the following result (Franklin, 1972) is obtained for diffusion via the single-vacancy mechanism

$$D_{0m} = \beta f r_0^2(T_0) v_D(T_0) \alpha^2(T, T_0) \sigma(T, T_0) F(T, T_0) \tag{121}$$

where σ and α are given by $\sigma = 1 + \sigma_0(T - T_0)$, $\alpha = 1 + \alpha_0(T - T_0)$, and T_0 is the temperature on which the parameters are based. The quantities σ and α are the temperature dependences of the frequency and the thermal expansion, respectively. The temperature dependences of the nearest neighbor distance r_0 and the Debye frequency v_D are included utilizing α and σ, respectively. The temperature dependent factor $F(T)$ contains the expanded partition function components for expansions of both the potential, the force constants in the SPC, and only the former in the EC. Anharmonic terms are included here in both the EC and SPC. Then $F(T)$ is given by

$$F(T) = [1 + \alpha^2(\mu_4 Z_{22}^S/\sigma^2 + \mu_2{}^2{}_{11}Z_{11}^S/\sigma^3 + \mu_3{}^2{}_{21}Z_{21}^S/\sigma^4) \\ + (\eta/\sigma^4)(\mu_4 Z_4^S + \mu_3{}^2{}_3 Z_3^S/\sigma^2)]/[1 + (\eta/\mu_2{}^2)(\mu_4 Z_4 + \mu_3{}^2{}_3 Z_3/\mu_2)] \tag{122}$$

where $\eta = T/T_0$ and the temperature dependence of the force constants is contained in μ_2, μ_3, and μ_4, which are given by $\mu_i = 1 + \mu_{i0}(T - T_0)$, where μ_{i0} is the thermal coefficient of the ith force constant and where $i = 2, 3, 4$ for harmonic, cubic, and quartic force constants, respectively. The Z's and Z^S's are the EC and SPC partition function components. It is not the purpose of our discussion here to derive these partition function components since that has been done before (Franklin, 1969) and will be discussed briefly below. However, the temperature dependence of D_{0m} must include that of the partition function components in $F(T)$.

Numerical evaluation of the preexponential part of the diffusivity for migration in the single-vacancy mechanism was done for Cu in the high temperature regime using Eq. (121) and the values of the constants are given in Table II. The results are shown in Fig. 13 for given values of μ_{30} and μ_{40}. It can be seen that D_{0m} is not constant with temperature except for $\mu_{30} = 3 \times 10^{-4}/°C$. For lower values of μ_{30}, $F(T)$ increases with increasing temperature and for larger values, it decreases. Consequently, as

TABLE II

VALUES OF CONSTANTS USED TO EVALUATE EQ. (128)

$\beta = 1.0$	$f = 0.781$
$T_0 = 20°C$	$r_0(20°C) = 2.556$ Å
$v_D(20°C) = 7.05 \times 10^{12}/\text{sec}$	$\alpha = 1.7 \times 10^{-5}/°C$
$\mu_{20} = -3.0 \times 10^{-14}/°C$	$\sigma_0 = -1.5 \times 10^{-4}/°C$

a result of this calculation, it can be seen that D_{0m} depends more strongly on μ_{30} than on μ_{40} and that the value of the temperature dependence of the cubic force constants, as well as the temperature dependence of the other quantities in Eq. (121), gives rise to a temperature variation of D_{0m} which is significant only at high temperatures for certain cases depending on the temperature dependence of the third- and, to a lesser extent, fourth-order force constants. If a temperature dependence of D_0 at high temperatures is observed which can be assigned to anharmonic effects, it might be possible to obtain a value for μ_{30} from diffusion measurements if μ_{40} is, indeed, negligible. It is important to note, however, that the temperature dependence of D_0 which is attributable to lattice dynamics is small, which shows why the Arrhenius plot is reasonably valid.

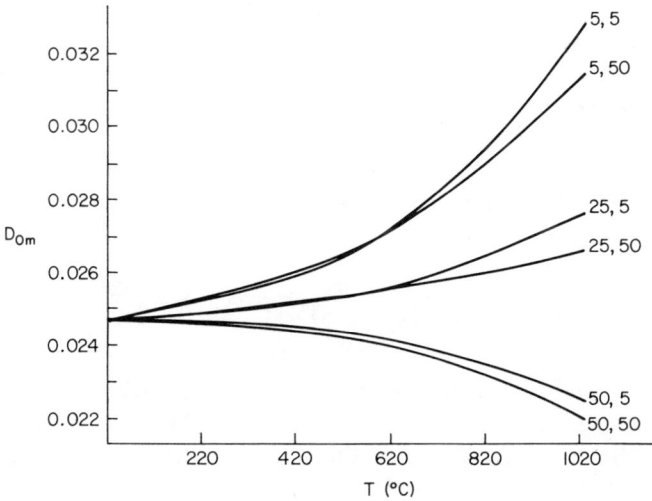

Fig. 13. The variation of the preexponential diffusivity for vacancy migration D_{0m} in Cu with temperature due to the temperature dependence of the anharmonic terms. The temperature dependence of the cubic and quartic force constants is given by μ_{30} and μ_{40}. In the figure, the first number by each curve refers to $\mu_{30} \times 10^5$ and the second number to $\mu_{40} \times 10^5$.

The calculation just completed is important in the sense that it shows explicitly why the Arrhenius plot has worked so well for rate processes, despite the fact that a large degree of anharmonicity is involved in the basic migration event. A long history of both experimental and theoretical work has shown that the temperature dependence of elementary jump events in rate processes is described quite well by $\exp(T^{-1})$ behavior. Other factors, such as different atomic mechanisms which have different activation energies, do, of course, affect the temperature dependence but the simple Arrhenius plot is adequate, even for a highly anharmonic material, for a single atomic mechanism over most of the temperature range. This was shown first (Franklin, 1972) in a quantitative manner for Cu and is reproduced, in part, above.

VI. Mass Dependence—Classical Case

Since Gosar (1964) and Prigogine and Bak (1959) have predicted a theoretical mass dependence for the diffusivity which is proportional to m^{-1} and m^{-2}, respectively, whereas the observed mass dependence is close to $m^{-1/2}$ it is important to show explicitly that a correct anharmonic derivation does, in fact, give the observed $m^{-1/2}$ mass dependence in the classical case. This has been done by Franklin (1972) and the results will be discussed briefly in this section. In addition, the harmonic lattice derivation, given first by Vineyard (1957), will be shown. Quantum mechanical considerations of the mass dependence are deferred until Section VII.

The product rule in molecular spectroscopy gives the product of the frequencies in terms of the product of $(m_i)^{-1/2}$ for $i = 1, 2, \ldots, 3N$ and the determinant of the force constants, $\phi^{1/2} = (\det \phi_{ij})^{1/2}$, such that

$$\prod_i^{3N} v_i = [\phi^{1/2}/(2\pi)^{3N}] \prod_i^{3N} m_i^{-1/2} \qquad (123)$$

This stems directly from the characteristic equation of the normal mode frequencies of a lattice, as shown by Vineyard (1957). If we let Eq. (123) denote the frequencies of the EC, then the product of frequencies of the lattice in its SPC is, separating the Nth mode,

$$\prod_i^{3N-1} v_i' = [(\phi')^{1/2}/v_{3N}'(2\pi)^{3N}] \prod_i^{3N} m_i^{-1/2} \qquad (124)$$

Following Vineyard, we can write $v_{3N}' = C(m^*)^{-1/2}$ where the constant C

1. CLASSICAL AND QUANTUM THEORY OF DIFFUSION

depends on ϕ_{ij}, and m^* is the effective mass for motion

$$m^* = \sum_{i=1}^{3N} m_i c_i^2 \tag{125}$$

where c_i are the direction cosines. Combining Eqs. (123) and (124) in the Vineyard equation, we obtain, for the jump frequency,

$$\Gamma = C'(m^*)^{-1/2} e^{-\Delta E/kT} \tag{126}$$

where C' is another constant which does not depend on the masses.

The weak point in the above analysis is that Eq. (124) is not correct for the mass dependence of defect modes which depend, in a complicated way, on isotopic or impurity mass. It is impossible, particularly for the SPC, to say that each mode of the defect system is proportional to $m^{-1/2}$. A better method of analysis is to study the defect modes of both the EC and SPC and analyze their mass dependence explicitly. Then, through the frequency products in the numerator and denominator of the Vineyard equation, together with the mass dependence of the frequencies in the anharmonic correction factors, the total mass dependence of the defect modes can be accounted for properly. Only a few of the $3N$ modes in an N-body system, namely the localized, gap, and resonance modes, will have a mass dependence which differs from $m^{-1/2}$, but these will make a significant effect on the mass dependence of D_0.

Regarding anharmonic effects, it has been shown (Franklin, 1972) that there is no explicit mass dependence in the anharmonic terms which might provide mass corrections to the harmonic frequencies of the EC and SPC. Therefore, the mass dependence of the anharmonic terms arises solely from secondary effects—effects which stem from deviations of the frequencies from $m^{-1/2}$ dependence in those terms. The fact that there is no explicit mass dependence in the anharmonic terms can be seen in Eqs. (135) and (136) which appear in the jump frequency in the classical case, which is given by Eq. (142). If a $m^{-1/2}$ mass dependence is assumed in these equations for the frequencies and if the deBroglie relation is used for $\hbar$ in Eq. (136c), the jump frequency is seen to have no mass dependent contribution from the anharmonic terms. However, the defect mode frequencies which occur in the EC and SPC may not have the perfect lattice $m^{-1/2}$ mass dependence and these will result in a mass dependence of Eqs. (135), (136), and (142) due to anharmonic terms, which will be in addition to the harmonic effects in the Vineyard equation.

The theoretical calculation of the deviations of defect mode frequencies from the $m^{-1/2}$ mass dependence could be accomplished by the Green's function solution of the defect mode problem and analysis of the mass dependence of the eigenfrequencies. This has been done in one-dimensional problems by Maradudin *et al.* (1971) and by Maradudin (1966b). However, it has not yet been done for diffusion-type defects in three dimensions. There appears to be no simple way to introduce the effects on the mass dependence of the coupling between atoms which are involved in the defect modes which contribute to diffusion. The only simple analysis is one in which each of the modes has a $m^{-1/2}$ dependence. If the normal modes of the lattice with a vacancy in the EC are close to those of the perfect lattice, such as those found by Land and Goodman (1967), then there would not be much error in using the perfect lattice approximation for the mass dependence of the EC. However, the mass dependence of the SPC is expected to deviate considerably from that of the perfect lattice so that a solution of the SPC frequencies, together with their mass dependence, would give an answer to this difficult problem.

The experimental analysis of the mass dependence involves the product of the correlation factor f and the mass factor ΔK, which are defined by

$$f^\alpha = (D^\alpha/D^\beta - 1)/(\Gamma^\alpha/\Gamma^\beta - 1) \qquad (127)$$

$$\Delta K = (\Gamma^\alpha/\Gamma^\beta - 1)/[(m^\beta/m^\alpha)^{1/2} - 1] \qquad (128)$$

where α and β denote two different isotopes with differing mass. Then the product of f and ΔK, which is what is measured, is given by

$$f^\alpha \Delta K = (D^\alpha/D^\beta - 1)/[(m^\beta/m^\alpha)^{1/2} - 1] \qquad (129)$$

The mass factor ΔK gives a measure of the deviation of the jump frequency from $m^{-1/2}$ mass dependence. Thus, ΔK depends on the detailed analysis of the mass dependence of the localized and resonance modes of the SPC and even the EC, if these differ significantly from those of the perfect lattice.‡

VII. Quantum and Anharmonic Effects

In this section, the anharmonic quantum mechanical form of the jump frequency is developed. Quantum statistics, including the zero-point energy, are considered here, beginning with the Hirschfelder–Wigner equation.

The approaches taken by different authors to the aspects of quantum statistics and tunneling in rate processes have been varied. Hirschfelder and Wigner (1939) pointed out that, in quantum considerations, because of the uncertainty principle, the transition state can be defined only if the region around the SPC is reasonably "flat." Also, they pointed out that

---

‡ In Chapter 3, Peterson deals in detail with the experimental study of the mass dependence of diffusion.

tunneling and reflection cause differences from classical rate theory and they presented the results of a one-dimensional quantum calculation including these effects. Subsequently, Alefeld (1964) considered the effects of quantum statistics, especially the zero-point energy, on the rate theory of Rice (1958). In Alefeld's theory, the jump frequency was developed utilizing Bose–Einstein statistics in the quantized partition functions. He noted, among other things, that the mean square displacement, denoted by $\langle u_i u_j \rangle$ in Eq. (111), gives rise to an effective temperature which depends upon isotopic mass. Gosar (1964) presented a theoretical development of the mobility of interstitials utilizing quantized lattice vibration theory and Bose–Einstein statistics to develop the current of interstitials in terms of the retarded Green's function. This theory did not have to assume a transition state since it followed a theory developed for viscous materials by Kramers (1940). Gosar (1964) obtained the proper order of magnitude for the frequency term for Li^+ diffusion in Ge and Si. However, the theory predicted a m^{-1} mass dependence for the diffusivity, which is known to be incorrect.

LeClaire (1966) wrote the partition functions for the EC and the SPC in quantized form and expanded them, keeping the first two terms in the power series. Ebisuzaki et al. (1967b) showed that retention of all the terms in the series led to significant differences from LeClaire's truncation, which were especially important for the diffusion of small masses such as hydrogen and deuterium. An overview of quantum effects, especially in terms of the interaction of the migrating particle with thermal phonons, was presented by Sussmann (1967). A temperature dependence for the jump rate of T^{-7} for intermediate quantum temperatures for symmetric perturbation potentials (T^{-3} for antisymmetric) was predicted. Flynn and Stoneham (1970) later showed the derivation of the T^{-7} temperature dependence explicitly utilizing the theory of polarons applied to diffusion. Other calculations using the phonon–particle interaction theory were given by Pirc et al. (1966).

Weiner (1968, 1970), in his quantum mechanical development of migration, developed the theory of minimum-uncertainty wave packets, which is analogous to the theory of coherent states developed by Glauber (1963) in the field of quantum electronics. Using quantum statistics, Weiner obtained a low temperature quantum frequency factor which was directly proportional to temperature. Subsequent work by Weiner and Partom (1969, 1970) dealt with the effects of tunneling as well as quantum statistics in one- and N-dimensional models. This work predicted a decrease of the jump frequency with respect to the Arrhenius plot at low temperatures due to quantum statistics and an increase due to tunneling. Experimental determination of the effects of quantum statistics and of tunneling at low temperatures is needed since, as Flynn (1972) points out, "It appears that the correct low temperature diffusion rate has not yet been established with certainty for any light interstitial" and since checks on the various theoretical predictions are needed.

An interesting application of quantum tate theory has been made by Löwdin (1963) to proton tunneling in DNA. Löwdin (1963) conjectured that mutations and related genetic changes might occur by proton tunneling between two states of different stability in the hydrogen bonds between base pairs.

In solids the diffusion of H_2, D_2, and T_2 in single crystals of Ni and Cu has been studied by Katz, et al. (1971). Diffusion coefficients for these three isotopes were measured from 723 to 1073°K and no systematic departures from the Arrhenius relationship was found. However, the temperature dependence of the diffusivity ratios for different isotopes was not explainable with classical harmonic theory whereas reasonable agreement was obtained with a quantized anharmonic theory. The calculated anharmonic corrections depended on the proton screening parameter, which is proportional to the density of electron states at the Fermi level. The interesting results of this investigation warrant further development of the theory together with experimental extensions to temperatures less than 723°K (quantum effects would be more pronounced) and to different matrix materials.

At least three discernible quantum effects can, under certain conditions, be observed in rate processes in solids. These are quantum statistics (Bose–Einstein distribution of phonons), tunneling, and the effects of zero-point energy. All of these quantum effects increase with decreasing temperature while the classical approximation is valid for high temperatures. Different approaches have been utilized to introduce quantum statistics into rate theory and some differences in the predictions of experimental results will be pointed out in this section.

Following Hirschfelder and Wigner (1939), we can define the jump rate by

$$\Gamma = \gamma f_D Z^S/Z \tag{130}$$

where γ, f_D, Z^S, and Z are the tunneling coefficient, the frequency of oscillation of the system in a generalized coordinate associated with the diffusion direction for the motion of the system, and the partition functions in the SPC and EC, respectively. In Eq. (130), f_D can be written as the ratio of the rms velocity in the diffusion coordinate to the de Broglie wavelength and the partition functions can be expressed by the trace over the exponentials in $\beta H^S(3N - 1)$ and $H(3N)$, where $H^S(3N - 1)$ and $H(3N)$

1. CLASSICAL AND QUANTUM THEORY OF DIFFUSION

are the Hamiltonians of the system in the SPC and EC, respectively, and $\beta = 1/kT$.‡ Then

$$\Gamma = \gamma \frac{[\langle \tfrac{1}{2}\dot{u}_1^2 \rangle]^{1/2}}{\lambda} \frac{\text{Tr} \exp\{-\beta H^S(3N-1)\}}{\text{Tr}\, e^{-\beta H(3N)}} \qquad (131)$$

In the quantum mechanical approach taken here, the partition functions must include the total Hamiltonian and not just the configuration part. The latter approach is usually taken in the classical derivation.

If anharmonicity and the effects of defects on the lattice vibration frequencies and eigenvectors is included in the Hamiltonians H and H^S, we obtain a reasonably accurate picture of atomic migration in the equilibrium rate process theory. The anharmonic Hamiltonian $H(3N)$ is given by Eq. (57) and the associated partition function is given by Eqs. (59) and (60). For the SPC, the partition function can be written, including effects of deformed force constants (Franklin, 1969),

$$Z^S(3N-1)\exp[\beta \phi_0(r_S)] = Z_2^S \{1 + \lambda^2 (Z_{22}^S/Z_2^S + Z_4^S/Z_2^S + Z_{11}^S * Z_{11}^S/Z_2^S + Z_{21}^S * Z_{21}^S/Z_2^S + Z_3^S * Z_3^S/Z_2^S)\} \qquad (132)$$

In both Z and Z^S, the partition function components are given by

$$Z_2^\alpha = \sum_n \exp(-\beta E_n^\alpha) \qquad (133a)$$

$$Z_{ij}^\alpha = -\beta \sum_n \exp(-\beta E_n^\alpha)\langle n|\phi_{ij}^\alpha|n\rangle \qquad (133b)$$

$$Z_{ij}^\alpha * Z_{ij}^\alpha = \beta \sum_{nn'} \exp(-\beta E_n^\alpha)\langle n|\phi_{ij}^\alpha|n'\rangle\langle n'|\phi_{ij}^\alpha|n\rangle / (E_{n'}^\alpha - E_n^\alpha) \qquad (133c)$$

where $\alpha = 0$ or S for the EC and SPC terms, respectively. In Eq. (133b) $j = 0$ for the Z_4 and Z_4^S terms. Equation (133b) accounts for the Z_4, Z_4^S, and Z_{22}^S partition function components, whereas Eq. (133c) accounts for the $Z_3 * Z_3$, $Z_{11}^S * Z_{11}^S$, $Z_{21}^S * Z_{21}^S$, and $Z_3^S * Z_3^S$ components. The subscripts ij used in Eqs. (133) denote the indices of partition function components. They are not tensor component indices.

The intent of the theory presented here (and, for the classical case, in the previous section and in Section II) is to present the unknown saddle-point force constants and, using the dynamical matrix, eigenfrequencies in terms of the known force constants and frequencies of the EC. Hence, we utilize the method of Born and Huang (1956) for the deformed lattice of the SPC. However, we also assume that the presence of defects causes perturbations in the perfect lattice properties, so we use the defect lattice

‡ Note that β is also used as the geometrical factor in the diffusivity.

representation for the EC as well as for the SPC. To be explicit, we use Eq. (33) for the transformation of the displacement $u(l\kappa)$ for both the SPC and EC. In terms of this transformation, the rms velocity, which is needed in Eq. (131), is taken for a single atom per unit cell from

$$\langle (\dot{u}_\alpha{}^l)^2 \rangle = (\hbar/2m) \sum_s \omega_s B_\alpha^{l*}(s) B_\alpha^l(s) \coth \tfrac{1}{2}\beta\hbar\omega_s \tag{134}$$

where l and s denote the lth unit cell and s degenerate quantum state, respectively. In the high temperature nearest neighbor approximation, the partition function components of Eq. (60) are given by (Maradudin et al., 1961)

$$Z_4/Z_2 = -kT\Phi^{IV}/16(\phi'')^2 \tag{135a}$$

$$Z_3 * Z_3/Z_2 = 172kT(\phi''')^2/9216(\phi'')^3 \tag{135b}$$

and those of the SPC, in the Debye approximation, by (Franklin, 1969)

$$Z_{22}^S/Z_2^S = -3\phi^{IV} \sum_l^{nn} \bar{u}^2(ll_0)/4m\omega_D'^2 \tag{136a}$$

$$Z_4^S/Z_2^S = -9\phi^{IV}kT/m^2\omega_D'^4 \tag{136b}$$

$$Z_{11}^S * Z_{11}^S/Z_2^S = 3(\phi'')^2 \ln(\omega_D'/\omega_0) \sum_l^{nn} [\bar{u}^2(ll_0)/16mh\omega_D'^3] \tag{136c}$$

$$Z_{21}^S * Z_{21}^S/Z_2^S = 9(\phi''')^2 \sum_l^{nn} [\bar{u}^2(ll_0)/256m^2\omega_D'^4] \tag{136d}$$

$$Z_3^S * Z_3^S/Z_2^S = (\phi''')^2 \frac{kT[54 + (81/4)\ln \omega_D'/\omega_0]}{144m^3\omega_D'^6} \tag{136e}$$

where ϕ'', ϕ''', and ϕ^{IV} are the harmonic, cubic, and quartic nearest neighbor force constants, $\bar{u}(ll_0)$ is the critical displacement between nuclei l and l_0 in the SPC, m is the mass, ω_D is the Debye frequency, and ω_0 is the minimum SPC frequency. The dynamical matrix, Eq. (37), can be used with the Debye approximation for ϕ'' in Eqs. (135a-b), in order to obtain the type of representation used in Eqs. (136a-136e), if desired. We have denoted the SPC frequencies by ω_s' and those of the EC by ω_s, since there is a difference in the wavevectors of the EC and SPC. However, for long wavelengths, this difference will be slight and the approximation is made for this case that $\omega_s' = \omega_s$. This is not a bad approximation since ω_s' represent the frequencies of the SPC for *undeformed* force constants.

The numerical value of each of the partition function components, given by Eq. (135) for the EC and Eq. (136) for the SPC, was calculated utilizing the data given in Tables I and III. The results are shown in

TABLE III

NEAREST NEIGHBOR FORCE CONSTANTS AND OTHER EXPERIMENTAL DATA FOR Cu[a]

Born–Mayer constants	$A = 0.0728$ eV, $B = 12.67$
Harmonic force constant	$\phi''(r_0) = AB^2/r_0^2 = 2.876 \times 10^4$ ergs/cm^2
Cubic force constant	$\phi'''(r_0) = -AB^3/r_0^3 = -14.29 \times 10^{12}$ ergs/cm^3
Quartic force constant	$\phi^{IV}(r_0) = AB^4/r_0^4 = 71.0 \times 10^{20}$ ergs/cm^4
Temperature dependence of ϕ''	$(d\phi''(r_0)/dT)_p = -8.61$ dyn/cm °K
Debye frequency ($\theta_d = 339$°K)	$v_d = 7.05 \times 10^{12}$ sec^{-1}
Lattice constant	$a_0(20°C) = 3.6153$ Å
Linear expansion coefficient	$\alpha = 17 \times 10^{-6}/°C$
Temperature dependence of v	$\sigma = -1.5 \times 10^{-4}/°C$

[a] The force constant data[b] are obtained from the Born–Mayer potential, $\phi(r) = A \exp[-B(r/r_0 - 1)]$, which is good for repulsive situations such as those encountered in diffusion.
[b] Hiki, Y., Thomas, J. F., and Granato, A. V. (1967). *Phys. Rev.* **153**, 764.

Table IV for both the classical and quantum mechanical cases. For the Z_4/Z_2 and $Z_3 * Z_3/Z_2$ EC terms, a factor of 11/12 was used since atoms neighboring a vacancy have only 11 nearest neighbors instead of 12. For the $Z_{11}^S * Z_{11}^S/Z_2^S$ term, the temperature dependence of the harmonic force constant was included. The temperature dependence of the anharmonic

TABLE IV

NUMERICAL VALUES OF THE PARTITION FUNCTION COMPONENTS AND TOTALS AND THE CALCULATED JUMP FREQUENCY FOR MIGRATION IN Cu[a]

	293°K		793°K		1293°K	
	C	QM	C	QM	C	QM
Z_{22}^S/Z_2^S	−0.293	−0.326	−0.342	−0.346	−0.407	−0.409
Z_4^S/Z_2^S	−0.060	−0.074	−0.222	−0.228	−0.513	−0.520
$Z_{11}^S * Z_{11}^S/Z_2^S$	5.42	6.00	4.94	5.02	4.35	4.37
$Z_{21}^S * Z_{21}^S/Z_2^S$	−0.0005	—	−0.0006	—	−0.0009	—
$Z_3^S * Z_3^S/Z_2^S$	0.004	—	0.018	—	0.050	—
Z_4/Z_2	−0.020	−0.024	−0.054	−0.056	−0.083	−0.084
$Z_3 * Z_3/Z_2$	0.006	—	0.016	—	0.026	—
Z_{tot}^S/Z_2^S	6.07	6.60	5.39	5.46	4.48	4.49
Z_{tot}/Z_2	0.99	0.98	0.96	0.96	0.94	0.94
$\Gamma_m^0 \times 10^{-13}$	3.25	3.86	2.75	2.82	2.14	2.15

[a] Values are given for 293, 793, and 1293°K for both the classical (C) and quantum mechanical (QM—where they differ from C) cases. The isotropic Debye and nearest neighbor approximations were assumed.

force constants was not accounted for in this calculation. The temperature dependent effects in this calculation included, therefore, the inherent dependence of the cubic and quartic terms [Eqs. (135), (136b), and (136e)], the thermal expansion which appears in $\bar{u}^2(ll_0)$, and the temperature dependence of the harmonic force constant which appears in $Z^S_{11} * Z^S_{11}/Z_2^S$. The figures shown in Table IV differ slightly from those given before (Franklin, 1969) for the Z_4^S/Z_2^S, $Z^S_{21} * Z^S_{21}/Z_2^S$, and $Z_3^S * Z_3^S/Z_2^S$ terms. The corrected equations (Franklin, 1972) which give rise to the values shown in Table IV are given by Eqs. (136).

We now arrive at the final form for the jump frequency by writing out the components of Eq. (131) explicitly. Since the eigenvectors are normalized, Eq. (134) becomes, in expanded form, letting $\alpha = 1$ (the diffusion coordinate),

$$\langle \dot{u}_1^2(l) \rangle = (2kT/m)\{1 + \tfrac{1}{48}(\hbar\omega_1/kT)^2 - (32 \cdot 360)^{-1}(\hbar\omega_1/kT)^4 + \cdots\} \quad (137)$$

In the high temperature approximation, Eq. (137) gives $2kT/m$, which is the classical result. The partition function in the SPC can be written, using the harmonic approximation first for simplicity,

$$\operatorname{Tr} Z_2^S = \operatorname{Tr} \exp[-\beta H_2^S(3N-1)] = \sum_{n_s=0}^{\infty} \prod_s^{3N-1} \exp[-\hbar\omega_s'(n_s + \tfrac{1}{2})/kT]$$

$$= \prod_s^{3N-1} (kT/\hbar\omega_s')\{1 - \tfrac{1}{24}(\hbar\omega_s'/kT)^2 + (7/5760)(\hbar\omega_s'/kT)^4 - \cdots\} \quad (138)$$

and, in a similar manner, the harmonic partition function for the EC is given by

$$\operatorname{Tr} Z_2 = \prod_s^{3N} (kT/\hbar\omega_s)\{1 - \tfrac{1}{24}(\hbar\omega_s/kT)^2 + (7/5760)(\hbar\omega_s/kT)^4 - \cdots\} \quad (139)$$

Now we can use $\lambda = h/(mkT)^{1/2}$ for the Broglie wavelength and write for the jump frequency, in the classical harmonic approximation,‡

$$\Gamma^{(0)}_C = \frac{(kT/m)^{1/2}}{h/(mkT)^{1/2}} \frac{\prod^{3N-1} kT/\hbar\omega_s'}{\prod^{3N} kT/\hbar\omega_s} = \frac{\prod^{3N} v_s}{\prod^{3N-1} v_s'} \quad (140)$$

and, in the harmonic quantum mechanical case, when $x_s < 1$ and $x_s' < 1$,

$$\Gamma^{(0)}_Q = \gamma\left(\frac{x_1}{4} \coth \frac{x_1}{4}\right) \frac{\prod^{3N}(2v/x)_s \sinh x_s/2}{\prod^{3N-1}(2v'/x')_s \sinh x_s'/2} \quad (141)$$

where $x = \hbar\omega/kT$. We note that in the high temperature limit, $x \to 0$ and the tunneling transmission coefficient approaches 1, so that Eq. (141) reduces

‡ The subscript zero in Γ_0 is raised to a superscript here and the subscripts C, Q, and A are used to denote classical, quantum mechanical, and anharmonic, respectively.

to the classical case given by Eq. (140). We also note that Eq. (141) differs from that of others (LeClaire, 1966; Ebisuzaki et al., 1967b; Weiner, 1970), since the final Vineyard equation for Γ [Eq. (10)] was used by them to begin with. Here we started with the initial Hirschfelder–Wigner (1939) equation and included the quantum form for $\langle u_t^2 \rangle^{1/2}/\lambda$ as well as for the partition functions. We note that v_i and v_i' represent frequencies in the EC and SPC, respectively, which can be widely different from each other since we did not use the representation here in which the saddle-point force constants are expressed in terms of those of the EC.

In the anharmonic case, factors A_s and A_s' can simply be inserted for each of the frequencies in the EC and SPC, respectively. We note, however, that two expansions are used—one for the potential energy and another for the force constants. The latter expansion gives rise to the ϕ_{11} term in Eqs. (16) and (136c) which contains the second-order force constant and is, therefore, much larger than the other terms in the partition functions which involve the cubic or quartic anharmonic force constants. This term is not truly anharmonic but we include it in our anharmonic representation nonetheless. Since anharmonic terms are significant primarily at high temperatures, the classical approximation, which is given by Eq. (140) for the harmonic case, can be made so that, together with the expansion of the partition functions, we have

$$\Gamma_{CA}^{(0)} = v_{3N} \frac{1 + (Z_2^S)^{-1}(Z_{22}^S + Z_4^S + Z_{11}^S * Z_{11}^S + Z_{21}^S * Z_{21}^S + Z_3^S * Z_3^S)}{1 + (Z_2)^{-1}(Z_4 + Z_3 * Z_3)} \quad (142)$$

where the assumption has been made that the frequencies of the SPC, after expanding the force constants in terms of those of the EC, are equal to those of the EC. The partition function components have been given for the classical case in Eqs. (135) and (136). Consequently, the evaluation of the jump frequency is complete in terms of quantities which are known for certain materials.

When the Debye method of frequency averaging is used, the full quantum mechanical anharmonic jump frequency for migration is given by

$$\Gamma_Q^{(0)} = 0.75 v_d \frac{Z_{tot}^S/Z_2^S}{Z_{tot}/Z_2} \frac{\sinh x_d/2}{x_d/2} \left(\frac{x_1}{4} \coth \frac{x_1}{4} \right)^{1/2} \quad (143)$$

where Z_{tot}^S/Z_2^S and Z_{tot}/Z_2 are given by the numerator and denominator, respectively, of Eq. (142). The tunneling factor was omitted in Eq. (143) since it is negligible except for very small masses at very low temperatures. Each of the partition function components given in Eqs. (135) and (136) contains quantum factors (Franklin, 1969, 1972) and the quantum factor shown in Eq. (143) is in addition to these. The temperature dependence of the total statistical quantum factor including quantum corrections for the velocity, harmonic and anharmonic terms is shown in Fig. 14.

The preexponential portion of the diffusivity is given, in the classical anharmonic case, by

$$D_0 = r_0^2 \, e^{\Delta S_f/k} \Gamma_{CA}^{(0)} \qquad (144)$$

where ΔS_f is the entropy of vacancy formation. It is important to note that we have been interested in the theoretical construct of the entropy of migration ΔS_m and that the formation entropy can be related to ΔS_m by

$$\Delta S_f = \Delta S_m + \Delta S_b \qquad (145)$$

where ΔS_b is the binding entropy of the vacancy to the source of vacancies, such as a dislocation or vacancy cluster. If ΔS_b is known, the

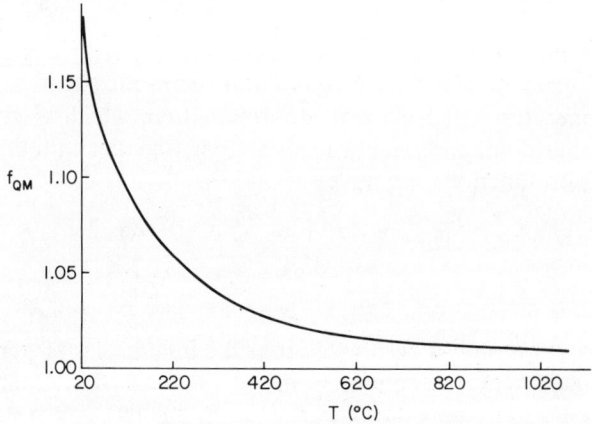

Fig. 14. The temperature dependence of the statistical quantum factor (tunneling not included) for Cu calculated using Eq. (143) including the quantum mechanical terms for the partition function components shown (in classical form) in Eqs. (135) and (136). Note that quantum effects are greater than 10% only for temperatures less than approximately the Debye temperature, which is 66°C for Cu. See Franklin (1972) for the quantum mechanical forms of the components given in Eqs. (135) and (136).

calculation of ΔS_m takes on added significance since the total entropy is given then by $2 \Delta S_m + \Delta S_b$. Using the above representations for the jump frequency, we find the entropy of migration to be temperature dependent for both the quantum and anharmonic cases. This can be seen by extracting the jump frequency for migration, $\Gamma_m = v \exp(\Delta S_m/k)$, from the basic equation for D_{0m} in the thermodynamic rate theory, which is given by

$$D_{0m} = \beta r_0^2 v e^{\Delta S_m/k} \qquad (146)$$

1. CLASSICAL AND QUANTUM THEORY OF DIFFUSION

Then, using Eq. (141), we find the quantum mechanical representation of the entropy of migration to be

$$\frac{\Delta S_m}{k} = \ln\left[\frac{\gamma}{v}\left(1 + \frac{x_1^2}{48} - \cdots\right) \frac{\prod^{3N-1} (v_s')^{-1}(1 - x_s'^2/24 + 7x_s'^4/5760 - \cdots)}{\prod^{3N} (v_s)^{-1}(1 - x_s^2/24 + 7x_s^4/5760 - \cdots)}\right] \tag{147}$$

We note that the value of ΔS_m depends on v, which is an unknown quantity. Consequently, if ΔS_m alone is desired rather than the combination of ΔS_m and v, the value of v must be postulated. One method of treating this problem is to let the saddle-point frequencies, v_i', be expanded in terms of those of the EC, as we did in Eq. (25). Then we simply set $v = v_{3N}$. This is allowable since v is not a well-known or specified quantity. In this manner, we define v as being the maximum frequency of the system in its EC. In order to accomplish this, we define v_i^0 and $v_i^{0'}$ as the harmonic frequencies, without quantum corrections, in the EC and SPC, respectively. Then we rewrite Eq. (147) as

$$\frac{\Delta S_m}{k} = \ln\left[\frac{\gamma}{v}\left(1 + \frac{x_1^2}{48} - \cdots\right) \frac{\prod^{3N-1} (v_s^{0'})^{-1} A_s'(1 - x_s'^2/24 + \cdots)}{\prod^{3N} (v_s^0)^{-1} A_s(1 - x_s^2/24 + \cdots)}\right] \tag{148}$$

where A_s' and A_s are the numerator and denominator, respectively, of Eq. (142). Now, if we let $v_s^{0'} = v_s^0$ for $s = 1, 2, \ldots, 3N - 1$ and $v_{3N}^0 = v$, we obtain

$$\frac{\Delta S_m}{k} = \ln\left[\gamma\left(1 + \frac{x_1^2}{48} - \cdots\right) \frac{\prod^{3N-1} A_s'(1 - x_s'^2/24 + \cdots)}{\prod^{3N} A_s(1 - x_s^2/24 + \cdots)}\right] \tag{149}$$

In the classical case, Eq. (149) becomes

$$(\Delta S_m/k)_C = \sum^{3N-1} \ln A_s' - \sum^{3N} \ln A_s \tag{150}$$

Eqs. (149) and (150) represent the culmination of the lattice vibration theory applied to the entropy of migration. In the derivation, the tremendous complexities of the Green's function solution of the defect modes of the EC and SPC were circumvented through the use of perturbation theory, which was shown in Section V to be tenable, even for grossly anharmonic lattices. In Eq. (149), the quantum effects of tunneling are included via γ, of the particle velocity by the expansion $(1 + x_1^2/48 - \cdots)$ and of the partition functions for the SPC and EC by the expansions in numerator and denominator, respectively, corresponding to the eigenfrequencies v_s' and v_s. Anharmonic effects are included via the factors A_s' and A_s. Besides the anharmonic expansion of the potential, A_s' includes the effects of expanding the SPC force constants in terms of those of the

EC so that, as can be seen from the dynamical matrix, the eigenfrequencies of the SPC are known in terms of those of the EC. In this manner, the full quantum mechanical anharmonic theory is soluable in terms of known quantities.

An approximation to Eqs. (149) and (150) can be made which is quite reasonable, as can be seen from the values of the partition function components given for Cu in Table IV, by choosing the largest term in Eq. (142), which is the $Z_{11}^S \overset{*}{*} Z_{11}^S$ term. In addition, the frequencies and quantum expansions can be cancelled in numerator and denominator in Eq. (149), letting $v_s^{0'} = v_s^0$, so that we obtain

$$\frac{\Delta S_m}{k} = \ln\left[\gamma \frac{1 + x_1^2/48 - x_1^4/32 \cdot 360 + \cdots}{1 - x_{3N}^2/24 + 7x_{3N}^4/5760 - \cdots} \prod_s^{3N-1} \zeta_s\right] \quad (151)$$

and, in the classical approximation,

$$(\Delta S_m/k)_C = \sum_s^{3N-1} \ln \zeta_s = \ln(1 + Z_{11}^S \overset{*}{*} Z_{11}^S/Z_2^S) \quad (152)$$

where we have used

$$\prod_s^N \zeta_s = \prod_s^N (1 + a_s) = 1 + \sum_s^N a_s + \sum_{sp}^N a_s a_p + \cdots \approx 1 + \sum_s^N a_s \quad \text{for} \quad a_s \ll 1 \quad (153)$$

and where

$$\sum_s^{3N-1} a_s = Z_{11}^S \overset{*}{*} Z_{11}^S/Z_2^S \quad (154)$$

The procedure shown here has been used in the derivation and evaluation of the partition function components shown in Eqs. (136) and (142). A different procedure can be used (or higher terms in the expansion of Eq. (153) can be included) if all of the a_s are not small.

The entropy of migration given in quantum mechanical and classical form by Eqs. (149) and (150), respectively, and the approximations given by Eqs. (151) and (152) represent the force constant lattice dynamics result for the jump frequency of migration. In the case of impurity diffusion rather than self-diffusion, the force constants involved in the calculation would be altered to account for the properties of the impurity. In the case of di- or multiple-vacancy diffusion, the calculation could be restructured to account for the differences in lattice structure. The calculation presented here takes into account the strain in the SPC through the utilization of computer calculations of the positions of the nuclei in the SPC. The strain of the EC was neglected since its contribution to migration calculations is small. Thermal variations of all parameters involved in the calculation except the cubic and quartic force constants, whose temperature dependence is not known, were included. The significance of anharmonicity to migration is evident from the magnitudes of the partition function components shown in Table IV. All of the components shown in Table IV, except $Z_{11}^S \underset{*}{*} Z_{11}^S$, contain anharmonic force constants and the relative magnitudes of these terms and the $Z_{11}^S \underset{*}{*} Z_{11}^S$ term provide the desired comparison to show the effects of anharmonicity on migration. A simple comparison is provided by subtracting one from Z_{tot}^S/Z_2^S and comparing with the $Z_{11}^S \underset{*}{*} Z_{11}^S$ term. When this is done for the classical case, it can be seen that the anharmonic terms contribute the following percentages to the entropy of migration at the indicated temperatures: 7% at 293°K, 13% at 793°K, and 29% at 1293°K. Another comparison is to relate the ratio of Z_{tot}^S/Z_2^S and Z_{tot}/Z_2 to ζ. When this is done, anharmonicity is seen to contribute the following percentages: $4\frac{1}{2}$% at 293°K, 6% at 793°K, and 12% at 1293°K. Since the temperature dependence of the anharmonic force constants has not been accounted for, an additional thermal effect due to this cause would affect the results, too.

VIII. Lattice Vibration Theory and Diffusion Experiments

The principal thrust of the theory in Sections II, V, VI, and VII is the development of the jump frequency $\Gamma^{(0)}$ for migration and the development of effects on the jump frequency. The dependences of $\Gamma^{(0)}$ on the variables of temperature, mass, and Debye frequency, as well as on the effects of anharmonic and quantum terms, are related to experimental measurements through the preexponential portion of the diffusivity D_0. In this section, some brief comments are given regarding $\Gamma^{(0)}$ in order to provide insights into experimental relationships to the theory.

The effects on $\Gamma^{(0)}$ of the following quantities, which are not independent except in limiting cases, can be measured experimentally and related to the theory of Sections V, VI, and VII: (1) anharmonicity, (2) quantum statistics, (3) mass dependence of the lattice frequencies, and (4) temperature dependence. Anharmonic effects could probably be studied in impurity diffusion by choosing impurities having different degrees of anharmonicity and relating the results to the partition function components shown in Table IV. In the application of the Born–Mayer potential (which was used in the calculations for Cu given in Table IV) to anharmonicity, high values of the B/r_0 ratio (see Table III) indicate a large degree of anharmonicity. Quantum statistics affect $\Gamma^{(0)}$ via both the explicit terms shown in Eq. (141) and the quantum terms in each of the partition function components which are shown in classical form in Eqs. (135) and (136) (see Franklin,

1969, for the quantum effects in these terms). From Table IV it is important to note that the magnitude of the $Z_{11}^S \underset{*}{*} Z_{11}^S$ term makes its quantum effect significant—even more significant than the terms shown in Eq. (141). The effects of this term, which is not really an anharmonic term, have not been included in experimental analyses of quantum effects before. Regarding the mass dependence, the classical anharmonic terms were shown in Section VI to make no contribution, whereas the quantum terms in both Eq. (141) and in the partition function components of Eqs. (135) and (136) do contribute. Again, the effect of the frequency in the $Z_{11}^S \underset{*}{*} Z_{11}^S$ term is significant as a result of the magnitude of this term. Deviations of the mass dependence from that predicted by the quantum mechanical anharmonic theory provide a measure of the mass dependence of defect mode frequencies—a quantity which is very difficult to derive theoretically for a three-dimensional lattice. The deviation of the temperature dependence of $\Gamma^{(0)}$ for a single atomic mechanism from the Arrhenius relation is due, at high temperatures, to anharmonic effects as shown in Section V and, at low temperatures, to quantum effects as discussed above. The effect of the temperature dependence of the anharmonic force constants, which is unknown, would, if known, enter the partition function components in Table IV which contain anharmonic force constants. While this effect is certainly small, it may be experimentally detectable at high temperatures.

The effects of the introduction of localized and gap modes in a lattice on the diffusion process involve changes in the lattice spectrum, as shown by Eqs. (55) and (56), and changes in the mass dependence of these frequencies. The increase of the localized mode frequencies in a three-dimensional lattice due to effects of altered masses and force constants will follow the trends indicated by the one-dimensional calculation shown in Section IV.C. Internal degrees of freedom, such as the extra degrees of freedom added by an interstitial molecular impurity, can be treated as shown in Section IV.F. Further work of a theoretical nature is needed to show the mass dependence of defect modes involved in the migration process and to compare with the perturbation approach.

REFERENCES

ALEFELD, G. (1964). *Phys. Rev. Lett.* **12**, 372.
BAK, T., and PRIGOGINE, I. (1959). *J. Chem. Phys.* **31**, 1368.
BARDEEN, J., and HERRING, C. (1952). "Imperfections in Nearly Perfect Crystals." Wiley, New York.
BELLOMONTE, L., and PRYCE, M. (1966). *Proc. Phys. Soc.* **89**, 967, 973.
BORN, M., and HUANG, K. (1956). "Dynamical Theory of Crystal Lattices." Oxford Univ. Press, London.

BRICE, D. (1965). *Phys. Rev.* **140A**, 1211.
BURTON, J. J. (1969). *Phys. Rev.* **182**, 885
BURTON, J. J. (1970). *Comments Solid State Phys.* **3**, 82.
BURTON, J. J., and JURA, G. (1967). *J. Phys. Chem. Solids* **28**, 705.
BURTON, J. J., and LAZARUS, D. (1970). *Appl. Phys. Lett.* **16**, 131.
COMPAAN, K., and HAVEN, Y. (1956). *Trans. Faraday Soc.* **52**, 786.
COMPAAN, K., and HAVEN, Y. (1958). *Trans. Faraday Soc.* **54**, 1498.
COWLEY, R. A. (1963). *Adv. Phys.* **12**, 421.
DIVINCENZO, D., and GIRIFALCO, L. A. (1971). Personal communication.
EBISUZAKI, Y., KASS, W. J., and O'KEEFE, M. (1967a). *J. Chem. Phys.* **46**, 1373.
EBISUZAKI, Y., KASS, W. J., and O'KEEFE, M. (1967b). *Phil. Mag.* [8], **15**, 1071.
EYRING, H. (1935). *J. Chem. Phys.* **3**, 107.
FEIT, M. D. (1971). *Phys. Rev.* **B3**, 1223.
FEIT, M. D. (1972). *Phys. Rev.* **B5**, 2145.
FLINN, P. A., and MARADUDIN, A. A. (1963). *Ann. Phys.* (N.Y.) **22**, 223.
FLYNN, C. P. (1968). *Phys. Rev.* **171**, 682.
FLYNN, C. P. (1971a). *Comments Solid State Phys.* **3**, 159.
FLYNN, C. P. (1971b). *Z. Naturforsch.* **26a**, 99.
FLYNN, C. P. (1972). "Point Defects and Diffusion." Oxford Univ. Press, London.
FLYNN, C. P., and STONEHAM, A. M. (1970). *Phys. Rev.* **B1**, 3966.
FRANKLIN, W. M. (1967). *J. Phys. Chem. Solids* **28**, 829.
FRANKLIN, W. M. (1969). *Phys. Rev.* **180**, 682.
FRANKLIN, W. M. (1972). *J. Chem. Phys.* **57**, 2659.
FRANKLIN, W. M., and GRADDICK, W. F. (1970). *Phys. Rev.* **B2**, 2906.
FRANKLIN, W. M., and SENGUPTA, P. (1972). *IEEE J. Quantum Electron.* **QE-8**, 393.
GESZTI, T. (1967). *Phys. Status Solidi* **20**, 165.
GIRIFALCO, L. A. (1962). *Phys. Rev.* **128**, 2630.
GLAUBER, R. J. (1963). *Phys. Rev.* **131**, 2766.
GLYDE, H. R. (1967). *Rev. Mod. Phys.* **39**, 373.
GOSAR, P. (1964). *Nuovo Cimento* **31**, 781.
GRIMES, H. (1968). Personal communication, based on computer calculations.
HIKI, Y., THOMAS, J. F., and GRANATO, A. V. (1967). *Phys. Rev.* **153**, 764.
HIRSCHFELDER, J. O., and WIGNER, E. (1939). *J. Chem. Phys.* **7**, 616.
HUNTINGTON, H. B., SHIRN, G. A., and WAJDA, E. S. (1955). *Phys. Rev.* **99**, 1085.
JOHNSON, R. A. (1966a). Department of Materials Science, Univ. of Virginia, private communication.
JOHNSON, R. A. (1966b). *Phys. Rev.* **145**, 423.
KATZ, L., GUINAN, M., and BORG, R. J. (1971). *Phys. Rev.* **B4**, 330.
KRAMERS, H. A. (1940). *Physica* **7**, 284.
LAND, P. L., and GOODMAN, B. (1967). *J. Phys. Chem. Solids* **28**, 113.
LAZARUS, D. (1960). *Solid State Phys.* **10**, 71.
LAWSON, A. W., RICE, S. A., CORNELIUSSEN, R. D., and NACHTRIEB, N. H. (1960). *J. Chem. Phys.* **32**, 447.
LECLAIRE, A. D. (1949). *Progr. Metal Phys.* **1**, 306.
LECLAIRE, A. D. (1953). *Progr. Metal Phys.* **4**, 265.
LECLAIRE, A. D. (1966). *Phil. Mag.* **14**, 1271.
LEIBFRIED, G., and LUDWIG, W. (1961). *Solid State Phys.* **12**, 275.
LIFSCHITZ, I. M. (1956). *Nuovo Cimento Suppl.* **3**, 716.
LIFSCHITZ, I. M., and KOSEVITCH, A. M. (1968). *Rep. Progr. Phys.* **31**, 235.
LORD, A. E., Jr., and BESHERS, D. N. (1966). *Acta Met.* **14**, 1659.
LÖWDIN, P. (1963). *Rev. Mod. Phys.* **35**, 724.

LUDWIG, W. (1963). *Ergeb. Exakten Naturw.* **35**, 1.
LUDWIG, W. (1966). *In* "Theory of Crystal Defects" (B. Gruber, ed.), p. 57. Academic Press, New York.
MANLEY, O. P., and RICE, S. A. (1960). *Phys. Rev.* **117**, 632.
MANNING, J. R. (1968). "Diffusion Kinetics for Atoms in Crystals." Van Nostrand-Reinhold, Princeton, New Jersey.
MARADUDIN, A. A. (1965). *Rep. Progr. Phys.* **28**, 331.
MARADUDIN, A. A. (1966a). *Solid State Phys.* **18**, 273.
MARADUDIN, A. A. (1966b). *Solid State Phys.* **19**, 1.
MARADUDIN, A. A., and FLINN, P. A. (1961). *Ann. Phys.* (N.Y.) **15**, 360.
MARADUDIN, A. A., FLINN, P. A., and COLDWELL-HORSFALL, R. A. (1961). *Ann. Phys.* (N.Y.) **15**, 337.
MARADUDIN, A. A., IPATOVA, I. P., MONTROLL, E. W., and WEISS, G. H. (1971). *Solid State Phys. Suppl.* 3(2nd ed.), 1.
MITANI, Y., and TAKENO, S. (1965). *Progr. Theor. Phys.* **33**, 779.
MONTROLL, E. W., and POTTS, R. B. (1955). *Phys. Rev.* **100**, 525.
PEGEL, B. (1967a). *Phys. Status Solidi* **22**, 223.
PEGEL, B. (1967b). *Phys. Status Solidi* **23**, 335.
PIRC, R., ŽEKŠ, B., and GOSAR, P. (1966). *J. Phys. Chem. Solids* **27**, 1219.
Pocono Manor Symposium. (1952). "Imperfections in Nearly Perfect Solids." Wiley, New York.
PRIGOGINE, I., and BAK, T. A. (1959). *J. Chem. Phys.* **31**, 1368.
PROKHOROV, A. M. (1972). *Phys. Today* Mar., p. 27.
RICE, S. A. (1958). *Phys. Rev.* **112**, 804.
RICE, S. A., and FRISCH, H. L. (1960). *J. Chem. Phys.* **32**, 1026.
ROTHMAN, S. J., and PETERSON, N. L. (1969). *Phys. Status Solidi* **35**, 305.
SCHOTTKY, G. (1965). *Phys. Status Solidi* **8**, 357.
SEEGER, A., and MEHRER, H. (1968). *Phys. Status Solidi* **29**, 231.
SEEGER, A., and MEHRER, H. (1969a). *Phys. Status Solidi* **35**, 313.
SEEGER, A., and MEHRER, H. (1969b). "Vacancies and Interstitials in Metals." North-Holland Publ., Amsterdam.
SEEGER, A., and MEHRER, H. (1970). *Phys. Status Solidi* **39**, 647.
SLATER, N. B. (1959). "Theory of Unimolecular Reactions." Methuen, London.
SUSSMANN, J. A. (1967). *J. Phys. Chem. Solids* **28**, 1643.
VARMA, C. M. (1971). *Phys. Rev.* **A4**, 313.
VINEYARD, G. (1957). *J. Phys. Chem. Solids* **3**, 121.
WAGNER, M. (1963). *Phys. Rev.* **131**, 2520.
WAGNER, M. (1964). *Phys. Rev.* **A133**, 750.
WEINER, J. H. (1968). *Phys. Rev.* **169**, 570.
WEINER, J. H. (1970). *Proc. 6th U.S. Nat. Congr. Appl. Mech.*, Harvard Univ., Cambridge, Massachusetts.
WEINER, J. H., and PARTOM, Y. (1969). *Phys. Rev.* **187**, 1134.
WEINER, J. H., and PARTOM, Y. (1970). *Phys. Rev.* **B1**, 1533.
WERT, C. A. (1950). *Phys. Rev.* **79**, 601.
WERT, C. A., and ZENER, C. (1949). *Phys. Rev.* **76**, 1169.
WIGNER, E. (1933). *Phys. Rev.* **43**, 252.
WYNBLATT, P. (1969). *J. Phys. Chem. Solids* **30**, 2201.
ZENER, C. (1951). *J. Appl. Phys.* **22**, 372.

2

Exact Defect Calculations in Model Substances

CHARLES H. BENNETT

IBM THOMAS J. WATSON RESEARCH CENTER
YORKTOWN HEIGHTS, NEW YORK ‡

and

SOLID STATE SCIENCES DIVISION
ARGONNE NATIONAL LABORATORY
ARGONNE, ILLINOIS §

I. Introduction	74
II. The Molecular Dynamics, Monte Carlo, and Molecular Statics Methods	76
A. Molecular Dynamics	78
B. Monte Carlo Method	79
C. Monte Carlo Calculation of Free Energy Differences and Equilibrium Defect Concentrations	81
D. Molecular Statics	84
E. Induced Defect Migration	88
III. Discussion of Molecular Dynamics and Monte Carlo Results on Point Defects at Thermal Equilibrium	91
A. Defect Jump Dynamics and Correlations	92
B. Comparison of Hard Sphere, Lennard-Jones, and "α-Iron" Systems.	96
C. Comparisons of Molecular Dynamics and Monte Carlo Results with Molecular Statics	101
D. Suggestions for Future Calculations	103
Appendix A. Monte Carlo Acceptance Ratio Method for Free Energy Differences	104
Appendix B. Lennard-Jones Vacancy Jump Calculations	105
Appendix C. Molecular Dynamics Calculation of the Isotope Effect.	107
References	112

‡ Present Address, at which work was performed under IBM auspices.
§ Previous Address, at which work was performed under U.S. Atomic Energy Comission auspices.

I. Introduction

In the temperature range where diffusion is important, the atoms of most solids can be regarded as classical point masses, moving under the influence of forces arising from the Born–Oppenheimer electronic ground state. An obvious but useful way to study a system of this sort is the technique called "molecular dynamics," in which one uses a fast digital computer to calculate the simultaneous classical trajectories of several hundred atoms, according to Newton's laws of motion, starting from an arbitrary set of initial positions and velocities. The technique is particularly suited to the study of point defects because, while they are small enough to be adequately simulated in an environment of only a few hundred atoms, their motion involves such strongly anharmonic forces that the motion cannot be treated by the more familiar method of lattice dynamics.

In any molecular dynamics calculation, the interatomic forces must be assumed, because they are not known with precision for any real material (though, in principle, they could be found by solving the many-electron Schrödinger equation). For this reason, the technique is generally unable to make quantitative predictions about particular real materials. Instead, it is best applied to qualitatively realistic model systems, such as the Lennard-Jones "substance," to gain a qualitative insight into the microscopic basis of macroscopic phenomena. With regard to point defects, molecular dynamics can fruitfully be used to investigate such questions as the preferred configuration of a defect, the extent to which it locally distorts or melts the lattice, the duration of diffusive jumps, and the extent to which these jumps depend on cooperative motion of atoms other than the jumping atom. More experimentally accessible quantities, such as equilibrium defect concentrations, jump frequencies, and isotope effects, can also be computed for the model systems, and their dependence on crystal structure and interatomic forces can be systematically investigated. Molecular dynamics calculations also yield data on the model system's equation of state and phase diagram, allowing model systems to be compared with one another much as if they were real materials.

The problem of finding correct or even reliable interatomic potentials to simulate particular real materials is a complicated one which will not be treated here; it is discussed from many viewpoints by Gehlen *et al.* (1972). This chapter aims instead to describe the available techniques for computing point defect properties in model systems, and to review the qualitative insights these techniques have yielded so far, into the behavior of point defects in crystals at thermal equilibrium. [The more complicated problem of defect production during radiation damage has also

2. EXACT DEFECT CALCULATIONS IN MODEL SUBSTANCES

been the object of much computer simulation, which will not be reviewed here (see, e.g., Beeler, 1972).]

Besides molecular dynamics, two other computational methods can be used to determine point defect properties from assumed interatomic forces: the Monte Carlo method of Metropolis et al. (1953) and the quasidynamic or molecular statics method of Gibson et al. (1960). The Monte Carlo method is a procedure for sampling a Gibbs canonical ensemble of configurations of the model system at a preselected volume and temperature; it yields data equivalent to molecular dynamics data on defect configurations and energies, but by itself cannot compute dynamic properties such as the

TABLE I

Exact Computational Techniques for Point Defects

Computational technique	Molecular dynamics	Monte Carlo	Hamiltonian manipulation[e]	Molecular statics[f]
Computer time required	Many hours	Many hours	Many hours	Few minutes
Fixed state variables[a]	E, V, N	T, V, N	T, V, N	$T = 0$ only, V, N
Calculated state variables[b]	T, p	E, p	Free energy differences between states	E, p
Statistical uncertainty	Present	Present	Present	Absent
Calculated thermodynamic defect properties[c]	E^F, V^F	E^F, V^F	χ	E^F, V^F, S^F at $T = 0$
Calculated dynamic defect properties[d]	Γ and jump dynamics at high temperature	None	Γ and jump dynamics at low temperatures Isotope effect	E^M, V^M, v^* at $T = 0$

[a] Thermodynamic variables fixed by initial conditions of run: E energy, V volume, N number of atoms, T temperature.

[b] Thermodynamic variables whose values are calculated during run.

[c] E^F, defect energy of formation; V^F, volume of formation; χ, equilibrium concentration; S^F, entropy of formation.

[d] Γ, jump frequency; E^M, V^M, energy and volume of defect motion.

[e] Molecular dynamics and Monte Carlo techniques in which the potential energy function $U(\mathbf{r}_1 \cdots \mathbf{r}_N)$ is manipulated.

[f] v^*, harmonic jump attempt frequency.

jump frequency. The third method, molecular statics,‡ may be viewed as molecular dynamics with a damping force, and is used to find configurations of static equilibrium (i.e. potential minima and saddle points) under the assumed forces. Because all thermal disorder is damped out, the results of molecular statics are strictly valid only at zero temperature; however, the method is computationally so much faster than the other two that it has been used in the majority of attempts to infer effective interatomic forces from experimentally measured point defect formation and motion energies. The relative slowness of Monte Carlo and dynamical calculations in fact stems precisely from their faithful simulation of the model system's thermal fluctuations, which must be averaged over hours of computing time.

Table I summarizes the capabilities of the three computational methods. The fourth category, "Hamiltonian manipulation," refers to techniques in which the potential energy function is manipulated during a molecular dynamics or Monte Carlo computation so as to facilitate the occurrence of events (such as the spontaneous formation of point defects) which are so rare on an atomic scale that there would be little hope of observing them in a direct simulation.

The computational methods and their application to point defects are discussed in some detail in Section II. Section III reviews the few exact (i.e., dynamic and Monte Carlo) studies that have been performed to date on point defects in thermal equilibrium.

II. The Molecular Dynamics, Monte Carlo, and Molecular Statics Methods

Before treating the computational methods separately, it is well to consider a problem faced by all three—the need to simulate a macroscopic specimen of matter by means of a model system containing a relatively small number of "atoms." When available computer memories were smaller, this was a serious limitation, but experience has shown (Alder et al., 1968; Tsai et al., 1970) that systems of a few hundred atoms, easily handled by present day computers, are sufficient to simulate a bulk crystal, provided periodic boundary conditions are used and extended defects (such as phase boundaries, grain boundaries, and dislocations) are absent. The periodic boundary condition consists of confining the atoms in a parallepiped whose opposite faces are considered physically identical, so that an atom leaving

‡ This method is to be distinguished from lattice statics (Kanzaki, 1957) which, like lattice dynamics, is based on an harmonic-approximation normal mode analysis of atomic displacements. Neither lattice statics nor lattice dynamics will be discussed in this chapter.

2. EXACT DEFECT CALCULATIONS IN MODEL SUBSTANCES

through one face reenters simultaneously, with the same velocity, at the corresponding point on the opposite face. The interaction potential is also propagated across the boundary, so that each atom experiences the same force and makes the same contribution to the total potential energy, as it would in an infinite system, consisting of the original parallelepiped periodically replicated to fill all of space. This prescription abolishes any physical distinction between the boundary of the parallelepiped and its interior. Periodic boundary conditions conserve energy and linear (but not angular) momentum and impose a fixed volume and number of atoms. In a crystalline system, they also fix the lattice orientation and the number of lattice sites.

Another stratagem for simulating an infinite crystal, formerly popular in molecular statics calculations, involved using a system consisting of a small spherical crystallite embedded in an infinite elastic continuum. Because of the conceptual difficulties in joining the lattice to the continuum, this stratagem now appears inferior to the use of a moderately large system with periodic boundary conditions, where the interaction between a defect and its own periodic images, if not negligible, would at least approximate the interaction between a typical defect and its neighbors in a real crystal.

The majority of static, dynamic, and Monte Carlo calculations have been done on model systems having pairwise–additive interaction potentials, of which the Lennard-Jones potential

$$u(r_{ij}) = 4\varepsilon[(\sigma/r_{ij})^{12} - (\sigma/r_{ij})^6] \tag{1}$$

is the classic example. (Here r_{ij} is the distance between a pair of atoms, σ is a unit of length, and ε is a unit of energy. The Lennard-Jones "substance" with $\sigma = 3.4$ Å and $\varepsilon/k = 120°$K corresponds roughly to argon.) To speed computation, the interaction is generally truncated at a range of several atomic diameters, and the potential energy and pressure are corrected for contributions from those atoms, in an infinite perfect crystal, which would be outside of the truncation range. Intrinsically long range (e.g. coulombic) interactions cannot be so truncated and thus slow the computation down considerably, as do noncentral or nonpairwise–additive forces. On the other hand, the density dependent (but configuration independent) potential term often used to simulate the electron–gas contribution to the cohesive energy of metals, because it has no effect on atomic motions at constant volume, can be incorporated afterward as a correction to the equation of state computed in its absence. However, the density dependent term must be taken into account in a more subtle manner in computing the formation volumes of point defects (Benedek and Ho, 1974).

A. MOLECULAR DYNAMICS

Molecular dynamics—the numerical solution of a model system's classical equations of motion—is a method of great generality. In principle, it can answer almost any physical question about any classical system whose Hamiltonian is known. In practice, of course, the size and speed of computers limit the number of degrees of freedom that can be handled and the time over which their motion can be followed. As indicated earlier, only the latter limitation is important in point defect studies.

The input data for a typical molecular dynamics calculation include the volume V of the periodic box, the number of "atoms" N, their initial coordinates $\{\mathbf{r}_i(0),\ i = 1, \ldots, N\}$ and initial velocities $\{\dot{\mathbf{r}}_i(0)\}$, and finally the Hamiltonian governing their motion. Vacancies are studied by omitting one or more atoms from an otherwise perfectly crystalline configuration, interstitials by including extra atoms.

The simultaneous equations of motion are solved by any of several techniques which involve extrapolating the atomic trajectories over successive short time intervals Δt. After each extrapolation, the forces on all the atoms are recalculated at their new positions and the resulting accelerations used in the next extrapolation (Beeler and Kulcinski, 1972; Gear, 1966; Rahman and Stillinger, 1971). For this procedure to work (i.e., conserve energy), the timestep Δt must be shorter than the reciprocal of the high frequency cutoff of the dynamical system's velocity spectrum; in monatomic systems a typical timestep is $2 \cdot 10^{-14}$ sec.

Computationally, the most expensive part of each timestep is not the extrapolation but the recalculation of the net forces on all the atoms. This involves computing and summing as many terms as there are simultaneously interacting pairs of atoms—typically 10,000 in a 250-atom system with a pairwise-additive potential of moderate range—and on a fast computer requires a few hundred milliseconds. ‡ Long range, non-spherically symmetric, or nonpairwise-additive potentials would all slow down the force calculation, though not necessarily to an intolerable extent. Indeed, molecular dynamics calculations have been done on such complicated systems as "beryllium fluoride" and "water" (Rahman and Stillinger, 1971; Rahman et al., 1972).

The preceding comments have referred to continuous potentials. For the hard sphere and square well potentials, an entirely different computational technique is used, which proceeds directly from one binary collision to the next (Alder and Wainwright, 1959). At crystalline densities,

---

‡ Throughout this chapter, estimates of computing time refer to fast machines available in 1973, such as the IBM model 360-195.

computational technique proceeds about an order of magnitude faster than in a comparable Lennard-Jones system.

The principal output of a dynamic computation is the set of atomic trajectories $\{r_i(t)\}$ from which all other properties are calculated. The frequency of defect jumps can be observed directly, while equation of state data are gathered by time averaging appropriate functions of the positions and velocities. The temperature, for example, is obtained by time averaging the kinetic energy (which undergoes considerable fluctuations, owing to the small number of degrees of freedom, even though the total energy is conserved), while the pressure is obtained, via the virial theorem, from the time-averaged interparticle momentum current (Alder and Wainwright, 1960; Verlet, 1967). In taking these averages, the beginning of the run is omitted, to allow the system time to relax to thermal equilibrium from its (generally nonequilibrium) initial conditions.

Molecular dynamics faces two serious difficulties in elucidating point defect properties.

1. The periodic boundary conditions and the small size of the system prevent the spontaneous formation or destruction of vacancies or interstitials; hence, their equilibrium concentrations (or, equivalently, their free energies of formation) cannot be directly observed. With long runs, equation of state differences between perfect and defect-containing crystals can be measured, but these yield only the (isobaric–isothermal) volume and energy of defect formation, not the entropy or free energy.

2. Good statistics on defect motion cannot be gathered if the defect in question does not jump, or jumps only a few times, during the time available for computation. Monovacancy jumps, for example, cannot be observed at temperatures much below the melting point.

Both these difficulties can be largely overcome by auxiliary techniques which intervene in the dynamical system in a physically impossible but statistically well defined way, to create a path between physically inaccessible states or to facilitate the occurrence of an infrequent event. These techniques will be described in detail in succeeding sections.

B. Monte Carlo Method

The Monte Carlo method (Metropolis *et al.*, 1953; Wood, 1968) does not simulate molecular motion; it merely generates a large number of configurations (i.e., snapshots of the N-particle, constant volume, periodic system) representative of the Gibbs canonical ensemble at temperature T. Equation of state data (from which defect formation energies and volumes can be calculated) are then obtained by appropriate averages over the generated configurations, rather than over time as in a dynamics calculation. The configurations are generated by a (Markovian) random walk through configuration space, a walk whose single step transition probabilities, between any two configurations I and II, satisfy the Boltzman relation

$$p(\text{I} \to \text{II})/p(\text{II} \to \text{I}) = e^{-[U(\text{II})-U(\text{I})]/kT} \qquad (2)$$

where $U(\text{I})$ and $U(\text{II})$ are the configurations' potential energies; this, in turn, is a sufficient condition for convergence to a canonical distribution over the accessible configurations. It may be noted that for $N > 100$, the canonical equation of state differs negligibly from the microcanonical one measured by molecular dynamics.

Some ingenuity is required to make the transition probabilities satisfy (2), since the potential energy U is generally a complicated function of all the coordinates. In the usual procedure, new configurations are initially generated by a random process not obeying (2), but some of them are then thrown away, by a second random process, explicitly depending on the potential energy. The resulting two-stage selection process then obeys (2). The following is a typical procedure for generating the next configuration in a Monte Carlo chain.

1. Starting from any configuration I, an atom is selected at random and shifted to a random point in a small cubic neighborhood centered on its earlier location. The result is called configuration II. Clearly, $p(\text{I} \to \text{II}) = p(\text{II} \to \text{I})$ at this stage. Next, the potential energy difference $\Delta U = U(\text{II}) - U(\text{I})$ is calculated.

2. Configuration II is accepted if a random number chosen from the interval [0, 1] is less than $e^{-\Delta U/kT}$; otherwise II is rejected and the earlier confirmation I is counted again, as its own successor. The trial move's acceptance probability, as a function of $\Delta U/kT$, is thus given by the "Metropolis" function

$$M(\Delta U/kT) \equiv \min\{1, \exp(-\Delta U/kT)\} \qquad (3)$$

and relation (2) is satisfied.

At first sight, it might appear that by making the steps of the Monte Carlo walk large instead of small, one could sample configuration space much more rapidly than in a molecular dynamics calculation, where successively generated configurations necessarily differ only slightly. Unfortunately, almost any large random change in the configuration of a nongaseous many-atom system results in a large positive ΔU due to core–core overlaps; hence, if large steps are attempted, they are almost certain to be rejected. Empirically, it is found that the best cubic neighborhood is one which results in acceptance of about half the attempted moves.

Under many conditions, molecular dynamics and Monte Carlo techniques appear to be about equally efficient for computing the equation of state. In either case, since the dependent state variables are obtained as averages of fluctuating quantities, long runs are necessary to reduce the statistical uncertainty of the results. Monte Carlo programs are easier to write and may be more efficient with soft potentials at low temperatures, where the dynamical behavior is dominated by weakly damped normal mode vibrations. On the other hand, they provide no information on dynamical properties such as defect jump frequencies, which must be studied by molecular dynamics.

C. Monte Carlo Calculation of Free Energy Differences and Equilibrium Defect Concentrations

Since defects such as vacancies and interstitials cannot spontaneously form or be destroyed in a small system with periodic boundary conditions, all Monte Carlo and molecular dynamics procedures for determining their concentrations depend on calculating free energies of formation. However, because the entropy is not the microcanonical or canonical average of any function of positions and velocities, neither molecular dynamics nor the Monte Carlo method just described is able to calculate the free energy of a system directly.

Fortunately, there are ingenious modifications of the basic molecular dynamics and Monte Carlo techniques by which free energy differences between states can be determined. Like calorimetric procedures for measuring entropy differences in the laboratory, they depend on conducting the system from one state to another in a thermodynamically reversible manner. The great power of these computer techniques lies in the fact that, unlike laboratory procedures, they are under no obligation to be physically realistic. For example, the free energy of vacancy formation is commonly determined by a technique in which one atom is reversibly decoupled from all the others in an initially perfect crystal, leaving behind a crystal with vacancy plus one atom of ideal gas.

Two generally applicable Monte Carlo techniques for calculating free energy differences will now be described, followed by examples of their use in determining defect free energies of formation. Later, it will be shown that these same techniques, in conjunction with molecular dynamics, can be used to calculate isotope effects and jump frequencies at arbitrarily low temperatures. The techniques are described in some detail, because they are not widely known and are, in the author's opinion, potentially very useful in the study of point defects.

The most convenient way of thermodynamically manipulating a Monte Carlo system on a computer is by varying its potential energy function $U(\mathbf{r}_1 \cdots \mathbf{r}_N)$, rather than its state variables N, V, or T. At constant (N, V, T), the Helmholtz free energy A is a functional of U:

$$A = \text{const} - kT \ln \int_{V^N} \exp[-U(\mathbf{r}_1, \mathbf{r}_2 \cdots \mathbf{r}_N)/kT]\, d\mathbf{r}_1, d\mathbf{r}_2 \cdots d\mathbf{r}_N \quad (4)$$

Now if U itself depends on a parameter λ, we have

$$A(\lambda) = \text{const} - kT \ln \int_{V^N} \exp[-U_\lambda/kT] \quad (5)$$

where explicit reference to the atomic coordinates $(\mathbf{r}_1 \cdots \mathbf{r}_N)$ has been suppressed. One procedure for finding free energy differences rests on the fact that while A is not the canonical (i.e., Monte Carlo) average of any function of $(\mathbf{r}_1 \cdots \mathbf{r}_n)$, its derivative with respect to λ is. For a continuous potential,

$$dA(\lambda)/d\lambda = \left\{\int_{V^N} (\partial U_\lambda/\partial \lambda) \exp[-U_\lambda/kT]\right\} \bigg/ \left\{\int_{V^N} \exp[-U_\lambda/kT]\right\} = \langle \partial U_\lambda/\partial \lambda \rangle_\lambda \quad (6)$$

the angle bracket $\langle \ \rangle_\lambda$ denoting a canonical average taken with the potential U_λ. The quantity $dA/d\lambda$ may be thought of as a force which the system exerts against the parameter λ, and it can be numerically integrated to give the free energy difference between two states determined by potentials U_0 and U_1. Let U_λ be chosen so that it varies continuously from U_0 to U_1 as λ varies from 0 to 1. Then

$$A_1 - A_0 = \int_0^1 [dA(\lambda)/d\lambda]\, d\lambda \quad (7)$$

This procedure is somewhat cumbersome, requiring a separate Monte Carlo run to measure $dA(\lambda)/d\lambda$ at each of several intermediate λ values.

A newer approach avoids the integration by use of the following relation, derived in Appendix A, which holds in principle for any two potential functions:

$$A_1 - A_0 = -kT \ln \{\langle M[(U_1 - U_0)/kT]\rangle_0 / \langle M[(U_0 - U_1)/kT]\rangle_1\} \quad (8)$$

Here M denotes the Metropolis function of Eq. (3), and $\langle \ \rangle_0$ and $\langle \ \rangle_1$ denote canonical averages taken with potentials U_0 and U_1, respectively. Physically, the numerator of (8) is the mean acceptance probability for a new kind of Monte Carlo move which, instead of attempting to change the *configuration*, attempts to suddenly switch the *potential* from U_0 to U_1, while the denominator is the analogous probability for a switch from U_1 to U_0. Equation (8) may thus be viewed as a generalization of Eq. (2), which relates the acceptance probability ratio for ordinary Monte Carlo moves to the potential energy difference between two configurations.

This acceptance ratio method is preferable to the continuous integration when U_0 and U_1 are fairly similar (e.g. when one atom is being decoupled from the rest). On the other hand, when U_0 and U_1 differ considerably, the numerator and denominator of (8) may be subject to large statistical errors, and the integration method becomes perferable.

The development here has been in terms of canonical (Monte Carlo) averages and continuous potentials; however analogous arguments can be developed for discontinuous potentials and for microcanonical averages. When the potential is discontinuous with respect to $(\mathbf{r}_1 \cdots \mathbf{r}_N)$, the integrand in Eq. (6) may be ill defined; however $dA/d\lambda$ generally remains well defined, and the numerical integration (7) can be performed using $\Delta A/\Delta \lambda$ values obtained by applying the acceptance ratio method to small intervals $\Delta \lambda$.

We are now in a position to determine the free energy of vacancy formation, by reversibly decoupling one atom in a crystal from all the rest. Let two equilibrium states of N atoms at volume V and temperature T, under periodic boundary conditions allowing N lattice sites, be defined by the potentials

$$U_0(\mathbf{r}_1 \cdots \mathbf{r}_N) = \sum_{i<j\leq N} u(r_{ij}) \qquad (9)$$

$$U_1(\mathbf{r}_1 \cdots \mathbf{r}_N) = \sum_{i<j\leq N-1} u(r_{ij}) \qquad (10)$$

where u is a pair potential. U_0 and U_1 differ only in the coupling of the Nth atom. Passing from U_0 to U_1 causes the Nth atom to "evaporate" from its crystal environment, leaving behind an $N-1$ atom crystal with a vacancy plus a noninteracting atom with zero potential energy. The free energy of "evaporation" can be measured either by λ integration or by acceptance ratios. The former method, used by Squire and Hoover (1969) involves making a series of Monte Carlo runs in which the Nth atom is partly coupled to all the others, with the coupling parameter λ varying between zero and one. The latter method, used by Bennett and Alder (1971a), involves determining Metropolis acceptance probabilities for attempts to suddenly couple or decouple the Nth atom from all the others (in the hard sphere systems studied by these authors, only the former probability had to be measured explicitly, since the purely repulsive nature of the pair potential guaranteed acceptance of all decoupling attempts).

With either method, statistical errors are greatly reduced by confining the uncoupled (or weakly coupled) Nth atom to a small volume v in the neighborhood of the vacancy, instead of letting it roam through the whole system volume V.

The reversible path just described creates a vacancy by transferring one atom to an ideal gas state of volume v, whereas in nature, the atom removed from a vacancy appears as an additional occupied lattice site and makes the normal atomic contribution to the system's volume and free energy. The desired Helmholtz free energy of vacancy formation at constant volume must therefore include, in addition to the free energy of "evaporation," terms representing the work of forming an extra occupied lattice site from the one-atom ideal gas and the work of compressing the resulting $N + 1$ site crystal back to its original volume. These terms are readily computed from the entropy and equation of state of the perfect crystal (see Squire and Hoover, 1969; Bennett and Alder, 1971a).

Free energies of formation for interstitials can be computed by analogous decoupling procedures, but there would be little point in doing so for self-interstitials whose *equilibrium* concentrations are so low as to be physically negligible.

The equilibrium concentrations of associated defects such as divacancies can be determined either by the decoupling procedure (e.g. by decoupling one of the neighbors of a vacancy already present in the crystal) or by directly observing the equilibrium between association and dissociation in a molecular dynamics system containing several simple defects (Bennett and Alder, 1971a). These two methods are complementary: the former requires accurate configurational averages, which are easier to measure at low temperatures (because of the lesser variety of accessible configurations); while the latter requires frequent defect migration, which occurs only at high temperatures. Equilibrium concentrations of vacancies and divacancies in the hard sphere and Lennard-Jones systems, computed by both the decoupling and the association-constant methods, are reviewed in Section III.B.

D. Molecular Statics

Unlike the other computational techniques so far described, molecular statics is incapable, even in principle, of yielding exact thermodynamic or dynamical properties of a model system, except at absolute zero. Its widespread use in defect studies stems from its small demand of computer time, its ability to handle large systems containing extended defects such as dislocations, and the liklihood, for most real materials, that errors due to ignorance of the interatomic potential will be more serious than those incurred by extrapolating molecular statics results to finite temperatures.

Molecular statics may be thought of as molecular dynamics with a damping force which progressively draws energy out of the system until it arrives at a configuration of stable static equilibrium—i.e., a local minimum of the potential energy function $U(\mathbf{r}_1 \cdots \mathbf{r}_N)$. Convergence is very rapid in crystalline systems, particularly if the "individual damping" method is used (Beeler and Kulcinski, 1972). This consists, for each atom, of setting its velocity to zero whenever the scalar product of its velocity and acceleration becomes negative. The equilibrium configuration and its potential energy are obtained to high precision, free from the statistical fluctuations which in the Monte Carlo and molecular dynamics methods necessitate averaging over long runs. Comparison of energy minima for perfect and defect-containing crystals yields defect energies of formation, while volumes of formation can be obtained if the energy minimizations are carried out with respect to volume as well as configuration. Defect entropies of formation in the low temperature limit where the harmonic approximation is valid can be computed exactly from the normal mode frequencies of the perfect and defective crystals (Burton, 1972); in the past, when computations with large matrices were more difficult than they are today, it was customary to estimate defect entropies from shifts in the Einstein frequencies of a few atoms in the immediate neighborhood of the defect (Huntington et al., 1955).

Besides computing energy minima, molecular statics can be used to find saddle-point configurations, from which energies, volumes, and entropies of defect motion can be estimated. One general technique for finding saddle points will now be described; it involves carrying out energy minimizations subject to the constraint that an appropriately chosen function of the atomic positions, termed a "reaction coordinate," remain constant. The reaction coordinate may be thought of as a parameter which, when externally varied while all other degrees of freedom are allowed to relax to the local energy minimum, causes the system to move from one stable configuration, through a saddle point, to another stable configuration. In particular, a function $\xi(r_1 \cdots r_N)$ is a suitable reaction coordinate for detecting a saddle point between two stable configurations C_1 and C_2, if it has the following properties.

1. ξ is a continuous function of the configuration (i.e., of the N atomic positions $\mathbf{r}_1 \cdots \mathbf{r}_N$), taking on the value ξ_1 at C_1 and a larger value ξ_2 at C_2.

2. Starting from C_1, if the system is subjected to a series of small displacements of the atomic positions in the direction of *increasing* ξ, each

displacement being followed by an energy minimization *at constant* ξ, then when ξ reaches the value ξ_2, the system will be in configuration C_2. Let the "reaction path" between C_1 and C_2 (i.e., the relaxed configuration as a function of ξ) be denoted by $C^*(\xi)$ and its energy by $E^*(\xi)$. Thus, $C^*(\xi_1) = C_1$ and and $C^*(\xi_2) = C_2$.

3. $C^*(\xi)$ is continuous where $E^*(\xi)$ reaches its maximum, although it may be discontinuous elsewhere.

If all these conditions are satisfied, the configuration at which E^* reaches its maximum is a saddle point of the unconstrained system, since the energy is simultaneously maximized with respect to ξ and minimized with respect to all other degrees of freedom. Although the reaction path is clearly an artifical construct, depending on the particular choice of reaction coordinate, the maximum of E^* is a true physical saddle point, and would not be affected by a small change in the definition of a successful reaction coordinate. A larger change, on the other hand, might lead to violation of condition 3, and the reaction path would show hysteresis around the saddle point instead of passing smoothly through it. A still larger change might shift the reaction path so that it passed through a different saddle point entirely.

As an example of the reaction coordinate technique in molecular statics, consider the jumping of a vacancy in an fcc lattice. In order to reach a nearest neighbor vacant site, the jumping atom, which will be designated atom 1, must pass through a "gate" of four other atoms (designated 2–5). Hence, a reasonable reaction coordinate would be

$$\xi \equiv [\mathbf{r}_1 - \tfrac{1}{4}(\mathbf{r}_2 + \mathbf{r}_3 + \mathbf{r}_4 + \mathbf{r}_5)] \cdot \mathbf{j} \tag{11}$$

where $\mathbf{j}$ is a unit vector in the jump direction (a 110 direction). In the stable configurations before and after the jump, the reaction coordinate takes on the approximate values $-\tfrac{1}{2}$ and $+\tfrac{1}{2}$ an atomic diameter, respectively. From the lattice symmetry, one may infer that the configuration at $\xi = 0$ is either a saddle point or possibly a shallow potential minimum; actual calculation, in the Lennard-Jones case, reveals a saddle point, and confirms the continuity of the reaction path through it, with this choice of reaction coordinate.

As an example of an unsatisfactory reaction coordinate, consider the perhaps more obvious choice

$$\xi = \mathbf{r}_1 \cdot \mathbf{j} \tag{12}$$

which involves only the jumping atom. An attempt to push the system into the saddle point by manipulating this reaction coordinate would merely drag the whole crystal in the $\mathbf{j}$ direction, with atom 1 acting as a sort of handle. Even if this were prevented by using a system with fixed rather than periodic boundaries, the reaction coordinate might (depending on the interatomic potential and the remoteness of the boundaries) be unable to stabilize the saddle-point configuration against a deformation in which the gate atoms would be pushed ahead or behind the jumping atom, a situation which would manifest itself by hysteresis and discontinuous energy maxima on the reaction path.

When the migrating defect is of lower symmetry (e.g. a divacancy), finding a suitable reaction coordinate may be more difficult, and it will generally not be possible to guess the ξ value corresponding to the saddle point, as was done in the fcc vacancy case. One must therefore proceed by trial and error, inventing reaction coordinates, testing them for lack of hysteresis, and searching for saddle points (of which there may be several), and possible metastable configurations [relative minima of $E^*(\xi)$ along the reaction path]. Once a saddle point has been found, its energy, volume, and normal mode frequencies (one of which is imaginary, corresponding to the unstable degree of freedom) can be calculated and compared to the analogous quantities for the stable defect to obtain the energy, volume, and entropy of defect "motion" at zero temperature.

The basic assumption made in extrapolating molecular statics results to finite temperatures is the harmonic approximation, i.e. the assumption that the system's potential energy is a purely quadratic function of the atomic displacements from the equilibrium (or saddle-point) configuration. A harmonic crystal, of course, undergoes no thermal expansion, and its defect formation and motion energies and entropies would be independent of temperature. The usual way of including thermal expansion in molecular statics calculations is to perform the calculations with an expanded crystal, whose lattice constant is constrained to agree with experiment or with some high temperature anharmonic model, rather than being allowed to relax to the equilibrium zero-temperature value. Defect jump frequencies, in the harmonic approximation, are computed by rate theory, using expressions such as Vineyard's (1957).

$$\Gamma_j = \left(\prod_{i=1}^{3N-3} v_i \right) \bigg/ \left(\prod_{i=1}^{3N-4} v_i' \right) \exp[(E - E')/kT] \qquad (13)$$

which gives Γ_j, the jump frequency per jump direction, in terms of the energies and normal mode frequencies of the stable (unprimed) and saddle-point (primed) configurations. In computing the preexponential frequency factor (which is often designated v^*), the one unstable normal mode of the saddle-point configuration is excluded, as are six zero-frequency modes corresponding to translations. Both the stable defect configuration and the saddle-point configuration should be computed using the same thermally expanded lattice constant. The quantity $E' - E$ is thus the motion energy at constant volume.

The main drawbacks of the molecular statics method, when extrapolated to finite temperatures, are the weakness of the harmonic approximation, the rather ad hoc way in which thermal expansion is introduced, and, most important, the method's inability to give a realistic picture of dynamical jump trajectories. (It should be emphasized that dynamical trajectories, i.e., paths along which a system might move during a spontaneous jump at finite temperature, are quite different from the artificial reaction paths mentioned earlier in connection with finding the saddle point. The latter, for example, pass exactly through the saddle point, but may be discontinuous elsewhere, while the typical dynamical trajectory is continuous everywhere but only passes near, not through, the saddle point.) The generation of realistic dynamical trajectories for defect jumps is the subject of Section II.E. A comparison of high temperature extrapolations from molecular statics with exact (Monte Carlo and molecular dynamics) high temperature results will be made in Section III.C.

E. Induced Defect Migration

In Section II.C, the potential function $U(\mathbf{r}_1 \cdots \mathbf{r}_N)$ was manipulated to bring about a physically impossible event—the evaporation of an atom from the interior of a crystal. Similar manipulation can be used to bring about events, such as defect jumps, which, while physically possible, occur too infrequently to be observed conveniently by conventional molecular dynamics techniques. Excessively low jump frequencies, for example, have been a problem in molecular dynamics studies of vacancy motion, limiting them to temperatures near or above the melting point (under periodic boundary conditions the crystal phase can be superheated considerably above its thermodynamic melting point) where an adequate number of jumps can be collected in the available computer time. Similarly, even at high temperatures, molecular dynamics cannot in practice determine the jump frequency of an isolated impurity or isotopic atom, since almost all the jumps would be made by host atoms.

One rather naive way of inducing a jump in a molecular dynamics system would be to start the calculation in an exact saddle-point configuration, with some arbitrary Maxwellian set of initial velocities. In the starting configuration the jumping atom would be midway between two "half vacancies," but very soon it would fall into one or the other, and one would have observed the molecular motions corresponding to the last half of a jump event—i.e., the descent from the saddle point. To observe the first (ascending) half of the jump, it would only be necessary to return to the initial configuration and compute the motion of the time reversed system obtained by reversing all the initial velocities.

The scheme just described has two serious flaws.

1. During a typical spontaneous jump, the system only passes *near*, not *through*, the exact saddle-point configuration of molecular statics. Therefore the induced trajectories, all of which pass through the exact saddle point, will not be representative of spontaneous jumps.

2. Even if one could choose a representative set of near saddle-point starting configurations and so induce realistic jumps, one would not know how often such configurations arise spontaneously; thus, one could not compute the absolute jump frequency, except, of course, in the harmonic approximation, by rate theory.

Both these problems can be solved by appropriate manipulation of the potential energy function. To generate an ensemble of near saddle-point configurations representative of those that would occur spontaneously under the potential $U(\mathbf{r}_1 \cdots \mathbf{r}_N)$, one starts with an exact saddle-point configuration and performs a molecular dynamics or Monte Carlo calculation using a modified potential function U^*, which is identical to U in a selected neighborhood of the saddle point but infinite elsewhere. U^* thus constrains the system to remain near the saddle point, but does not otherwise alter the relative probabilities of configurations. The U^* configurations can therefore be used to generate realistic induced trajectories through the saddle-point neighborhood, thereby overcoming the first problem. The second problem—that of determining the spontaneous jump frequency—can be overcome by using the methods of Section II.C to measure the free energy difference $A^* - A$, i.e., the reversible work required to pass from the potential U to the potential U^*. The factor $\exp[(A^* - A)/kT]$ then represents the extent by which the constraining potential has enhanced the system's probability of being in the saddle-point neighborhood, and the frequency of spontaneous trajectories through this neighborhood can be determined.

Not all trajectories through the saddle-point neighborhood lead to successful jumps, because sometimes the jumping atom may proceed only a short distance beyond the saddle point and then return to the site from which it came. However, the fraction of successful jumps can be easily determined by examining the induced trajectories, which will include both successful and unsuccessful jump attempts, in the same ratio as would spontaneous trajectories through the saddle-point neighborhood. The induced trajectory technique will be described in considerable detail, because it has not, to the author's knowledge, been used before in many body systems. (An analogous technique, called "combined phase space/trajectory analysis", has been used successfully in rate computations for gas phase chemical reactions; see, e.g., Jaffe et al., 1973; Anderson, 1973.)

The heart of the induced trajectory method is the correct choice of the confining potential U^*, which should restrict the system to a region of phase space that intersects essentially all successful jump trajectories between the two sites in question, but relatively few unsuccessful trajectories. This can be conveniently accomplished by defining U^* in terms of a reaction coordinate of the type discussed in Section II.D. For example, for a monovacancy jumping in an fcc Lennard-Jones crystal, an appropriate confining potential would be

$$U^*(\mathbf{r}_1 \cdots \mathbf{r}_N) = U(\mathbf{r}_1 \cdots \mathbf{r}_N) + \begin{cases} 0 & \text{if } |\xi| < 0.1\sigma \\ \infty & \text{if } |\xi| > 0.1\sigma \end{cases} \quad (14)$$

where ξ, the reaction coordinate defined earlier [Eq. (11)], measures the position of the jumping atom (atom 1) relative to the four-atom gate between two "half vacancies." The infinite potential barriers in U^* serve to hold the system near the saddle-point configuration in which atom 1 and the gate atoms are all coplanar. Strictly speaking, a system under U^* could still escape from the intended saddle-point neighborhood by a spontaneous jump of an atom other than atom #1 into one of the "half vacancies." Such unintended jumps can be prevented by including in both U and U^* a "single-occupancy" term, defined to be infinite if any atom other than atom 1 is outside its own Wigner–Seitz cell and zero otherwise; however, in practice, this is only necessary near the melting point, where spontaneous jumps have a significant chance of occurring during a run of normal length. Whether or not the single-occupancy restriction is included in the dynamics program, it must be formally included in calculating the free energies A and A^*, so that the free energy difference $A^* - A$ refers to the activation free energy from a *pair* of adjacent stable defect states to the *single* saddle-point neighborhood between them.

Figure 1 shows a typical plot of the reaction coordinate versus time during a molecular dynamics run on a 255-atom one-vacancy Lennard-Jones system, under the confining potential U^*. As one can see, the system frequently attempted to leave the high potential confinement region $|\xi| < 0.1\sigma$ but was reflected back by the infinite barriers. Induced trajectories were generated by dynamically continuing the atomic motion forward and backward in time beyond the reflections, using the unperturbed Lennard-Jones potential U; one such continuation, representing a successful jump, is shown by the dotted line. The free energy of activation to the saddle-point neighborhood, $A^* - A$, was computed as described in Appendix B and used to determine the spontaneous jump frequency under

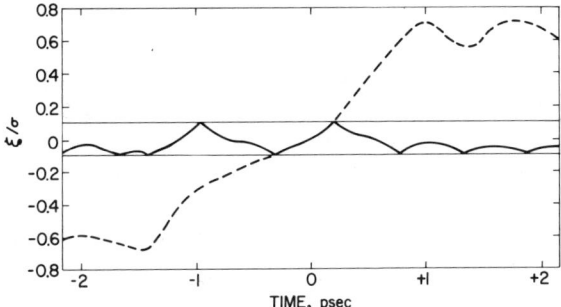

Fig. 1. The solid curve shows the evolution in time of the reaction coordinate ξ, defined in the text, for a 255-atom Lennard-Jones dynamical system moving under the potential U^*, which prevents atom 1 from leaving the saddle-point neighborhood between two vacant sites, by means of infinite potential barriers at $\xi = \pm 0.1$. The dashed lines show the evolution of ξ during a continuation of the motion forward and backward in time beyond the barriers. Together with the intervening solid segment, they constitute a typical trajectory for a successful vacancy jump. The density and temperature of the system, described in Appendix B, correspond roughly to argon at 80°K.

the unperturbed potential U. These results are discussed further in Sections III.A and III.B.

It should perhaps be remarked that the implicit assumption of an equilibrium ensemble distribution for the spontaneously arising saddle-point configurations and velocities is not an approximation (it does not, for example, require that the jumping atom spend a long time in the saddle-point neighborhood). The equilibrium distribution in the saddle-point neighborhood (or indeed in any part of phase space) follows rigorously as long as the system as a whole (including the N atoms and the occasionally jumping vacancy) is in a state of thermal equilibrium.

In addition to making possible the determination of defect jump frequencies at arbitrarily low temperatures, the induced trajectory method can be used for isotope effect studies, by making atom 1 more or less massive than all the others and observing the effect on the induced trajectories and the frequency of successful jumps. The free energy difference $A^* - A$ does not need to be calculated, since it is independent of the atomic masses. Recent isotope effect calculations by the author on the Lennard-Jones system are discussed in Appendix C.

III. Discussion of Molecular Dynamics and Monte Carlo Results on Point Defects at Thermal Equilibrium

Equilibrium Monte Carlo and molecular dynamics point defect calculations have so far been performed on three models: the fcc hard sphere system, the fcc Lennard-Jones system, and a bcc system intended to represent α iron (Table II), but for none of the systems can the picture of point defect properties be said to be complete. Comparison among the systems is further hindered by statistical errors, the variety of truncation conventions that have been applied to the Lennard-Jones potential, and the lack of equation of state data on the α-iron model. In view of these difficulties, discussion will be limited to the following: (A) the microscopic dynamics of vacancy jumps, as observed by molecular dynamics; (B) a rather crude comparison of equilibrium defect concentrations and jump frequencies in the three systems, in terms of a corresponding states principle; (C) an assessment of the validity of molecular statics for estimating defect concentrations and jump frequencies at high temperatures; and (D) suggestions for future computer experiments. The reader is referred to the original papers for other discussion and for comparisons with experimental data on real materials.

A. Defect Jump Dynamics and Correlations

Molecular dynamics is uniquely suited to observing the microscopic dynamical processes attending defect migration, but in practice this endeavor has been somewhat impeded, at least with regard to monovacancies, by their low jump frequency at realistic solid temperatures. The next two paragraphs will report some new results obtained by the author (see also Appendix B) on vacancy jumps in a fcc Lennard-Jones crystal ($N = 255$, $V/N\sigma^3 = 1.003$, $kT/\varepsilon = 0.51$) comparable to solid argon at 60°K. The jumps were generated by the induced trajectory technique described earlier, which overcomes the limitation of low jump frequency.

Figure 2 shows the reaction coordinate ξ as a function of time for eight successful jumps and one U-turn. This reaction coordinate, defined by Eq. (11), measures the position of the jumping atom relative to the four-atom "gate" through which it must pass; for any given dynamical trajectory the condition, $\xi = 0$ will correspond roughly, but not necessarily exactly, to the jumping atom's position of maximum potential energy. The starting configurations (and velocities) were randomly selected from among the $\xi = 0$ crossings occurring at equilibrium in a constrained system, having infinite potential barriers at $\xi/\sigma = \pm 0.1$; the trajectories were then dynamically continued forward and backward in time for about 2 psec, as shown.

The trajectories are noteworthy for their diversity, ranging from fast jumps (e.g., A, B) during which the jumping atom scarcely decelerates, to slow jumps (C, D, E) in which the jump velocity $d\xi/dt$ may pass through two

TABLE II
Some Point Defect Calculations in Model Systems

Pair potential	Computational method[a]	Crystal structure	Defect species[b]	Thermodynamic properties[c]	Dynamic properties[d]	Investigators
Chang's "α iron" (1968)	MD	bcc	1V, 2V, 1I	—	Γ	Tsai et al. (1970)
Hard spheres	MS	bcc	1V, 2V, 1I	E^F	E^M	Bennett and Alder (1971a, b)
	MD (HM)	fcc	1V, 2V, vacancy clusters	V^F, χ	Γ	
	MD	fcc	1V, 2V	—	Jump correlations[e]	
Lennard-Jones	MC (HM)	fcc	1V	χ	—	Squire and Hoover (1969)
Lennard-Jones	MD	fcc	1V	E^F	—	Cotterill and Pedersen (1972)
Lennard-Jones	MD (HM)	fcc	1V, 2V	—	Γ, $\xi(t)^f$	Bennett (Appendix B)
Lennard-Jones	MS	fcc	1V, 2V	$E^F S^F$	E^M, v^*	Glyde (1967), Burton (1969), Burton (1972), many others

[a] MD, molecular dynamics; MC, Monte Carlo; MS, molecular statics; HM, Hamiltonian manipulation.
[b] 1V, monovacancy; 2V, divacancy; 1I, interstitial, etc.
[c] E^F (V^F), energy (volume) of formation; χ, equilibrium concentration.
[d] Γ, jump frequency; E^M, energy of motion; v^*, harmonic jump attempt frequency.
[e] Correlations in time and direction between successive jumps of the vacancy or divacancy.
[f] $\xi(t)$, reaction coordinate as a function of time during jumps.

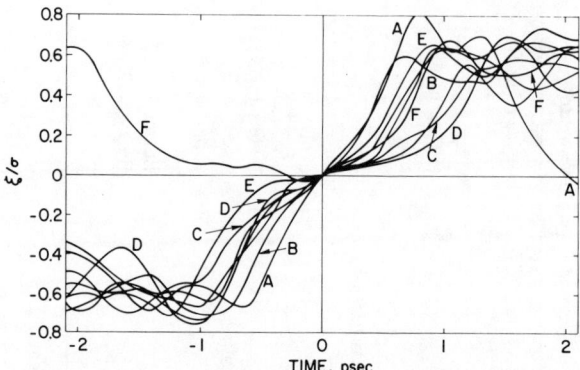

Fig. 2. Reaction coordinate as a function of time during eight successful vacancy jumps (*A* through *E* and the three unlabeled curves) and one unsuccessful attempt (*F*) in a 255-atom fcc Lennard-Jones system. The trajectories were generated by dynamically continuing nine representative $\xi = 0$ crossings occurring in an equilibrated system under the confining potential U^*. The density and temperature correspond to argon at 60°K (see Appendix B).

minima (*D*), suggesting a double saddle point. (This might occur if the four-atom "gate" happened to be decidedly nonplanar when the jumping atom passed through.) Memory (or anticipation) of the jump extends only about 1.5 psec before and after the $\xi = 0$ crossing; by ± 2 psec, the oscillations of $\xi(t)$ have become typical of those that would occur in the long intervals (about 500 psec) between spontaneous jumps. Trajectory *A* is an exception—the jumping atom makes another jump almost immediately, apparently without dissipating the kinetic energy acquired during the first jump. Unsuccessful trajectories, such as *F*, were also relatively uncommon and accounted for only about 10% of $\xi = 0$ crossings.

Figure 3 shows the mean $\xi(t)$ curve for 15 successful jumps and their time reverses. (This lumping together of positive and negative times improves statistics and is justified by the time reversal invariance of the Hamiltonian which guarantees the antisymmetry of $\langle \xi(t) \rangle$ at thermal equilibrium.) The dotted line represents the thermal mean crossing velocity $\langle V^2 \rangle / \langle |V| \rangle$, obtained by assuming Maxwellian velocity distributions for all five atoms involved in the reaction coordinate:

$$\langle V^2 \rangle / \langle |V| \rangle = \int_0^\infty v^2 \exp(-\mu v^2/2kT)\, dv \bigg/ \int_0^\infty v \exp(-\mu v^2/2kT)\, dv$$

$$= (\pi kT/2\mu)^{1/2} \quad (15)$$

Here μ is the reduced mass of the reaction coordinate, $\tfrac{4}{5}$ of the atomic mass of this case. It can be seen that successful jumping atoms have very nearly thermal velocity while they are in the saddle-point neighborhood,

2. EXACT DEFECT CALCULATIONS IN MODEL SUBSTANCES

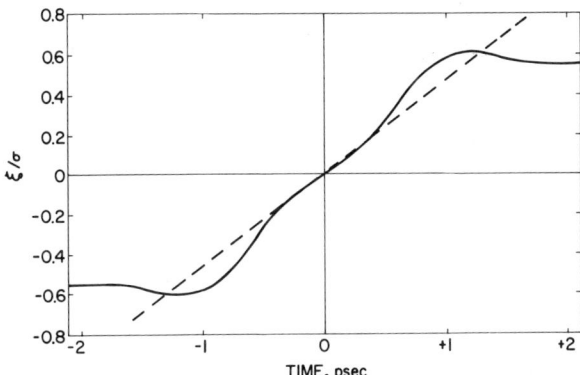

Fig. 3. Average of 15 successful induced vacancy jump trajectories $\xi(t)$ and their time inverses $-\xi(-t)$, in a Lennard-Jones system under the conditions of Fig. 2. The dashed line represents the mean thermal crossing velocity $(5\pi kT/8m)^{1/2}$ calculated from kinetic theory. Its agreement with the observed crossing velocity reflects the fact that most $\xi = 0$ crossings lead to successful jumps.

but about twice this velocity during the ascent and descent. Comparison of Figs. 2 and 3 shows that averaging the trajectories has smeared the velocity maximum somewhat and that the typical maximum is closer to 2.5 times the velocity at the saddle point. Thus, a large part of the activation energy of motion (about $8kT$ at this temperature) appears, temporarily, as kinetic energy in the reaction coordinate. By determining the fraction of this energy in the jumping atom alone, one could estimate the ΔK factor and from that the isotope effect; however, a better approach, not involving the harmonic approximation, would be to observe the isotope effect directly, by varying the mass of the jumping atom (cf. Appendix C).

The duration of a vacancy jump (and presumably also of an interstitial jump) is about 2 psec or, more generally, about $(ma^2/kT)^{1/2}$, where m is the atomic mass and a is the jump distance. Therefore, whenever two consecutive jumps are separated by a time interval shorter than this, they may be expected to be strongly correlated, because they are to some extent happening simultaneously. If such correlated jumps occurred often, they could seriously affect the interpretation of tracer diffusion and isotope effect experiments, in which it is usually assumed that the path of each vacancy or interstitial is a random walk. Consecutive jump correlations have been extensively explored in the fcc hard sphere system; they have a significant effect on monovacancy diffusion only at metastable superexpanded densities, but influence the more frequently jumping divacancies even at stable solid densities (Bennett and Alder, 1971b). Two main types of correlation or "persistence" were noted. The first, which occurred only for monovacancies at metastable densities, involved a simultaneous colinear sliding of a row of two or more atoms and produced the effect of two or more jumps of the vacancy in the same direction, with zero time delay. The second, observed at all densities for both mono- and divacancies, consisted of an increased probability for making a 60° angle and a decreased probability for a 120° or a 180° (U-turn) angle, between successive jump directions, when the jumps occurred closer together in time than about $(ma^2/kT)^{1/2}$. The second type of persistence may be largely a geometric effect steming from the inability of two jumping atoms to be in the same place at the same time.

Persistence was not looked for in the bcc "α-iron" study (Tsai, et al., 1970) but may have been significant owing to the high jump rates.

B. Comparison of Hard Sphere, Lennard-Jones, and "α-Iron" Systems

The data on the three model systems near their melting points, summarized in Table III, are in accord with laboratory experience in that: (1) different fcc substances have similar defect concentrations and diffusion coefficients at the melting point, and (2) divacancies and interstitials are more mobile than vacancies, but are insufficiently numerous to contribute much to self-diffusion under equilibrium conditions. The data also illustrate a peculiar technical problem of computer "experiments": the ease of inadvertently choosing boundary conditions which place the crystal under a positive or negative pressure equivalent to several kilobars or in a metastable superheated state far above the equilibrium melting point. The possibility for superheating arises because, in a system of only a few hundred atoms, the nonnegligible interface tension destabilizes the system against coexistence of two phases and makes the melting transition irreversible. To determine the true thermodynamic melting point of a model system, one must perform separate equation of state computations on the solid and liquid phases, to find the temperature at which their free energies are equal. (The equilibrium melting point so determined may, of course, differ somewhat from the laboratory melting point of the material the model system is intended to represent.) The hard sphere and Lennard-Jones systems are well characterized thermodynamically (Hoover and Ree, 1968; Alder, et al., 1968; Hansen and Verlet, 1969), and the data in Table III indeed represent crystals near the thermodynamic melting point. The "α-iron" system's melting point, on the other hand, has not been determined, and the jump frequencies cited, at the laboratory melting temperature of iron, refer to a computer crystal that may be superheated and almost certainly is under a large positive pressure. (In real iron, this pressure, which arises from the predominantly repulsive pair interactions,

2. EXACT DEFECT CALCULATIONS IN MODEL SUBSTANCES

TABLE III
APPROXIMATE POINT DEFECT PROPERTIES NEAR THE MELTING POINT

Model system	pv_i/kT^a	$-\phi/kT^b$	Defect	Concentration	Jump frequency[c] Γ (sec^{-1})	Γ/ν
Hard Sphere fcc (Bennett and Alder, 1971a) Γ scaled to argon, 80°K	11.3	0	$1V$ $2V$	2×10^{-4} 10^{-6}	3.2×10^{10} 1.2×10^{11}	0.1 0.4
Lennard-Jones fcc (Appendix B) (Squire and Hoover, 1969) Γ scaled to argon, 80°K	-0.3	10.9	$1V$ $2V^d$	2×10^{-4} 10^{-6}	2.6×10^{10} 1.7×10^{11}	0.08 0.5
"α-Iron" bcc[e] (Tsai et al., 1970) Γ scaled to iron, 1810°K	≈ 5	5.0	$1V$ $1I$	— —	10^{12} 4×10^{12}	0.6 1.8

[a] p, pressure; v, volume per atom.
[b] ϕ, potential energy per atom.
[c] The jump frequency Γ is given both in absolute units of jumps per second per defect (summed over all jump directions) and in terms of a thermal scaling frequency $\nu = (kT/ma^2)^{1/2}$, based on the atomic mass m, the nearest neighbor distance a, and the temperature T. The approximate duration of a jump is $1/\nu$.
[d] The divacancy concentration in the Lennard-Jones system is computed from the monovacancy concentration (Squire and Hoover, 1969) assuming the same binding free energy, about $1.5kT$, as found in the hard sphere system (Bennett and Alder, 1971a).
[e] The pressure of the iron system was not reported, but pv/kT must have been at least 5 to protect the system from melting at such a small value of $-\phi/kT$.

would be offset by a positive volume dependence of the electronic part of the cohesive energy.)

Some sort of corresponding states principle must be adopted in order to compare data collected under diverse pressures and temperatures, on systems with different potentials. This is especially true for the hard sphere system which, because it lacks any cohesive forces, must always be held together by external pressure. A convenient corresponding states principle, which will be used here without any effort to justify it theoretically, is suggested by the near equality between the pressure volume term pv/kT in the hard sphere crystal at its melting point, and the cohesive potential energy $-\phi/kT$ in the Lennard-Jones crystal at its melting point (Table III). Henceforth, two computer crystals will be considered to be in corresponding states when they have equal values of the quantity w/kT, where $w \equiv pv - \phi$. The parameter kT/w may be likened to a dimensionless temperature; in both the hard sphere and Lennard-Jones systems, kT/w is close to 0.090 at the equilibrium melting point, while irreversible melting of the metastable superheated solid occurs near $kT/w = 0.11$.

The correspondence rule just described is especially suitable for discussing vacancies, because the work of forming an *unrelaxed* vacancy, in a crystal with pairwise-additive forces, is just $pv - \phi$. Because of relaxation, the true work of forming a vacancy (i.e., the isobaric Gibbs free energy of formation G^F) will always be less than the unrelaxed estimate w. Figure 4 plots the logarithm of the relaxational enhancement of the vacancy concentration $(w - G^F)/kT$ versus the corresponding states parameter kT/w, for the 107-atom hard sphere system of Bennett and Alder (1971a, curve) and the 107-atom Lennard-Jones system of Squire and Hoover (1969, points). An approximate extrapolation of the hard sphere data (dashed curve) is given, based on the assumption that the vacancy volume of formation decreases monotonically as the system is compressed to the close packed limit $kT/w = 0$.

Above the equilibrium melting point $kT/w = 0.90$, the hard sphere curve increases sharply. This increase is accompanied (Bennett and Alder, 1971a) by much steeper increases in the isobaric volume and entropy of vacancy formation, with V^F reaching $1.5v$ and S^F reaching $8k$ at $kT/w = 0.103$. Such large relaxations suggest extensive local melting around the vacancy and may indicate incipient vacancy-nucleated melting of the metastable bulk solid, which has a higher chemical potential than its melt. No temperature dependence of $(w - G^F)/kT$ is evident in the two Lennard-Jones data points.

Even for hard spheres, however, the temperature dependence is not strong below the equilibrium melting point, and it appears that in both the hard sphere and Lennard-Jones systems (and presumably also in other

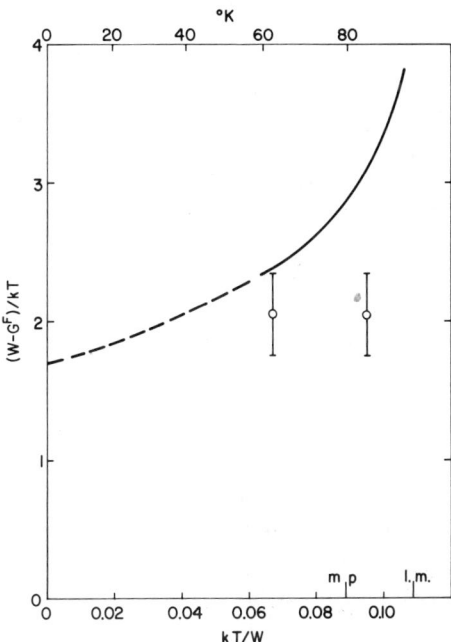

Fig. 4. Excess vacancy concentration, over the unrelaxed estimate $\exp(-w/kT)$, in hard sphere and Lennard-Jones fcc crystals. In the ordinate, G^F is the true free energy of formation determined by Monte Carlo decoupling techniques, while w is the work required to form a (hypothetical) unrelaxed vacancy. The solid curve represents the hard sphere data of Bennett and Alder (1971a); the two points the Lennard-Jones data of Squire and Hoover (1969). For both systems, the parameter kT/w is near 0.09 at the melting point (mp). The limit of metastability (l.m.) of the superheated hard sphere solid, above which even small systems quickly melt, is also shown. The upper scale shows the correspondence between kT/w and temperature, for an argon-scaled Lennard-Jones system under zero pressure.

fcc crystals under pairwise-additive potentials), the equilibrium vacancy concentration can be approximated to within a factor of e by the corresponding states expression $e^{2.0 - w/kT}$, throughout the physically interesting temperature range for diffusion, roughly $0.04 < kT/w < 0.09$.

Experimental determinations of the vacancy concentration in fcc materials at their melting points have generally yielded values between 10^{-3} and 10^{-4}, in rough agreement with the hard sphere and Lennard-Jones models. The one conspicuous exception is krypton, for which Losee and Simmons (1968) found a vacancy concentration about 10 times greater than that predicted by the Lennard-Jones model of Squire and Hoover. It is unlikely that this finding can be accounted for without appealing to short range nonpairwise-additive forces in the noble gas solid.

Point defect jump frequencies have been determined by molecular dynamics calculations on all three model systems (Table III); however, it is difficult to compare the bcc "α-iron" results of Tsai et al. (1970) with the other work, because these authors did not report the pressure, which must have been considerable in view of the high temperatures (relative to the cohesive potential energy $-\phi$) attained without melting. At the laboratory melting point of iron (about 1810°K), this system exhibits a considerably higher vacancy jump frequency (cf. Table III) than either of the fcc systems, and its interstitial jump frequency is four times higher still. Compared to the scaling time $v^{-1} \equiv (ma^2/kT)$, which represents the approximate duration of a jump, the interstitial jumps so often that its consecutive jump directions must be highly correlated.

In Fig. 5, the Lennard-Jones and hard sphere data on vacancy and divacancy jump frequencies are plotted against the corresponding states parameter w/kT. The behavior of divacancies is similar in the two systems, but the hard sphere monovacancy shows a much higher enthalpy of motion than the Lennard-Jones monovacancy. (The enthalpy of motion $-[\partial \ln \Gamma/\partial(1/kT)]_p$ is very nearly proportional to the slope in Fig. 5 for both the hard sphere and the Lennard-Jones system, owing to the weak temperature dependence of v and the nearly linear relation between w/kT

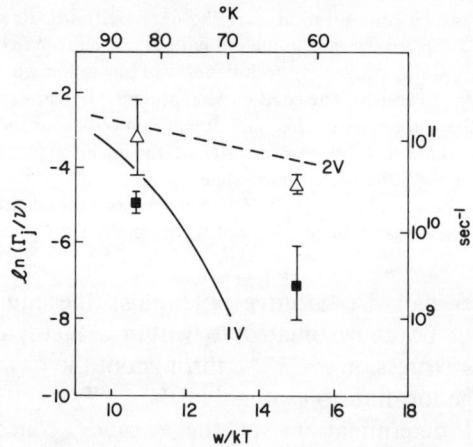

Fig. 5. Defect jump frequencies calculated by molecular dynamics. In the left ordinate, Γ_j is the jump frequency per jump direction and $v = (kT/ma^2)^{1/2}$ is the thermal scaling frequency. The lower abscissa is the corresponding states parameter w/kT, analogous to reciprocal temperature. The right and top scales give equivalent total jump frequencies ($12\Gamma_j$) and temperatures for an argon-scaled Lennard-Jones crystal under zero pressure. The plotted data are for hard sphere vacancies (—) and divacancies (---), and for Lennard-Jones vacancies (■) and divacancies (△). The Lennard-Jones data are from Appendix B; the hard sphere data from Bennett and Alder (1971a).

2. EXACT DEFECT CALCULATIONS IN MODEL SUBSTANCES

and $1/kT$ in both systems.) This discrepancy should not be too surprising, since the physical meaning of the enthalpy is entirely different in the two systems, being $p \cdot v$ work in one and potential energy in the other. However, it does show the limitations of the corresponding states argument, since it means that two fcc systems having the same vacancy concentration [given roughly by $\exp(2 - w/kT)$] can have very different jump frequencies if the pair potentials are sufficiently different.

The jump rate of divacancies is much less temperature dependent than that of monovacancies. One corollary of this which is perhaps not immediately apparent is a variation in mean diffusive lifetime of divacancies, observed in both the Lennard-Jones and hard sphere systems (Appendix B; Bennett and Alder, 1971a). At high temperatures (i.e., $w/kT < 12$), where the monovacancy jump rate is fairly high, a typical divacancy will make only a few jumps before it dissociates into two monovacancies (these may of course reassociate later, after a variable number of jumps). On the other hand, when the monovacancy jump rate is very low, the associative and dissociative jump rates tend to be low also, and the typical divacancy makes many jumps between the time it is formed and the time it dissociates.

C. Comparisons of Molecular Dynamics and Monte Carlo Results with Molecular Statics

In view of the ease and precision of molecular statics and the associated approximate methods for computing defect concentrations and jump frequencies, it would be very useful to know how far they can be trusted to reproduce the exact but expensive results of Monte Carlo and molecular dynamics. (These results, of course, are only "exact" in the sense of being relatively free of systematic errors; they are still subject to random statistical errors which can be reduced only by more computing time.) Worthwhile comparisons between exact and approximate results can be made for both the Lennard-Jones and "α-iron" system. It should perhaps be emphasized that such comparisons are pure tests of the mathematical validity of the harmonic approximation and do not depend on how "good" the potential is.

Table IV compares the monovacancy concentrations and jump frequencies predicted by static and exact methods for two high temperature Lennard-Jones crystals, corresponding roughly to argon at 60 and 80°K. Somewhat surprisingly, the static and exact jump frequencies are in good agreement, while the static estimate of the equilibrium vacancy concentration is too low by a factor of 5. This discrepancy probably results from the strongly anharmonic restoring forces experienced by the atoms

TABLE IV

Comparison between Static and Exact Estimates of the Equilibrium Vacancy Concentration and Jump Frequency in the Lennard-Jones FCC Solid

		Equivalent argon temperature: 60°K	80°K
[a] Reduced temperature	kT/ε	0.505	0.67
Reduced pressure	pv/kT	-0.42_8	-0.34_7
Nearest neighbor lattice parameter	a/σ	1.1215	1.1344
[b] Static formation energy	E^F/ε	8.4_1	8.17_{10}
[c] Static formation entropy	S^F/k	1.9_1	1.7_1
[d] Static motion energy	E^M/ε	4.36_{10}	3.7_1
[d] Log attempt frequency	$\ln[v^* \cdot (m\sigma^2/\varepsilon)^{1/2}]$	0.7_3	0.7_3
[e] Log vacancy concentration, $\ln \chi$	Static	-14.3_2	-10.2_2
	Exact	-12.9_5	-8.5_5
[f] Log jump frequency, $\ln[\Gamma_j(m\sigma^2/\varepsilon)^{1/2}]$	Static	-7.9_4	-4.8_4
	Exact	-7.5_{10}	-5.3_2

[a] The temperatures, pressures, and lattice constants are for the two Monte Carlo crystals in which Squire and Hoover (1969) determined the equilibrium vacancy concentration. The molecular dynamics calculations of Appendix B, in which the vacancy jump frequency was determined, reproduced these conditions. The subscripts denote error bounds, thus -0.42_8 means -0.42 ± 0.08.

[b] E^F and E^M were computed by the author on a 255-atom periodic system (see Appendix B) at zero temperature but with the same lattice parameter a/σ as was used in the high temperature runs. E^M is the motion energy at constant volume, while E^F is the formation energy at a *constant number density of lattice sites* (because the vacancy volume and atomic volume are nearly equal, this is very nearly the formation energy at constant pressure). E^M is much more strongly density dependent than E^F; e.g., at the equilibrium 0°K lattice spacing $(a/\sigma = 1.090)$, E^F/ε is 8.50_{10} while E^M/ε has increased to 6.50_{10}. The sublimation energy at 0°K is 8.61ε.

[c] Static Lennard-Jones vacancy formation entropy data of Burton (1972).

[d] The harmonic attempt frequency v^* was estimated by Glyde (1967) to be about 9×10^{-11} sec^{-1} and only weakly dependent on a/σ.

[e] The exact value is from Squire and Hoover's Monte Carlo calculation (1969). The error bound quoted by these authors, ± 0.25, has been increased to ± 0.5 to allow for the additional uncertainty resulting from their use of experimental Ar and Kr (rather than Lennard-Jones) solid phase entropy data in calculating the free energy of formation. The static estimate is computed by the formula $\ln \chi = S^F - E^F/kT - pv/kT$.

[f] The exact value is from molecular dynamics (Appendix B); the static estimate is given by $\Gamma_j = v^* \exp(-E^M/kT)$. There is no pv term in this formula because E^M and v^* are computed at constant volume (cf. footnote b).

adjacent to a vacancy, and is in accord with the finding by Cotterill and Pederson (1972) that molecular statics overestimates the formation energy by about $1.5 \pm 1.0\ kT$ near the melting point. (These authors give a "molecular statics" formation energy which is simply the sublimation energy, without any static relaxation. Accordingly, their static E^F values should be lowered by about 0.05×10^{-13} erg to agree with the definition of molecular statics employed in the present chapter.)

No defect concentration data are available for the "α-iron" model of Tsai et al. (1970), but static energies of motion (at constant volume) were computed and compared with dynamic values obtained from the temperature dependence of the jump frequency (also at constant volume). The dynamic and static values for interstitials were in good agreement at 0.23 eV but the dynamic E^M for vacancies, 0.42 eV, was significantly lower than the static value, 0.60 eV. Tsai et al. attributed this discrepancy to the double saddle point for vacancy migration in the bcc structure and tentatively suggested a "pumping" action in which the neighbors of the jumping atom at the first saddle point later help push it through the second saddle point. It should perhaps be noted that even at zero temperature, the vacancy in the bcc "α-iron" model is highly relaxed, having a static formation energy only 72% of the sublimation energy (cf. 98% for the fcc Lennard-Jones vacancy), and that the dynamic calculations were performed at rather high temperatures ($kT/\phi = 0.2$ or more, as opposed to 0.09 in the Lennard-Jones crystal at its melting point); both these factors could lead to a complex jump mechanism.

D. Suggestions for Future Calculations

Among the promising areas for future molecular dynamics and Monte Carlo calculations are the following:

a. isotope effect calculations, using the induced trajectory method, particularly in bcc systems, where experiments suggest smaller isotope effects than can be computed within the context of the harmonic approximation;

b. free energy of formation calculations for vacancies in bcc systems and for divacancies in fcc systems (The extensive relaxation of these defects makes them interesting, but may lead to large statistical fluctuations, necessitating longer runs than those needed in the fcc monovacancy case.);

c. further qualitative investigation of the detailed dynamics of jumps and the correlations between them, particularly in bcc systems;

d. calculations on both fcc and bcc structures using the same potential function to distinguish structure dependent phenomena from potential dependent phenomena;

e. efforts to verify and explain the apparent discrepancy between static and exact estimates of the vacancy concentration in the Lennard-Jones system, and to ascertain in detail the domain of reliability of molecular statics; and

f. work on systems with impurity diffusion, more complicated force laws, etc. Such work should be undertaken with considerable caution, in view of the inadequate understanding we now have of even the most simple systems.

Many of these calculations, e.g., (b) and (e), for which it has been difficult to achieve sufficient statistical accuracy on older computing equipment, can be expected to succumb to the greater speed of present and anticipated machines.

Appendix A. Monte Carlo Acceptance Ratio Method for Free Energy Differences

The following method for determining the Helmholtz free energy difference $A_1 - A_0$, between two systems having equal N, V, and T but different potential functions ($U_1 \neq U_0$), is new in detail but closely related to the methods of MacDonald and Singer (1967) and Valleau and Card (1972).

To derive the formula (8), we note that the definition of the Metropolis function, viz.,

$$M(x) = \min(1, e^{-x}) \qquad (3)$$

implies the following identity, for any values x and y:

$$e^{-y} M(x - y) = e^{-x} M(y - x) \qquad (A1)$$

Taking $x = U_1(\mathbf{r}_1 \cdots \mathbf{r}_N)/kT$ and $y = U_0(\mathbf{r}_1 \cdots \mathbf{r}_N)/kT$, and suppressing explicit reference to the atomic coordinates, one obtains the functional identity

$$\exp(-U_0/kT) M[(U_1 - U_0)/kT] = \exp(-U_1/kT) M[(U_0 - U_1)/kT] \qquad (A2)$$

Integrating both sides over all of configuration space yields

$$1 = \frac{\int \exp(-U_0/kT) M[(U_1 - U_0)/kT] \, d\mathbf{r}_1 \cdots d\mathbf{r}_N}{\int \exp(-U_1/kT) M[(U_0 - U_1)/kT] \, d\mathbf{r}_1 \cdots d\mathbf{r}_N} \qquad (A3)$$

which, when multiplied by the unknown constant $\exp[(A_0 - A_1)/kT]$,

2. EXACT DEFECT CALCULATIONS IN MODEL SUBSTANCES

becomes

$$\exp[(A_0 - A_1)/kT] = \frac{\int \exp[(A_0 - U_0)/kT] M[(U_1 - U_0)/kT] \, d\mathbf{r}_1 \cdots d\mathbf{r}_N}{\int \exp[(A_1 - U_1)/kT] M[(U_0 - U_1)/kT] \, d\mathbf{r}_1 \cdots d\mathbf{r}_N} \quad (A4)$$

Here, the numerator and denominator on the right side can be recognized as canonical averages, i.e. quantities that can be measured during ordinary Monte Carlo runs. Taking the logarithm and using the angle bracket notation for the canonical averages, one obtains the desired result

$$A_1 - A_0 = -kT \ln\{\langle M[(U_1 - U_0)/kT]\rangle_0 / \langle M[(U_0 - U_1)/kT]\rangle_1\} \quad (8)$$

whose intuitive interpretation in terms of Monte Carlo acceptance probabilities has already been mentioned.

In using the method, the numerator of (8) is determined, with more or less statistical error, by averaging the function $M[(U_1 - U_0)/kT]$ over the configurations generated by a Monte Carlo run under the potential U_0; the denominator is determined analogously by a run under the potential U_1. The statistical errors tend to be large when the potentials U_0 and U_1 differ greatly; in that case it may be necessary to pass from U_0 to U_1 via a sequence of intermediate potentials. In the limit of a large number of intermediate steps, this becomes equivalent to the older λ-integration method of Eq. (7).

Appendix B. Lennard-Jones Vacancy Jump Calculations

Table BI summarizes the results of some recent molecular dynamics calculations by the author on a 256-site fcc Lennard-Jones system, undertaken primarily to test the induced trajectory technique described in Section II.E. In all calculations, the potential was truncated at 2.5σ but the energies $E/N\varepsilon$ reported in the table include an infinite lattice correction for neighbors beyond that distance. The runs at $kT/\varepsilon = 0.67$ and 0.50 reproduce the conditions of the 80 and 60°K argon-scaled Monte Carlo runs of Squire and Hoover (1969).

Divacancy jump frequencies were determined directly, by observing spontaneous divacancy jumps in a 254-atom system. Monovacancy jump frequencies were determined both directly and by the induced jump technique of Section II.E, using the confining potential U^* of Eq. (14), whose effect is illustrated in Fig. 1. A representative collection of induced trajectories through the $|\xi| < 0.1\sigma$ neighborhood was collected, and the induced jump frequency $c \cdot \Gamma^*$ was determined. In this expression, Γ^* is the frequency of

TABLE BI

LENNARD-JONES VACANCY AND DIVACANCY JUMP FREQUENCIES

$E/N\varepsilon^b$	a/σ	kT/ε	Defect	$\ln \Gamma_j^a$ Direct	$\ln \Gamma_j^a$ Indirect
−6.28	1.134	0.67	1V	−5.3 ± 0.2	−5.0 ± 0.5
−6.24	1.134	0.67	2V	−3.0 ± 1.0	—
−6.90	1.121	0.50	1V	—	−7.5 ± 1.0
−6.87	1.121	0.50	2V	−4.0 ± 0.2	—

^{a} Γ_j is the jump frequency per jump direction in units of $(\varepsilon/m\sigma^2)^{1/2}$ or approximately 4.5×10^{11} sec^{-1} for the argon scaling. The relatively large statistical uncertainty in the divacancy jump rate at $kT/\varepsilon = 0.67$ stems from the fact that at this temperature the typical divacancy makes only a few jumps before dissociating into monovacancies.

^{b} Energies include truncation correction. The corrected pressure, pv/kT, was about −0.2 at $kT/\varepsilon = 0.67$ and about −0.7 at $kT/\varepsilon = 0.50$.

$\xi = 0$ crossings under U^* and c is the *conversion coefficient*, defined (Anderson, 1973) as the fraction of such crossings which, when dynamically extended under U, do not recross $\xi = 0$ before completing a successful jump. (The conversion coefficient is thus the ratio of successful jumps to $\xi = 0$ crossings; in the present case, it was found to be about 90% at both temperatures.)

In order to determine the spontaneous jump rate under U, one must know the factor P^* by which the potential U^* enhances the system's probability of being in the saddle-point neighborhood. This was computed with the help of potentials U_λ, analogous to U^* with the barriers at $\xi/\sigma = \pm \lambda$ instead of ± 0.1. Using a series of such potentials with $\lambda = 0.1, 0.2, \ldots, 0.8$, runs were made to determine the quantities $P(\lambda_1, \lambda_2)$ where, e.g., $P(0.3, 0.4)$ is the fraction of time a system, confined to $|\xi| < 0.4\sigma$ by the potential $U_{0.4}$, spontaneously spends with $|\xi|$ less than 0.3σ. These quantities, the microcanonical equivalent of Monte Carlo acceptance ratios (cf. Appendix A), represent volume ratios in phase space and can be multiplied together to give

$$P^* = P(0.1, 0.8) = P(0.1, 0.2) \cdot P(0.2, 0.3) \cdots P(0.7, 0.8) \qquad (B1)$$

This is the phase space volume ratio between the saddle-point neighborhood, $|\xi| < 0.1\sigma$, and the pair of equilibrium defect states it joins. In computing P^*, it is not necessary to consider λ values greater than 0.8, because, even in the absence of barriers, the jumping atom's neighbors prevent $|\xi|/\sigma$ from exceeding this value. Combining the results obtained so far, the spontaneous jump frequency per jump direction under the potential U is

2. EXACT DEFECT CALCULATIONS IN MODEL SUBSTANCES

simply

$$\Gamma_j = c \cdot \Gamma^* \cdot P^* \tag{B2}$$

while the total spontaneous jump frequency in all directions is 12 times as great (12 being the number of sites to which a monovacancy can jump on the fcc lattice). In determining jump frequencies by this method, the main statistical uncertainty comes from the P factors, which in the present study were determined rather imprecisely.

The table shows adequate agreement between the directly and indirectly determined jump frequencies, when the former is high enough to be observed, and demonstrates the suitability of the induced jump method at low temperatures, where the direct method would require an excessive amount of computer time.

To calculate the static formation and motion energy of monovacancies, a damped version of the same dynamics program was used. Computations were performed on 255- and 256-atom systems at the same lattice parameter a/σ used in the high temperature runs. The range of the potential was increased from 2.5σ to $2.7a$ so as to include, at each density, the first seven shells of neighbors. This reduced the magnitude and uncertainty of truncation corrections and permitted the formation and motion energies to be computed with an estimated error of 0.1ε or less. Damping was introduced by diminishing all velocities 5% after each timestep of 5×10^{-14} sec, and resulted in convergence of the configuration and potential energy within about 300 timesteps. To compute saddle-point configurations, an approximate saddle-point configuration was set up and allowed to relax subject to the constraint $\xi = 0$. The static energy of motion at constant volume was obtained by the formula

$$E^M = E'_{255} - E_{255} \tag{B3}$$

where E' and E are the saddle-point and equilibrium energies of the 255-atom system, respectively. The formation energy at constant number density of lattice sites is given by

$$E^F = E_{255} - (255/256)E_{256} \tag{B4}$$

where E_{256} is the relaxed energy of a 256-atom perfect crystal. The numerical values obtained for E^F and E^M are given in Table IV.

Appendix C. Molecular Dynamics Calculation of the Isotope Effect

Because it can be measured experimentally with considerable precision, and because it contains information on the diffusion mechanism, the isotope effect in tracer diffusion is an important target for theoretical calculation. The present appendix describes a molecular dynamics calculation of the relative jump rates into a vacancy of Lennard-Jones atoms with masses 1.00 and 1.05 times the host atom mass, under the conditions of the 80°K system of Appendix B (i.e., $N = 255$, $a/\sigma = 1.134$, $kT/\varepsilon = 0.67$). As before, the calculation involves generating trajectories forward and backward in time from an equilibrium ensemble of starting points in the saddle-point neighborhood. The isotope effect is found to be about 10% smaller than that calculated on the basis of the harmonic approximation, in agreement with most experimental results on fcc materials (Peterson, 1968). The chief error of harmonic treatments is their neglect of *unsuccessful* saddle-point crossings (e.g., U-turns), which become more prevalent the lighter the diffusing atom.

The spontaneous jump rate can be expressed (cf. Appendix B) as the product of three factors

$$\Gamma_j = P^* \cdot \Gamma^* \cdot c \tag{C1}$$

where P^* is the probability that the system will be found in a designated saddle-point neighborhood in configuration space; Γ^*, the probability per unit time that a system in this neighborhood will cross the dividing hypersurface $\xi = 0$; and c, the fraction of such crossings which lead to successful jumps. The individual factors are sensitive to the particular choice of saddle-point neighborhood and reaction coordinate ξ, but their product is not. Since P^* is independent of atomic masses, the change in jump rate caused by a differential change dm_1 in the mass of the jumping atom is given by

$$d\Gamma_j/\Gamma_j = d\Gamma^*/\Gamma^* + dc/c \tag{C2}$$

Any reaction coordinate which is linear in all the atomic coordinates

$$\xi = \text{const} + \sum_{i=1}^{N} \mathbf{a}_i \cdot \mathbf{r}_i \tag{C3}$$

can be characterized by a reduced mass μ,

$$\mu = 1/\Sigma(a_i^2/m_i) \tag{C4}$$

and, as m_1 is varied (while keeping $\mathbf{a}_i$ fixed), the crossing frequency Γ^* varies simply as $\mu^{-1/2}$. The m_1 dependence of Γ_j is then

$$d\Gamma_j/\Gamma_j = -\tfrac{1}{2} d\mu/\mu + dc/c$$
$$= -\tfrac{1}{2}(dm_1/m_1) \cdot (a_1^2 m_1^{-1}/\Sigma a_i^2 m_i^{-1}) + dc/c \tag{C5}$$

of which only the last term needs to be measured by molecular dynamics.

2. EXACT DEFECT CALCULATIONS IN MODEL SUBSTANCES

Before discussing the measurement of dc/c, we shall draw a connection between Eq. (C5) and the harmonic theory of the isotope effect developed by Vineyard (1957), Mullen (1961), and Le Claire (1966). Harmonic treatments are based on the assumption that there is a particular linear combination of atomic coordinates, namely, the unstable normal mode coordinate Q_1' about the exact saddle-point configuration, such that every crossing of the $Q_1' = 0$ hypersurface leads to a successful jump. This would indeed be so in the harmonic limit, where the motion of each normal mode is independent of all the others. In that case, the last term in Eq. (C5) vanishes (dc must vanish where c attains its maximum of unity), and the coefficient of $-\frac{1}{2}dm_1/m_1$ in the first term is simply the well-known ΔK factor, i.e., the fraction of the unstable mode's kinetic energy which resides in the jumping atom. In order best to reveal the extent of anharmonicity of the 80°K Lennard-Jones crystal, the isotope-effect calculations to be described below were all carried out using the harmonic reaction coordinate $\xi \equiv Q_1'$ based on the unstable mode of the equal mass $m_1 = 1.00$ system. This reaction coordinate has a reduced mass of 1 atomic mass and a ΔK factor of 0.98, but is otherwise quite similar to the simple reaction coordinate [Eq. (11)] used earlier, being zero at the saddle-point and about $\pm 0.6\sigma$ for the stable defect configurations before and after the jump.

The expected mass dependence of c is so small that there is no hope of determining it simply by measuring c in the $m_1 = 1.00$ and $m_1 = 1.05$ systems separately. Instead, a differential sampling method must be used to explore that subset of $\xi = 0$ crossings which would lead to successful jumps in the $m_1 = 1.05$ system but not in the $m_1 = 1.00$ system, or vice versa. Such a one-to-one comparison of events in the two systems is possible because, for every microstate of the $m_1 = 1.00$ system $(\mathbf{r}_1, \mathbf{r}_2 \ldots \mathbf{r}_N; \dot{\mathbf{r}}_1, \dot{\mathbf{r}}_2 \cdots \dot{\mathbf{r}}_N)$, there exists a corresponding equiprobable microstate of the $m_1 = 1.05$ system, namely, $(\mathbf{r}_1, \mathbf{r}_2 \cdots \mathbf{r}_N; \dot{\mathbf{r}}_1/(1.05)^{1/2}, \dot{\mathbf{r}}_2 \cdots \dot{\mathbf{r}}_N)$. The differential sampling procedure is outlined below:

1. A molecular dynamics calculation is performed on the $m_1 = 1.00$ system at equilibrium under the constraint $\xi = 0$. The microstates X generated by this calculation all lie on the $6N - 2$ dimensional hyperplane $\xi = \dot{\xi} = 0$ and will be called *surface states*.

2. Starting from the surface state X, a *crossing state* X_u is prepared by giving the reaction coordinate a positive thermal crossing velocity $\dot{\xi} = u$, drawn from the velocity-weighted Maxwell distribution

$$p(u) = (u/kT) \exp(-\tfrac{1}{2}u^2/kT) \tag{C6}$$

The states X_u so prepared are representative of the forward $\xi = 0$ crossings which would occur in an unconstrained system at thermal equilibrium.

3. The trajectory through the crossing state X_u is computed forward and backward in time to determine its success. A crossing is considered successful if and only if it is the *last* $\xi = 0$ crossing on a trajectory segment which starts with

$$\zeta < -0.2\sigma \quad \text{and ends with} \quad \xi > +0.2\sigma. \tag{C7}$$

4. Empirically, it is found that each surface state X has a threshold crossing velocity $\theta_{1.00}(X)$ such that the crossing X_u succeeds if and only if $u > \theta_{1.00}(X)$. The mean conversion coefficient for all $m_1 = 1.00$ crossings *through the surface state X*,

$$c_{1.00}(X) = \int_{\theta_{1.00}}^{\infty} p(u) \, du \tag{C8}$$

can therefore be determined quite accurately by zeroing in on the threshold velocity $\theta_{1.00}(X)$ by successive trajectory calculations.

5. From any crossing state X_u of the $m_1 = 1.00$ system, the corresponding crossing state X_u' of the $m_1 = 1.05$ system can be obtained by dividing the jumping atom's velocity by $(1.05)^{1/2}$. The X_u' states so generated are representative of spontaneous crossings in an unconstrained $m_1 = 1.05$ system; in each, the jumping atom has a greater momentum but a smaller velocity than in the corresponding X_u state. Like the X_u states, the X_u' states associated with a given X exhibit a threshold $\theta_{1.05}(X)$, such that X_u' succeeds if and only if $u > \theta_{1.05}(X)$. Typically, the 1.00 and 1.05 thresholds differ by a few parts in 10^3; either may be higher, depending on X. The anharmonic contribution to the isotope effect arises from u values between the two thresholds, i.e., from crossings which succeed for one isotope but fail for the other (cf. Fig. C1).

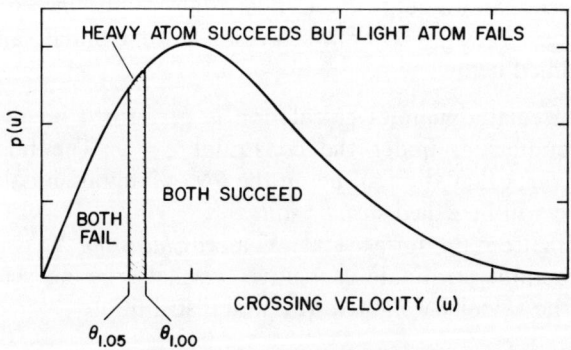

Fig. C1. Success of jump as a function of crossing velocity u through a typical point X on the $6N - 2$ dimensional saddle-point hyperplane. The curve $p(u)$ is the Maxwellian spectrum of spontaneous crossing velocities. The shaded area is equal to $c_{1.05}(X) - c_{1.00}(X)$ and contributes to the anharmonic isotope effect.

2. EXACT DEFECT CALCULATIONS IN MODEL SUBSTANCES

6. The difference in conversion coefficients, $c_{1.05} - c_{1.00}$, is obtained as an average over surface states X of the integral of $p(u)$ between the two thresholds.

$$c_{1.05} - c_{1.00} = \langle c_{1.05}(X) - c_{1.00}(X) \rangle_X = \left\langle \int_{\theta_{1.05}(X)}^{\theta_{1.00}(X)} p(u)\, du \right\rangle_X \quad (C9)$$

This may be found to the desired accuracy by measuring $\theta_{1.00}(X)$ and $\theta_{1.05}(X)$ for a sufficient number of representative surface states X, generated as described in paragraph 1 above.

The program outlined above has so far been carried out on 26 surface states of the 80°K Lennard-Jones system, with the results shown in Table C1.

TABLE C1

MASS DEPENDENCE OF THE CONVERSION COEFFICIENT

Sample size = 26	$c_{1.00}(X)$	$c_{1.05}(X) - c_{1.00}(X)$
Minimum	0.006	−0.0028
Maximum	1.000	+0.0166
Mean	0.80	+0.0018
Standard error of mean	±0.06	±0.0009

The positive sign of $c_{1.05} - c_{1.00}$ means that saddle-point crossings by the heavier isotope have a significantly higher chance of success; therby reducing the effective ΔK factor,

$$d(\ln \Gamma_j)/d \ln m_1^{-1/2} = \Delta K_{\text{harm}} - 40(c_{1.05} - c_{1.00})/c_{1.00} \quad (C10)$$

from its harmonic value of 0.98 to an anharmonic value of 0.89 ± 0.05, a finding which is in agreement with most experimental results on fcc materials (Peterson, 1968). Anharmonicity of the jump mechanism is also indicated by the fact that 20% of all $Q_1' = 0$ crossings fail, whereas in a harmonic crystal all would succeed.

Bcc systems are an important target for future isotope-effect calculations of the kind described here because experiments (e.g., Mundy, 1971, on sodium) suggest an isotope effect only half as large as reasonable harmonic estimates. A small and very mass dependent conversion coefficient is quite plausible in bcc systems as a result of double barrier ring, which, it may be imagined, sometimes traps a diffusing atom (particularly a light one) in the saddle-point neighborhood for some time, causing it to cross the $Q_1' = 0$ hyperplane several times without completing a successful jump.

Acknowledgments

The author wishes to thank B. J. Alder, W. W. Wood, P. Ho, J. J. Burton, and particularly A. Rahman for many helpful discussions. The work was performed at the Solid State Sciences Division, Argonne National Laboratory, under U.S. Atomic Energy Commission auspices, and at the IBM Thomas J. Watson Research Center.

References

Alder, B. J., Hoover, W. G., and Young, D. A. (1968). *J. Chem. Phys.* **49**, 3688.
Alder, B. J., and Wainwright, T. E. (1959). *J. Chem. Phys.* **31**, 459.
Alder, B. J., and Wainwright, T. E. (1960). *J. Chem. Phys.* **33**, 1439.
Anderson, J. B. (1973). *J. Chem. Phys.* **58**, 4684.
Beeler, J. R., Jr. (1972). In "Radiation-Induced Voids in Metals" (J. W. Corbett and L. C. Ianniello, eds.), pp. 739–756, U.S. Atomic Energy Commission, Oak Ridge, Tennessee.
Beeler, J. R., Jr., and Kulcinski, G. L. (1972). In "Interatomic Potentials and Simulation of Lattice Defects" (Gehlen *et al.*, eds.), p. 735. Plenum, New York.
Benedek, R., and Ho, P. (1974). "Lattice Statics of Interstitials in Tungsten" *J. Phys. F* **4**, 183.
Bennett, C. H., and Alder, B. J. (1971a). *J. Chem. Phys.* **54**, 4796.
Bennett, C. H., and Alder, B. J. (1971b). *J. Phys. Chem. Solids* **32**, 2111.
Burton, J. J. (1969). *Phys. Rev.* **182**, 885.
Burton, J. J. (1972). *Phys. Rev.* **B5**, 2948.
Chang, R. (1968). *Phil. Mag.* **16**, 1021.
Cotterill, R. M. J., and Pedersen, L. B. (1972). In "Interatomic Potentials and Simulation of Lattice Defects" (Gehlen *et al.*, eds.), p. 439. Plenum, New York.
Gear, C. W. (1966). Argonne Nat. Lab. Rep. ANL 7126, Argonne, Illinois.
Gehlen, P. C., Beeler, J. R., Jr., and Jaffee, R. I., eds. (1972). "Interatomic Potentials and Simulation of Lattice Defects," Plenum, New York.
Gibson, J. B., Goland, A. N., Milgram, M., and Vineyard, G. H. (1960). *Phys. Rev.* **120**, 1229.
Glyde, H. R. (1967). *Rev. Mod. Phys.* **39**, 373.
Hansen, J. P., and Verlet, J. (1969). *Phys. Rev.* **184**, 151.
Hoover, W. G., and Ree, F. H. (1968). *J. Chem. Phys.* **49**, 3606.
Huntington, H. B., Shirn, G. A., and Wajda, E. S. (1955). *Phys. Rev.* **99**, 1085.
Jaffe, R. L., Henry, J. M., and Anderson, J. B. (1973). *J. Chem. Phys.* **59**, 1128.
Kanzaki, H. (1957). *J. Phys. Chem. Solids* **2**, 24.
Le Claire, A. D. (1966). *Phil. Mag.* **14**, 1271.
Losee, D. L., and Simmons, R. O., (1968). *Phys. Rev.* **172**, 934.
McDonald, I. R., and Singer, K. (1967). *J. Chem. Phys.* **47**, 4766.
Metropolis, N., Rosenbluth, A. W., Rosenbluth, M. N., Teller, A. H., and Teller, E. (1953). *J. Chem. Phys.* **21**, 1087.
Mullen, J. G. (1961). *Phys. Rev.* **121**, 1649.
Mundy, J. N. (1971). *Phys. Rev.* **B3**, 2431.
Peterson, N. L. (1968). *Solid State Phys.* **22**, 409.

RAHMAN, A., AND STILLINGER, F. H. (1971). *J. Chem. Phys.* **55**, 3336.
RAHMAN, A., FOWLER, R. H., AND NARTEN, A. H. (1972). *J. Chem. Phys.* **57**, 3010.
SQUIRE, D. R., AND HOOVER, W. G. (1969). *J. Chem. Phys.* **50**, 701.
TASI, D. H., BULLOUGH, R., AND PERRIN, R. C. (1970). *J. Phys. C* **3**, 2022.
VALLEAU, J. P., AND CARD, D. N. (1972). *J. Chem. Phys.* **57**, 5457.
VERLET, L. (1967). *Phys. Rev.* **159**, 98.
VINEYARD, G. H. (1957). *J. Phys. Chem. Solids* **3**, 121.
WOOD, W. W. (1968). *In* "The Physics of Simple Liquids" (H. N. V. Temperley, J. S. Rowlinson, and G. S. Rushbrooke, eds.). North-Holland Publ., Amsterdam.

3

Isotope Effects in Diffusion‡

N. L. PETERSON
MATERIALS SCIENCE DIVISION
ARGONNE NATIONAL LABORATORY
ARGONNE, ILLINOIS

I. Preface	116
II. Introduction	117
A. General Nature of Correlation Effects	117
B. Correlation and the Isotope Effect	119
III. Self-Diffusion in Pure Metals	124
A. Face-Centered Cubic Metals	126
B. Body-Centered Cubic Metals	132
C. Hexagonal Close Packed Metals	136
D. Self-Diffusion in Germanium and Silicon	137
IV. Diffusion in Dilute Alloys	138
A. Jump Frequencies near an Impurity–Vacancy Pair	138
B. Rapid Impurity Diffusion in Metals	143
C. Effect of Resonant Vibrational Modes on Diffusion	145
V. Diffusion in Concentrated Alloys	146
A. Random Alloys	146
B. Ordered Alloys	147
VI. Diffusion in Alkali and Silver Halide Crystals	149
A. Vacancy-Pair Contribution to Cation Diffusion in NaCl	151
B. Interstitialcy Diffusion in Silver Halides	153
C. Sodium Diffusion in Rubidium Chloride	155
VII. Diffusion in Transition Metal Oxides	157
A. Cation Self-Diffusion in Transition Metal Oxides	158
B. Cobalt Impurity Diffusion in NiO	160
VIII. Correlation Effects in Grain Boundary Diffusion	163
References	167

‡ Work performed under the auspices of the U.S. Atomic Energy Commission.

I. Preface

Atomic migration in crystals is believed to occur by a series of jumps of individual atoms from site to site throughout the crystal. A number of mechanisms have been proposed for the elementary atomic jump. The generally accepted mechanisms involve the motion of one of two types of point defects, *vacancies* (monovacancies, divacancies, and higher order vacancy clusters) or *interstitials* (single interstitials, dumbbell interstitials, interstitialcies, and crowdions) (Peterson, 1973).

When the macroscopic diffusion coefficient D, as defined by Fick's laws, is related to the mobility of individual atoms, a term called the correlation factor is obtained. This correlation factor is a measure of the nonrandomness of the jump process and has a well-defined value for self-diffusion in a cubic crystal that is different for each of the possible mechanisms of diffusion. The correlation factor for diffusion in alloys is related to the relative jump frequencies of both A and B atoms. Thus, a measurement of the correlation factor for diffusion can frequently identify the mechanism of self-diffusion or provide considerable detail concerning the diffusive process in alloys.

A measurement of the mass effect (isotope effect) on diffusion is the most powerful method of determining information about the correlation factor in a broad spectrum of materials. When combined with other diffusion measurements, such as the temperature, pressure, and compositional dependence of the diffusion coefficient, the isotope effect has provided a detailed understanding of impurity and self-diffusion in a number of metallic and nonmetallic solids. The advances in our understanding of diffusion in solids obtained from measurements of the isotope effect for diffusion will be reviewed.

In Section II, the physical concepts of correlation effects and the relation between correlation and isotope effects are presented. The methods of calculating correlation factors have been extensively reviewed by LeClaire (1970) and will not be discussed in this chapter. Sections III–VIII will present results for self-diffusion in pure metals, diffusion in dilute alloys, diffusion in concentrated alloys, diffusion in alkali and silver halide crystals, diffusion in transition metal oxides, and correlation effects in grain boundary diffusion. Diffusion in amorphous and organic materials and interstitial diffusion in solids will not be extensively discussed in this chapter. Several examples of isotope-effect measurements in amorphous materials may be found in the works of Barr *et al.* (1972) and Shelby (1971); studies of organic materials have been reported by Fox and Sherwood (1971) and Chadwick and Sherwood (1972). The isotope effect for hydrogen diffusion in metals is reviewed in Chapter 5.

II. Introduction

A. General Nature of Correlation Effects

If the directions of successive jumps of an atom are random, the path followed by each atom during diffusion may be described as a random walk. A number of different approaches to random walk theory lead to the same equation for diffusion in an isotropic medium (Manning, 1968) first written explicitly by Einstein (1905),

$$D = \langle R^2 \rangle / 6t \qquad (1)$$

The term $\langle R^2 \rangle$ is the mean value over all possible diffusion paths of the square of the net vector displacement R of an atom in time t. If a total of N jumps takes place in time t and each jump is of the same length r, we may write

$$\langle R^2 \rangle = \left\langle \left(\sum_{}^{N} r \right)^2 \right\rangle = Nr^2 [1 + 2(\langle \cos \theta_1 \rangle + \langle \cos \theta_2 \rangle + \cdots)] \qquad (2)$$

where $\langle \cos \theta_j \rangle$ is the average value of the cosine of the angle between the first jump and the jth following jump. For a true random walk, where the direction of any jump is independent of the direction of preceding jumps, all the $\langle \cos \theta \rangle$'s in Eq. (2) are zero, and $\langle R^2 \rangle = Nr^2$.

For a number of diffusion mechanisms, the direction of a given tracer atom jump depends on the direction of the preceding jumps. In such cases, all the $\langle \cos \theta \rangle$'s are no longer zero. For these diffusion mechanisms, a correlation factor f equal to the term in brackets in Eq. (2) is introduced to correct for the fact that atomic jump directions are correlated with one another. Since the total number of jumps N equals the number of jumps per second Γ times the time t, Eqs. (1) and (2) become

$$D = \tfrac{1}{6} r^2 \Gamma f \qquad (3)$$

If a diffusion mechanism requires a defect to be next to an atom for that atom to execute a jump, correlation effects always arise. Let us consider the motion of a given tracer atom when diffusion occurs by the vacancy mechanism. After an atom has made an exchange with the vacancy, the vacancy is immediately available to effect a second jump of the atom in the reverse direction. Hence, the probability is greater than random that the next jump of the tracer atom will be a reversal of the first. Thus, $\langle \cos \theta_1 \rangle$ will be negative. After the reversal jump, the vacancy remains a neighbor of the tracer atom; therefore, the probability is greater than random that the next jump will be in the same direction as the first. Thus, $\langle \cos \theta_2 \rangle$ will be positive but numerically smaller than $\langle \cos \theta_1 \rangle$. The net effect is that the sum of the $\langle \cos \theta \rangle$'s will have a finite negative value rather than zero, and f [equal to the term in brackets in Eq. (2)] will be less than unity. For self-diffusion in a pure crystal, the vacancy moves randomly, but a given tracer atom does not.

Correlation effects for diffusion by a divacancy mechanism arise for the same reasons as those given for a monovacancy mechanism. However, if the two vacancies that make up the divacancy are allowed to dissociate to second-nearest neighbor positions [as is necessary for divacancy diffusion in the body-centered cubic (bcc) lattice], self-diffusion by the divacancy mechanism will require more than one unique jump frequency, and f may be temperature dependent. The value of f for the divacancy mechanism is usually smaller than that for the monovacancy mechanism (Howard, 1966; Mehrer, 1973).

Correlation effects also exist for diffusion by the interstitialcy mechanism. When a tracer atom jumps from an interstitial site to a normal lattice site, a lattice atom must move into another interstitial site. After this jump, the tracer atom (now on a lattice site) has an interstitial next to it; therefore, the second jump of the tracer atom has a greater than random probability of being a reversal of the first. Thus, $\langle \cos \theta_1 \rangle$ will be negative. The next jump of the tracer atom from its interstitial site will take place with equal probability in any of the allowed directions. For the interstitialcy mechanism, only alternate pairs of consecutive jumps are correlated; the jumps from the interstitial site to the normal lattice site are in random directions, but the jumps from the normal site to the interstitial site are correlated to the preceding jump.

The mathematical procedures for calculating correlation factors have been reviewed by LeClaire (1970). For self-diffusion in cubic crystals, f is generally a geometrical factor that depends only on the diffusion mechanism and the crystal structure. Values of f for self-diffusion in the face-

TABLE I

CORRELATION FACTORS FOR SELF-DIFFUSION IN THE
FACE-CENTERED CUBIC LATTICE

Mechanism	Correlation factor
Vacancy	0.781
Divacancy	0.475
Interstitial	1.000
Interstitialcy (AgCl)	
Collinear jump	0.667
Noncollinear jump	0.969

centered cubic (fcc) lattice by various diffusion mechanisms are given in Table I. Values of f for self-diffusion in other crystal structures have been determined by Compaan and Haven (1956, 1958), Mullen (1961b), and Howard (1966). The value of $f = 0.475$ for self-diffusion by the divacancy mechanisms, first obtained by Howard (1966), is supported by the recent calculations of Mehrer (1972) and Montet (1973); however, Bennett and Alder (1971) obtain a value of $f = 0.458$.

B. Correlation and the Isotope Effect

Because the correlation factor is not the same for different diffusion mechanisms (Table I), a determination of f often establishes the mechanism for self-diffusion and can provide detailed information about the relative jump frequencies near an impurity–vacancy pair for impurity diffusion. A measurement of the isotope effect for diffusion can provide information regarding the value of f for tracer diffusion in any material for which a suitable pair of radioisotopes is available.

When atomic motion of the diffusing species occurs by a unique jump frequency w, such as for self-diffusion or impurity diffusion by the monovacancy mechanism in an isotropic lattice, the diffusion of isotope α may be written as

$$D_\alpha = A w_\alpha f_\alpha \tag{4}$$

where A contains geometrical terms and defect concentrations that do not depend on the jump frequency w or on the mass of the tracer. If the correlation factor has the form

$$f_\alpha = u/(u + w_\alpha) \tag{5}$$

where u contains atom-vacancy exchange rates other than the tracer atom, we obtain from Eq. (4)

$$(D_\alpha/D_\beta) - 1 = f_\alpha[(w_\alpha/w_\beta) - 1] \tag{6}$$

where α and β refer to the properties of two isotopes of the same element with masses m_α and m_β.

Following the early work of Schoen (1958) and Tharmalingam and Lidiard (1959) who first derived Eq. (6), it was thought that Eqs. (5) and (6) were valid only when the jump direction was an axis of at least twofold rotational symmetry. Recently, Bakker (1971) has shown that Eqs. (5) and (6) are valid under much more general conditions. Although these more general conditions can only be stated in a complicated manner, Bakker has established that Eq. (5) is valid for self-diffusion and impurity diffusion by single vacancies and divacancies (if dissociation jumps are not considered) and for diffusion by vacancy–impurity pairs in fcc metals.

In some cases, f may not be of the form given by Eq. (5) even if there is only one tracer jump frequency. In general, several different possible tracer jump frequencies may be present, as for multiple interstitialcy diffusion (collinear and noncollinear interstitialcy jumps) or for diffusion in anisotropic crystals. For these cases, D will have the form (LeClaire, 1970)

$$D = \left(\sum_i A_i w_i\right) f \quad \text{or} \quad D = \sum_i A_i w_i f_i \tag{7}$$

where the A_i are constants, as in Eq. (4), and f_i is the partial correlation factor associated with the ith type of jump that has a jump frequency w_i. With D given by Eq. (4), LeClaire (1970) obtains

$$(D_\alpha/D_\beta) - 1 = (1 + w \, \partial \ln f/\partial w)[(w_\alpha/w_\beta) - 1] \tag{8}$$

If D is given by Eq. (7), he obtains the general form

$$\frac{D_\alpha}{D_\beta} - 1 = D_\beta^{-1} \sum_i A_i f_i w_i \left\{ 1 + \sum_j w_j \frac{\partial \ln f_i}{\partial w_j} \left[\frac{(w_{\alpha j} - w_{\beta j})/w_{\beta j}}{(w_{\alpha i} - w_{\beta i})/w_{\beta i}} \right] \right\} \frac{w_{\alpha i}}{w_{\beta i}} - 1 \tag{9}$$

To use Eq. (6) [or its more general form, Eq. (9)], the relative jump frequencies of the isotope must be expressed in terms of the relative isotopic masses. Wert (1950) has shown from elementary reaction-rate theory that w may be written in the form

$$w = v \exp(-g^m/kT) \tag{10}$$

where v is the vibration frequency of an atom about its equilibrium position in the jump direction and g^m is the difference in the Gibbs free energy of the crystal between the configuration where the atom is at the saddle point and the configuration where the atom is in a lattice site. If g^m is independent of isotopic mass, then $w_\alpha/w_\beta = v_\alpha/v_\beta$, and the isotope effect E may be written as

$$E \equiv [(D_\alpha/D_\beta) - 1]/[(m_\beta/m_\alpha)^{1/2} - 1] = f_\alpha \tag{11}$$

for diffusion mechanisms in which only one atom undergoes a jump displacement, provided the vibration frequencies follow the classic relationship

$$(v_\alpha/v_\beta) = (m_\beta/m_\alpha)^{1/2} \tag{12}$$

For mechanisms where n atoms move during a jump, the m_i's in Eqs. (11) and (12) must be replaced by $(n-1)m + m_i$, where m is the average mass of the host lattice.

The derivation of Eq. (10) considers that the atomic jump involves only motion of the diffusing atom; the movement of the jumping atom

3. ISOTOPE EFFECTS IN DIFFUSION

is thought to be completely decoupled from the lattice atoms. Deviations from Eqs. (10) and (12), which are caused by many-body interactions at the saddle point, have been considered by Vineyard (1957). Based on Vineyard's many-body treatment of the atomic jump process, Mullen (1961a) and LeClaire (1966) obtain the relation

$$E \equiv [(D_\alpha/D_\beta) - 1]/[(m_\beta/m_\alpha)^{1/2} - 1] = f_\alpha \Delta K \qquad (13)$$

where ΔK is the fraction of the kinetic energy at the saddle point, associated with motion in the jump direction, that belongs to the diffusing atoms. Hence, ΔK is bounded by zero and unity.

The correlation factor in Eq. (13) applies to the motion of isotope α. LeClaire (1966) has shown that f_α and f_β are related by the equation

$$f_\beta/f_\alpha = (w_\alpha/w_\beta)/[1 - f_\alpha + f_\alpha(w_\alpha/w_\beta)] \qquad (14)$$

if jumps that involve only the isotopes α and β are affected by the mass change. The right hand side of Eq. (14) is generally equal to unity within 1% unless w_α/w_β is unusually large.

In the derivation of Eq. (6), it was assumed that changing the mass of the tracer atom affects only tracer jumps w_α, and that all other jumps (u) remain unchanged. However, if ΔK is less than unity, the motion of a given atom involves the motion of several atoms, and u jumps can no longer be independent of the tracer mass. LeClaire (1966) and Bakker (1971) have made estimates of this effect on the validity of Eq. (13). For self-diffusion by single vacancies, LeClaire obtains

$$E = f \Delta K [1 + (2/9N)(f^{-1} - 1)\{(\Delta K)^{-1} - 1\}] \qquad (15)$$

where N is the number of nonmigrating atoms that participate in the jump. For impurity diffusion, the term $\frac{2}{9}$ in Eq. (15) is replaced by $[1 + (7w_3/2w_1)]^{-1}$, where w_3 and w_1 are solvent-vacancy exchange frequencies for a vacancy neighboring an impurity atom (see Fig. 7). Bakker (1971) estimates the maximum correction for self-diffusion by divacancies in the fcc lattice may be obtained from Eq. (15) with $\frac{2}{9}$ replaced by $\frac{1}{34}$. The correction term within brackets in Eq. (15) is unity if f or ΔK is unity. For sodium self-diffusion (the smallest measured ΔK), the correction term is 1.014 if it is assumed that $N = 6$. Hence, the correction introduced by Eq. (15) for diffusion by either monovacancies or divacancies is smaller than the several percent error in the measurement of E.

In deriving Eq. (13), it was also assumed that the activation energy for defect migration is independent of isotopic mass. By making this assumption, zero-point energy or quantum effects have been neglected. LeClaire (1966), Ebisuzaki et al. (1967a,b), and Franklin (1972) have considered this effect. The correction to Eq. (13) that arises from quantum-effect considerations is about 2% or less for the experiments considered in this chapter. This correction can be large when the temperature of the diffusion experiment is approximately the same as the Debye temperature. Contributions to diffusion from quantum-effect considerations are more fully discussed by Franklin in Chapter 1.

For impurity or self-diffusion by a single-defect mechanism, Eq. (13) accurately describes the isotope effect, and contributions from Eqs. (14) and (15), or from quantum effects, are small—generally less than several percent. In cases for which f is known for each possible diffusion mechanism, the experimental value of $f\Delta K$ deduced from the measurement of D_α/D_β is often consistent with only one diffusion mechanism. Thus, an isotope-effect measurement may identify the diffusion mechanism and provide a value of ΔK. Theoretical estimates of ΔK (Huntington et al., 1970; Achar, 1970; Brown et al., 1971a) are in reasonable agreement with the experimental value of ~ 0.9 for close packed lattices. Feit (1972a) has shown that the large theoretical value of ΔK found for all crystal structures (Huntington et al., 1970; Brown et al., 1971b) is a characteristic feature of the reaction-rate theory. Based on the dynamical theory of diffusion, Feit (1972b) estimates $\Delta K = 0.5$ for diffusion by the monovacancy mechanism in bcc sodium, which is in good agreement with experiment.

When two diffusion mechanisms are simultaneously operative, the value of the isotope effect will be a weighted average of these mechanisms. If both monovacancies and divacancies contribute to diffusion, the net isotope effect is given by (Bakker, 1969; Mehrer and Seeger, 1969)

$$E = f_{1V} \Delta K_{1V} [D_{1V}/(D_{1V} + D_{2V})] + g_{2V} \Delta K_{2V} [D_{2V}/(D_{1V} + D_{2V})] \quad (16)$$

where the subscripts $1V$ and $2V$ refer to properties of monovacancies and divacancies, respectively. The term $g_{2V} = f_{2V}$ ($f_{2V} = 0.475$ for self-diffusion in the fcc lattice) if the two vacancies that comprise the divacancy never dissociate, and g_{2V} is a multiple-termed function of several jump frequencies when dissociation of the divacancy is included (Mehrer, to be published). Since $f_{1V} > f_{2V}$ (see Table I) and ΔK_{2V} is probably $\leq \Delta K_{1V}$, the observed isotope effect should decrease as the temperature increases even if ΔK_{1V} and ΔK_{2V} are independent of temperature. [Both experiment (Peterson and Rothman, 1967a; Rothman and Peterson, 1969) and theory (Achar, 1970) suggest that ΔK is independent of temperature.] If anharmonic effects are present, ΔK may decrease with temperature; however, anharmonic effects are thought to be small in the determination of ΔK (Franklin, 1972).

For impurity diffusion isotope effects, it is useful to obtain f_i from the measured value of $f_i \Delta K$. Thus, it is desirable to relate ΔK for impurity diffusion to ΔK for self-diffusion, when the latter is known, and to obtain ΔK for solvent self-diffusion when suitable isotopes are not available for the measurement. This can be accomplished qualitatively from an empirical relation first proposed by Barr and Mundy (1965). They suggest that ΔK for self-diffusion is approximately equal to the activation volume for self-diffusion ΔV divided by the molar volume V.

A qualitative argument in favor of this empirical relation may be developed. When the relaxation around a vacancy is large, $\Delta V/V$ is small, and the atoms that create the saddle-point configuration will move appreciably during the jump process. The kinetic energy transferred by the jumping atom to the saddle-point atom will be larger (and, hence, ΔK smaller) when $\Delta V/V$ is smaller. This suggests that the activation volume for vacancy formation ΔV_f, rather than for vacancy motion ΔV_m, is the more important quantity in determining ΔK. Using a similar argument, LeClaire (1966) obtained the approximate relation

$$\Delta K \approx [1 + (n/3)|(1 - \Delta V_f/V)|]^{-1} \qquad (17)$$

where n is the number of atoms that relax during the decay of the saddle-point configuration.

This same qualitative argument may be used to suggest the following relation between impurity and self-diffusion:

$$\Delta K_{\text{self}}/\Delta K_{\text{imp}} \simeq \Delta V_{\text{self}}/\Delta V_{\text{imp}} \qquad (18)$$

Bonanno and Tomizuka (1965) measured $\Delta V_{\text{self}}/\Delta V_{\text{imp}}$ for indium and antimony diffusion in silver. They found this ratio differed from unity by 5% for indium and 10% for antimony. This suggests that $\Delta K_{\text{self}}/\Delta K_{\text{imp}}$ is ~ 1 if the impurity and solvent are near neighbors in the periodic table (i.e., have similar masses and chemical properties).

The major difficulty with isotope-effect measurements is the required accuracy of the experiment. This may be deduced from Eq. (13); m_β and m_α generally differ by about 5%, and $f\Delta K$ has a maximum value of unity. Hence, D_α and D_β will differ by only 2 or 3%. If $f\Delta K$ is desired to within 2%, the relative diffusion coefficients D_α/D_β must be determined to 0.04%. This accuracy may be obtained from experiments in which both isotopes diffuse simultaneously. The concentration C of tracer that has diffused initially from a thin layer on the sample surface to a penetration depth x after a diffusion anneal for time t is given by the equation

$$C = [S_0/(\pi Dt)^{1/2}] \exp(-x^2/4Dt) \qquad (19)$$

where S_0 is the tracer concentration per unit area at $x = 0$ and $t = 0$.

The relative concentration of the two isotopes as a function of penetration is given by

$$\ln(C_\alpha/C_\beta) = \text{const} - \ln C_\alpha[(D_\alpha/D_\beta) - 1] \tag{20}$$

Hence, the measurement of the relative concentration of the two isotopes, as a function of the concentration of one of the isotopes, yields the quantity $[D_\alpha/D_\beta) - 1]$ directly. Errors due to the time and temperature of the diffusion anneal and the errors in sectioning do not enter into the error in $[(D_\alpha/D_\beta) - 1]$. Careful radiotracer counting procedures and high purity tracers with significant differences in mass allow a determination of D_α/D_β to the desired accuracy.

From this general background on correlation and isotope effects, we will examine the various diffusion problems that have been studied by the isotope-effect technique. We will start with the simplest case of self-diffusion in pure monotonic solids, proceed to more difficult problems, and complete the discussion with diffusion along grain boundaries. Only the portion of each problem area that has obtained enhanced understanding from isotope-effect measurements will be discussed in detail.

III. Self-Diffusion in Pure Metals

Although the isotope-effect measurement provides significant information about the mechanism of self-diffusion, the maximum understanding of self-diffusion is obtained when the isotope-effect results are combined with results from measurements of the temperature and pressure dependence of the self-diffusion coefficient. A brief review of the pressure and temperature dependence of D will be presented before discussing the self-diffusion results for various materials.

The temperature dependence of the tracer diffusion coefficient is frequently found to obey the Arrhenius equation

$$D = D_0 \exp(-Q/kT) \tag{21}$$

From simple reaction-rate theory, one may obtain the expressions

$$Q = h^f + h^m \tag{22}$$

and

$$D_0 = a^2 f v \exp[(s^f + s^m)/k] \tag{23}$$

for self-diffusion by a single-defect mechanism. Here, h^f and h^m are the enthalpies, and s^f and s^m are the entropies of defect formation and migration, respectively; a is the edge length of the elementary cube; and

3. ISOTOPE EFFECTS IN DIFFUSION

v is the vibration frequency of an atom about its equilibrium position in the jump direction. More rigorous treatments of reaction-rate theory provide a slightly different interpretation of v, but the general form of Eqs. (21)–(23) remains unchanged for a harmonic solid (Vineyard, 1957).

If two mechanisms of diffusion are operative, Eq. (21) must be replaced by a sum of two exponential functions. For tracer self-diffusion in an fcc lattice by both monovacancies and nearest neighbor divacancies, one obtains

$$D = a^2 f_{1V} v_{1V} \exp[(s_{1V}^f + s_{1V}^m)/k] \exp[-(h_{1V}^f + h_{1V}^m)/kT]$$
$$+ 4a^2 f_{2V} v_{2V} \exp[(2s_{1V}^f + s_{2V}^m + \Delta s_{2V})/k] \exp[-(2h_{1V}^f + h_{2V}^m - h_{2V}^B)/kT] \quad (24)$$

The terms Δs_{2V} and h_{2V}^B are the divacancy binding entropy and enthalpy, respectively. A fit of Eq. (24) to experimental data may require the adjustment of five parameters: a preexponential factor D_0 and an activation energy Q for each of the two mechanisms, as well as a parameter α that describes the temperature dependencies of the activation enthalpies and entropies. This fitting procedure has been used by Seeger and Mehrer (1970) on the self-diffusion data for a number of metals.

The effect of hydrostatic pressure P on diffusion may be expressed as

$$(\partial \ln D/\partial P)_T = -\Delta V/kT + (\partial \ln a^2 v f/\partial P)_T \quad (25)$$

where ΔV is the activation volume for diffusion when only one mechanism of diffusion is operative. The second term on the right hand side of Eq. (25) is generally a small correction, on the order of several percent of ΔV (Mehrer and Seeger, 1972).

If more than one mechanism is operating, $(\partial \ln D/\partial P)_T$ will vary with pressure and temperature. Mehrer and Seeger (1972) have derived expressions for the pressure and temperature variation of $(\partial \ln D/\partial P)_T$ for the simultaneous operation of two diffusion mechanisms. An accurate determination of $(\partial \ln D/\partial P)_T$ as a function of temperature will generally permit a determination of the activation volume for the two diffusion mechanisms and may set limits on their relative contribution to the total diffusion coefficient.

The magnitude of ΔV may not uniquely define the diffusion mechanism, but it may add to the evidence in favor of one mechanism or another. General arguments (Peterson, 1973) as well as detailed calculations (Tewordt, 1958) suggest that ΔV_{1V} for diffusion by monovacancies is on the order of $\frac{1}{2}$ to $\frac{3}{4}$ of an atomic volume, ΔV_{2V} for diffusion by divacancies is slightly smaller than $2\Delta V_{1V}$, and the activation volume for diffusion by self-interstitials is thought to be much smaller than ΔV_{1V}. Hence, an experimental value of ΔV between $\frac{1}{2}$ and $\frac{3}{4}$ of an atomic volume is considered evidence in favor of a monovacancy mechanism, and a curved plot of $\ln D$ vs P, which can be resolved into two values of ΔV that differ by a factor of two, may be considered evidence in favor of diffusion by both monovacancies and divacancies.

A. FACE-CENTERED CUBIC METALS

The isotope effect has been measured for self-diffusion in four fcc metals, e.g., copper, silver, palladium, and iron. The measurements have been made over a significant temperature range in copper and silver. It is concluded that self-diffusion in fcc metals occurs predominantly by the monovacancy mechanism. A contribution from the divacancy mechanism may be observed at temperatures near the melting point. In the fcc metals, the isotope effect has been a powerful tool in identifying the predominant mechanism for diffusion and, in the case of copper and silver, the isotope-effect measurements have been useful in placing limits on the magnitude of the divacancy contribution to self-diffusion. A detailed discussion will be presented for for self-diffusion in copper (where the divacancy contribution is small) and silver (where a noticeable divacancy contribution is present).

1. *Self-Diffusion in Copper*

Self-diffusion in copper has been measured a number of times (Kuper et al., 1954; Beyeler and Adda, 1968; Mercer, 1955; Rothman and Peterson, 1969). However, the present discussion will concentrate on the data of Rothman and Peterson (1969) who also presented results on the temperature dependence of the isotope effect. Their high temperature results between 698 and 1061°C are well described by the Arrhenius equation with $D_0 = 0.78$ cm^2/sec, and $Q = 2.19 \pm 0.01$ eV (see Fig. 1). Their measurements of the simultaneous diffusion of ^{64}Cu and ^{67}Cu in single crystals of copper give an average value of $(D_{64}/D_{67}) - 1 = 0.0155 \pm 0.0003$ that is independent of temperature over the range of 894 to 1061°C. The known values of the correlation factor (Table I) and Eq. (13) allow one to obtain the following values of ΔK that are required to make each of the various diffusion mechanisms consistent with the observed value of $(D_{64}/D_{67}) - 1$: interstitial mechanism, $\Delta K = 0.68$; vacancy mechanism, $\Delta K = 0.87$; divacancy mechanism, $\Delta K = 1.45$; collinear interstitialcy mechanism, $\Delta K = 1.99$; and direct-exchange mechanisms, $\Delta K = 1.34$. Since ΔK is bounded by zero and unity, the only allowed diffusion mechanisms are the vacancy and the interstitial. High temperature equilibrium measurements show that the vacancy is the predominant defect in copper (Simmons and Balluffi, 1963); hence, the vacancy mechanism may be concluded to be the dominant mechanism for self-diffusion in copper with

3. ISOTOPE EFFECTS IN DIFFUSION

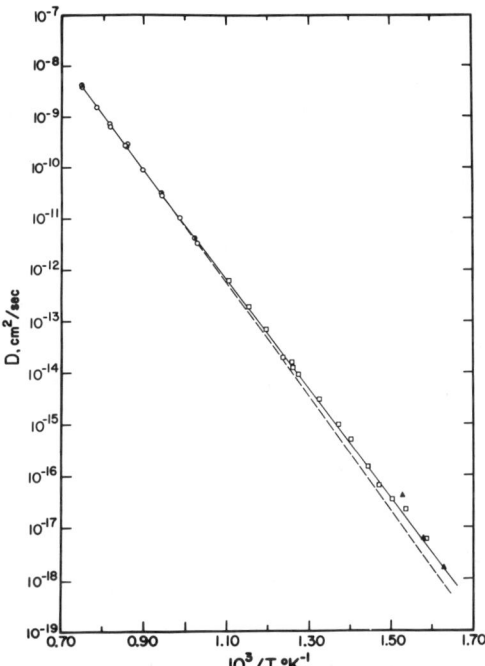

Fig. 1. Log D versus reciprocal absolute temperature for self-diffusion in copper. Data: ○ Rothman and Peterson (1969), ▲ Lam *et al.* (1974), □ Maier *et al.* (1973), — the analysis by Mehrer and Seeger (1969) of the results of Rothman and Peterson, – – – a linear least-squares analysis of the results of Rothman and Peterson.

$\Delta K = 0.87 \pm 0.02$. The fact that D_{64}/D_{67} is independent of temperature led Rothman and Peterson to conclude that the divacancy contribution to self-diffusion in copper must be small.

The pressure dependence of the copper self-diffusion coefficient, as determined by Beyeler and Adda (1968) and by McArdle (1968) give an activation volume of 0.9 atomic volume. This is consistent with diffusion by monovacancies or a combination of monovacancies and divacancies. The diffusion measurements at pressure are not sufficiently accurate to differentiate between one or two mechanisms of diffusion, but they are sufficiently accurate to reject interstitials as an important mechanism of diffusion in copper.

Mehrer and Seeger (1969) have combined the temperature dependence of the tracer diffusion coefficient and the isotope effect with the high temperature equilibrium defect data of Simmons and Balluffi (1963) to obtain what they believe to be an optimum consistent set of defect properties for copper. They obtain $h_{1V}^f = 1.03 \text{ eV}$, $s_{1V}^f \approx 0.3k$,

$h_{1V}^m = 1.06$ eV, $v_{1V} \exp(s_{1V}^m/k) = 1.4 \times 10^{14}$ sec^{-1}, $h_{2V}^m - h_{2V}^B \approx 0.54$ eV, $\Delta s_{2V} \approx 2k$, and $v_{2V} \exp(s_{2V}^f/k) \approx 1.6 \times 10^{14}$ sec^{-1}. The divacancy properties are clearly less accurate than corresponding monovacancy properties because of the dominance of single vacancies in the diffusion process.

The slight divacancy contribution to self-diffusion in copper deduced by Mehrer and Seeger (1969) requires a temperature dependent isotope effect that is just within the experimental error of Rothman and Peterson (1969) with $\Delta K_{1V} = \Delta K_{2V} = 0.95$. It also produces a curved Arrhenius plot that adequately fits the high temperature results of Rothman and Peterson, but is not required by these results. Recent low temperature tracer diffusion measurements (10^{-13}–10^{-18} cm^2/sec) by Lam et al. (1974) and Maier et al. (1973) (Fig. 1), as well as self-diffusion coefficients deduced by Bowden and Balluffi (1969) from electron-microscopy observations of void shrinkage support the Mehrer–Seeger interpretation of self-diffusion in copper.

Wynblatt (1971) has performed an analysis similar to that of Mehrer and Seeger on other copper self-diffusion data (Kuper et al., 1954; Beyeler and Adda, 1968). Wynblatt obtains slightly different defect parameters from those reported by Mehrer and Seeger, but his conclusions as to the mechanisms responsible for atomic transport are identical to those expressed by Rothman and Peterson (1969).

The isotope-effect measurements for self-diffusion in copper have played a significant role in the unambiguous deduction of the dominant mechanisms of diffusion and have placed some limits on the divacancy contribution. If the isotope-effect measurements had been extended to lower temperature (a difficult task with the short lived ^{64}Cu isotope), a more well-defined limit on the divacancy contribution may have been obtained.

2. Self-Diffusion in Silver

Although general agreement exists between the various investigations of self-diffusion in silver (Hoffman and Turnbull, 1951; Slifkin et al., 1952; Tomizuka and Sonder, 1956; Rothman et al., 1970), the data from the most recent high temperature investigation of Rothman et al. (1970) show a small systematic deviation with temperature from a linear Arrhenius plot. Recently, low temperature tracer diffusion measurements in the range of $10^{-13} > D > 10^{-18}$ cm^2/sec have been reported by Lam et al. (1972, 1973) that confirm the curvature in the Arrhenius plot suggested by the high temperature measurements ($D > 10^{-11}$ cm^2/sec) of Rothman et al. (1970) (see Fig. 2). This slight curvature in the Arrhenius plot suggests that a second mechanism of diffusion may be important. The temperature dependence of the isotope effect for self-diffusion in silver (Rothman et al., 1970) (see Fig. 3) provides additional evidence in favor of two mechanisms of diffusion. The values of the isotope effect E at low temperatures

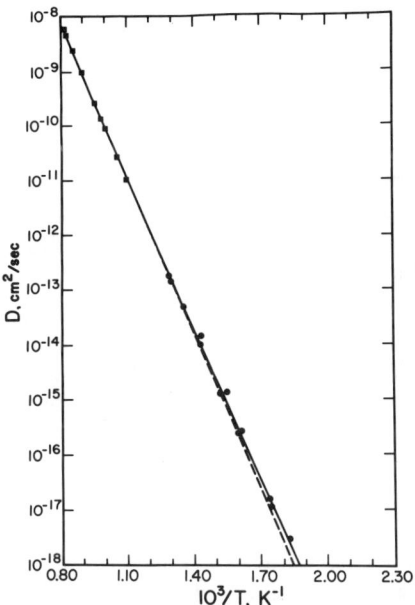

Fig. 2. Log D versus reciprocal absolute temperature for self-diffusion in silver from the data of Rothman *et al.* (1970) and Lam *et al.* (1972, 1973). - - - the analysis by Mehrer and Seeger (1970) of the results of Rothman *et al.* (1970).

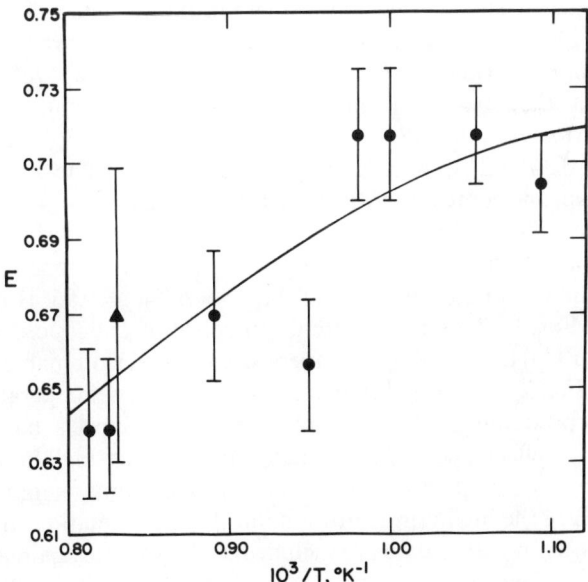

Fig. 3. The isotope effect versus reciprocal absolute temperature for self-diffusion in silver from the work of Rothman *et al.* (1970) and Peterson *et al.* (1973); — is from the analysis by Mehrer and Seeger (1970) of the results of Rothman *et al.* (1970).

are consistent with diffusion by monovacancies and $\Delta K = 0.92$, and the decrease in E at higher temperatures is consistent with an increase in the importance of divacancy diffusion. [Reimers and Bartdorff (1972) have reported less accurate measurements of E as a function of temperature that are in complete agreement with those shown in Fig. 3.]

Two limiting cases for the divacancy contribution to self-diffusion in silver may be deduced from the data in Fig. 3. First, the difference between the measured value of E and the value of f_{1V} may be assigned entirely to the divacancy contribution at all temperatures, i.e., $\Delta K_{1V} = 1.0$. Assuming $\Delta K_{1V} = \Delta K_{2V} = 1.0$ and $f_{2V} = 0.475$ yields the maximum divacancy contribution of $D_{2V}/D_{1V} = 0.872$ at high temperature ($E = 0.639$) and $D_{2V}/D_{1V} = 0.263$ at low temperature ($E = 0.718$). The second limiting case is obtained if it is assumed that diffusion at low temperatures takes place only by monovacancies, i.e., $\Delta K_{1V} = 0.918$, and the divacancy contribution is responsible for the further decrease of E at higher temperatures. If ΔK is independent of temperature, and the minimum divacancy contribution at the melting point is obtained if $\Delta K_{2V} = 0$, then $D_{2V}/D_{1V} = 0.12$. Unless $f\Delta K$ is temperature dependent, the data in Fig. 3 require the simultaneous operation of two diffusion mechanisms.

Mehrer and Seeger (1970) have used the temperature dependencies of the diffusion coefficient ($D > 10^{-11}$ cm^2/sec) and the isotope effect, the fraction of vacant lattice sites at the melting point deduced by Simmons and Baluffi (1960), and limited information from nonequilibrium (quenching and annealing) experiments (Doyama and Koehler, 1962; Quéré, 1960) to deduce what they believe to be an optimum consistent set of defect properties for silver. They obtain $h_{1V}^f = 0.99$ eV, $h_{1V}^m = 0.86$ eV, $s_{1V}^f \approx 0.5k$, $h_{2V}^m - h_{2V}^B = 0.34$ eV, $h_{2V}^m = 0.58$ eV, and $\Delta S_{2V} \approx 2.6k$. The properties that pertain to divacancies are clearly less accurate than corresponding properties of monovacancies because of the dominance of the monovacancies in the diffusion process. The temperature dependence of the isotope effect, calculated from the above parameters and $\Delta K_{1V} = \Delta K_{2V} = 0.96$, is shown as a solid line in Fig. 3. The temperature dependence of D deduced by Mehrer and Seeger (1970) from the high temperature data of Rothman et al. (1970) is shown in Fig. 2 as a dotted line extrapolated to low temperatures.

The interpretation in the two preceding paragraphs is based entirely on the isotope-effect and diffusion data obtained at high temperatures. Although the low temperature diffusion data differ only slightly from an extrapolation of the high temperature data, the interpretation must change when the low temperature data are included in the analysis. Lam et al. (1973) find that when their data and the data of Rothman et al. (1970) are fitted to Eqs. (16) and (24) with a temperature independent value of $f_{2V} = 0.475$, an optimum fit is obtained only with the physically unreasonable result that $\Delta K_{2V} > 1$. This value of f_{2V} was calculated under the assumption that the two vacancies which comprise the divacancy are always nearest neighbors (Howard, 1966). If the divacancy is allowed to dissociate to at least second-nearest neighbor positions and if the two vacancies still attract each other at second-nearest neighbor positions, f_{2V} will depend on more than one type of jump and may be temperature dependent. If jumps that involve nearest neighbor and next-nearest neighbor divacancy configurations are considered, f_{2V} will increase and $f_{2V} = 0.475$ will be a minimum value. Mehrer has fitted the data of Figs. 2 and 3 using this more complete evaluation of f_{2V} (Lam et al., 1973). He finds a satisfactory fit to the data with an acceptable value of ΔK. Lam et al. (1973) conclude that migration processes other than the simple, completely associated, divacancy mechanism contribute to divacancy migration at high temperatures; this other process is believed to be either dissociation of the divacancy to the next-nearest neighbor configuration or a reorientation of the divacancy by a next-nearest neighbor jump of one vacancy.

Bonanno and Tomizuka (1965) and Tomizuka (1961) have found the pressure dependence of the silver self-diffusion coefficient $(\partial \ln D/\partial P)_T$ to be independent of pressure with $\Delta V = 0.9$ atomic volume. Although these measurements do not require two mechanisms of diffusion, it can be shown that two mechanisms of diffusion do not become evident in a $(\partial \ln D/\partial P)_T$ plot at high kT unless an unusually large range in pressure is investigated. The data of Tomizuka can be fitted under the assumption that two mechanisms of diffusion are important, with $D_{2V} = D_{1V}$ and $\Delta V_{2V} = 2\Delta V_{1V}$. These parameters give a good fit to the data of Tomizuka (1961) and yield $\Delta V_{1V} = 0.75$ atomic volume. Thus, the pressure data for self-diffusion in silver is consistent with diffusion by both monovacancies and divacancies at high temperatures. Although the accuracy of the divacancy properties deduced from the diffusion measurements may be questioned (Burton and Froozan, 1973), the mechanisms of diffusion in silver are well established to be monovacancies with an allowed range of contributions from divacancies at high temperature.

3. Self-Diffusion in Palladium and γ Iron

Isotope-effect measurements for self-diffusion in palladium and γ iron are sufficiently accurate to permit some statements about diffusion mechanisms, but the temperature dependence of E has not been determined with sufficient accuracy to allow a detailed analysis of the properties of the defects.

Peterson (1964) has measured the isotope effect for self-diffusion in palladium and found $E = 0.79 \pm 0.04$ over a small range in temperature near the melting point. Since ΔK cannot exceed unity, self-diffusion in palladium must occur by monovacancies with $(1 - \Delta K) \leq 0.02$. Although a divacancy contribution would go undetected because of the small temperature range investigated, the magnitude of E requires that any divacancy contribution must be extremely small.

Heumann and Imm (1968), Walter and Peterson (1969), and Graham (1969) have measured the isotope effect for self-diffusion in γ iron. The results of Heumann and Imm and Walter and Peterson are in good agreement and support a monovacancy mechanism of diffusion. The values of E are sufficiently smaller than f_{1V} to allow a significant divacancy contribution to diffusion or a ΔK as small as 0.68. The apparent temperature dependence of E observed by Heumann and Imm is consistent with an appreciable divacancy contribution to self-diffusion in γ iron. The measurements of Graham are apparently erroneous.

B. Body-Centered Cubic Metals

Self-diffusion in bcc metals is controversial, and no unique explanation of the diffusion behavior exists for most of them. The self-diffusion coefficient at the melting point is $\sim 10^{-7}$ cm^2/sec for most bcc metals compared with $\sim 10^{-8}$ cm^2/sec for most fcc metals. The Arrhenius plots occasionally show significant curvature. A detailed discussion will be presented for sodium, which is the most thoroughly studied bcc metal. The evolving concept of self-diffusion in sodium is more complex than for the fcc metals, but may well be typical of many bcc metals. The isotope-effect measurements for self-diffusion in bcc iron will be compared with the more complete results for sodium.

1. Self-Diffusion in Sodium

The temperature, pressure, and mass dependence of the self-diffusion coefficient have been measured for sodium. The temperature dependence of the tracer diffusion coefficient has been measured over five orders of magnitude and clearly demonstrates a curved Arrhenius plot (Mundy et al., 1966; Mundy, 1971) (Fig. 4). The data may be fit by a sum of two or three exponential functions with no apparent statistical preference for one or the other. The plot of log D vs P is clearly nonlinear (Mundy, 1971; Nachtrieb et al., 1952) and may be fit using two or three mechanisms of diffusion. However, if three mechanisms are operating, the two with the largest activation volume occur at higher temperatures and have the same activation volume. Also, the higher temperature mechanisms have an activation volume that is twice as large as the low temperature mechanism. The isotope effect varies significantly with temperature (Mundy et al., 1966; Mundy, 1971) (Fig. 5) and requires three diffusion mechanisms if $f \Delta K$ is a constant for each mechanism.

3. ISOTOPE EFFECTS IN DIFFUSION

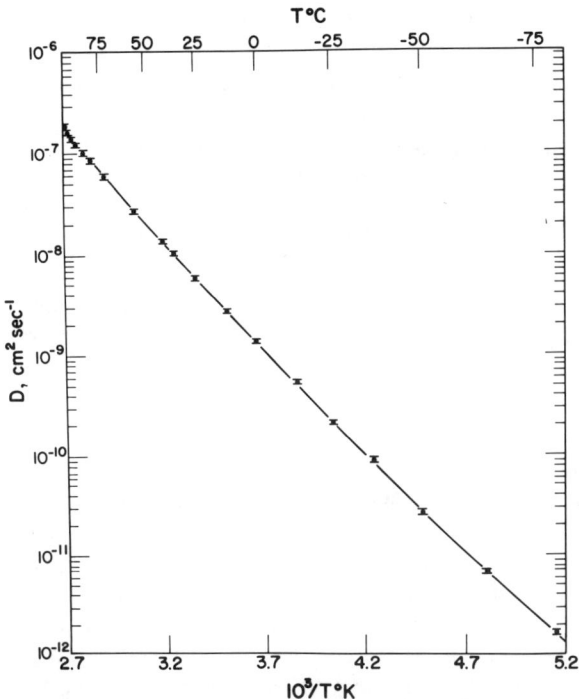

Fig. 4. Log D versus reciprocal absolute temperature for self-diffusion in sodium from the work of Mundy (1971).

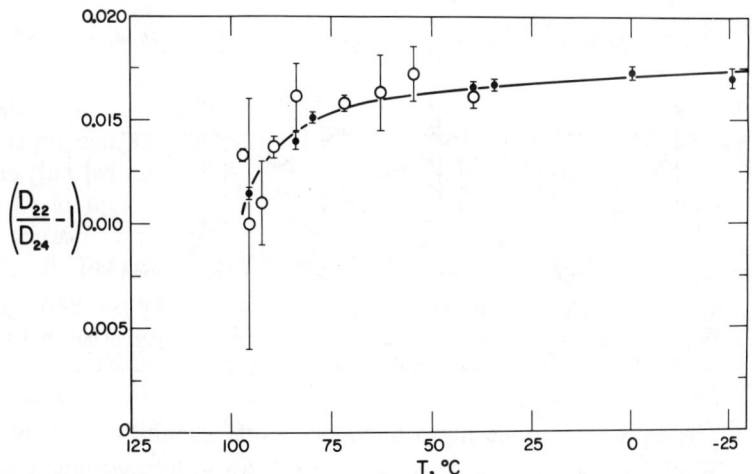

Fig. 5. The isotope effect versus temperature for self-diffusion in sodium. Data: ● from Mundy (1971) and ○ from Mundy et al. (1966).

The model that appears to be most successful in quantitatively fitting the experimental data was first proposed by Seeger (1970). In this model, the low temperature diffusion mechanism is the monovacancy mechanism, and the two high temperature diffusion mechanisms suggested by the temperature dependence of $f\Delta K$ arise from the two types of divacancy jumps that must be considered in the bcc lattice. A divacancy that initially consists of two nearest neighbor vacancies can move by nearest neighbor jumps only if the divacancy partly dissociates into two vacancies at next-nearest neighbor sites. If the most stable configuration of the divacancy is two vacancies at next-nearest neighbor sites, it can diffuse either by jumping to a metastable nearest neighbor configuration or by dissociating further. The correlation factor for divacancy diffusion from these two configurations will be different and will lead to an apparent f_{2V} that is temperature dependent, if further dissociation is allowed; but the activation volumes for diffusion by the two divacancy configurations would be similar, in agreement with the experimental data of Mundy (1971).

Using the values of f_{2V}, calculated as a function of temperature on the basis of Seeger's model by Kunz (1971), Ho (1972) found a fair fit between the model and the data with $\Delta K_{1V} = \Delta K_{2V} = 0.6$. Mehrer (to be published) has improved the fit of this model with the data by developing a more rigorous relation between $(D_\alpha/D_\beta) - 1$ and $f\Delta K$ for diffusion by the two divacancy configurations, and by considering divacancy configurations beyond next-nearest neighbors in the evaluation of f_{2V}. Mehrer considers the divacancy configurations and jump frequencies shown in Fig. 6 for his calculation

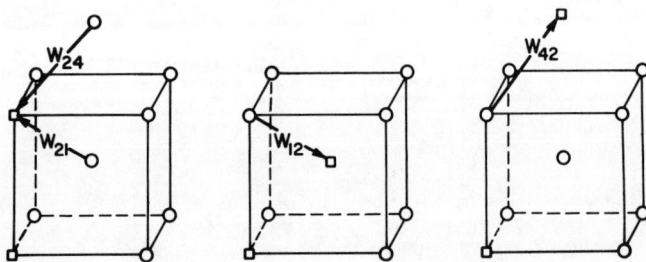

Fig. 6. Divacancy jumps in the bcc lattice. The squares represent vacant lattice sites. The w_{ij} are atomic jump frequencies that change the divacancy configuration from an ith nearest neighbor to a jth nearest neighbor configuration during that jump.

of f_{2V}. The w_{ij}'s are atomic jump frequencies that change the divacancy configuration from an ith nearest neighbor to a jth nearest neighbor configuration during that jump. In thermal equilibrium, the concentration of

3. ISOTOPE EFFECTS IN DIFFUSION

divacancies in the second-nearest neighbor configuration C_{2V}^{2n} must be related to the concentration of divacancies in the first- and fourth-nearest neighbor configurations (C_{2V}^{1n} and C_{2V}^{4n}) by the relations (Mehrer, 1973)

$$3C_{2V}^{1n} w_{12} = 4C_{2V}^{2n} w_{21} \tag{26}$$

and

$$12 C_{2V}^{2n} w_{24} = 3 C_{2V}^{4n} w_{42} \tag{27}$$

Using these relations, the tracer diffusion coefficient for motion by divacancies can be written in terms of the concentration of divacancies in one configuration,

$$D_{2V} = 2a^2 C_{2V}^{2n} (w_{21} + w_{24}) f_{2V} \tag{28}$$

where f_{2V} is a function of w_{21}/w_{24}. The optimum solution suggests that the most stable configuration of a divacancy consists of two vacancies at second-nearest neighbor sites and all four jump frequencies shown in Fig. 6 must be included in the determination of f_{2V} (Mehrer, to be published).

Although this model of self-diffusion in sodium by monovacancies and the complicated motion of divacancies may not be unique, no other mechanism or group of mechanisms has obtained as good agreement with the data. If this interpretation is correct, one must accept the rather small values of $\Delta V_{1V} = 0.33$, $\Delta V_{2V} = 0.66$, and $\Delta K = 0.6$ for a bcc metal compared with $\Delta V_{1V} \sim 0.8$ and $\Delta K \sim 0.9$ for fcc metals. This smaller value of ΔK for the bcc lattice is not without theoretical support (Feit, 1972b).

2. Self-Diffusion in Body-Centered Cubic Iron

Walter and Peterson (1969), Graham (1969). deGonzalez and deReca (1971), and Irmer and Feller-Kniepmeier (1972) have measured the isotope effect for self-diffusion in α iron; and Walter and Peterson (1969) and Graham (1969) have made a similar measurement in δ iron. Agreement among most of the investigators suggests that Graham's measurements are erroneous, as they were in γ iron. The magnitude of E in α and δ iron is the same as in sodium at comparable temperatures relative to the melting point. This suggests that atomic migration in iron and sodium occurs by the same mechanism or mechanisms.

The model developed for sodium may be applicable to self-diffusion in many bcc metals. Experiments are now in progress at Argonne National Laboratory to determine if the temperature dependence of the isotope effect for self-diffusion in chromium and niobium is similar to that for sodium and bcc iron. Although the mysteries of atomic migration in bcc metals are far from solved, progress is being made in developing a more complete understanding of these technologically important metals.

C. Hexagonal Close Packed Metals

For self-diffusion in a noncubic crystal, f_{1V} may no longer be a purely geometrical factor. For diffusion in the hexagonal close packed (hcp) lattice, two types of atomic jumps must be considered; jumps in the basal plane (A jumps with jump frequency w_A) and jumps out of the basal plane (B jumps with jump frequency w_B). The diffusion coefficient may be determined experimentally both parallel to the basal plane D_x and perpendicular to the basal plane D_z. The B jumps have a component parallel to the basal plane; hence, D_x contains contributions from both A- and B-type jumps.

Three partial correlation factors exist for self-diffusion by monovacancies in the hcp lattice (Mullen, 1961b; Howard, 1966): f_{Bz} is the correlation factor for B jumps in the z direction, and f_{Ax} and f_{Bx} are the correlation factors for A and B jumps in the x direction. The measurable diffusion coefficients may be related to the correlation factors and jump frequencies by the relations

$$D_x = \tfrac{1}{2}na^2(3w_A f_{Ax} + w_B f_{Bx}) \quad (29)$$

and

$$D_z = \tfrac{3}{4}nc^2 w_B f_{Bz} \quad (30)$$

where n is the atomic fraction of vacancies, and a and c are the dimensions of the unit cell. The ratio D_x/D_z may be written as

$$(c/a)^2(D_x/D_z) = (2/f_{Bz})(w_A f_{Ax}/w_B + \tfrac{1}{3}f_{Bx}) \quad (31)$$

Mullen (1961b) has calculated f_{Ax}, f_{Bx}, and f_{Bz} as functions of w_A/w_B. This dependence may be used to calculate $(c/a)^2(D_x/D_z)$ as a function of w_A/w_B from Eq. (31). Hence, an experimental determination of D_x/D_z may be used to obtain f_{Ax}, f_{Bx}, and f_{Bz} from Mullen's calculations.

Since diffusion parallel to the x direction involves two types of jumps, Eq. (9) must be used to express the isotope effect in terms of the various jump frequencies and correlation factors. Using the expressions derived for the partial correlation factors by Huntington and Ghate (1962) and Ghate (1964), Eq. (9) yields the expression (Peterson and Rothman, 1967a)

$$E_x = [3w_A f_{Ax}(f_{Ax}\Delta K_A) + w_B f_{Bx}(f_{Bx}\Delta K_B)]/(3w_A f_{Ax} + w_B f_{Bx}) \quad (32)$$

where ΔK_A and ΔK_B are the kinetic energy factors for w_A and w_B jumps, respectively. Since only w_B jumps contribute to D_z, Eq. (13) may be used to obtain

$$E_z = f_{Bz} \Delta K_B \quad (33)$$

3. ISOTOPE EFFECTS IN DIFFUSION

Peterson and Rothman (1967a) have measured D_x, D_z, E_x, and E_z as a function of temperature for self-diffusion in zinc. From the measurements of D_x/D_z, the partial correlation factors are obtained; ΔK_B and ΔK_A are then obtained from Eqs. (33) and (32). Peterson and Rothman found $\Delta K_B = 0.93 \pm 0.03$ and $\Delta K_A = 0.96 \pm 0.03$ independent of temperature in the range of 289 to 418°C. The apparent anisotropy is not considered significant. Batra (1967) obtained results at two temperatures that are about 5% smaller than those of Peterson and Rothman. The linearity of the Arrhenius plot, the temperature independence of the isotope effect, and the rather large value of ΔK are consistent with self-diffusion by monovacancies with no noticeable divacancy contribution.

Chhabildas and Gilder (1972) have found the activation volume to be slightly temperature dependent for zinc. They interpret this to mean that the thermal coefficient of expansion of an activated vacancy is 15 times greater than the normal coefficient of expansion of the perfect lattice. If the isotope effect were not temperature independent, the temperature dependent activation volume could be interpreted as support for a divacancy contribution to diffusion in zinc.

D. SELF-DIFFUSION IN GERMANIUM AND SILICON

A number of the parameters for self-diffusion are distinctly different for germanium and silicon than for most metals (fcc, hcp, or bcc metals); the diffusion coefficient at the melting point is about 10^4 smaller, D_0 is larger, and the Q is distinctly larger for germanium and silicon. These differences lead Seeger and Chik (1968) to propose a more complex combination of diffusion mechanisms for these materials.

Seeger and Chik propose that both the vacancy and the interstitialcy mechanisms contribute significantly to self-diffusion in silicon and that the interstitialcy mechanism becomes dominant at higher temperatures. They propose an extended vacancy mechanism as the dominant mechanism for self-diffusion in germanium at high temperatures. Although the extended vacancy is not well defined, they suggest, on the grounds of entropy considerations, that this defect may be viewed as 10 atoms occupying the same volume as 11 atoms in the crystalline state. Seeger and Chik further suggest that, if the extended vacancy is responsible for self-diffusion in germanium, the translational kinetic energy during the jump will be distributed over many atoms, and ΔK will be small.

Campbell (1973) has recently made four measurements of the isotope effect for self-diffusion in germanium between 900 and 925°C, and finds that $f\Delta K$ lies between 0.25 and 0.30. Assuming diffusion occurs by monovacancies ($f_{1V} = 0.5$), Campbell's measurements give $\Delta K = 0.5$–0.6.

(A divacancy mechanism is most unlikely because of the small defect concentration.) Although this value of ΔK is small compared with typical values of 0.9 for close packed metals, it is similar in value to $\Delta K \sim 0.5$ for bcc metals. Although Campbell's value of ΔK is not inconsistent with the proposal of Seeger and Chik, it does not provide a unique proof that the extended vacancy mechanism is the only mechanism of self-diffusion in germanium.

IV. Diffusion in Dilute Alloys

A number of experiments have been performed on impurity diffusion in pure metals and tracer diffusion in dilute alloys. Three types of isotope-effect experiments that have enhanced our understanding of the diffusion process will be discussed. First, the effect of perturbations on the atomic jump frequencies produced by an isolated substitutional impurity atom is reviewed. Second, the mechanisms for the rapid metallic solute diffusion in metals are discussed. Third, the effect of local or resonance vibrational modes on the diffusion of heavy impurities in a light matrix is considered.

A. Jump Frequencies near an Impurity–Vacancy Pair

For diffusion in a pure cubic metal, all vacancy-atom exchanges occur with equal probability, and the correlation factor for self-diffusion by monovacancies is a numerical constant independent of temperature. The presence of an impurity atom will change the jump frequencies of the neighboring solvent atoms relative to the values in the absence of the impurity. If the effect of the impurity on the solvent jump frequencies is short ranged, four jump frequencies for the vacancy near an impurity atom may be defined for the fcc lattice. As shown in Fig. 7, w_1 is the

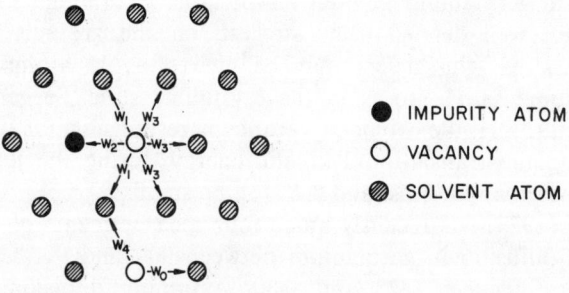

Fig. 7. Vacancy jumps near an impurity atom in an fcc crystal.

3. ISOTOPE EFFECTS IN DIFFUSION

frequency of exchange of a vacancy neighboring an impurity atom with any of the four solvent atoms that are also neighbors of the impurity; w_2 is the frequency of exchange of the impurity and the vacancy; w_3 is the frequency of exchange of a vacancy neighboring an impurity with any of the seven solvent atoms adjacent to the vacancy but not neighbors of the impurity (dissociation jump); w_4 is the frequency of the association jump (reverse of a w_3 jump); and other jumps are assumed to occur with a frequency w_0, which is the frequency of the solvent-vacancy exchange in the pure solvent. Manning (1964) has derived the following expression for the correlation factor for impurity diffusion f_i in the fcc lattice in terms of these various jump frequencies:

$$f_i = (w_1 + \tfrac{7}{2}Fw_3)/(w_1 + w_2 + \tfrac{7}{2}Fw_3) \tag{34}$$

where F is the fraction of vacancies making w_3 jumps that effectively do not return to the site from which the w_3 jump was made, and F is a known function of w_4/w_0 (Manning, 1964). Since the temperature dependence of the various w_i's will follow Eq. (10), and the various g_i^m's will, in general, be different, f_i will vary with temperature.

Actually three dissimilar w_3 jumps occur that displace the vacancy to second-, third-, or fourth-nearest neighbor positions of the impurity. It has been assumed in the derivation of Eq. (34) that all three occur with the same w_3 frequency. Other vacancy jumps at distances farther from the impurity may have frequencies different from w_0; however, the experimental data to date may be adequately explained by the above formulation.

As illustrated in Eq. (34), f_i is a function of a number of jump frequencies. Thus, the analysis of f_i is not always straightforward, and the extent to which information can be obtained regarding jump frequency ratios depends on the amount of information available from other measurements. First, the isotope effect for impurity and self-diffusion may be measured in the pure solvent. The self-diffusion isotope effect will yield a value of ΔK if only one known diffusion mechanism is operative. [ΔK can also be estimated by Eq. (17).] The impurity diffusion isotope effect will give f_i if ΔK is assumed to be the same for both solvent and impurity diffusion. The near equality of ΔK for impurity and self-diffusion is supported by theory if the masses of the solvent and impurity are similar (Feit, 1972b).

A knowledge of f_i alone allows some general conclusions to be drawn. If f_i is close to unity, the impurity jump frequency w_2 must be small compared with the solvent jump frequencies. If f_i is quite small, then $w_2 \gg w_1, w_3, w_4, w_0$. This appears to be the case for sodium diffusion in potassium for which Barr et al. (1967) find $E = 0.17$.

If the self-diffusion coefficient $D_s(0)$ and the impurity diffusion coefficient D_i are known along with f_i, some information about limiting values of certain jump frequency ratios may be obtained. The ratio $D_i/D_s(0)$ may be written as

$$D_i/D_s(0) = (w_2/w_0)(f_i/f_0) \exp(-\Delta g/KT) = (w_2/w_0)(f_i/f_0)w_4/w_3 \quad (35)$$

where f_0 is the correlation factor for self-diffusion in the pure solvent and $-\Delta g$ is the impurity–vacancy binding energy. From Eqs. (34) and (35), one may obtain the relation

$$w_1/w_3 = [D_i/D_s(0)][f_0/(1-f_i)](w_0/w_4) - 7F/2 \quad (36)$$

Since F is a known function of w_4/w_0, experimental values of D_i, $D_s(0)$, and f_i give w_1/w_3 as a function of w_4/w_0 from Eq. (36). Equation (36) also sets a lower limit for w_0/w_4 because w_1/w_3 must be greater than zero. For each value of w_0/w_4 above its minimum value, one can obtain corresponding values of w_1/w_3 and, from Eq. (35), values of w_2/w_3. The isotope effect for iron diffusion in silver and copper (Mullen, 1961a) and for chromium diffusion in nickel (Heumann and Reerink, 1966; LeClaire, 1967) have been analyzed in this manner using Eq. (36). As an example for chromium diffusion in nickel, LeClaire (1967) obtains a minimum value of $w_0/w_4 \sim 1$ and a corresponding minimum value of $w_2/w_3 \sim 2.6$.

Several measurements of the isotope effect for impurity diffusion in bcc and hcp metals have also been made. Coleman et al. (1968) have measured the isotope effect as a function of temperature for iron diffusion in vanadium. They observe that $f \Delta K$ decreases as the temperature increases from a value of 0.7 below 1350°C to 0.3 at 1800°C. The value of the isotope effect, as well as its temperature dependence and a significant curvature in the Arrhenius plot, strongly suggests that two mechanisms of diffusion are operative, probably monovacancies and divacancies. Batra (1967) measured the isotope effect for cadmium diffusion in zinc; he was able to show that the cadmium–vacancy binding energy is 0.09 eV and that the cadmium–vacancy exchange frequency is 3.7 times larger than the zinc–vacancy exchange frequency. Mao (1972) reports preliminary studies of the isotope effect for zinc diffusion in cadmium. He concludes that the results rule out the interstitial mechanism and are consistent with the monovacancy mechanism of diffusion.

To obtain experimental values of the jump frequency ratios, information in addition to D_i, $D_s(0)$, and f_i must be obtained. This information can be obtained from measurements of the effect of impurity additions on self-diffusion in dilute alloys. As impurity atoms are added to the solvent, some of the solvent atoms will jump with a frequency w_1, w_3, or w_4

rather than w_0. Lidiard (1960) was the first to derive correctly the average solvent-atom jump frequency as a function of impurity content c in terms of the various w_i's. He found that the self-diffusion coefficient in an alloy of composition c, $D_s(c)$, could be written as

$$D_s(c) = D_s(0)(1 + bc) \qquad (37)$$

and

$$f_i = 1 - [4f_0/(b + 18)](D_i/D_s(0)) \qquad (38)$$

where b is a constant for a given alloy system at a given temperature. A measurement of D_i, $D_s(0)$, and b will give a value of f_i for impurity diffusion in the pure solvent within the framework of the Lidiard theory.

In deriving Eq. (38), Lidiard assumed that all solvent jumps in the alloy have a correlation factor f_0 as in the pure solvent. Howard and Manning (1967) have removed this assumption and found that, for a given set of values of D_i, $D_s(0)$, and b a range of possible values of f_i exist, and that unique values of w_4/w_0, w_3/w_1, and w_2/w_1 correspond to each value of f_i. Hence, given experimental values of D_i, $D_s(0)$, b, and f_i, a unique set of jump frequency ratios may be obtained.

Experimental data on these four quantities exist for zinc diffusion in silver and copper. The terms D_i, $D_s(0)$, b, and E have been measured for zinc diffusion in silver (Rothman and Peterson, 1967) and copper (Peterson and Rothman, 1967b, 1970). The term ΔK was assumed to be the same for the impurity as for the pure solvent. The data were analyzed by Manning to obtain the following jump frequency ratios for a vacancy near a zinc atom in silver and copper (see Table II).

TABLE II

JUMP FREQUENCY RATIOS FOR A VACANCY NEAR A
ZINC ATOM IN SILVER AND COPPER

Solvent	T (°C)	w_4/w_0	w_3/w_1	w_2/w_1
Ag	747	1.15	0.27	1.53
	880	1.3	0.39	1.54
Cu	894	1.2	0.5	2.5
	947	1.5	0.9	3.6

The apparent temperature dependence of these jump frequencies should not be taken too seriously because they are sensitive to the value of f_i; an increase in f_i of 0.01 decreases w_3/w_1 and w_2/w_1 by 15%. Thus, the jump frequency ratios are known only to 30%, even though the relative diffusion coefficients of ^{65}Zn and ^{69}Zn were determined to within 0.03%. However, the jump frequency ratios are of the expected magnitude. Since zinc diffuses faster than solvent atoms in copper and silver, w_2/w_1 is expected to be greater than unity. Similarly, since zinc is believed to attract vacancies in copper and silver, w_4/w_0 is expected to be greater than unity.

Rothman and Peterson (1967) also measured the isotope effect for zinc diffusion in silver as a function of temperature in the range of 697 to 952°C. The experimental values of f_i are shown as points with error bars in Fig. 8. Starting with any one of the above values for w_4/w_0 or w_3/w_1 for zinc diffusion in silver, Rothman and Peterson calculated w_4/w_0, w_3/w_1, and w_2/w_1 as a function of temperature using Eqs. (34) and (35) and the experimental values of f_i (Fig. 8), D_i, and

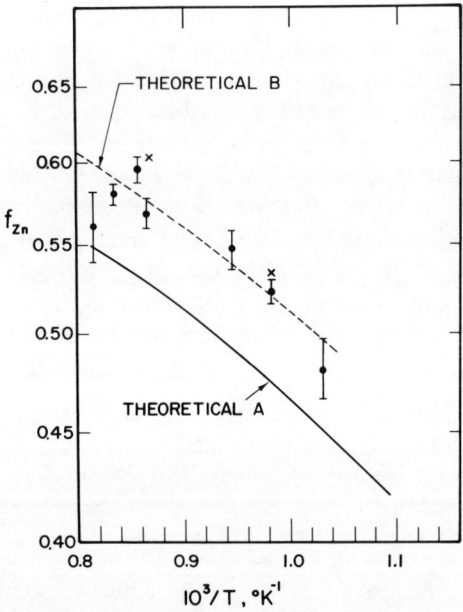

Fig. 8. Correlation factor for the diffusion of zinc in silver versus reciprocal absolute temperature from the work of Rothman and Peterson (1967). The points with error bars are data from isotope-effect experiments. The crosses are values calculated from Lidiard's theory. See text for descriptions of the two curves labeled "theoretical."

$D_s(0)$. They found that the temperature dependence of these jump frequency ratios was well represented by the LeClaire theory (1962). The curve labeled "theoretical B" in Fig. 8 shows f_i calculated from Eq. (34) using the smooth values of the jump frequency ratios. The curve labeled "theoretical A" shows f_i calculated entirely from the LeClaire theory. In this theory, w_4 jumps are not included [F in Eq. (34) is taken as unity] and it is assumed, in the expression for w_i [Eq. (10)], that $v_1 = v_3 = v_0$. The analysis by Rothman and Peterson which includes w_4 jumps shows that $v_4 = 0.99$, v_0 and $v_3 = 0.9v_0$, but $v_1 = 2.9v_0$; this difference between v_1 and v_0 accounts for most of the difference between curves A and B. Good agreement is found between f_i from isotope-effect measurements and that determined by the LeClaire theory for zinc diffusion in copper (Peterson and Rothman, 1970) if w_4 jumps are included in the LeClaire theory.

The crosses in Fig. 8 show f_i determined from Eq. (38). The good agreement between f_i from Eq. (38) and from the isotope effect for zinc diffusion in silver is not found for zinc diffusion in copper. Equation (38) gives $f_i = 0.56$ at 894°C and 0.60 at 947°C, whereas the isotope-effect technique gives $f_i = 0.47$ at both temperatures for zinc diffusion in copper (Peterson and Rothman, 1970). The 25% difference between the values of f_i from the two techniques is much larger than the estimated error of 4% in either value of f_i.

B. Rapid Impurity Diffusion in Metals‡

It has been known for nearly 80 years that gold diffuses rapidly in lead at moderate temperatures (Roberts-Austen, 1896a,b). However, it has only been during the last 6 years that the subject of rapid metallic impurity diffusion in metals has received significant attention. It now appears that the noble metals diffuse, at least partially, by the interstitial mechanism in a large number of materials. Since this topic is covered extensively in Chapter 4, only two systems, cadmium and silver diffusion in lead for which isotope-effect measurements have been made, will be discussed here.

The rapid diffusion and low activation energy for cadmium diffusion in lead suggests an interstitial mechanism. However, it was found that the self-diffusion of lead is somewhat enhanced by additions of cadmium, i.e., b in Eq. (37) is greater than zero (Miller, 1969a), but b is smaller than the minimum value required for diffusion by the monovacancy mechanism. Since b is thought to be zero for interstitial impurities, Miller (1969b) proposed a mechanism in which a mobile interstitial cadmium ion is tightly bound to a neighboring vacancy and moves

---

‡ This topic is dealt with extensively by Warburton and Turnbull in Chapter 4.

essentially as an interstitial–vacancy pair. This model predicts $b \approx f_0 \, D_i/D_s(0)$, which is in agreement with the data. In addition, the model predicts that f_i for cadmium diffusion in lead should be small in contrast to a value of $f \, \Delta K$ equal to unity for interstitial diffusion by means of a simple random walk. Miller and Edelstein (1969) have measured the isotope effect for cadmium diffusion in lead and find $f_i \, \Delta K = 0.12 \pm 0.03$. Since the values of $D_i/D_s(0)$, b, and $f \, \Delta K$ are all consistent with the proposed close-pair mechanism, Miller and Edelstein conclude that cadmium dissolves dissociatively into both interstitial and substitutional sites in lead and diffuses primarily by the interstitial–vacancy-pair mechanism.

The interstitial mechanism for silver diffusion in lead appears to be more complicated than for cadmium diffusion in lead. The correct mechanism of diffusion must be consistent with the following experimental data: $D_i/D_s(0) \approx 500$ at the melting point and increases with a decrease in temperature (Dyson et al., 1966); small additions of silver to lead enhance lead self-diffusion ($b = 138$ at 300°C) (Miller, 1970) but do not enhance silver diffusion (Miller et al., 1973); $f_i \, \Delta K = 0.25 \pm 0.05$ over a fair temperature range (150–300°C) (Miller et al., 1973; Herzig et al., 1971); and internal friction measurements (Turner et al., 1972) show that noble metal impurities produce defects with symmetry lower than that of the lead lattice, which is not true for an isolated silver atom in a normal octahedral interstitial site.

Miller and co-workers (1973) suggest a reasonable explanation of the above data for silver diffusion in lead that requires two mechanisms of diffusion. First, the fact that b is not zero requires that one of the diffusion mechanisms must couple impurity and solvent jumps. Substitutional or interstitial impurity atoms associated with vacancies will produce an enhancement. The maximum contribution of the coupled mechanism to the total impurity diffusion, which is consistent with the experimental value of b, is 9% for diffusion by vacancies and $\sim 20\%$ for diffusion by interstitial–vacancy pairs. Since the binding energy between an interstitial impurity atom and a neighboring vacancy must be large, Miller et al. suggest that the interstitial impurity–vacancy-pair mechanism is preferred for 20% of the diffusion, and the remainder proceeds by an interstitial mechanism. Miller et al. suggest that this interstitial mechanism is an impurity–solvent split interstitial; the impurity atom jumps from association with one lead atom to association with a neighboring lead atom. This jump sequence requires the movement of both the original and neighboring lead atoms. Qualitative arguments suggest that $f_i \, \Delta K$ should lie between 0.20 and 0.34 for this mechanism of interstitial motion. This split interstitial can migrate without causing concomitant enhancement of self-diffusion, and it has a geometry that is consistent

3. ISOTOPE EFFECTS IN DIFFUSION

with the internal friction measurements. Although the mechanism for silver diffusion in lead may not be uniquely established, the data clearly require interstitial motion that is more complex than simple random walks.

C. EFFECT OF RESONANT VIBRATIONAL MODES ON DIFFUSION

Resonant vibrational modes may develop for an impurity that could make ΔK_i for impurity diffusion significantly different from ΔK_0 for self-diffusion. Impurities with mass greater than the solvent lattice atoms will give rise to resonance modes that are characterized by a large amplitude of vibration of the impurity atom or of those atoms which interact directly with the impurity. Achar (to be published) has developed, from the dynamical theory, the effect of resonance modes on ΔK and finds a significant change in ΔK_i for impurities with appreciably greater mass than the host atoms. The effect of impurity mass on the value of ΔK_i has also been determined from the dynamical theory by Feit (1972b). In Feit's approach, the resonance modes are not treated explicitly, and the variation of ΔK_i is determined relative to ΔK_0.

To test the theory, Mundy and McFall (1973) have measured the isotope effect for the heaviest convenient impurity (silver) and an intermediate impurity (sodium) diffusing in the lightest convenient solvent (lithium). The values of $\Delta K_i/\Delta K_0$ from Achar's calculations are 0.72 and 0.145 for sodium and silver diffusion in lithium, respectively. Corresponding values from Feit's theory depend on the value of ΔK_0. Taking $\Delta K_0 = 0.5$ for lithium (the approximate value for sodium self-diffusion), Feit's theory gives $\Delta K_i/\Delta K_0$ values of 0.26 and 0.05 for sodium and silver. Although the two dynamical theories predict different values of $\Delta K_i/\Delta K_0$, they both give the same value of the experimental quantity $(\Delta K)_{Na}/(\Delta K)_{Ag} \sim 5$.

The experimental values from the measurements of Mundy and McFall are $(f\Delta K)_{Na} = 0.19 \pm 0.01$, and $(f\Delta K)_{Ag} = 0.26 \pm 0.01$. Since sodium and silver diffuse at the same rate in lithium [their Arrhenius plots are indistinguishable within experimental error (Mundy and McFall, 1973)], it is reasonable to assume that $f_{Na} \sim f_{Ag}$ within 10 or 20%. Thus, the experimental results give

$$(\Delta K)_{Na}/(\Delta K)_{Ag} \approx (f\,\Delta K)_{Na}/(f\,\Delta K)_{Ag} = 0.73 \pm 0.07 \tag{39}$$

This suggests that, unless f_{Ag} is at least 20% larger than f_{Na}, the value of $(\Delta K)_{Ag}$ is measurably larger than $(\Delta K)_{Na}$ and not five times smaller, as predicted by theory (Feit, 1972b; Achar, to be published). Thus, one must question the assumptions used in the application of the dynamical theory to the mass dependence of the isotope effect for impurity diffusion; the basic dynamical theory of diffusion is neither supported nor rejected by these experiments.

It should be noted that any potential contribution from quantum effects to diffusion should be a maximum in lithium because of the high Debye temperature relative to normal diffusion temperatures in this material. The contribution from quantum effects to ΔK_i is not reliably known at this time. However, it would be surprising to find that quantum effects could account for the factor of 5 in the disagreement between theory and experiment.

V. Diffusion in Concentrated Alloys

Tracer diffusion in concentrated alloys cannot be treated as rigorously as for dilute alloys. The atomistic approach to diffusion in concentrated alloys has been limited to the two extreme cases, a completely random alloy or a completely ordered alloy. The isotope-effect technique has been used to study diffusion in both random and ordered binary alloys. These studies will be considered separately in Section V.

A. Random Alloys

Manning (1967, 1971) has treated the kinetics of diffusion in a concentrated alloy using a rather simplified model. He assumes that vacancy exchanges with tracer atoms A and B occur with jump frequencies w_A and w_B independent of the configuration of the surrounding atoms, and all vacancy exchanges with nontracer atoms are assumed to occur with an average jump frequency W given by

$$W = w_A N_A + w_B N_B \tag{40}$$

where N_A and N_B are the molar fractions of A and B. This model is expected to give a more accurate description of correlation effects in concentrated alloys than in dilute alloys, because vacancy binding to the solute becomes more important in the latter.

On the basis of this model, the correlation factor f_A for diffusion of A atoms in a random A–B alloy may be written as (Manning, 1967)

$$f_A = [(M_0 + 2)(N_A D_A + N_B D_B) - 2D_A]/[(M_0 + 2)(N_A D_A + N_B D_B)] \tag{41}$$

where M_0 is a constant equal to 7.15 for the fcc lattice and 5.33 for the bcc lattice. The equation for f_B is similar, with $-2D_A$ replaced by $-2D_B$.

3. ISOTOPE EFFECTS IN DIFFUSION

A direct comparison between the correlation factors calculated from Eq. (41) and the results from an isotope-effect experiment cannot be made because we do not know ΔK. However, the ratio f_A/f_B calculated from measured values of D_A and D_B can be compared with $(f\Delta K)_A/(f\Delta K)_B$ if one assumes that $\Delta K_A = \Delta K_B$, an assumption that should be valid within the framework of Manning's model.

Peterson and Rothman (1967b, 1970) have measured the isotope effect for zinc and copper diffusion in a series of copper–zinc alloys that contain 4, 30, and 49% zinc. In α-phase CuZn, Peterson and Rothman's measurements of $(f\Delta K)_{Zn}/(f\Delta K)_{Cu}$ show rather poor agreement with predicted f_{Zn}/f_{Cu} values from Eq. (41) at 4 at. % zinc (0.557 ± 0.020 vs 0.456 ± 0.050) but better agreement at 30 at. % zinc (0.706 ± 0.023 vs 0.650 ± 0.025). Agreement in the disordered β-CuZn at 49% zinc was superb (0.738 ± 0.035 vs 0.732 ± 0.023), whereas agreement is not as good in ordered β-CuZn (0.615 ± 0.050 vs 0.721 ± 0.023). Manning's theory provides values of f_{Zn}/f_{Cu} in good agreement with the experimental values of $(f\Delta K)_{Zn}/(f\Delta K)_{Cu}$ for those cases in which the theory should be applicable (30 at. % zinc and disordered β-CuZn). The agreement with experiment is poorer for those cases in which the theory is less applicable (4 at. % zinc and ordered β-CuZn).

If the agreement between the experimental and theoretical ratios of the correlation factors implies that f_{Zn} and f_{Cu} from Eq. (41) are correct, then $\Delta K_{Zn} = \Delta K_{Cu} = 0.76 \pm 0.03$ in a copper–30 at. % zinc alloy and $\Delta K_{Zn} = \Delta K_{Cu} = 0.39 \pm 0.01$ in disordered β-CuZn. These values are in general agreement with values of ΔK for fcc and bcc metals, but do show that alloying pure copper ($\Delta K = 0.87$) with zinc lowers ΔK.

Fishman et al. (1970) have measured $(f\Delta K)_{Fe}$ and $(f\Delta K)_{Co}$ in disordered equiatomic FeCo. They obtain $(f\Delta K)_{Fe}/(f\Delta K)_{Co} = 0.92 \pm 0.10$ compared with $f_{Fe}/f_{Co} = 0.95 \pm 0.05$ from Eq. (41). They also obtain $\Delta K = 0.88 \pm 0.10$ in the fcc phase and 0.73 ± 0.10 in the bcc phase of FeCo. This value of ΔK for the bcc phase is rather large compared with 0.50 reported for sodium (Mundy, 1971) and bcc iron (Walter and Peterson, 1969).

B. Ordered Alloys

The lattice of an ordered CsCl-type alloy may be considered as two interpenetrating sublattices and each sublattice of the AB alloy is occupied by atoms of one kind. The nearest neighbor positions of a site on one sublattice all lie on the other sublattice. Random vacancy motion between nearest neighbor sites is not possible without disrupting the ordered arrangements of the atoms. The energy required to put an atom on the wrong sublattice is often considered to be too large to allow random vacancy motion by means of nearest neighbor jumps.

Elcock and McCombie (1958) and Elcock (1959) have suggested the possibility of a six-jump vacancy cycle that allows diffusion to take place by means of nearest neighbor vacancy jumps. After the complete cycle of six jumps, the vacancy has moved a given distance and the order of the system is still preserved.

The actual sequence may have many intervening jumps that must be retraced. Beeler and Delaney (1963) followed the order–disorder process by a Monte Carlo computation on a planar lattice. They observed that $<1\%$ of the Elcock–McCombie loops are actually completed in six consecutive jumps. At least 90% of the vacancy jumps were consumed in retracted excursions from the basic loop. However, net transport by pure and inefficient Elcock–McCombie loops was always observed when the long range order parameter was unity.

Since Elcock–McCombie loops are a highly correlated set of jumps and retracted intervening jumps are expected, the correlation factor for diffusion by this mechanism should be small compared with random vacancy motion in a disordered alloy. Peterson and Rothman (1967b, 1970) have measured the isotope effect for copper and zinc diffusion in ordered and disordered β-CuZn. They obtain the moderately large value of $f\Delta K = 0.325 \pm 0.010$ for copper diffusion in both the ordered and disordered states. Peterson and Rothman conclude that the sequence of atomic jumps in the ordered and disordered alloys is similar, and that the Elcock–McCombie loop mechanism is not the dominant mechanism in the ordered state. However, it should be noted that at the lowest temperature, where D is sufficiently large to allow a copper isotope-effect experiment to be done ($\sim 410°C$), the order in β-CuZn is far from perfect ($S = 0.7$). Thus, it is not surprising that Elcock–McCombie loops are not dominant at $410°C$ because these loops are an inefficient means of transport, and with $S = 0.7$ (approximately $\frac{3}{4}$ of the atoms have at least one "wrong" neighbor), these loops are not required.

Fishman et al. (1970) have measured the isotope effect for iron diffusion in ordered and disordered FeCo. They find that $f\Delta K$ decreases as the temperature decreases and the corresponding degree of long range order in the compound increases. Although the errors in $f\Delta K$ are large in the ordered phase ($D < 10^{-13}$ cm^2/sec), a distinct decrease in $f\Delta K$ in the ordered state relative to the disordered state is observed. Fishman et al. conclude that the correlation factor decreases as the degree of order increases and the most plausible jump mechanism in the highly ordered region is the six-jump Elcock–McCombie mechanism.

VI. Diffusion in Alkali and Silver Halide Crystals

For diffusion in a pure ionic crystal, experimental methods, other than the isotope effect, are available for measuring the correlation factor. These methods exist because the tracer self-diffusion coefficient D_T of the ions contains a correlation factor, but the contribution σ to the conductivity does not. Since the comparison of diffusion and ionic conductivity can be a powerful tool in identifying the diffusion mechanism and can play an important role in the isotope-effect experiments described in Section VI, we shall describe this technique before discussing the isotope-effect studies.

The derivation of the relation between D_T and σ will begin with the general Nernst–Einstein equation

$$D_q = kT\mu/q \tag{42}$$

where μ and D_q are the mobility and diffusion coefficients, respectively, of the charge carriers of charge q. The contribution of the defects to the conductivity is

$$\sigma = n_d q\mu \tag{43}$$

where n_d is the number of defects per unit volume. Since the defects diffuse by a random walk [see Eq. (3)],

$$D_q = \tfrac{1}{6} r_q^2 \Gamma_q \tag{44}$$

where r_q is the charge displacement by one jump. The tracer follows a correlated diffusion path so that D_T has a form of Eq. (3)

$$D_T = \tfrac{1}{6} r_T^2 \Gamma_T f \tag{45}$$

where r_T is the ion displacement by one jump. Combining Eqs. (42)–(45) yields

$$D_T/\sigma = (kT/q^2)(\Gamma_T/n_d \Gamma_q)(r_T/r_q)^2 f \tag{46}$$

If the charge carriers are single vacancies, $r_T = r_q$ and $\Gamma_T = n_d \Gamma_q/N$, where N is the number of ions per unit volume, and

$$D_T/\sigma = (kT/q^2 N)f \tag{47}$$

If the charge carriers are interstitial ions, Eq. (47) may be slightly modified. If the interstitial ion diffuses by simple interstitial jumps, Eq. (47) remains the same with f equal to unity. If the interstitial ions move by interstitialcy jumps, two ions move for each defect jump, and

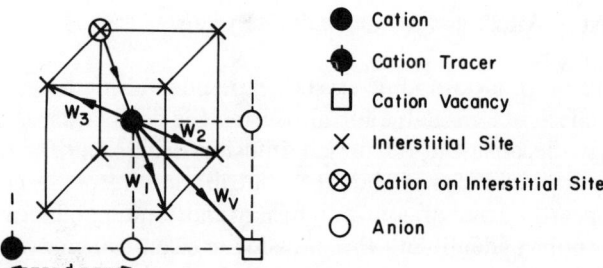

Fig. 9. Cation jump processes in the AgBr structure. A cation tracer may move from one substitutional site to another by exchanging places with a cation vacancy w_v, or an interstitial cation may push the cation tracer from its normal lattice site into one of the neighboring interstitial sites (w_1, w_2, or w_3), and the initial interstitial atom will then occupy the normal lattice sites. [Reprinted with permission from Chen and Peterson, *J. Phys. Chem. Solids* (1972), Pergamon Press.]

$\Gamma_T = 2n_d \Gamma_q/N$. For a direct interstitialcy jump (w_1 jump in Fig. 9), the charge q is displaced twice as far as a tracer ion during a jump ($r_q = 2r_T$), and

$$D_T/\sigma = (kT/q^2 N)(\tfrac{1}{2} f_{I_1}) \tag{48}$$

For indirect interstitialcy jumps, r_T/r_q depends on the geometry of the jump; for the w_2 jump in Fig. 9, $r_q = r_T(\tfrac{8}{3})^{1/2}$, and $\tfrac{1}{2}$ in Eq. (48) becomes $\tfrac{3}{4}$.

When two or more types of mobile defects contribute to diffusion and conductivity, Eq. (46) includes a sum of contributions. For pure crystals with Frenkel disorder (equal number of interstitials and vacancies transporting the same charge g),

$$D_T/\sigma = \{[(D_{T_i}/D_{q_i})\mu_i + (D_{T_v}/D_{q_v})\mu_V]/n_d(\mu_i + \mu_V)\}(kT/q^2) \tag{49}$$

where the subscripts i and V refer to properties of interstitials or vacancies. If both anions and cations are mobile, the sum of D_T for both species must be used for comparison with the total conductivity. We see from the preceding discussion that the experimental quantity D_T/σ is critically related to the type of defects responsible for ionic transport through its dependence on r_T/r_q and f. A given experimental value of D_T/σ is often consistent with only one explanation of the mechanism of ionic and mass transport in a pure ionic conductor.

The first isotope-effect experiment discussed in Section VI explores the contribution from neutral vacancy pairs to sodium diffusion in NaCl. The second experiment investigates the mass term in Eq. (13) for the simultaneous motion of two atoms (interstitialcy jumps) in AgBr and AgCl. The third experiment discussed uses isotope-effect measurements in a study of the sodium ion migration energy in RbCl.

A. VACANCY-PAIR CONTRIBUTION TO CATION DIFFUSION IN NaCl

A comparison of the sum of the cation and anion tracer diffusivity with the total conductivity by means of Eq. (47) yields a correlation factor close to that for diffusion by single vacancies in NaCl, but perfect agreement between theory and experiment is not evident (Downing and Friauf, 1970). It appears that neutral defects, such as cation vacancy–anion vacancy pairs, contribute to D_T but not to σ. It is difficult to deduce the magnitude of the contribution from neutral pairs to the diffusion process from measurements of D_T and σ only; the pair contribution to the diffusivity of the anions and cations may be different, the correlation factor for diffusion by pairs will be a function of the relative jump frequencies of cations w_c' and anions w_a' with the vacancy pair, and the experimental accuracy required to detect small contributions from neutral pairs is difficult to obtain.

Three attempts have been made in recent years to determine the contribution of cation–anion vacancy pairs to the diffusion of sodium ions in NaCl. Nelson and Friauf (1970) measured the simultaneous diffusion and drift of sodium tracers in NaCl in an electric field. In this experiment, the neutral pairs will contribute to diffusion (spreading of the tracer profile) but not to the drift. By obtaining both the diffusion and drift in one experiment, the errors were minimized. Nelson and Friauf obtained a pair contribution of $\sim 40\%$ near the melting point. Although the contact between the two crystals in the drift experiment was far from perfect, the results of Nelson and Friauf should be more accurate than those deduced from separate measurements of D_T and σ.

Beniere et al. (1970) obtained the contribution of vacancy pairs to anion diffusion in NaCl from measurements of D_{Cl} in NaCl crystals doped with divalent cation impurities. (Divalent cation impurities reduce the number of anion vacancies, but do not affect the concentration of vacancy pairs; hence, the anions will migrate predominantly by vacancy pairs if sufficient divalent cation impurities are added.) Using a theoretical result that cation diffusion by pairs is the same as anion diffusion by pairs in NaCl, Tharmalingam and Lidiard (1961) and Beniere et al. (1970) conclude that the vacancy-pair contribution to sodium ion diffusion in NaCl is about 10% at the melting point. Using this value and their own measurements of D_{Cl^-}, D_{Na^+}, and σ, Beniere et al. reach the unusual conclusion that f_{1V} equals unity.

Rothman et al. (1972) measured the isotope effect as a function of temperature for sodium diffusion in NaCl. Their results, together with an earlier measurement by Barr and LeClaire (1964), are shown in Fig. 10. Although some scatter exists in the values of E at lower temperatures,

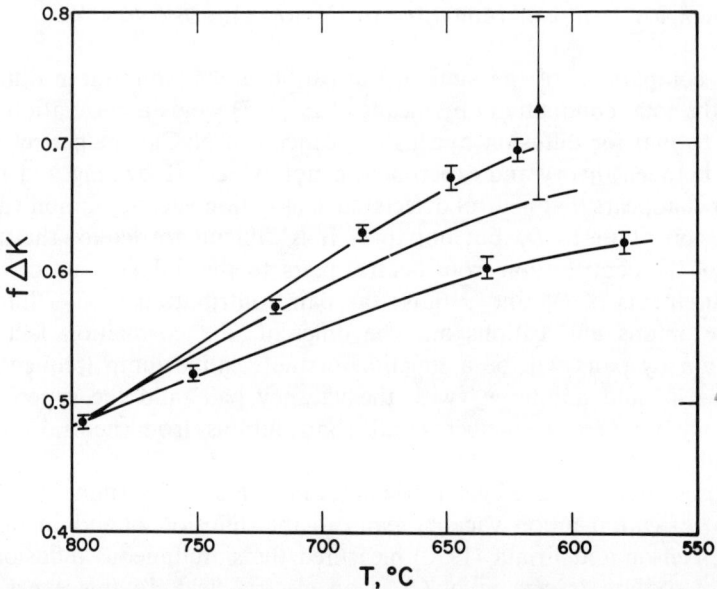

Fig. 10. The isotope effect versus temperature for sodium diffusion in NaCl from the work of Rothman et al. (1972). The triangle represents a point from Barr and LeClaire (1964). The three lines represent possible fits to the data as discussed in the text. [Reprinted with permission from Rothman et al., J. Phys. Chem. Solids (1972), Pergamon Press.]

some limiting values of the vacancy-pair contribution can be deduced from the decrease in $f\Delta K$ with an increase in temperature. The minimum vacancy-pair contribution may be obtained from Fig. 10 using Eq. (16) and the minimum allowed values of $(f\Delta K)_{1V}$ and $(f\Delta K)_{2V}$. Taking $(f\Delta K)_{1V} = 0.624$ (the lowest value at low temperatures) and $(f\Delta K)_{2V} = 0$ gives a minimum pair contribution at the melting point of 22.5%.

Rothman et al. (1972) place further limits on the vacancy-pair contribution in the following manner: from each of the three lines in Fig. 10, one can calculate a set of values of D_{2V}/D_{1V}, as a function of temperature, for an assumed pair of values of E_1 (i.e., ΔK_1) and E_2. A least-squares fit of log (D_{2V}/D_{1V}) to T^{-1} then gives a value of $\Delta Q \equiv Q_{2V} - Q_{1V}$ and a value of D_{02}/D_{01}, the ratio of the D_0's for the two mechanisms. Using this ratio, ΔQ, and the experimental values of D at the ends of the temperature range, one can obtain the values of D_{01} and Q_1 that correspond to the values of E_1 and E_2 assumed above. All pairs of values of E_1 and E_2 are acceptable that (1) give an Arrhenius dependence of D_{2V}/D_{1V} and (2) yield a plot of log D vs T^{-1}, calculated from the above values of D_{01},

D_{02}, Q_{1V}, and Q_{2V}, which is not much more curved than the experimentally determined Arrhenius plot. No matter which line in Fig. 10 is used, only values of $(f \Delta K)_{2V}$ less than 0.2 yield a good fit according to the above criteria. Rothman et al. estimate that ΔK_{2V} is not less than 0.8; this means that $f_{2V} \leq 0.25$. Since f_{2V} is a function of the relative jump frequencies of cations w_c' and anions w_a' with the vacancy pair (Howard, 1966; Compaan and Haven, 1956, 1958), w_a'/w_c' must be less than or equal to 0.1. If the middle line of Fig. 10 is accepted as best representing the data, one obtains $\Delta K_1 = 0.9$, $\Delta Q = 0.82$–0.90 eV, $Q_{1V} = 1.89$–1.96 eV, and the vacancy-pair contribution near the melting point is between 30 and 45%.

The isotope-effect results of Rothman et al. clearly support a significant vacancy-pair contribution to sodium ion diffusion in NaCl near the melting point, as proposed by Nelson and Friauf. The isotope-effect results do not support the value of $f_{1V} = 1.0$ or a value of $w_a'/w_c' \geq 1$ proposed by Beniere et al. (1970). It would appear that the results of Beniere et al. and Rothman et al. are completely consistent if the assumption that D_{Na^+} (pairs) = D_{Cl} (pairs) used by Beniere et al. is replaced by D_{Na^+} (pairs)/D_{Cl^-} (pairs) = 5. (This ratio of 5 corresponds to $w_a'/w_c' = 0.1$, which is consistent with the isotope-effect measurements and the measurements of Nelson and Friauf.)

B. Interstitialcy Diffusion in Silver Halides

In the derivation of the general isotope-effect equation [Eq. (13)], it was assumed that only one atom changes lattice sites during the jump processes. If n atoms simultaneously move to new lattice sites during the jump, m_i in Eq. (13) is assumed to be replaced by $(n - 1)m_0 + m_i$, where m_0 is the average mass of the nontracer atoms. Since interstitialcy migration is believed to be a near simultaneous motion of two atoms, an isotope-effect measurement for diffusion by the interstitialcy mechanism would test the validity of this assumption.

Teltow (1949), Ebert and Teltow (1955), Kurnik (1952), and Abbink and Martin (1966) have found from measurements of ionic conductivity in AgBr, a series of AgBr–CdBr$_2$ alloys, AgCl, and a series of AgCl–CdCl$_2$ alloys, that an appreciable fraction of the silver ions move by an interstitial-type mechanism in pure AgBr and pure AgCl. They also report values of $\phi = \mu_i/\mu_V$ as a function of temperature.

Friauf (1957) has made simultaneous measurements of D_T and σ in AgBr as a function of temperature and found that D_T/σ is temperature dependent and is not consistent with simple interstitial diffusion or any combination of interstitials and vacancies for the migration of silver. (Anion transport in the silver halides is completely negligible to cation transport.) If all three interstitialcy jumps in Fig. 9 occur, Eq. (49) may be written as

$$\frac{D_T}{\sigma} = \frac{kT}{N_q^2} \left\{ \frac{\frac{1}{2}f_I[(w_1 + 3w_2 + 3w_3)/(w_1 + 2w_2 + w_3)]\phi + f_V}{(\phi + 1)} \right\} \quad (50)$$

where f_I is the overall correlation factor for interstitialcy jumps. Friauf fitted his results to Eq. (50), assuming that w_3 jumps do not occur and using the experimental values of ϕ for AgBr. Since f_I is a known function of $\kappa = w_2/w_1$, Eq. (50) was solved to give w_2/w_1 as a function of temperature. The excellent fit between Friauf's results and Eq. (50) presents convincing evidence that transport in AgBr occurs by Frenkel defects and that the interstitials move by both direct and indirect interstitialcy jumps with $\kappa = 0.026 \exp(-0.15/kT)$.

As divalent impurities are added to AgBr, cation vacancies are created and cation interstitials are depressed, and usually a temperature range exists in which the interstitial concentration is so depressed that diffusion is almost entirely due to vacancies. From measurements of ionic conductivity and tracer diffusion, Miller and Maurer (1958) have shown that all silver ions move by a vacancy mechanism in AgBr doped with 0.03 to 0.08 mole % Cd at temperatures between ~ 100 and 200°C. Results similar to those of Friauf have been obtained by Weber and Friauf (1969) and Gracey and Friauf (1969) on AgCl.

Peterson et al. (1973) have measured the isotope effect for silver diffusion in cadmium-doped AgBr (to obtain ΔK for vacancy diffusion in pure AgBr), in pure AgBr at six temperatures in the range of 175 to 337°C (to obtain ΔK for both direct and indirect interstitialcy jumps), and in pure AgCl.

Peterson et al. analyze their isotope-effect results using the equation

$$(D_\alpha/D_\beta) - 1 = \alpha_1 \Delta K_1 + \alpha_2 \Delta K_2 + \alpha_V \Delta K_V \quad (51)$$

where the subscript V refers to properties of vacancy jumps, and the subscripts 1 and 2 refer to properties of w_1- and w_2-type interstitialcy jumps, respectively. The α's are known but complex functions of ϕ, κ, and atomic mass (LeClaire, 1971). Using the literature values of ϕ (Teltow, 1949; Abbink and Martin, 1966) and κ (Friauf, 1957; Weber and Friauf, 1969; Gracey and Friauf, 1969), Peterson et al. obtain $\Delta K_V = 0.66 \pm 0.02$, $\Delta K_1 \sim 1$, and $\Delta K_2 \sim 0.1$. Simple interstitial jumps and w_3-type interstitialcy jumps were ignored in the analysis.

The value of unity for ΔK_1 indicates that the direct interstitialcy type of jump occurs with little or no disturbance of the surrounding lattice. However, the small value of ΔK_2 means that an appreciable concomitant motion of neighboring nonmigrating atoms occurs during the indirect interstitialcy jump. This seems reasonable on qualitative geometrical grounds, because one might expect the linear motion of the w_1 jump to provide for a ΔK_1 closer to unity than the value of ΔK_2 for the dogleg type of w_2 jump. The striking difference in the ΔK value is consistent with the activation energy being smaller for w_1 jumps than for w_2 jumps [0.058 eV compared with 0.274 eV for AgBr (Weber and Friauf, 1969)] and with the larger preexponential factor for w_2 jumps (1090 cm^2/sec) compared with w_1 jumps (6.1 cm^2/sec). This indicates a large activation entropy for w_2 jumps, which is consistent with a many-body process.

The above analysis is clearly consistent with the simultaneous activation of two atoms during an interstitialcy jump; however, this is not required by the results. If the two atoms jump consecutively rather than as a single entity, the values of ΔK_1 and ΔK_2 would be roughly halved. Although the larger values of ΔK_1 and ΔK_2 are preferred, the present measurements do not provide direct confirmation that two atoms move simultaneously during an interstitialcy jump.

C. Sodium Diffusion in Rubidium Chloride

Tosi and Doyama (1966) have reported calculations of the energies of formation (h^f) and motion (h^m) of cation vacancies in NaCl, KCl, and RbCl, the energies of motion of Na$^+$, K$^+$, and Rb$^+$ as impurities in these chlorides, and the impurity–vacancy binding energy h^b for these impurities. They calculate an unusually small value of h^m for sodium ion migration in RbCl (0.22 eV) compared with $h^m = 1.02$ eV for rubidium ion migration.

The relations between these energies and the measurable activation energies Q, for diffusion are as follows: for impurity diffusion in the intrinsic region,

$$Q_{\text{int}} = \tfrac{1}{2}h^f + h^m - k(\partial \ln f / \partial T^{-1})_{\text{int}} + h^b \tag{52}$$

and in the extrinsic region,

$$Q_{\text{ext}} = h^m - k(\partial \ln f / \partial T^{-1})_{\text{ext}} + h^b \tag{53}$$

An experimental measurement of Q and the temperature dependence of the correlation factor in both the intrinsic and extrinsic regions will give $\tfrac{1}{2}h^f$ and $h^m + h^b$ that may be compared with the theoretical values of Tosi and Doyama.

Peterson and Rothman (1969) have measured the diffusion of sodium ions and the isotope effect for sodium diffusion in RbCl as a function of temperature. Their temperature dependence of $f\Delta K$ for sodium diffusion in RbCl is shown in Fig. 11. From their experimental values of $Q_{\text{int}} = 2.06$

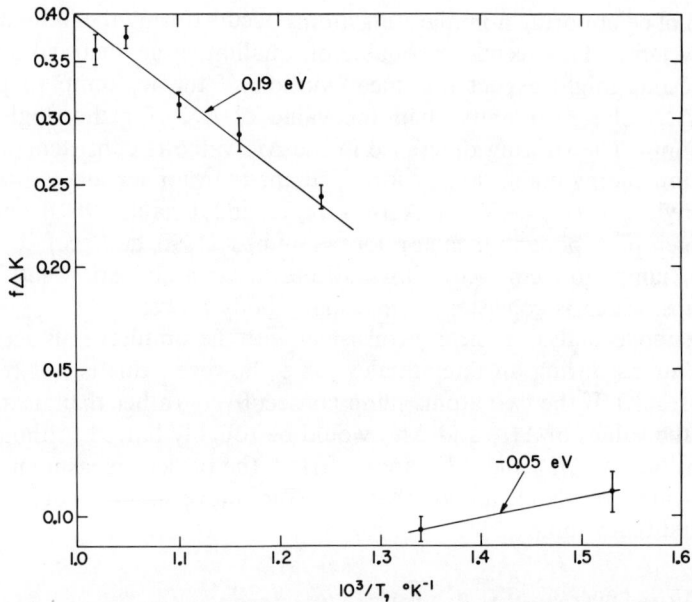

Fig. 11. The isotope effect versus reciprocal absolute temperature for sodium diffusion in RbCl in the intrinsic region (left) and the extrinsic region (right), from the work of Peterson and Rothman (1969).

$\pm$ 0.02 eV, $Q_{\text{ext}} = 0.59 \pm 0.01$ eV, $(\partial \ln f/\partial T^{-1})_{\text{int}} = -0.19 \pm 0.02$ eV, and $(\partial \ln f/\partial T^{-1})_{\text{ext}} = 0.05 \pm 0.03$ eV, Peterson and Rothman obtain $h^m + h^b = 0.65$ eV and $\frac{1}{2}h^f = 1.22$ eV. Tosi and Doyama calculated that Na$^+$ repels vacancies in RbCl with a repulsive energy of 0.1 eV. If the 0.1 eV value for h^b is used, one obtains $h^m = 0.55$ eV compared with the theoretical value of $h^m = 0.22$ eV. The motional energy for Na$^+$ is much smaller than for Rb$^+$ (but not as small as the theoretical value by a factor of 2).

The experimental value of $\frac{1}{2}h^f = 1.22$ eV is in poor agreement with the theoretical values of Tosi and Doyama (1966) (1.05 eV) and Boswarva and Lidiard (1967) (as large as 1.12 eV). The fact that the experimental value of $\frac{1}{2}h^f$ is larger than the theoretical value is in agreement with the divergence between experiment and calculations based on the usual ionic model in other alkali halides, Boswarva and Lidiard point out that the disagreements between theory and experiment are probably due to an overestimate of the polarization energy of the ions neighboring the vacancy.

In deducing $\frac{1}{2}h^f$ from their experimental data, Peterson and Rothman assumed that $h^m + h^b$ is the same in both the intrinsic and extrinsic regions, i.e., the energy required for a sodium ion-vacancy exchange plus the sodium

3. ISOTOPE EFFECTS IN DIFFUSION

ion-vacancy binding energy is the same for vacancies that are bound to polyvalent impurity ions and for free vacancies. No direct measurement is available to verify the assumption.

The reversal in sign of $\partial \ln f / \partial T^{-1}$ between the intrinsic and extrinsic regions was qualitatively explained by Peterson and Rothman. In the intrinsic region, Na^+ exchanges with the vacancy much more rapidly than do any of the Rb^+ neighbors. At 575°C, the Na^+-vacancy exchange rate w_2 is about 10 times greater than the Rb^+-vacancy exchange rate w_0 if $w_1 = w_3 = w_4 = w_0$. Also w_2 increases less rapidly with an increase in temperature than $w_0 [h^m(Na^+) < h^m(Rb^+)]$. This produces a small value of f from Eq. (34) that increases as the temperature increases.

In the extrinsic region, a number of the vacancies may be bound to polyvalent impurity ions. If the polyvalent impurity exchanges with the vacancy less frequently than does the Na^+, the correlation factor will be much smaller than in the intrinsic region (Howard, 1966). If h^m for the polyvalent impurity ions lies between h^m for Na^+ and for Rb^+, the correlation factor can be temperature independent or even increase as the temperature decreases. This is consistent with the observed temperature dependence and small value of $f \Delta K$ in the extrinsic region.

VII. Diffusion in Transition Metal Oxides

In contrast to most ionic compounds, such as alkali halides, many metal oxides exist in nonstoichiometric compositions. These oxides contain nonstoichiometric defects that vary in concentration, depending on temperature and ambient oxygen partial pressure. The NaCl-structured oxides to be considered in this section, NiO, CoO, and FeO, can be expressed in a formula $M_{1-x}O$ in which the nonstoichiometric defect has been generally recognized as a cation vacancy. To preserve electroneutrality in the crystal, an equivalent number of electron holes must be present to compensate for missing cations. The transition elements nickel, cobalt, and iron can be oxidized to 2+ or 3+ states; thus, the electron holes can be localized to the lattice cations. At high temperatures, the localized electron holes can move from one cation site to another with sufficient frequency to produce p-type semiconductivity.

If negligible association exists between charged defects, the accomodation of an excess oxygen ion or the creation of a cation vacancy in $M_{1-x}O$ can be expressed in the following reaction:

$$\tfrac{1}{2}O_2(g) + \alpha M^{2+} \rightarrow O^{2-} + V_c^{\alpha\prime} + \alpha M^{3+} \qquad (54)$$

where M^{2+} and O^{2-} are a cation and an oxygen ion on their respective lattice sites, $V_c^{\alpha\prime}$ is a cation vacancy with α negative charge relative to the lattice, and M^{3+} is a lattice cation with a localized electron hole. From Eq. (54), the concentrations of electron holes and cation vacancies are expected to be proportional to $p_{O_2}^{1/n}$ at a given temperature, where p_{O_2} is the equilibrium oxygen partial pressure, and $n = 2(\alpha + 1)$. The electrical conductivity and the cation self-diffusivity are expected to be functions of $p_{O_2}^{1/n}$ if the electrical conduction and the cation diffusion are solely by means of electron holes and cation vacancies, respectively.

The first set of experiments considered in Section VII explores the mechanism of self-diffusion and the charge on the vacancy in NiO, CoO, and FeO. Because of its large deviation from stoichiometry, FeO is discussed separately from NiO and CoO. The second group of experiments is concerned with the effect of small additions of solute (Co) on solute and cation self-diffusion in NiO. In both groups of experiments, isotope-effect measurements contributed significantly to the total understanding of the problem.

A. Cation Self-Diffusion in Transition Metal Oxides

1. *Self-Diffusion in* NiO *and* CoO

The effect of equilibrium oxygen partial pressure on cation self-diffusion has been determined in CoO by Carter and Richardson (1954) and in NiO by Volpe and Reddy (1970). The value of n in $D \propto p_{O_2}^{1/n}$ was found to 2.9, 3.3, and 3.6 at 1000, 1150, and 1350°C, respectively, for CoO, and 6.25 and 5.0 at 1245 and 1380°C, respectively, for NiO. An experimental value of n other than 4 or 6 indicates that a single-defect model, which involves only positive holes and singly or doubly charged cation vacancies, is insufficient for complete interpretation. A noninteger value of n can be understood in terms of the coexistence of more than one type of ionized cation vacancy. Since the mobility of an electron hole is large relative to that of a cation vacancy, the ionization state of a cation vacancy may change many thousands of times within the lifetime of a vacancy at a given lattice site. When more than one type of ionized cation vacancy is in equilibrium, a migrating vacancy can be treated as having an α-effective negative charge, where α need not be an integer. From this viewpoint, it is impractical to distinguish the mobilities of neutrals, or singly or doubly charged vacancies in these materials.

Recent measurements of the isotope effect for cation self-diffusion have provided additional evidence to establish the diffusion mechanism in CoO and NiO crystals. Chen *et al.* (1969) measured the simultaneous diffusion of ^{55}Co and ^{60}Co in CoO and obtained $f\Delta K = 0.58 \pm 0.01$, independent of temperature in the range of 1081 to 1411°C. They conclude that cation self-diffusion in CoO occurs by only one mechanism, the monovacancy mechanism and $\Delta K = 0.75$.

Volpe et al. (1971) determined the simultaneous diffusion of ^{57}Ni and ^{66}Ni in NiO crystals and obtained $f \Delta K = 0.61 \pm 0.02$, independent of temperature, between 1201 and 1678°C. In their earlier measurements of cation self-diffusion and electrical conductivity in NiO as a function of oxygen pressure, Volpe and Reddy (1970) reported that a small temperature dependent contribution to diffusion by an interstitial mechanism may occur along with diffusion by vacancies. The isotope-effect measurements require that cation self-diffusion in NiO occurs by only one mechanism, and the monovacancy mechanism is concluded to be the most likely mechanism with $\Delta K = 0.78$.

2. Self-Diffusion in $Fe_{1-x}O$

Wüstite has a large range of nonstoichiometry; the composition of nearest approach to stoichiometry is $Fe_{0.95}O$, and the deviation from stoichiometry may be as high as $Fe_{0.84}O$ at 1400°C (Darken and Gurry, 1945). The apparent excess in oxygen results from vacancies on cation sites with some of the tetrahedral sites occupied by cations in a virtually complete fcc oxygen sublattice. In view of this large defect concentration, a single-defect model that does not include defect clustering would appear to be inappropriate for this oxide. From neutron and X-ray diffraction studies, several models of the defect cluster have been proposed that include cation vacancies in octahedral sites and a smaller number of interstitial cations in tetrahedral sites (Roth, 1960; Koch and Cohen, 1969). In view of the significant defect clustering, a simple linear relationship between the cation self-diffusivity and the deviation from stoichiometry x in $Fe_{1-x}O$ would not be anticipated at all temperatures.

Tracer diffusion of Fe in $Fe_{1-x}O$ has been studied by a number of investigators (Carter and Richardson, 1954; Himmel et al., 1953; Desmarescaux and Lacombe, 1963; Leroy et al., 1970; Hembree and Wagner, 1969; Chen and Peterson, 1973). The results of Chen and Peterson (1973) on $Fe_{1-x}O$ single crystals and those of Himmel et al. (1953) at 983°C are shown in Fig. 12. In contrast to earlier results (Himmel et al., 1953; Desmarescaux and Lacombe, 1963; Hembree and Wagner, 1969), the studies of Chen and Peterson on single crystals show that the diffusivity decreases with an increase in the deviation from stoichiometry at 802°C, is insensitive to the O/Fe ratio at 1003°C, and increases slightly with an increase in the deviation from stoichiometry at 1200°C.

A plausible explanation of the data in Fig. 12 is that iron ions migrate by "free mobile vacancies" which are in equilibrium with defect clusters; however, the mobility or concentration of free vacancies may decrease with an increase in deviation from stoichiometry. Recent measurements of the isotope effect for the diffusion of ^{52}Fe and ^{59}Fe in $Fe_{1-x}O$

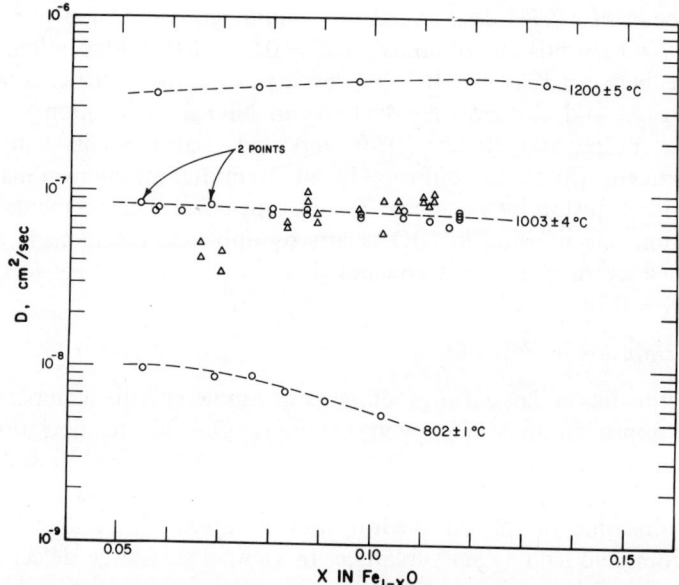

Fig. 12. Log D as a function of deviation from stoichiometry for cation self-diffusion in $Fe_{1-x}O$. $\triangle$ is ^{55}Fe at 983°C from Himmel et al. (1953). $\bigcirc$ is ^{59}Fe from Chen and Peterson (1973).

crystals (Chen and Peterson, 1973) give values of $f\,\Delta K$ that are similar to those observed for self-diffusion in CoO and NiO. This strengthens the concept that iron diffusion in $Fe_{1-x}O$ takes place by simple mobile vacancies in equilibrium with less mobile defect clusters. Recent Mössbauer studies of $Fe_{1-x}O$ (Greenwood and Howe, 1972) indicate that the average cation jump frequency decreases with an increase in deviation from stoichiometry at 800°C. However, it is premature to formulate a detailed account of the diffusion process in $Fe_{1-x}O$.

B. Cobalt Impurity Diffusion in NiO

In an attempt to study the detailed process of cation impurity diffusion and the effect of impurity-defect interactions on cation self-diffusion in oxides, Chen and Peterson (1972) have measured the isotope effect for cobalt diffusion in NiO and the dependence of cobalt and nickel diffusivities on dilute cobalt concentrations in $(Ni_{1-c}Co_c)O$ crystals. The experimental values of $(f\Delta K)_{Co}$, deduced from measurements of the simultaneous diffusion of ^{55}Co and ^{60}Co in NiO, increase from 0.45 ± 0.01 at 1179°C to 0.61 ± 0.01 at 1649°C. The correlation factor f_{Co} was determined from the product $(f\Delta K)_{Co}$ by assuming that ΔK for cobalt impurity diffusion

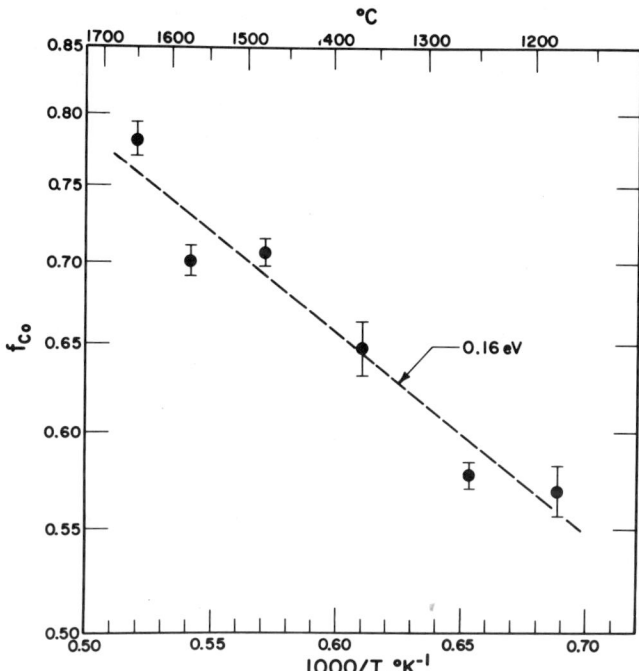

Fig. 13. Log f_{Co} versus reciprocal absolute temperature for cobalt diffusion in NiO from Chen and Peterson (1972).

is the same as that for nickel self-diffusion in NiO ($\Delta K = 0.78$), and f_{Co} is plotted as a function of temperature in Fig. 13.

The diffusion coefficients for both cobalt and nickel diffusion in $(Ni_{1-c}Co_c)O$ crystals increase linearly with cobalt concentrations (up to $C = 0.015$) in agreement with Eq. (37). The enhancement factors in Eq. (37) for $D_{Ni}(b_1)$ and $D_{Co}(b_2)$ were found to be equal within the experimental errors ($b_1 = 16.2 \pm 1.6$, $b_2 = 16.4 \pm 2.6$ at 1382°C, and $b_1 = 10.9 \pm 1.2$, $b_2 = 10.6 \pm 2.1$ at 1496°C). The equal enhancement for both solvent and solute diffusion is further demonstrated by the fact that the ratios of D_{Co}/D_{Ni}, obtained from the simultaneous diffusion of ^{60}Co and ^{57}Ni, are independent of composition at a given temperature. The ratio D_{Co}/D_{Ni} equals 2.00 ± 0.01 at 1382°C and 1.84 ± 0.01 at 1496°C.

Lidiard's theory [Eqs. (37) and (38)] has successfully explained the effect of solute additions on self-diffusion in many metallic systems; however, it does not predict the enhancement of solute diffusivity. By extending Lidiard's analysis, Miller (1969b) has shown that the association of a vacancy with two solute atoms may result in a linear enhancement for solute diffusion, and will add a quadratic term to the expression for the enhancement of the solvent diffusivity. Thus both solvent and solute diffusion may be enhanced by additions of solute, but the enhancement effect for solvent diffusion should be greater than that for solute diffusion. This prediction is contrary to the experimental results for the $(Ni_{1-c}Co_c)O$ crystals.

In addition to the effects of solute additions on self-diffusion considered by Lidiard for metallic systems, an effect in oxide systems arises from charge neutrality conditions. Since the third ionization energy for Co^{3+} is lower than that for Ni^{3+} (Johnston et al., 1959), the concentration of cation vacancies must vary at a given temperature in a dilute oxide solid solution to conserve charge neutrality. This will cause the free vacancy concentration to vary with cation composition as expressed by the relation $[V_c^{\alpha'}](c) = [V_c^{\alpha'}](o)(1 + b_V c)$. The combined effect of the vacancy–solute interactions given by Lidiard and the change in free vacancy concentration results in the enhancement factors $b_1 = b + b_V$ and $b_2 = b_V$ for the diffusion of solvent and solute, respectively. Since the experimental results show $b_1 = b_2$, the enhancement of both cobalt and nickel diffusion is primarily caused by the increased concentration of cation vacancies; the vacancy–solute interactions and their contribution to the enhancement of solvent diffusion are negligible ($b \approx 0$).

These conclusions suggest that $\Delta g \ll kT$ and $w_0 = w_1 = w_3 = w_4 \neq w_2$.

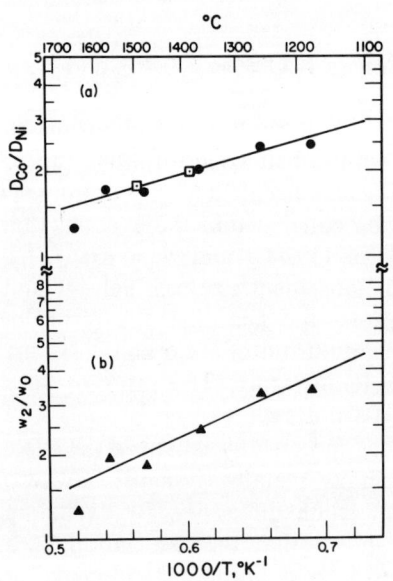

Fig. 14. Temperature dependence of (a) D_{Co}/D_{Ni} (●) and (b) w_2/w_0 (▲) as calculated from measured values of f_{Co}. Experimental data from the simultaneous diffusion of ^{60}Co and ^{57}Ni (□) are also shown. [Reprinted with permission from Chen and Peterson, J. Phys. Chem. Solids (1972), Pergamon Press.]

Using these conditions, w_2/w_0 and D_{Co}/D_{Ni} in pure NiO were calculated as a function of temperature from only the measured values of f_{Co} using Eqs. (34) and (35). The calculated values of w_2/w_0 and D_{Co}/D_{Ni} are plotted versus temperature in Fig. 14. The energy term in the temperature dependence of D_{Co}/D_{Ni} is $\Delta Q = Q_2 - Q_1$ where Q_2 and Q_1 are the activation energies for cobalt-tracer diffusion and nickel-tracer diffusion in NiO, respectively. The direct experimental values of Q_2 (Chen and Peterson, 1972) and Q_1 (Volpe and Reddy, 1970) give $\Delta Q = -0.29$ eV in good agreement with -0.274 from the line in Fig. 14. Also, the experimental values of D_{Co}/D_{Ni} determined from the simultaneous diffusion of ^{60}Co and ^{57}Ni in NiO at 1382 and 1496°C are in good agreement with the line in Fig. 14 deduced only from values of f_{Co}. Thus, a self-consistent view of cobalt diffusion in NiO can be developed with $w_0 = w_1 = w_3 = w_4 \neq w_2$ and a solute–vacancy interaction energy $\Delta g \ll kT$.

VIII. Correlation Effects in Grain Boundary Diffusion

A correlation factor can be defined for tracer diffusion along a dislocation or grain boundary just as it can be defined for diffusion through a lattice. However, since the regularity of the lattice is perturbed in the vicinity of the dislocation or grain boundary, the actual calculation of the correlation factor is more difficult for grain boundary diffusion. Nevertheless, Robinson and Peterson (1972) have measured the isotope effect for silver diffusion along silver grain boundaries and have obtained a correlation factor that provides some interesting information about the diffusion process along grain boundaries.

Mathematical solutions of the grain boundary diffusion equation have been obtained for two sets of boundary conditions, a constant concentration source, and an instantaneous source of tracer. The Whipple constant concentration solution (Whipple, 1954; LeClaire, 1963) and the Suzuoka instantaneous source solution (Suzuoka, 1961, 1964) are

Suzuoka: $\quad D'\delta = (\partial \ln c/\partial Y^{6/5})^{-5/3} (4D/t)^{1/2} (0.72\beta^{0.008})^{5/3}$ \quad (55)

Whipple: $\quad D'\delta = (\partial \ln c/\partial Y^{6/5})^{-5/3} (4D/t)^{1/2} (0.78)^{5/3}$ \quad (56)

where D' is the grain boundary diffusivity, D is the lattice diffusivity, δ is the grain boundary width, t is the annealing time, c is the specific activity in a section whose center is Y centimeters from the original surface, and $\beta = D'\delta/2D(Dt)^{1/2}$. Both solutions give equivalent results when β is large.

For the simultaneous diffusion of two isotopes α and β, Eqs. (55) and (56) may be rearranged to give

$$\partial \ln(c_\alpha/c_\beta)/\partial \ln c_\beta = [1 - (D_\alpha/D_\beta)^{0.286} (D_\beta'/D_\alpha')^{0.592}] \qquad (57)$$

from the Suzuoka solution, and

$$\partial \ln(c_\alpha/c_\beta)/\partial \ln c_\beta = [1 - (D_\alpha/D_\beta)^{0.300} (D_\beta'/D_\alpha')^{0.600}] \qquad (58)$$

from the Whipple solution. The grain boundary width δ has been assumed to be independent of isotopic mass in deriving Eqs. (57) and (58). If D_α/D_β is known from lattice diffusion measurements, D_β'/D_α' is easily obtained from experimental measurements of $\partial \ln(c_\alpha/c_\beta)/\partial \ln c_\beta$ using either Eq. (57) or (58). The value of D_β'/D_α' from the Whipple and Suzuoka solutions are generally the same to within several hundredths of a percent. The effects of the differences in the exponents for the two solutions tend to cancel, and the differences in D_β'/D_α' that do exist are well within experimental error. It is interesting to note that negative values of $\partial \ln(c_\alpha/c_\beta)/\partial \ln c_\beta$ are possible for grain boundary diffusion even though D_α/D_β and D_α'/D_β' are greater than unity. That is, the heavier tracer may appear to diffuse faster than the lighter tracer under certain circumstances.

An exact expression that relates D_β'/D_α' to the correlation factor for grain boundary diffusion is not known. Robinson and Peterson assume that an equation similar to Eq. (13) is valid

$$f' \Delta K' = [(D_\beta'/D_\alpha') - 1]/[(m_\alpha/m_\beta)^{1/2} - 1] \qquad (59)$$

where f' is the correlation factor for grain boundary diffusion, and $\Delta K'$ has the same definition as for lattice diffusion. It should be noted that Eq. (59) may not be precisely valid because the symmetry conditions required for the derivation of Eq. (13) are not fulfilled for grain boundary diffusion.

Robinson and Peterson (1972) measured the isotope effect for silver diffusion as a function of temperature along random grain boundaries and along the boundary of a 16° misorientation symmetric $\langle 100 \rangle$ tilt bicrystal. The values of $f' \Delta K'$ for the random grain boundaries are shown as a function of temperature in Fig. 15. The experiments on the bicrystals gave the same results within experimental error. The ratio of lattice diffusivities D_α/D_β was assumed to be independent of temperature and equal to the value measured by Rothman *et al.* (1970) at 640°C.

Robinson and Peterson developed a model for diffusion along dislocations by a vacancy mechanism. Six jump frequencies different from w_0 are defined for vacancy jumps in the compression region of an edge dislocation. The correlation factor for diffusion along a dislocation was then solved in terms of ratios of jump frequencies using the method

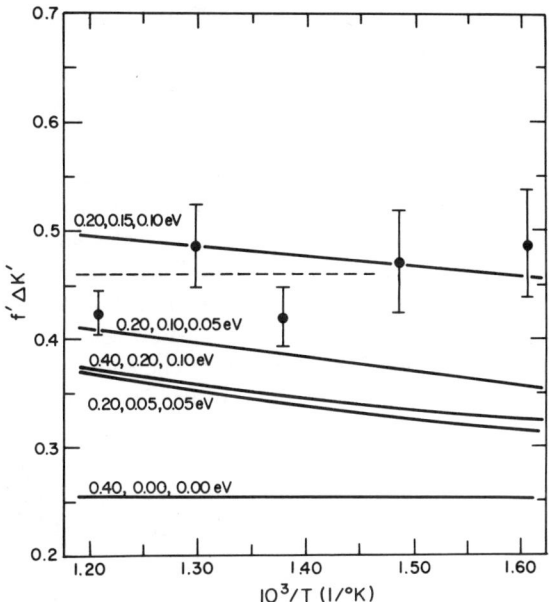

Fig. 15. Temperature dependence of the isotope effect for grain boundary self-diffusion in silver from the work of Robinson and Peterson (1972). The points with error bars are experimental data; - - - represents the average value of $f' \Delta K'$; the solid curves represent theoretical values of $f' \Delta K'$ for various binding conditions; and $\Delta K' = 0.75$. [Reprinted with permission from Robinson and Peterson, *J. Phys. Chem. Solids* (1972), Pergamon Press.]

of Howard (1966), which is applicable to atomic jumps that are along directions of low crystal symmetry (Robinson and Peterson, 1973). The jump frequency ratios used in the calculation of f' involved ratios of vacancy jumps toward the dislocation versus away from the dislocation or toward the dislocation versus along the dislocation. The vacancy dislocation binding energy differs from site to site around the dislocation and allows a numerical evaluation of the jump frequency ratios and of f'. Assuming a constant entropy factor and migration energy, the jump frequency ratio w_i/w_j may be written as

$$w_i/w_j = \exp(-(\Delta E_i - \Delta E_j)/kT) \tag{60}$$

where ΔE_i is the change in vacancy energy due to a jump of the type i. One may then assume a value of the vacancy dislocation binding energy as a function of distance from the dislocation and solve for f' as a function of temperature. Figure 15 shows $f' \Delta K'$ as a function of temperature for various binding conditions; $\Delta K'$ was assumed equal to 0.75 solely for the graphical representation in Fig. 15. The binding condition

0.2, 0.1, and 0.05 eV was taken from the calculation of Doyama and Cotterill (1968). This corresponds to a vacancy dislocation attraction of 0.2 eV when the vacancy is on the dislocation core, 0.1 eV when the vacancy is one jump distance from the core, and 0.05 eV when the vacancy is two jump distances from the core. The other vacancy dislocation binding conditions were chosen to represent the tight and loose binding one might expect when diffusion occurs predominantly along the dislocation core or by a nonlocalized vacancy mechanism.

The binding condition 0.40, 0.00, 0.00 eV is unacceptable because it requires the nonphysical condition $\Delta K' > 1$. Thus, dislocation pipe diffusion does not occur predominantly by atomic motion along a single row of atoms; three-dimensional motion is required. The calculations do not rule out the case of a constant vacancy dislocation binding energy at all sites. This condition gives $f' = 0.82$ and $\Delta K' = 0.56$. Whereas these results give a physically acceptable $\Delta K'$, the assumption of uniform binding is not reasonable. The results of Doyama and Cotterill (1968) demand a vacancy dislocation binding energy that decreases fairly rapidly with distance from the dislocation.

The binding conditions 0.4, 0.2, 0.1 eV and 0.20, 0.05, 0.05 eV require values of $\Delta K'$ greater than unity at low temperatures to match the calculated values of f' to the experimental values of $f'\Delta K'$. The binding conditions 0.20, 0.10, 0.05 eV and 0.20, 0.15, 0.10 eV both give physically acceptable values for $\Delta K'$; the former binding condition gives values of $\Delta K'$ between 0.85 and 0.97, the latter binding condition gives values of $\Delta K'$ between 0.70 and 0.76. It may be argued that $\Delta K'$ should be somewhat less than ΔK. For this reason, Robinson and Peterson (1973) state a preference for the binding condition 0.20, 0.15, 0.10 eV.

The Robinson and Peterson calculation of f' is limited in the number of different vacancy jumps. The binding condition 0.60, 0.55, 0.50 eV gives essentially the same value of f' as the condition 0.20, 0.15, 0.10 eV. Thus, the measurements and model of Robinson and Peterson require a vacancy dislocation binding energy that decreases with distance from the dislocation at the rate of 0.05 to 0.10 eV per nearest neighbor spacing; however, the results do not specify the absolute value of the binding energy at any given plane.

Robinson and Peterson (1972) also present a value of $f'\Delta K'$ for silver diffusion in grain boundaries of copper polycrystals (0.18 ± 0.07) that is considerably lower than for silver diffusion in silver polycrystals (Fig. 15). This result is reasonable. Since the solubility of silver in the copper lattice is small, silver tracer atoms in copper will tend to segregate to the grain boundary and diffusion will be more correlated.

In terms of the grain boundary diffusion model presented above, this means that silver atom jumps away from the boundary are much more difficult than silver atom jumps toward or along the boundary in copper.

ACKNOWLEDGMENTS

The author wishes to thank B. N. N. Achar, D. R. Campbell, W. K. Chen, N. Q. Lam, H. Mehrer, J. W. Miller, J. N. Mundy, J. T. Robinson, and S. J. Rothman for providing data or copies of their manuscripts before publication, and M. F. Adams for editorial assistance in preparing this article. This work was performed under the auspices of the U.S. Atomic Energy Commission.

REFERENCES

ABBINK, H. C., AND MARTIN, D. S. (1966). *J. Phys. Chem. Solids* **27**, 205.
ACHAR, B. N. N. (1970). *Phys. Rev.* **B2**, 3848.
ACHAR, B. N. N., to be published.
BAKKER, H. (1969). *Phys. Status Solidi* **31**, 271.
BAKKER, H. (1971). *Phys. Status Solidi* **44**, 369.
BARR, L. W., AND LECLAIRE, A. D. (1964). *Proc. Brit. Ceram. Soc.* **1**, 109.
BARR, L. W., AND MUNDY, J. N. (1965). "Diffusion in bcc Metals," p. 171. Amer. Soc. Metals, Metals Park, Ohio.
BARR, L. W., MUNDY, J. N., AND SMITH, F. A. (1967). *Phil. Mag.* **16**, 1139.
BARR, L. W., MUNDY, J. N., AND ROWE, A. H. (1972). "Amorphous Materials," p. 243. Wiley, New York.
BATRA, A. P. (1967). *Phys. Rev.* **159**, 487.
BEELER, J. R., AND DELANEY, J. A. (1963). *Phys. Rev.* **130**, 962.
BENIERE, F., BENIERE, M., AND CHEMLA, M. (1970). *J. Phys. Chem. Solids* **31**, 1205.
BENNETT, C., AND ALDER, B. (1971). *J. Phys. Chem. Solids* **32**, 2111.
BEYELER, M., AND ADDA, Y. (1968). *J. Phys.* **29**, 345.
BONANNO, F. R., AND TOMIZUKA, C. T. (1965). *Phys. Rev.* **137**, A1264.
BOSWARVA, I. M., AND LIDIARD, A. B. (1967). *Phil. Mag.* **16**, 805.
BOWDEN, H. G., AND BALLUFFI, R. W. (1969). *Phil. Mag.* **19**, 1001.
BROWN, R. C., WORSTER, J., MARCH, N. H., PERRIN, R. C., AND BULLOUGH, R. (1971a). *Z. Naturforsch.* **26a**, 77.
BROWN, R. C., WORSTER, J., MARCH, N. H., PERRIN, R. C., AND BULLOUGH, R. (1971b). *Phil. Mag.* **23**, 555.
BURTON, J. J., AND FROOZAN, F. (1973). *Phil. Mag.* **27**, 473.
CAMPBELL, D. R. (1973). Private communication.
CARTER, R. E., AND RICHARDSON, F. D. (1954). *J. Metals* **6**, 1244.
CHADWICK, A. V., AND SHERWOOD, J. N. (1972). *J. Chem. Soc., Faraday Trans.* **1**(68), Pt. 1, 47.
CHEN, W. K., AND PETERSON, N. L. (1972). *J. Phys. Chem. Solids* **33**, 881.
CHEN, W. K., AND PETERSON, N. L. (1973). *J. Phys. (Paris)* **34**, C9-303. A less accurate measurement of $f \Delta K$ for cation diffusion in $Fe_{1-x}O$ may be found in Leroy *et al.* (1970).

Ann. Chim. **5**, 275. A more extensive account of the work of Chen and Peterson will be published in 1974.
CHEN, W. K., PETERSON, N. L., AND REEVES, W. T. (1969). *Phys. Rev.* **186**, 887.
CHHABILDAS, L. C., AND GILDER, H. M. (1972). *Phys. Rev.* **B5**, 2135.
COLEMAN, M. G., WERT, C., AND PEART, R. F. (1968). *Phys. Rev.* **175**, 788.
COMPAAN, K., AND HAVEN, Y. (1956). *Trans. Faraday Soc.* **52**, 786.
COMPAAN, K., AND HAVEN, Y. (1958). *Trans. Faraday Soc.* **54**, 1498.
DARKEN, L. S., AND GURRY, R. W. (1945). *J. Amer. Chem. Soc.* **67**, 1398.
DEGONZALEZ, C. O., AND DERECA, N. E. W. (1971). *J. Phys. Chem. Solids* **32**, 1067.
DESMARESCAUX, P., AND LACOMBE, P. (1963). *Mem. Sci. Rev. Met.* **60**, 899.
DOWNING, H. L., Jr., AND FRIAUF, R. J. (1970). *J. Phys. Chem. Solids* **31**, 845.
DOYAMA, M., AND COTTERILL, R. M. J. (1968). *Suppl. Jap. Inst. Metals* **9**, 55.
DOYAMA, M., AND KOEHLER, J. S. (1962). *Phys. Rev.* **127**, 21.
DYSON, B. F., ANTHONY, T., AND TURNBULL, D. (1966). *J. Appl. Phys.* **37**, 2370.
EBERT, I., AND TELTOW, J. (1955). *Ann. Phys.* **15**, 268.
EBISUZAKI, Y., KASS, W. J., AND O'KEEFFE, M. (1967a). *Phil. Mag.* **15**, 1071.
EBISUZAKI, Y., KASS, W. J., AND O'KEEFFE, M. (1967b). *J. Chem. Phys.* **46**, 1373.
ELCOCK, E. W. (1959). *Proc. Phys. Soc.* (London) **73**, 250.
ELCOCK, E. W., AND MCCOMBIE, C. W. (1958). *Phys. Rev.* **109**, 605.
EINSTEIN, A. (1905). *Ann. Phys.* **17**, 549.
FEIT, M. D. (1972a). *Phil. Mag.* **25**, 769.
FEIT, M. D. (1972b). *Phys. Rev.* **B5**, 2145.
FISHMAN, S. G., GUPTA, D., AND LIEBERMAN, D. S. (1970). *Phys. Rev.* **B2**, 1451.
FOX, R., AND SHERWOOD, J. N. (1971). *Trans. Faraday Soc.* **67**, 3364.
FRANKLIN, W. M. (1972). *J. Chem. Phys.* **57**, 2659.
FRIAUF, R. J. (1957). *Phys. Rev.* **105**, 843.
GHATE, P. B. (1964). *Phys. Rev.* **133**, A1167.
GRACEY, J. P., AND FRIAUF, R. J. (1969). *J. Phys. Chem. Solids* **30**, 421.
GRAHAM, D. (1969). *J. Appl. Phys.* **40**, 2386.
GREENWOOD, N. N., AND HOWE, A. T. (1972). *J. Chem. Soc.* (London) *Dalton Trans.*, p. 122.
HEMBREE, P., AND WAGNER, J. B., Jr. (1969). *Trans. Met. Soc. AIME* **245**, 1547.
HERZIG, CH., HEUMANN, TH., AND WOLTER, D. (1971). *Z. Naturforsch.* **26a**, 1477. These authors obtain a larger value of $f\Delta K$; however, the results of Miller *et al.* (1973) are preferred by this author.
HEUMANN, TH., AND IMM, R. (1968). *J. Phys. Chem. Solids* **29**, 1613.
HEUMANN, TH., AND REERINK, W. (1966). *Acta Met.* **14**, 201.
HIMMEL, L., MEHL, R. F., AND BIRCHENALL, C. E. (1953). *Trans. AIME* **197**, 827.
HO, P. S. (1972). Materials Science Center, Cornell Univ. Rep. Nos. 1748 and 1823.
HOFFMAN, R. E., AND TURNBULL, D. (1951). *J. Appl. Phys.* **22**, 634.
HOWARD, R. E. (1966). *Phys. Rev.* **144**, 650.
HOWARD, R. E., AND MANNING, J. R. (1967). *Phys. Rev.* **154**, 561.
HUNTINGTON, H. B., FEIT, M. D., AND LORTZ, D. (1970). *Cryst. Lattice Defects* **1**, 193.
HUNTINGTON, H. B., AND GHATE, P. B. (1962). *Phys. Rev. Lett.* **8**, 421.
IRMER, V., AND FELLER-KNIEPMEIER, M. (1972). *J. Appl. Phys.* **43**, 953.
JOHNSTON, W. D., MILLER, R. C., AND MAZELSKY, R. (1959). *J. Amer. Chem. Soc.* **63**, 198.
KOCH, F., AND COHEN, J. B. (1969). *Acta Crystallog.* **B25**, 275.
KUNZ, P. (1971). Diplomarbeit, Universitat Stuttgart, Germany.
KUPER, A., LETOW, H., Jr., SLIFKIN, L., SONDER, E., AND TOMIZUKA, C. T. (1954). *Phys. Rev.* **96**, 1224.

Kurnik, S. W. (1952). *J. Chem. Phys.* **20**, 218.
Lam, N. Q., Rothman, S. J., and Nowicki, L. J. (1972). *Bull. Am. Phys. Soc.* **17**, 244.
Lam, N. Q., Rothman, S. J., Mehrer, H., and Nowicki, L. J. (1973). *Phys. Status Solidi* **57**, 225.
Lam, N. Q., Rothman, S. J., and Nowicki, L. J. (1974). *Phys. Status Solidi*, **63**, K29.
LeClaire, A. D. (1962). *Phil. Mag.* **7**, 141.
LeClaire, A. D. (1963). *Brit. J. Appl. Phys.* **14**, 351.
LeClaire, A. D. (1966). *Phil. Mag.* **14**, 1271.
LeClaire, A. D. (1967). *Acta Met.* **15**, 573.
LeClaire, A. D. (1970). *In* "Physical Chemistry" (H. Eyring, D. Henderson, and W. Jost, eds.), Vol. 10, Chapter 5. Academic Press, New York.
LeClaire, A. D. (1971). *In* "Atomic Transport in Solids and Liquids" (A. Lodding and T. Lagerwall, eds.), p. 265. Verlag der Zeitschrift für Naturforschung, Tübingen.
Leroy, B., Beranger, G., and Lacombe, P. (1970). *Ann. Chim.* **5**, 275.
Lidiard, A. B. (1960). *Phil. Mag.* **5**, 1171.
Maier, K., Bassani, C., and Schüle, W. (1973). *Phys. Lett.* **44A**, 539.
Manning, J. R. (1964). *Phys. Rev.* **136**, A1758.
Manning, J. R. (1967). *Acta Met.* **15**, 817.
Manning, J. R. (1968). "Diffusion Kinetics for Atoms in Crystals." Van Nostrand-Reinhold, Princeton, New Jersey.
Manning, J. R. (1971). *Phys. Rev.* **B4**, 1111.
Mao, C. W. (1972). *Phys. Rev.* **B5**, 4693.
McArdle, P. B. (1968). *Bull. Am. Phys. Soc.* **13**, 489. (1969). Ph.D. Thesis, Univ. of Arizona, Tucson, Arizona.
Mehrer, H. (1972). *J. Phys. F* **2**, L11.
Mehrer, H. (1973). *J. Phys. F* **3**, 543.
Mehrer, H., to be published.
Mehrer, H., and Seeger, A. (1969). *Phys. Status Solidi* **35**, 313.
Mehrer, H., and Seeger, A. (1970). *Phys. Status Solidi* **39**, 647.
Mehrer, H., and Seeger, A. (1972). *Cryst. Lattice Defects* **3**, 1.
Mercer, W. L. (1955). Ph.D. Thesis, Leeds Univ., England.
Miller, A. S., and Maurer, R. J. (1958). *J. Phys. Chem. Solids* **4**, 196.
Miller, J. W. (1969a). *Phys. Rev.* **181**, 1095.
Miller, J. W. (1969b). *Phys. Rev.* **188**, 1074.
Miller, J. W. (1970). *Phys. Rev.* **B2**, 1624.
Miller, J. W., and Edelstein, W. A. (1969). *Phys. Rev.* **188**, 1081.
Miller, J. W., Mundy, J. N., Robinson, L. C., and Loess, R. E. (1973). *Phys. Rev.* **B8**, 2411.
Montet, G. L. (1973). *Phys. Rev.* **B7**, 650.
Mullen, J. G. (1961a). *Phys. Rev.* **121**, 1649.
Mullen, J. G. (1961b). *Phys. Rev.* **124**, 1723.
Mundy, J. N. (1971). *Phys. Rev.* **B3**, 2431.
Mundy, J. N., and McFall, W. D. (1973). *Phys. Rev.* **B7**, 4363.
Mundy, J. N., Barr, L. W., and Smith, F. A. (1966). *Phil. Mag.* **14**, 785.
Nachtrieb, N. H., Weil, J. A., Catalano, E., and Lawson, A. W. (1952). *J. Chem. Phys.* **20**, 1189.
Nelson, V. C., and Friauf, R. J. (1970). *J. Phys. Chem. Solids* **31**, 825.
Peterson, N. L. (1964). *Phys. Rev.* **136**, A568.
Peterson, N. L. (1973). "Diffusion." Amer. Soc. Metals, Metals Park, Ohio. Gives a review of the established mechanisms for self-diffusion in pure metals.
Peterson, N. L., and Rothman, S. J. (1967a). *Phys. Rev.* **163**, 645.

PETERSON, N. L., AND ROTHMAN, S. J. (1967b). *Phys. Rev.* **154**, 558.
PETERSON, N. L., AND ROTHMAN, S. J. (1969). *Phys. Rev.* **177**, 1329.
PETERSON, N. L., AND ROTHMAN, S. J. (1970). *Phys. Rev.* **B2**, 1540.
PETERSON, N. L., BARR, L. W., AND LECLAIRE, A. D. (1973). *J. Phys. C* **6**, 2020.
QUÉRÉ, Y. (1960). *C. R. Acad. Sci.* (France) **251**, 367.
REIMERS, P., AND BARTDORFF, D. (1972). *Phys. Status Solidi* **50**, 305.
ROBERTS-AUSTEN, W. C. (1896a). *Phil. Trans. Roy. Soc.* **187**, 404.
ROBERTS-AUSTEN, W. C. (1896b). *Proc. Roy. Soc.* **59**, 281.
ROBINSON, J. T., AND PETERSON, N. L. (1972). *Surface Sci.* **31**, 586.
ROBINSON, J. T., AND PETERSON, N. L. (1973). *Acta Met.* **21**, 1181.
ROTH, W. L. (1960). *Acta Crystallog.* **13**, 140.
ROTHMAN, S. J., AND PETERSON, N. L. (1967). *Phys. Rev.* **154**, 552.
ROTHMAN, S. J., AND PETERSON, N. L. (1969). *Phys. Status Solidi* **35**, 305.
ROTHMAN, S. J., PETERSON, N. L., AND ROBINSON, J. T. (1970). *Phys. Status Solidi* **39**, 635.
ROTHMAN, S. J., PETERSON, N. L., LASKAR, A. L., AND ROBINSON, L. C. (1972). *J. Phys. Chem. Solids* **33**, 1061.
SCHOEN, A. H. (1958). *Phys. Rev. Lett.* **1**, 138.
SEEGER, A. (1970). Private communication.
SEEGER, A., AND CHIK, K. P. (1968). *Phys. Status Solidi* **29**, 455.
SEEGER, A., AND MEHRER, H. (1970). *In* "Vacancies and Interstitials in Metals" (A. Seeger, D. Schumacher, W. Schilling, and J. Diehl, eds.), p. 1. North-Holland Publ., Amsterdam.
SHELBY, J. E. (1971). *Phys. Rev.* **B4**, 2681.
SIMMONS, R. O., AND BALLUFFI, R. W. (1960). *Phys. Rev.* **119**, 600.
SIMMONS, R. O., AND BALLUFFI, R. W. (1963). *Phys. Rev.* **129**, 1533.
SLIFKIN, L., LAZARUS, D., AND TOMIZUKA, C. T. (1952). *J. Appl. Phys.* **23**, 1032.
SUZUOKA, T. (1961). *Trans. Jap. Inst. Metals* **2**, 25.
SUZUOKA, T. (1964). *J. Phys. Soc.* (Japan) **19**, 839.
TELTOW, J. (1949). *Ann. Phys.* **5**, 63, 71.
TEWORDT, L. (1958). *Phys. Rev.* **109**, 61.
THARMALINGAM, K., AND LIDIARD, A. B. (1959). *Phil. Mag.* **4**, 899.
THARMALINGAM, K., AND LIDIARD, A. B. (1961). *Phil. Mag.* **6**, 1157.
TOMIZUKA, C. T. (1961). "Progress in Very High Pressure Research," p. 266. Wiley, New York.
TOMIZUKA, C. T., AND SONDER, E. (1956). *Phys. Rev.* **103**, 1182.
TOSI, M. P., AND DOYAMA, M. (1966). *Phys. Rev.* **151**, 642.
TURNER, T. J., PAINTER, S., AND NIELSON, C. H. (1972). *Solid State Commun.* **11**, 577.
VINEYARD, G. H. (1957). *J. Phys. Chem. Solids* **3**, 121.
VOLPE, M. L., AND REDDY, J. (1970). *J. Chem. Phys.* **53**, 1117.
VOLPE, M. L., PETERSON, N. L., AND REDDY, J. (1971). *Phys. Rev.* **B3**, 1417.
WALTER, C. M., AND PETERSON, N. L. (1969). *Phys. Rev.* **178**, 922.
WEBER, M. D., AND FRIAUF, R. J. (1969). *J. Phys. Chem. Solids* **30**, 407.
WERT, C. (1950). *Phys. Rev.* **79**, 601.
WHIPPLE, R. T. (1954). *Phil. Mag.* **45**, 1225.
WYNBLATT, P. (1971). *J. Phys. Chem. Solids* **32**, 15.

4

Fast Diffusion in Metals

W. K. WARBURTON AND D. TURNBULL

DIVISION OF ENGINEERING AND APPLIED PHYSICS
HARVARD UNIVERSITY
CAMBRIDGE, MASSACHUSETTS

I. Introduction	172
II. Experience on Fast Diffusion	174
A. Semiconductor Hosts	174
B. Groups III and IV Metal Hosts	177
C. Alkali Metal Hosts	191
D. Lanthanide and Actinide Hosts	193
E. Occurrence of Fast Diffusion	193
III. Corroboration of Existence of Interstitial-Type Defects	196
A. Liquid State Properties of Fast Diffusing Systems	196
B. Electron Channeling Experiments	198
C. Mössbauer Data	199
D. Supersaturated Alloys	199
E. Centrifuge Experiments	199
F. Precipitation Experiments	200
G. Resistivity Measurements	201
H. Internal Friction Measurements	202
IV. Fast Diffusion Mechanisms	206
A. Introduction	206
B. The Vacancy Mechanism	207
C. Interstitial-Type Mechanisms	211
D. Interstitial–Vacancy Pairs	216
E. Choosing among the Mechanisms	219
V. Interpretation of Fast Diffusion Behavior of Particular Systems	221
A. Pb(Au)	221
B. Polyvalent Solvent (Noble Metal) Alloys	223
C. Pb(Cd) and Pb(Hg)	225
References	226

I. Introduction

It is well established that self-diffusion in most face-centered cubic metals, as well as in many body-centered cubic metals, occurs predominantly by vacancy mechanisms. The transport of substitutional impurities in these same hosts also appears to be effected principally by vacancies (Peterson, 1968). At high dilutions, the impurity diffusivities in these systems rarely differ from the self-diffusivities by factors greater than 10–20. This rather small diffusivity dispersion probably reflects, primarily, the high efficiency of screening of point charges in metals, which limits the impurity–vacancy binding energies to values which are small relative to the vacancy formation energies.

In contrast with the substitutional behavior, the diffusivities of impurities having a substantial interstitial component in the metal host are often many orders of magnitude greater than the host self-diffusivities. At thermal equilibrium, any impurity will be distributed over both interstitial and substitutional sites of the host. Assuming equivalence of interstitial sites, as well as of substitutional sites, the equilibrium may be represented by the equation

$$s \rightleftarrows i + v \tag{1}$$

where s, i, and v represent, respectively, substitutional, interstitial, and vacancy, and the corresponding atom fractions are denoted by x_s, x_i, and x_v. Provided defects are in equilibrium locally and in the dilute solution regime, the concentrations must fulfill the condition

$$x_i x_v / x_s = K(P, T) = \exp[-(g_v^0 + g_i^0)/kT] \tag{2}$$

where g_v^0 and g_i^0 are the *standard* Gibbs free energies of forming, respectively, a vacancy and an interstitial from a substitutional impurity atom. If, in addition, the vacancies are in equilibrium with the external surface of the specimen, we have $x_v = x_v'$, $x_i = x_i'$, $x_s = x_s'$, and

$$x_i'/x_s' = K(P, T)/x_v' = \exp(-g_i^0/kT) \tag{3}$$

where the primes label equilibrium concentrations subject to the condition that the vacancy chemical potential is fixed at its external surface value. These concentrations are designated the free surface equilibrium (fse) values.

In those cases where the system is only in local equilibrium, the distribution of impurity between interstitial and substitutional sites can be conveniently expressed, by combining Eqs. (2) and (3), as

$$x_i/x_s = [x_v'/(x_v' + \Delta x_v)] x_i'/x_s' \tag{4}$$

where Δx_v is the departure of the vacancy-atom fraction from its fse value x_v'. Thus, the actual distribution ratio can greatly exceed the fse ratio x_i'/x_s', when $\Delta x_v < 0$, as in a perfect crystal. In contrast, the ratio would be suppressed to below x_i'/x_s' where there is a supersaturation of vacancies, i.e., $\Delta x_v > 0$.

The equilibrium distribution x_i'/x_s' between interstitial and substitutional sites had been thought to correlate well with the ratio of atom sizes λ_2/λ_1 of the impurity to the host, where the atom size may be taken to be, e.g., the Wigner–Seitz diameter of the particular element in its own metal structure. More particularly, according to Hägg's (1929, 1930) empirical rule, x_i' is substantial only in systems where $\lambda_2/\lambda_1 \lesssim 0.59$. Qualitatively, this rule is justified on the supposition that the repulsive part of the interatomic pair potential rises steeply with decreasing interatomic separation r_{12} in the range less than the average Wigner–Seitz diameter λ_{12}. Thus, interstitial solution would be essentially limited to impurities with atom sizes not greatly exceeding the size of the holes delineated by the Wigner–Seitz diameter in the host structure.

However, a number of systems have been recognized (see earlier reviews by Peterson, 1968; Anthony, 1970a; Miller, 1971) in which the impurity diffusivity is several orders of magnitude greater than that of the host, even though the atom size ratio greatly exceeds the Hägg limit and sometimes reaches values of ~ 0.85, considered favorable for substantial substitutional solution. This "fast diffusion" behavior is thought to be exceptional because the diffusivities lie far above the range considered normal for vacancy controlled mechanisms, while the relatively large atom size ratio would seem to preclude interstitial occupancies sufficient to contribute significantly to the atomic mobilities. The earliest recognized and most widely studied system which exhibits fast diffusion is the solution of gold in lead; indeed, this system has become the textbook example for showing that diffusion in the solid state really occurs. The atom size ratio $\lambda_{Au}/\lambda_{Pb} = 0.83$, but several experimental investigations, beginning with that of Roberts-Austen (1896, 1900), generally agree that the diffusivity of Au in Pb is, at all temperatures, many orders of magnitude greater than the lead self-diffusivity; e.g., at 175°C, $D_{Au-Pb}/D_{Pb-Pb} \sim 10^5$. That this extraordinarily fast diffusion is an intrinsic or bulk property was shown by experiments which indicated only negligible short-circuiting by grain boundaries (Ascoli, 1960) and by dislocations (Kidson, 1966).

Our main object in this paper is to survey the fast diffusion phenomena in metal systems and their interpretation. As we shall see, the diffusivities indicate that, in some systems at least, transport is effected principally by the motion of interstitial-type defects. Therefore, we also will review some studies, besides those on diffusivity, which have provided important information on the existence and nature of these defects.

Various aspects of fast diffusion in metals have been reviewed earlier by Peterson (1968), Turnbull (1966), Anthony (1970a), and Miller (1971). More particularly, Anthony reviewed the developments in the field through 1968 quite comprehensively. The present article overlaps Anthony's somewhat but its main emphasis is on the more recent developments.

For a helpful perspective, we shall begin with a discussion of the parallel and well-documented fast diffusion phenomena in semiconductors. The concepts growing out of these studies, in particular the dissociation model of diffusion, have strongly influenced the interpretation of fast diffusion behavior in metallic systems. From a chemical viewpoint, this behavior is quite similar to that in metals in that it occurs in hosts which are group IV or compound semiconductors, and the fast diffusing impurities are mostly noble or transition atoms. While the total content of these impurities in semiconductors rarely exceeds atom fractions of 10^{-7}, the results indicate a more or less equal, within two orders of magnitude, distribution of impurities between interstitial and substitutional sites. These rather substantial interstitial components have been attributed to the open structure of the hosts rather than to any special interaction between host and impurity atoms.

II. Experience on Fast Diffusion

A. Semiconductor Hosts

The phenomenon of fast diffusion in germanium and in silicon hosts was reviewed in 1959 by Reiss and Fuller (1959). More recently, Hall and Racette (1964) surveyed much of the earlier literature on the fast diffusion of Cu in Ge and Si and in various compound semiconductors, including GaAs, InAs, InSb, AlSb, CdS, PbS, Bi_2Te_3, and Ag_2Se. Among the fast diffusing impurities in Ge and Si are noble, transition, and Li metal atoms. The most recent review is by Seeger and Chik (1968).

1. *Dissociative Mechanism*

Fast diffusion in semiconductors seems to be fairly well typified by the extensively studied behavior of Cu in Ge (Woodbury and Tyler, 1957; Letaw *et al.*, 1956; Millea, 1966). At 700°, the diffusivity of Cu is more than 10^9 times the Ge self-diffusivity. When Ge crystals with a very high degree of perfection became available, it was found that the acceptor Cu diffusivity decreased markedly with decreasing dislocation content of the Ge. Also, it was noted that in the early stages of diffusion the *p*-type copper had penetrated the Ge deeply only in regions surrounding dislocations

4. FAST DIFFUSION IN METALS

and that the dimension of the radial diffusion zone around the dislocation center was, along the entire length of the dislocation, about the same as the depth of penetration δ of Cu from the free surface into a perfect section of the crystal. Even dislocations remote from, and not apparently connecting with, the free surface from which the Cu diffused were encased by a radial diffusion zone having this same radius δ.

Frank and Turnbull (1956) interpreted these observations by supposing that the Cu is distributed among both interstitial and substitutional sites and that the substitutional Cu (p-type) is transported primarily by dissociation of the substitutional atom, according to Eq. (1), followed by migration of the interstitial thereby created, until it meets another vacancy. Thus, the transport of *substitutional* copper is limited by the rate at which the migrating interstitials encounter vacancies. These vacancies have to be supplied by dislocations and free surfaces and it is this requirement which accounts for the striking effects of crystal perfection on the diffusion rate. Van der Maesen and Brenkman (1954) had proposed an interstitial mechanism for the diffusion of substitutional copper, but they did not recognize the necessary provision of vacancies and the resulting key role played by vacancy sources in the transport process.

Thus, in crystals with a *normal* content of dislocations, there is, in general, a sufficient abundance of vacancy sources to keep the vacancies everywhere in equilibrium, so $\Delta x_v = 0$. Then, if transport by vacancies is negligible, the effective diffusivity of substitutional solute in an intrinsic crystal would be

$$D_{\text{eff}} = D_i x_i'/(x_i' + x_s') \tag{5}$$

where D_i is the solute diffusivity in its interstitial state. On the other hand, in regions of a crystal far removed from vacancy sources, Δx_v is essentially unconstrained. Thus, if vacancies in this region are annihilated, as, for example, by absorption of interstitials, Δx_v may easily become negative and equal in magnitude to x_v'. Under these conditions (4) shows that x_i/x_s may become large and, by comparison to (5), $D_{\text{eff}} \approx D_i$. Thus, when a Ge crystal which is perfect, except for a small density of isolated dislocations, is put in contact with a Cu source, it will be penetrated rapidly to great depths by interstitially sited and migrating Cu atoms, which act as donors in the Ge host. That this occurs is shown by the rapid penetration of radiocopper into perfect regions of Ge crystal and by the quick encasement, already alluded to, of deeply embedded dislocations by an outwardly migrating zone of p-type Cu solution. This outward migration of substitutional Cu from dislocation centers, as well as from the free surface, will be limited by the rate of migration of vacancies produced at these sources. More particularly, Frank and Turnbull showed that it should be described by the diffusivity

$$D_d = D_v\left(\frac{x_v}{x_v + x_s}\right) = D_v\left[\frac{x_v'}{x_v' + (x_s'/x_i')x_i}\right] \quad (6)$$

which is the solvent vacancy diffusivity D_v multiplied by the probability that the vacancy is dissociated from the solute.

2. Interstitial-Vacancy Pairs

According to the theory developed by Longini and Greene (1956) and Shockley and Moll (1960) both the interstitial and substitutional populations of an electrically active impurity may depend strongly on the concentrations of other acceptor or donor impurities in the host. Since interstitial copper behaves as a donor in Ge, its concentration, and thus x_i'/x_s', should increase with concentration of acceptor impurities. Therefore, the effective diffusivity of Cu in Ge [see Eq. (5)] also should increase with acceptor content c_p and approach the diffusivity D_i of interstitial Cu in the limit where c_p is so high that $x_i' \gg x_s'$. The experiments on the solution and diffusivity of Cu in Ge, Si, and GaAs, reported by Hall and Racette (1964), are in general accord with this and other predictions of the theory.

The diffusivity of Au in Ge, in contrast with that of Cu, is quite low and it is not affected appreciably by large changes in the concentrations of other impurities, whether they are p- or n-type. This insensitivity to doping is inconsistent with a vacancy mechanism. It would be consistent with an interstitial mechanism if $x_i'/x_s' \gg 1$ in intrinsic Ge, but such a distribution is highly unlikely in view of the low diffusivity. Millea (1966) shows that the behavior can be explained more plausibly by some form of a direct interchange, Au ↔ Ge, mechanism. He suggests that this interchange may be effected by the excitation of substitutional Au to an interstitial position where it stays bound to the vacancy so created (the resulting defect will be called an i–v pair), followed by exchange of the vacancy with an adjacent Ge atom and the collapse of i-Au into the displaced vacancy. This model has been extensively developed to describe fast diffusion in certain metal systems, particularly Pb(Cd) and Pb(Hg), as discussed in Sections IV.D and V.C.

3. Interstitial Correlation Factors

It seems plausible that the correlation factor f contained in the diffusivity D_i of freely migrating interstitials should be unity. Pell's (1960) measurements of the relative diffusivities in Si of two lithium isotopes, ^{6}Li and ^{7}Li which migrate interstitially, appear to be in accord with this expectation. More specifically, his results indicate that $f \Delta K = 0.94 \pm 0.25$, where

4. FAST DIFFUSION IN METALS

ΔK ($\lesssim 1$) is the kinetic energy factor, which is thoroughly discussed by Peterson (Chapter 3; 1968).

B. Groups III and IV Metal Hosts

1. *Noble Metals in Lead*

Using a resistometric method, Rossolimo (1971) determined that the solubility of Au in Pb, at 1 atm pressure and over the temperature range 150–215°C, is specified satisfactorily by

$$x_e = (4.42 \pm 0.87) \times 10^4 \exp[-(16.92 \pm 0.20 \text{ kcal})/RT] \qquad (7)$$

where x_e is the atom fraction of Au in solution in equilibrium with the intermetallic phase (probably $AuPb_3$). Earlier determinations of x_e by Seith and Etzold (1934), in which saturation was achieved by solid state diffusion, are in satisfactory accord with the relation.

As we noted, several investigations have confirmed the exceptionally rapid diffusion rate of Au in Pb. Ascoli's (1960) finding that this rate is as high in single as in polycrystalline Pb specimens rules out the possibility that the fast diffusion only reflects grain boundary short-circuiting. Kidson (1966) showed that the single-crystal diffusivity was not changed significantly by the introduction, through plastic bending and annealing, of 6×10^6 dislocations/cm^2 emergent in the surface from which the Au diffused. Also, he demonstrated that the amount of Au taken in by the lead was not appreciably enhanced by the introduction of these dislocations. From these and other results, Kidson (1966) and Dyson et al. (1966) concluded that the fast diffusion is, indeed, a bulk property rather than a manifestation of short-circuiting by either one- or two-dimensional imperfections. This conclusion is also supported by Rossolimo's (1971) finding that the isothermal increase (at 200°C) of resistivity of a coarsely crystalline lead wire during dissolution of a thin Au coating on the wire occurs at the rate given by the fast diffusion coefficient (Fig. 1).

The investigations of Ascoli (1960), Kidson (1966), and Warburton (1973a) are in good agreement on the actual magnitude and temperature dependence of the diffusivity of radiotracer gold, at high dilutions, into single-crystal lead. Figure 1 shows the log diffusivity versus $1/T$ relation composed from these studies; it has the Arrhenius form with the constants, D_0 and Q, listed in Table I. The results reported earlier by both Seith and Keil (1933, 1935) and Seith and Etzold (1934) approach this curve at the highest temperatures but fall below it by a factor which increases with decreasing temperature to 4–5 at the lowest temperature of measurement,

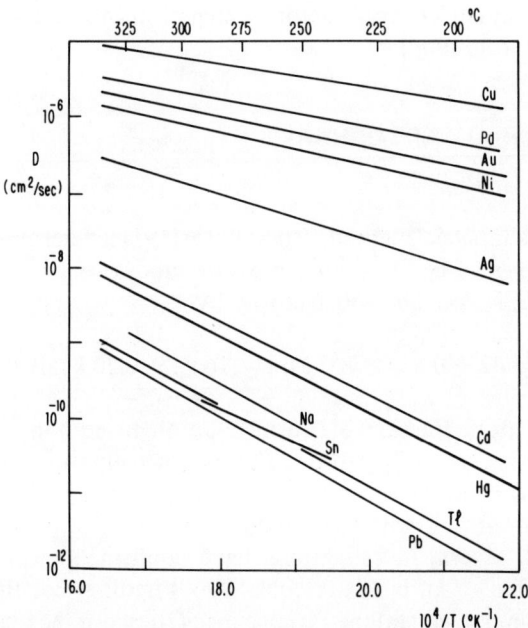

Fig. 1. Diffusivities of various elements in lead as a function of inverse temperature. The curves shown here are representative of the values listed in Table I.

as shown in Fig. 2. Rossolimo and Turnbull (1973) suggested that this behavior may indicate that the Au diffusivity decreases with increase of Au concentration, which was much higher in the experiments of Seith *et al.* than in those from which the relation was composed. Since the ratio of Au diffusivity to Pb self-diffusivity, at high dilution, ranges from $\sim 10^4$ to $\sim 10^6$ as the temperature of measurement goes from its highest to its lowest value, transport of Au by an interchange mechanism, whether direct or indirect, would require a marked minimum enhancement of the self-diffusivity of Pb by dissolution of Au, as will be discussed in Section IV.B. For example, at 175°C, dissolving 0.1 at. % Au in Pb should increase the Pb diffusivity by a factor of at least 100. It was such large increases in Pb diffusivity, demanded by the interchange interpretations of the Au diffusivity, which motivated the enhancement studies of Seith and Keil (1933), Dyson *et al.* (1966), and Miller (1970). In contrast with the predictions, Seith and Keil and Dyson *et al.* found no appreciable enhancement of D_{Pb} by Au dissolution. The more precise measurements of Miller did reveal a definite enhancement, but it was two orders of magnitude smaller than the minimum factor required by the interchange mechanisms. Enhancement data are usually represented by

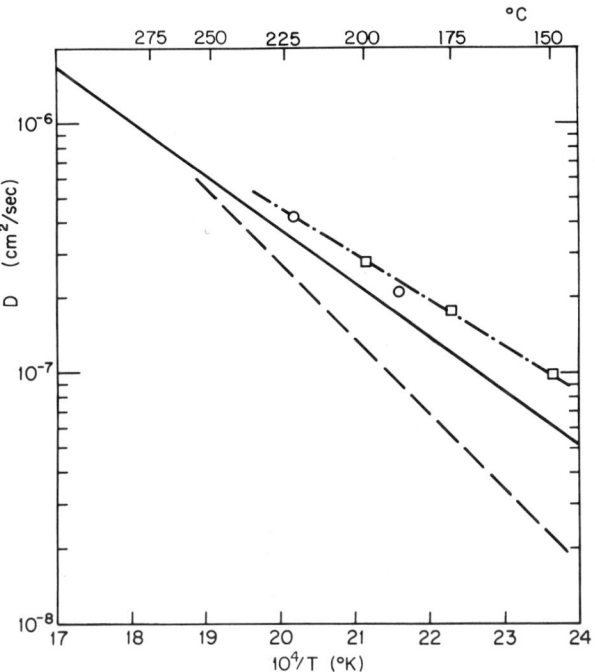

Fig. 2. Diffusivity of gold into lead reported by several researchers, as a function of inverse temperature. Data: (—) Kidson (1966), ^{198}Au; ○ Kidson (1966), ^{199}Au; □ Warburton (1973a), ^{195}Au; (- - -) Seith and Keil (1933). Kidson's ^{199}Au samples, as discussed in the text, are much more dilute in Au than his ^{198}Au samples.

polynomial expansions of isothermal diffusivity as a function of alloy composition

$$D_i(x_2)/D_i(0) = D_i/D_i^0 = 1 + b_{i1}x_2 + b_{i2}x_2^2 + b_{i3}x_2^3 \tag{8}$$

where up to third-order terms may be necessary for an accurate fit. The subscript i may equal 1 or 2, referring, respectively, to the host or impurity. The effect Miller observed was essentially linear in Au concentration and is thus represented by the values b_{11} listed in Table I. The conclusion from these enhancement studies (Seith and Keil, 1933; Wagner, 1938; Kidson, 1966; Dyson et al. 1966; Miller, 1970) is that the interchange mechanisms, direct or indirect (e.g., vacancy, interstitialcy) make only negligible contributions to the actual transport of Au in Pb. Rather, this transport must be effected, primarily, by Au migrating as an interstitial-type defect.

TABLE I

DIFFUSION PROPERTIES OF VARIOUS IMPURITIES IN GROUP III AND IV METAL HOSTS†

System	D_0 (cm²/sec)	Q_0 (kcal/mole)	$D_2(0)/D_1(0)$	b_{11}	b_{min}	b_{21}	$f\Delta K$	$\Delta V^*/V_0$	T (°C)	Ref.
Pb(Cu)	$(7.9 \pm 2.0) \times 10^{-3}$	8.0 ± 0.4	10^4 (at mp)						225–mp	a
							~ 0.2	0.04 ± 0.03	218–530	b
										c
Pb(Ag)	$(4.6 \pm 1.0) \times 10^{-2}$	14.4 ± 0.5	6×10^2 (at mp)						125–mp	a
	7.5×10^{-2}	15.2							113–300	d
									300.3	e
			870	137	1675		$0.25 \pm .06$		300	f
							$0.32 \pm .02$		178	f
								0.34 ± 0.04	25–350	g
Pb(Au)	4.1×10^{-3}	9.35	2×10^3 (at mp)						190–300	h
	8.7×10^{-3}	10.0							190–320	i
			0.89×10^5	4.3×10^3	1.7×10^5				215.2	e
			1.53×10^5	5.7×10^3	3.0×10^5				199.4	e
			6.7×10^5			-1.3×10^3			175.0	j
			2.0×10^6			-2.1×10^3			149.7	j
								0.28 ± 0.03	25–350	g
Pb(Zn)	$(1.6 \pm 0.4) \times 10^{-2}$	11.3 ± 0.2							182–299	ll
Pb(Cd)	$(4.1 \pm 1.0) \times 10^{-1}$	21.2 ± 0.2	20 (at mp)						150–mp	k
			29.1	30.0 ± 3.1	38.5	14			248.2	k
									248	l
							0.12 ± 0.04		248	m
Pb(Hg)	$1.05 {}^{+0.27}_{-0.21}$	22.7 ± 0.2	10 (at mp)						193–300	n
			17.6	27.2 ± 2.3	16.1	19.4 ± 3.2			251.8	n
									248.9	n

4. FAST DIFFUSION IN METALS

Pb(Tl)	0.5 ± 0.2	24.3 ± 0.4	2		206–323	o
Pb(Sn)	1.6×10^{-1}	23	2		245–285	p
Pb(Na)	6.3	28.3 ± 3.0	3.5		249–313	q
Pb(Ni)	$(9.4 \pm 3.4) \times 10^{-3}$	10.6 ± 0.4		0.13 ± 0.04	208–591	jj
Pb(Pd)	$(3.8 \pm 1.6) \times 10^{-3}$	8.6 ± 0.4		0.04 ± 0.04	209–588	kk
Pb(Pb)	1.4 ± 0.5	26.1 ± 0.4		0.64	206–323	o
	0.7 ± 0.2	24.8 ± 0.6		0.71 ± 0.04	165–mp	r
				0.72		s
						t
	0.9 ± 0.3	25.5 ± 0.3			200–300	k
	$0.9 \; {+0.6 \atop -0.4}$	25.5 ± 0.5			214–295	n
Sn(Cu)		cm²/sec				
[100]	$\sim 2.0 \times 10^{-6}$	7.9			25	u
[001]	2.4×10^{-3}				135–223	u
Sn(Au)						
[001]	5.8×10^{-3}	11.0 ± 0.4			125–232	v
[100]	1.6×10^{-1}	17.7 ± 0.5			125–232	v
Sn(Ag)						
[001]	7.1×10^{-3}	12.3 ± 0.4			125–232	v
				0.01 ± 0.01		f
[100]	1.8×10^{-1}	18.4 ± 0.5		0.40 ± 0.01	125–232	v
						f
Sn(Zn)	$(5.9 \pm 1.2) \times 10^{-3}$	11.8 ± 0.1	(polycrystalline samples)		90–232	w
Sn(Cd)						
[001]	220	28.2 ± 0.6			190–225	x
[100]	120	27.0 ± 0.5			190–225	x

TABLE I (Continued)
Diffusion Properties of Various Impurities in Group III and IV Metal Hosts†

System	D_0 (cm²/sec)	Q_0 (kcal/mole)	$D_2(0)/D_1(0)$	b_{11}	b_{min}	b_{21}	$f \Delta K$	$\Delta V^*/V_0$	T (°C)	Ref.
Sn(Hg)										
[001]	$7.5 {}^{+6.4}_{-3.5}$	25.3 ± 0.6							175–232	y
[110]	$30 {}^{+20}_{-12}$	26.8 ± 0.5							175–232	y
Sn(In)										
[001]	12 ± 3	25.6 ± 0.5							181–221	z
[100]	34 ± 7	25.8 ± 0.5							181–221	z
Sn(Sb)										
[001]	$79 {}^{+21}_{-17}$	29.1 ± 0.2							192–226	aa
[100]	$77 {}^{+43}_{-28}$	29.4 ± 0.4							192–226	aa
Sn(Sn)										
[001]	8 ± 6	25.6 ± 0.8							178–222	bb
	7.7 ± 3.0	25.6 ± 1.2							160–223	cc
[100]	1.4 ± 0.5	23.3 ± 0.5							178–222	bb
	10.7 ± 1.0	25.2 ± 1.0							160–223	cc
								0.33		t
								0.33 ± 0.02		cc
In(Au)	9×10^{-3}	6.7 ± 0.9							20–150	dd
In(Ag)										
c	1.1×10^{-1}	11.5 ± 0.3							20–150	dd
a	5.2×10^{-1}	12.8 ± 0.3						0.41 ± 0.04	96–113	ee

4. FAST DIFFUSION IN METALS

In(In)						
c	2.7		18.7 ± 0.3		39-144	ff
a	3.7		18.7 ± 0.3		39-144	ff
				0.52 ± 0.03	118-148	gg
Tl(Ag)						
hcp-a	3.8×10^{-2}		11.8 ± 0.4		85-226	hh
hcp-c	2.7×10^{-2}		11.2 ± 0.4		85-226	hh
bcc	4.2×10^{-2}		11.9 ± 0.4		226-300	hh
Tl(Au)						
hcp-a	5.3×10^{-4}		5.2 ± 0.4		120-226	hh
hcp-c	2.0×10^{-5}		2.8 ± 0.4		120-226	hh
bcc	5.2×10^{-2}		6.0 ± 0.4		220-300	hh
Tl(Tl)						
hcp-a	0.4		22.6 ± 1.0		150-226	ii
hcp-c	0.4		22.9 ± 0.5		150-226	ii
bcc	0.7		20.0 ± 0.5		226-275	ii

† Included are: the diffusion constants D_0 and Q_0 describing the temperature dependence; the ratio of impurity diffusivity to host diffusivity, $D_2(0)/D_1(0)$; the first solvent enhancement coefficient b_{11}; the minimum value of this parameter consistent with a vacancy mechanism b_{min}; the first solute enhancement coefficient b_{21}; the value $f\Delta K$ obtained from an isotope experiment; the ratio of the activation volume of diffusion to the atomic volume, $\Delta V^*/V_0$; and the temperatures of the reported measurements.

[a] Dyson et al. (1966).
[b] Candland et al. (1972).
[c] Miller (1972).
[d] Seith and Keil (1933).
[e] Miller (1970).
[f] Miller et al. (1972).
[g] Weyland et al. (1971).
[h] Ascoli (1960).
[i] Kidson (1966).
[j] Warburton (1973a).
[k] Miller (1969a).
[l] Miller (1969c).
[m] Miller and Edelstein (1969).
[n] Warburton (1973c).
[o] Resing and Nachtrieb (1961).
[p] Seith and Laird (1932).
[q] Owens and Turnbull (1972).
[r] Hudson and Hoffman (1961).
[s] Nachtrieb et al. (1959).
[t] Devries et al. (1963).
[u] Dyson et al. (1967).
[v] Dyson (1966).
[w] Bergner and Lange (1966).
[x] Huang and Huntington (1972).
[y] Warburton (1972).
[z] Sawatzky (1958).
[aa] Enderby et al. (1966).
[bb] Meakin and Klokholm (1960).
[cc] Coston and Nachtrieb (1964).
[dd] Anthony and Turnbull (1966).
[ee] Ott (1971a).
[ff] Dickey (1959).
[gg] Ott and Norden-Ott (1971).
[hh] Anthony et al. (1968).
[ii] Shirn (1955).
[jj] Candland and Vanfleet (1973).
[kk] Vanfleet et al. (1973).
[ll] Ross et al. (1974).

Warburton (1973a) measured the Au diffusivity in Pb(Au) alloys between 150 and 200°C and found a striking decrease in D_{Au} with increasing Au concentration. The values of b_{21} associated with these measurements are given in Table I. Following Rossolimo and Turnbull (1973) (see Section III.F), he explained this result in terms of the formation, at low temperatures and high concentrations, of a slowly diffusing Au defect containing two Au atoms. It should also be noted that Warburton's values for ^{195}Au diffusivity, extrapolated to infinite dilution, lie above Kidson's ^{197}Au values in this temperature region (see Fig. 2), just as Kidson's values lie above the earlier Seith and Keil work in concentrated alloys. Kidson's results from two ^{199}Au/Pb couples, which were much more dilute, show the same elevation. In light of the negative enhancement results, the depression of Kidson's ^{197}Au values is presumably due to the quantity of tracer Au he used.

As is indicated by the log D vs 1/T relations displayed in Fig. 1 (Arrhenius constants, Table I), Cu and Ag are also fast diffusers in Pb. We note that the isothermal order of the noble metal diffusivities in Pb is $D_{Cu} > D_{Au} > D_{Ag}$. This same order is found for most of the other metal matrices in which fast diffusion occurs. Some possible interpretations of this order are discussed by Anthony (1970a).

Dyson et al. (1966) found no appreciable enhancement of the lead self-diffusivity by copper dissolution. Miller (1970) found a linear increase of D_{Pb} with silver concentration but the enhancement coefficient, b_{11} (see Table I) was smaller than b_{11} for Au by a factor of ~ 40 and it was also negligibly small in comparison with the minimum b_{11} demanded if Ag were transported in Pb principally by an interchange mechanism. From these results, we conclude that the transport of Cu and Ag in Pb, like that of Au, occurs mainly by migration of interstitial-type defects.

From measurements of the isotope effect on diffusion, Miller et al. (1972) calculated that for the diffusion of Ag in Pb, $f \Delta K = 0.25 \pm 0.06$ at 300°C and 0.32 ± 0.02 at 178°C (see Table I). The interpretation of this result will be discussed in Section V.B.1.

The effect of pressure on the diffusivities of the noble metals in Pb has been measured and it seems to be satisfactorily characterized, for each impurity, by a single volume of activation ΔV^* which is listed in Table I. Values for lead self-diffusion are included for comparison and completeness.

2. Cd and Hg in Pb

In contrast with the fast diffusing noble metal impurities, thallium, which is presumed to be trivalent, and radiotracer Pb diffuse in Pb at the much slower "normal" rates expected for vacancy controlled diffusion. This

4. FAST DIFFUSION IN METALS

comparison suggested that the occurrence of fast diffusion might be favored by a low impurity valence and it motivated the studies of the diffusivities of the divalent impurities, Cd (Miller, 1969a,c) and Hg (Warburton, 1973c) in Pb. The log D vs $1/T$ relations found for these impurities, at high dilutions, are displayed in Fig. 1 and the corresponding Arrhenius constants are given in Table I. We note that these diffusivities, while well below those of the noble metals, are still 1–2 orders of magnitude above the self-diffusivity of Pb. Accordingly, the activation energies are much higher than those for the noble metals but 12–20% below that for Pb self-diffusion.

The self-diffusivity of Pb is markedly enhanced by dissolution of Cd or Hg, although not nearly so much as other systems showing similar diffusivity ratios. For satisfactory description of the isothermal enhancement over the entire concentration range (several atomic percents) of the measurements, the b_{1j} coefficients of Eq. (8) must be specified through the cubic term, b_{13}. These coefficients for Cd and Hg are listed in Table II, together with some values from more normal systems, and typical isothermal courses of Pb and Hg diffusivity with Hg impurity concentration are shown in Fig. 3.

Increasing impurity concentration also leads to increasing diffusivity of

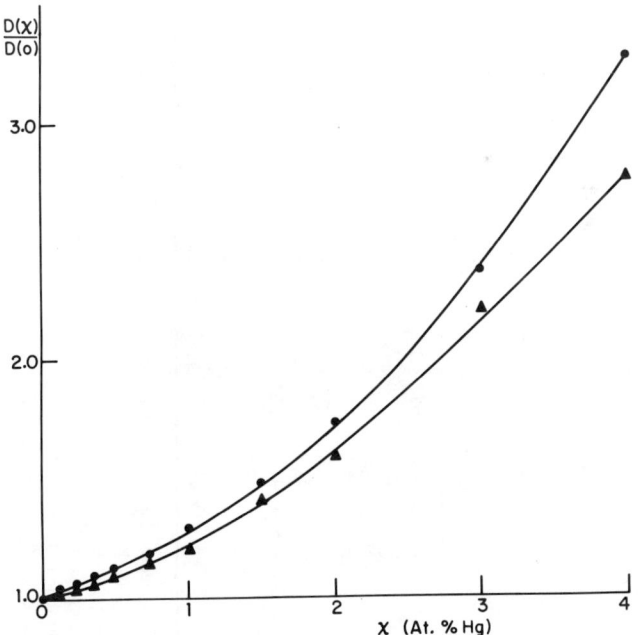

Fig. 3. Enhancement of the diffusivity of ^{210}Pb ● and ^{203}Hg ▲ in lead, as a function of mercury concentration, at 275 and 225°C, respectively.

TABLE II

Enhancement Coefficients in Pb(Cd) and Pb(Hg) Systems†

System	T(°C)	$\frac{D_2(0)}{D_1(0)}$	b_{11}	b_{min}	b_{12}	b_{13}	T(°C)	b_{21}	b_{22}	b_{23}	Ref.
Pb(Cd)	198.7	44.9	45.2 ± 2.5	69.3	440						a
	248.2	29.1	30.0 ± 3.1	38.5	410						a
	300.5	19.9	19.1 ± 2.9	20.8	1070	44,000	248.0	14	1000		b
Pb(Hg)	225.5	20.2	28.2 ± 3.2	21.2	120 ± 150	(22.0 ± 2.0) × 10³	224.0	18.1 ± 2.9	450 ± 190	(5.8 ± 3.4) × 10³	c
	251.8	17.6	27.2 ± 2.3	16.1	180 ± 95	(14.7 ± 1.0) × 10³	248.9	19.4 ± 3.2	370 ± 150	(3.9 ± 1.7) × 10³	c
	274.1	15.7	23.2 ± 1.6	12.5	490 ± 130	(9.0 ± 2.7) × 10³	272.0	25.7 ± 3.3	−700 ± 150	(6.6 ± 1.7) × 10³	c
	294.6	14.3	22.1 ± 3.0	9.8	390 ± 290	(9.3 ± 7.4) × 10³	300.2	29.2 ± 4.3	−740 ± 400	(24.8 ± 9.5) × 10³	c
Ag(In)		5.3	18	−7.4							d
Ag(Sb)		7.6	66	−3.2							d
Ag(Pb)		11	87	3.4							d

† Three systems exhibiting fast diffusion by a vacancy mechanism are shown for comparison.
[a] Miller (1969a).
[b] Miller (1969c).
[c] Warburton (1973c).
[d] Howard and Manning (1967).

4. FAST DIFFUSION IN METALS

the corresponding (Cd or Hg) impurity according to the isothermal relation

$$D_2(x_2)/D_2(0) = D_2/D_2^0 = 1 + b_{21}x_2 + b_{22}x_2^2 \tag{9}$$

This effect is adequately described by specification of the coefficients, which are given in Table II, through the square term.

Miller (1969c) noted, by recognizing the dependence of D_2 on x_2, that the form of the isothermal dependence of Pb diffusivity on Cd concentration given by Eq. (9) can be simplified to

$$D_1(x_2)/D_1(0) \cong 1 + [D_2(x_2)/D_1(0)]x_2 \tag{10}$$

which implies that, for Cd, the coefficients of Eqs. (8) and (9) are interrelated as follows:

$$b_{11} = D_2(0)/D_1(0), \qquad b_{12} = b_{11}b_{21}, \qquad b_{13} = b_{11}b_{22} \tag{11}$$

This enhancement is just that required by a direct interchange mechanism of diffusion; but at the lower temperatures, where

$$D_2(0)/D_1(0) > 18,$$

it is well below the minimum enhancement [see Eq. (28)] demanded by a vacancy mechanism.

The diffusivity behavior and enhancement effects of Hg in Pb are generally similar to those of Cd in Pb. However, Warburton (1973c) found that to describe the enhancement of Pb diffusivity by Hg, it is necessary to introduce a constant factor $\alpha > 1$ into the second term of Eq. (10); thus

$$D_1(x_2)/D_1(0) \approx 1 + \alpha[D_2(x_2)/D_1(0)]x_2 \tag{12}$$

This expression applies to both Hg and Cd but for Cd solutions $\alpha \cong 1$. The enhancement of Pb diffusivity by Hg lies just above the minimum enhancement permitted by the vacancy mechanism over the temperature range of the measurements, but it extrapolates to values below the minimum (see Fig. 4) at low temperature.

Miller and Edelstein (1969) measured the ratio of the diffusivities of ^{109}Cd and ^{115}Cd at high dilutions. From this they calculated that the product of the correlation factor and kinetic factor for the diffusion of Cd in Pb is $f\Delta K = 0.12 \pm 0.04$ (see Table I), which will be discussed in Sections IV.D and V.C.

3. *Other Metal Impurities in* Pb

Thallium and tin diffuse in Pb (see Arrhenius constants in Table II) at rates which differ little from that of Pb self-diffusion and which are considered normal for vacancy controlled diffusion. In contrast with the other

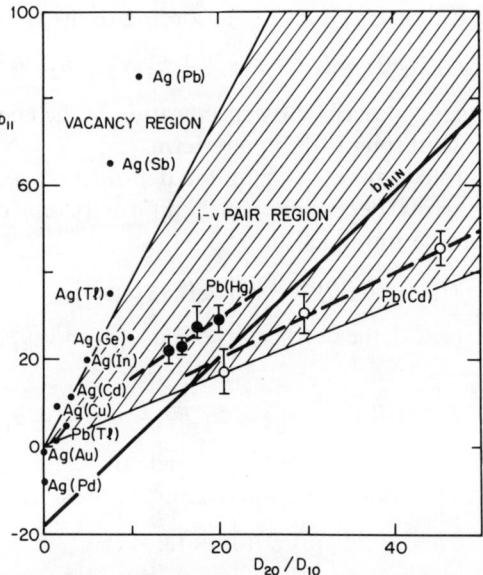

Fig. 4. Values of the coefficient of enhancement of self-diffusion, b_{11}, plotted for various systems as a function of the ratio of impurity diffusivity to host diffusivity at the same temperature.

monovalent impurities studied, Na (see Table I) also appears to diffuse in Pb at a normal rate (Owens and Turnbull, 1972). Its diffusivity is, over the temperature range studied, about 3–4 times that of Pb, but this factor may be due, in large part, to the small mass of Na relative to Pb. From this result, it is apparent that there may be factors in addition to impurity valence which play a determining role in the occurrence of fast diffusion.

Candland and Vanfleet (1973) have measured the diffusion of both Ni and Pd in Pb at pressures up to 50 kbar. Both impurities are extremely rapid diffusers with diffusivities approximately similar to Au in Pb (see Table I). Both also show limited (order 0.1 at. %) solubilities in Pb, (Hansen, 1958), but whether they will prove to be similar to the diffusion of noble metals in other respects remains to be seen.

To further delineate the occurrence of fast diffusion, studies of the diffusivities of Li, Mg, and Zn in Pb would be especially helpful.

Note added in proof. Ross et al. (1974) have measured the diffusivity of Zn in Pb and it is indeed a fast diffuser, unlike the other IIB metals Cd and Hg. The order of impurity diffusivities in Pb is now Cu > Au > Pd > Ni ≈ Zn > Ag > Cd > Hg > Tl ≈ Pb.

4. Diffusion in Sn

The noble metals (Cu, Ag, Au) are also fast diffusers in crystalline tin (Dyson, 1966; Dyson et al., 1967). More particularly, their diffusivities range, depending on the element, temperature, and crystallographic direction, from 1 to 10 orders of magnitude greater than the self-diffusivity of tin (Fig. 5 and Table I). These diffusivities exhibit a striking crystalline anisotropy being $1\frac{1}{2}$–$2\frac{1}{2}$ orders of magnitude, depending on the element and temperature, higher along the "c" (fourfold symmetry) axis than in the orthogonal directions. This anisotropy effect contrasts sharply with that in normal diffusion in which (e.g., self-diffusion of Sn) diffusion is fastest in the directions transverse to the c axis, but only by factors of 1–2.

The atoms in the white tin structure delineate a set of relatively open square channels along the c direction with passages transverse to c which are more restricted. It has been shown that diffusion in this structure

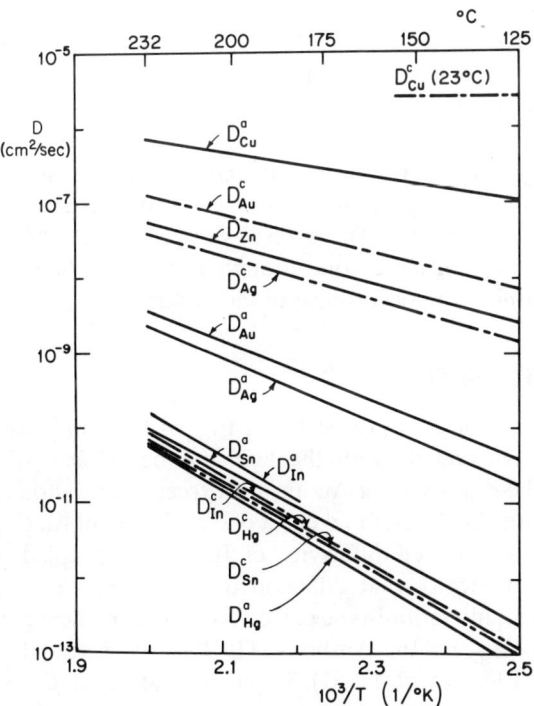

Fig. 5. Diffusivities of various elements along tin a and c axes as a function of inverse temperature. (– – –) are for diffusion along the c axis, (—) for diffusion along the a axis. Zinc was diffused into polycrystalline specimens; Cu diffusivity along the c axis was only measured at one temperature, 23°C.

by a vacancy mechanism should be fastest in directions perpendicular to c, which is accordant with the normal diffusion behavior (Coston and Nachtrieb, 1964). In contrast, it appears that interstitial atoms migrating in the tin structure would encounter the least resistance when moving along the c channels. Thus, the striking inverse anisotropy observed in fast, relative to normal, diffusion seems to indicate the dominance of a mechanism controlled by the motion of interstitial-type defects.

The observed order of the isothermal diffusivities, $D_{Cu} > D_{Au} > D_{Ag}$ is the same in Sn as in Pb, but the scaling of these quantities is quite different in the two hosts. Thus, the average D_{Cu}/D_{Au} is about 3 orders of magnitude higher in Sn than in Pb, while D_{Au}/D_{Ag} is about an order of magnitude lower.

Miller et al. (1972) measured the relative diffusivity of two Ag isotopes in Sn single crystals. The variation of D with isotopic mass was considerable, and consistent with $f\Delta K = 0.40 \pm 0.01$, for diffusion perpendicular to the c axis, but it was hardly significant, corresponding to $f\Delta K = 0.01 \pm 0.01$, for diffusion along the c axis. A similar effect also appears to exist in the case of Cu diffusion in Sn (Miller, 1972). Such a small mass effect would be quite unexpected, though explicable, in interstitially controlled diffusion.

The diffusivities of Cd and Hg in Sn (Table I and Fig. 5), in contrast with their behavior in Pb, differ rather little from the host self-diffusivity. However, the existing information, obtained from polycrystalline specimens with no control of the crystallographic orientation variable, indicates that Zn is a fast diffuser in Sn. These results suggest that the impurity/host radius ratio may play some role, though not one defined by the Hägg limit, in determining the occurrence of fast diffusion.

5. Diffusion in In and Tl

It has been established (Table I and Fig. 6) that both Ag and Au are fast diffusers in In and in both the hcp and bcc phases of Tl. In these systems, also, the diffusivity of Au always exceeds substantially that of Ag. Powell and Braun (1964) found no effect of dissolution of Au to the solubility limit at 152°C on the self-diffusivity of In which, coupled with the fast diffusion result, indicates that diffusion of Au in In is controlled mainly by migration of an interstitial-type defect. As further evidence for the formation of interstitials in In, Anthony (1970a) discusses the Mössbauer experiments of Flinn et al. (1967), which indicate that Co also occupies interstitial sites in In.

Additionally, it should be noted that in metals exhibiting a phase transition between a bcc and a close packed structure, the self-diffusivity is generally at least an order of magnitude higher on the bcc side of the

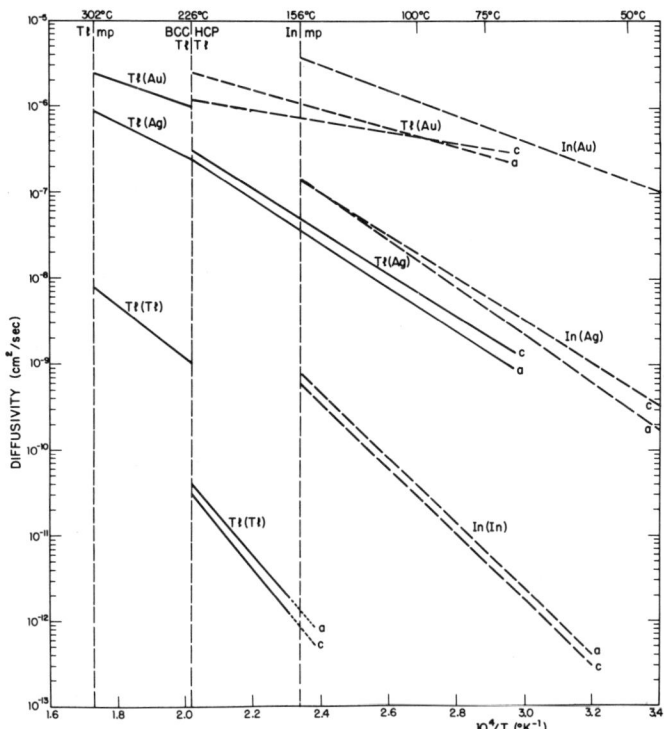

Fig. 6. Self-diffusivities and diffusivities of silver and gold in indium and thallium as a function of inverse temperature.

transition than on the close packed structure side. For example, the self-diffusivity of Tl increases about $1\frac{1}{2}$ orders of magnitude in the hcp → bcc transition at 495°K (see Fig. 6). However, the diffusivities of Ag or Au in Tl actually decrease by a small amount as a result of this transition. An important contributing factor to this changed structural effect may be that the proportion of interstitial impurity decreases in the transition (hcp → bcc) owing to the accompanying decrease in the volume of the structural interstices. For a more complete discussion of the results of these systems, see the review by Anthony (1970a).

C. Alkali Metal Hosts

It has been shown that some of the noble metals are fast diffusers in alkali metal hosts, all of which have bcc crystal structure (see Table III). An outstanding example of this behavior is the diffusivity of Au in Na which is 2–3 orders of magnitude higher than the self-diffusivity of Na.

TABLE III
Fast Diffusion in Alkali Metals†

System	D_0 (cm²/sec)	Q_D (kcal/mole)	$f \Delta K$‡	$\Delta V^*/V_m$	T (°C)	Ref.
Li(Cu)	0.47 ± 0.11	9.2 ± 0.2			50–181	a
Li(Ag)	0.12 ± 0.05	12.6 ± 0.2			67–160	b
			1.12 ± 0.25		169.8	c
			1.50 ± 0.40		77.9	c
Li(Au)	0.21 ± 0.08	11.0 ± 0.2			45–150	d
			0.50 ± 0.25		167.9	c
			2.00 ± 0.50		37.2	c,e
Li(Zn)	$0.57 {+0.31 \atop -0.20}$	12.98 ± 0.24			57–173	f
			0.88 ± 0.25		153.8	c
Li(Na)	0.41 ± 0.09	12.6 ± 0.15			50–176	p
			0.50 ± 0.25		175.0	c
			0.87 ± 0.37		69.9	c
Li(Li)	0.12 ± 0.05	12.6 ± 0.2			60–170	g
			3.38 ± 0.38		$T > 110$	h
			4.38 ± 0.38		$T < 110$	h
K(Au)	$(1.29 \pm 0.6) \times 10^3$	3.23 ± 0.3		0.28 ± 0.03	35–80	i
K(K)	0.16 ± 0.01	9.36 ± 0.05			6–53	j
					−52–62	k
				0.55 ± 0.01	15–30	l
Na(Au)	3.3 ± 1.0	2.20 ± 0.2			0–77	m
Na(Na)	0.145 ± 0.015	10.09 ± 0.07			0–97	n
			0.35 ± 0.05	0.41 ± 0.02	−50 – −42	i
Na(Na)$_{vacancy}$	0.0057 ± 0.0004	8.5 ± 0.2			−79 – +97	o
Na(Na)$_{divacancy}$	0.72 ± 0.05	11.5 ± 0.5		0.30 ± 0.03	−79 – +97	o
				0.70 ± 0.07	−79 – +97	o

The Au diffusivity in Li is relatively slower than in Na but it is still 10–20 times the Li self-diffusivity. In Li, the order of the diffusivities is, again, $D_{Cu} > D_{Au} > D_{Ag}$. The maximum solubilities of the noble in the alkali metals are almost an order of magnitude smaller (0.01 compared with 0.1 at. %) than in Pb and Sn. This would make it difficult to carry out meaningful experiments on the effects of impurity concentration on solvent self-diffusivity. However it is thought that fast diffusion in these systems can be accounted for most plausibly by a mechanism where the tracer at least intermittently occupies interstitial sites (Ott, 1969, 1970; Ott and Lodding, 1970; Lodding and Ott, 1971; Ott et al., 1969).

An important conclusion from these results is that fast diffusion can occur in monovalent, as well as in polyvalent, hosts.

D. Lanthanide and Actinide Hosts

Dariel and his associates (1969a,b,c) found that fast diffusion can occur in praseodymium, lanthanum and uranium hosts (see Table IV and Fig. 7 for a summary of results). The noble metal diffusivity behavior in the hexagonal and bcc phases of Pr proved to be quite similar to that in the Groups III and IV metal hosts. Co was found to be a fast diffuser in both Pr and U and to show anomalous behavior in Pu; in the Pr structures, it diffuses even a little more rapidly than Cu. The diffusivity of Zn in Pr, though lower than the noble metal diffusivities, is much higher than the self-diffusivity of Pr.

E. Occurrence of Fast Diffusion

We consider here what experience indicates about the requirements on the properties of the host and impurity for the occurrence of fast diffusion. Our considerations are guided by the evidence that fast diffusion, at least in some systems, is effected by migration of interstitial-type defects.

Footnotes to Table III:

† See Table I for an explanation of the headings.

‡ The values for Li are actually $[(^6D_N/^7D_N) - 1]/[(7/6)^{1/2} - 1]$, where the tracer (N) has been diffused in isotopically pure ^{6}Li and ^{7}Li, respectively, having masses 6 and 7.

^{a} Ott (1969).
^{b} Ott and Norden-Ott (1968).
^{c} Ott and Lodding (1970).
^{d} Ott (1968).
^{e} Ott (1971b).
^{f} Mundy et al. (1969).
^{g} Ott et al. (1968).
^{h} Lodding et al. (1970).
^{i} Hultsch and Barnes (1961).
^{j} Smith and Barr (1970).
^{k} Mundy et al. (1971).
^{l} Kohler and Ruoff (1965).
^{m} Barr et al. (1969).
^{n} Mundy et al. (1966).
^{o} Mundy (1971).
^{p} Mundy et al. (1967).

TABLE IV Fast Diffusion in Lanthanides and Actinides[†]

System		D_0 (cm²/sec)	Q_D (kcal)	T (°C)	Ref.
Lanthanides					
La(Au)	[PC]	$\left(2.2{+0.8 \atop -0.6}\right) \times 10^{-2}$	18.1 ± 0.6	610–800	a
La(La)	[PC]	$1.5{+1.2 \atop -0.7}$	45.1 ± 1.2	660–850	a
Pr(Cu)					
dhcp	[PC]	$\left(8.4{+2.5 \atop -1.9}\right) \times 10^{-2}$	18.1 ± 0.5	653–786	b
bcc	[PC]	$\left(5.7{+2.1 \atop -1.5}\right) \times 10^{-2}$	17.8 ± 0.7	813–914	b
Pr(Co)					
dhcp	[PC]	$\left(4.7{+2.3 \atop -1.6}\right) \times 10^{-2}$	16.4 ± 0.8	612–766	c
Pr(Au)					
dhcp	[PC]	$\left(4.3{+1.6 \atop -1.1}\right) \times 10^{-2}$	19.7 ± 0.6	610–750	c
bcc	[PC]	$\left(3.3{+1.1 \atop -0.8}\right) \times 10^{-2}$	20.1 ± 0.6	800–910	c
Pr(Ag)					
dhcp	[PC]	$\left(1.4{+0.4 \atop -0.3}\right) \times 10^{-1}$	25.4 ± 0.5	610–770	c
bcc	[PC]	$\left(3.2{+1.1 \atop -0.8}\right) \times 10^{-2}$	21.5 ± 0.7	800–920	c
Pr(Zn)					
dhcp	[PC]	$\left(1.8{+0.9 \atop -0.6}\right) \times 10^{-1}$	24.8 ± 0.9	603–767	d
bcc	[PC]	$\left(6.3{+3.2 \atop -2.1}\right) \times 10^{-1}$	27.0 ± 0.9	822–922	d
Pr(Pr)					
bcc		$\left(8.7{+5.6 \atop -3.4}\right) \times 10^{-2}$	29.4 ± 1.1	825–925	e
Actinides					
γ-U(Co)		$\left(3.5{+1.0 \atop -0.8}\right) \times 10^{-4}$	12.6 ± 0.6	780–1080	f
γ-U(Fe)		$\left(2.7{+0.4 \atop -0.4}\right) \times 10^{-4}$	12.0 ± 0.3	780–1080	f
γ-U(Ni)		$\left(5.4{+0.9 \atop -0.7}\right) \times 10^{-4}$	15.7 ± 0.4	780–1080	f
γ-U(Mn)		$\left(1.8{+2.0 \atop -0.9}\right) \times 10^{-4}$	13.9 ± 1.7	780–1080	f
γ-U(Cr)		$\left(5.4{+1.0 \atop -0.9}\right) \times 10^{-3}$	24.5 ± 0.4	780–1080	f
γ-U(Cu)		$\left(2.0{+0.4 \atop -0.3}\right) \times 10^{-3}$	24.1 ± 0.4	780–1080	f
γ-U(U)		$\left(1.2{+0.6 \atop -0.4}\right) \times 10^{-3}$	26.7 ± 1.0	780–1080	f
γ-U(Nb)		$\left(4.9{+1.2 \atop -0.9}\right) \times 10^{-2}$	39.7 ± 0.5	780–1080	f

[†] "PC" denotes polycrystalline samples. See Table I for an explanation of the headings.
[a] Dariel et al. (1969a).
[b] Dariel (1971).
[c] Dariel et al. (1969b).
[d] Dariel (1970).
[e] Dariel et al. (1969c).
[f] Peterson and Rothman (1964).

4. FAST DIFFUSION IN METALS

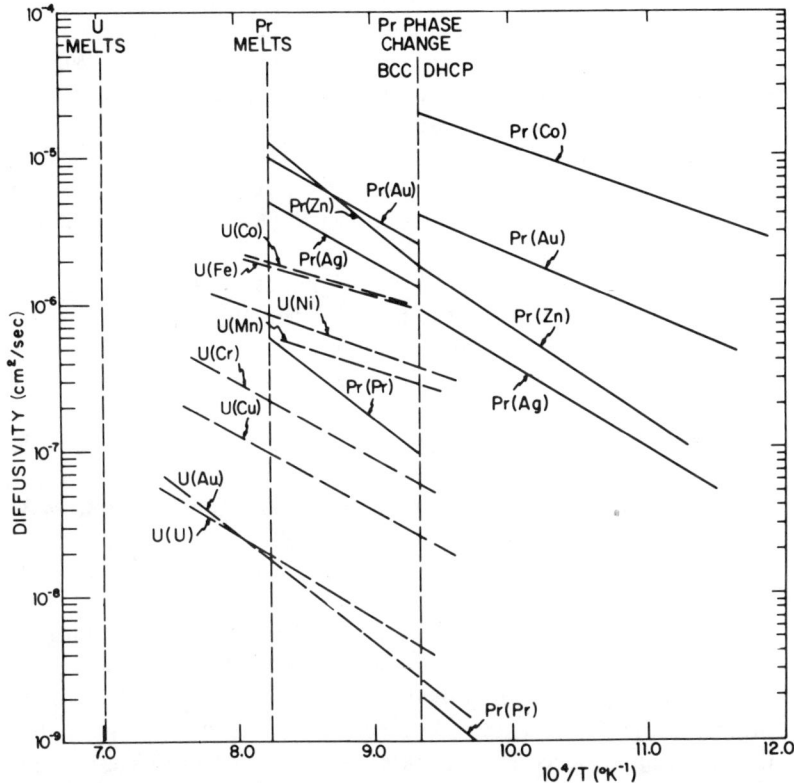

Fig. 7. Diffusivities of various elements in uranium and praseodymium as a function of inverse temperature.

We note first that the impurity/host Wigner–Seitz radius ratios of the systems which exhibit fast diffusion range from ~0.62 to 0.93. These ratios are all higher, and generally much higher than the limiting value $\lesssim 0.59$, specified by Hägg's rule for substantial interstitial solution. Dyson *et al.* (1966) suggested that a necessary, but not sufficient, condition for fast diffusion is that there be no ion–ion overlap when the impurity atom is placed in the largest interstice of the host structure. With the generally accepted values for the metallic valences and corresponding ionic radii, it appears that this condition is fulfilled by all of the fast diffusing systems. This means that the most likely hosts for fast diffusion (we might call them "voluminous hosts") are those having relatively large Wigner–Seitz radii and relatively small ion-core radii. Apparently, fast diffusion is compatible with a wide range of host valences, 1–4, at least.

All of the fast diffusing impurities recognized so far are elements from the noble (Cu, Ag, Au), transition, or IIB (Zn, Cd, Hg) metal groups. All are thought to be either mono- or divalent in the metallic state. We have noted already that the tendency toward fast diffusion decreases sharply with increasing impurity valence. For example, the order of the diffusivities in Pb of the elements leading up to Pb in the periodic table is the reverse of their metallic valence order, i.e., $D_{Au} \gg D_{Tl} > D_{Pb}$; further Au(1+) and Hg(2+) are fast while Tl(3+) and radiotracer Pb(4+) are normal diffusers in Pb.

That factors besides impurity valence are important in determining the occurrence and rate of fast diffusion, even when the necessary ion size condition is fulfilled, is shown by the findings that Na(1+) is a normal diffuser in Pb and that the diffusivity of Au is, in all cases, substantially higher than that of Ag. The behavior of Na might be correlated with that of the other impurities by amending the necessary condition for fast diffusion to require that the volume of the largest interstice formed by the host ions exceed the Wigner–Seitz, as well as the ionic, volume of the impurity. We note, however, that the diffusion of the noble metals in Al apparently is normal, even though fast diffusion would be allowed by the amended condition.

All of the fast diffusing impurities identified so far have filled, or partly filled, electronic d levels with energies not far below those of occupied s levels. Consequently, their ions ought to have relatively high polarizabilities and arguments have been made which connect the tendency toward interstitial solution, and hence fast diffusion behavior, with this ease of ionic polarization (Anthony, 1967; Warburton, 1972). Such a connection might account for the faster diffusion of Au relative to Ag in all fast diffusing systems so far studied since the polarizability of the Au ion is higher than that of Ag, while the Wigner–Seitz volumes differ little.

III. Corroboration of Existence of Interstitial-Type Defects

The presence of substantial concentrations of interstitial-type defects ought to be reflected by certain other properties, as well as by the diffusion behavior, of fast diffusing systems. Indeed, there have been a number of studies of such properties that have indicated some interstitial dissolution or particular alloy interactions which would favor such solution. These studies are surveyed in this section.

A. Liquid State Properties of Fast Diffusing Systems

Interactions which cause some interstitial solution of impurity in a crystalline host might be expected to lead to host–impurity nearest neighbor separations and solution volumes in the molten state which are substantially smaller than calculated by summing the corresponding properties of the pure molten constituents. These expectations seem to be borne out by the diffraction and volumetric studies in the molten state of some fast diffusing systems which we now describe.

Enderby et al. (1966) studied neutron scattering from liquid Cu_6Sn_5, changing the scattering powers of the alloys by isotopic substitution of ^{63}Cu and ^{65}Cu for natural copper. In this manner they obtained the partial structure factors for Sn–Sn, Cu–Cu, and Cu–Sn, which are essentially the Fourier transforms of the respective radial distribution functions. The positions of the extrema in these functions are given in Table V, where

TABLE V

First Extrema, $r_{\alpha-\beta}$, in Partial Radial Distribution Functions between Type α Atoms and Type β Atoms in Liquid Sn(N) Alloys, Where N is a Noble Metal[†]

System	Ref.	Measurement	r_{Sn-Sn} (Å)	r_{N-N} (Å)	r_{Sn-N} (Å)	$r^{H.S.}_{Sn-N}$ (Å)
Cu_6Sn_5	a	Neutron scattering	3.30	2.65	2.65	2.98
Sn(Cu)	b	x-ray scattering	3.14	2.60	2.70	2.87
Sn(Ag)	c	x-ray scattering	3.18	2.88	2.98	3.03
Sn(Au)	d	x-ray scattering	3.16	2.84	2.88	3.00

[†] The hard sphere Sn–N separation, $r^{H.S.}_{Sn-N}$, calculated from the average of r_{Sn-Sn} and r_{N-N}, is given for comparison.
[a] Enderby et al. (1966).
[b] North and Wagner (1970).
[c] Halder and Wagner (1967).
[d] Wagner et al. (1969).

they may be compared with the data from the pure materials. The Cu–Cu and Sn–Sn data are in good agreement with the pure material data, while the Cu–Sn data look surprisingly like the Cu–Cu data. This implies that Cu–Sn separations are, on the average, nearly the same as Cu–Cu separations. Thus it appears that there is an attractive interaction between Sn and Cu atoms. We note that such an interaction is not necessarily inconsistent with the observed small solubility of Cu in solid Sn, where the additional requirement of not distorting the periodic lattice excessively is also imposed.

Wagner and colleagues (Wagner et al., 1969; North and Wagner, 1970; Halder and Wagner, 1967) have used x-ray techniques to obtain partial radial distribution functions (reproduced in Fig. 8) for the component

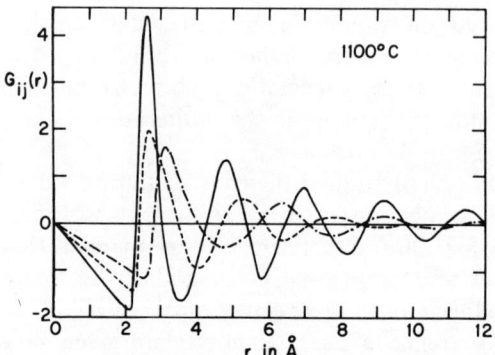

Fig. 8. Reduced partial distribution functions $G_{ij}(r) = 4\pi r \rho_0 \{g_{ij}(r) - 1\}$ in liquid Cu–Sn (Isothermal). [Reprinted from North and Wagner (1970). *Physics and Chemistry of Liquids* **2**, 98.]

atoms in Sn(Cu), Sn(Ag), and Sn(Au) alloys. The first peak positions, corresponding to nearest neighbor separations, are also given in Table V, together with the separations in the pure materials and their hard sphere averages for comparison. These results are qualitatively similar to those of the neutron scattering experiments and again seem to indicate an attractive interaction between Sn and noble metal atoms in the melt. It is interesting that the alloy of the system with the least mobile of the noble metal impurities, Sn(Ag), shows the most hard sphere-like behavior.

Another indication of such attractive behavior in liquid alloys is provided by Rossolimo's (1971) measurements of the specific volumes of Pb(Au) alloys between 0 and 28% Au at 345°C. The partial molar volume of Au was constant and equivalent to

$$\bar{V}_{Au} = (9.31 \pm 0.19) \quad \text{cm}^3/\text{mole} \tag{13}$$

If the volumes were strictly additive, $\bar{V}_{Au}$ would be equal to the molar volume of supercooled liquid gold at the same temperature which, extrapolating linearly from the molten state, is

$$\bar{V}_{Au} = 10.85 \quad \text{cm}^3/\text{mole} \tag{14}$$

Thus the experimental result is 14% smaller than strict additivity would predict and leads us to expect a foreshortening of the Pb–Au distance in liquid alloys similar to the effects observed for the Sn–(noble metal) systems.

B. Electron Channeling Experiments

Tomlinson and Howie (1968) observed the variation in intensity of electrons emitted near crystal symmetry directions from crystals doped with radioactive β emitters. Such intensity distributions are expected to show maxima in certain prominent directions if the doping atoms are sited substitutionally, but minima if sited interstitially. Details of the nature of the site may be determined by noting which directions are those showing the maxima or minima. ^{110m}Ag, emitting 87 and 530 keV electrons in a Ag lattice, for example, showed maxima in the $\langle 001 \rangle$ directions. ^{198}Au, emitting 963 keV electrons from a Pb lattice, however, shows a minimum in the $\langle 001 \rangle$ direction, suggesting that, for 0.06 wt. % alloys, the Au occupies interstitial sites. Since it is the $\langle 001 \rangle$ direction showing the maximum, the octahedral site $\frac{1}{2}[111]$ seems to be ruled out, the emitted electron seeming to come from *between* the (020) planes. It is not clear whether split interstitials are also thereby ruled out on the grounds that, for such a defect, two-thirds of the gold atoms would lie on (020) planes.

C. Mössbauer Data

Flinn *et al.* (1967) found that when Co is dissolved in In, the recoil free fraction f of γ rays from ^{57}Co's decay daughter, ^{57}Fe, is anomalously high. They suggest that this behavior may be due to Co in interstitial sites. They reasoned that an interstitial would experience larger restoring forces than a substitutional and thus have a larger f value. Anthony (1970a) notes that larger restoring forces for interstitials are consistent with the observed larger solution hardening effect of interstitials.

D. Supersaturated Alloys

Giessen and co-workers (Giessen *et al.*, 1971; Ray *et al.*, 1972) formed metastable solid solutions of Cu and Fe in Y and Gd containing up to 15 at. % solute by splat cooling with quench rates of 10^7–10^8 °C/sec. The alloy densities and unit cell volumes were measured and indicated that the alloys Y(Fe) and Y(Cu) are of the pure interstitial type, while in Gd(Cu) and Gd(Fe), the solute atoms replace solvent atoms with a ratio of 2 : 1, suggesting that they form disubstitutionals. The difference between the two hosts appears to be mainly one of atomic size: since the ionic radius of Gd^{3+} is larger than that of Y^{3+}, interstitial formation is favored in the latter, disubstitutional formation in the former.

E. Centrifuge Experiments

Anthony (1967, 1970b) measured the sedimentation of Au in In under the influence of high centrifugal fields. His samples were formed by diffusing plated ^{195}Au into In specimens at room temperature, etching away the excess Au, and recasting the In into appropriately shaped specimens. These specimens were subjected to centrifugal fields of approximately 10^5 G for times of the order of 1 day and then sectioned and counted. From the resulting redistribution of the ^{195}Au, the effective mass, which is essentially the difference between the mass of the diffusing atom and the mass of the atom normally occupying the same site, could be obtained. Since the mass of a Au atom is 197.0 amu, while In is 114.8 amu, the effective mass of a substitutional Au atom is 42% of the Au mass. Since no In atoms normally occupy interstitial sites, the effective mass of an interstitial or diinterstitial Au defect would then be one or two times its own mass, respectively, assuming no lattice relaxation occurred (Anthony, 1970b; Barr and LeClaire, 1969). For a disubstitutional, with two Au atoms replacing one In, the effective mass could be about 1.42 times a single Au mass. Anthony's experimental result, 1.7 ± 0.7, then suggests most strongly that the Au is dissolved disubstitutionally in In, though the possibilities of interstitials or diinterstitials cannot be ruled out.

Using similar techniques, Barr and Smith (1969) studied the equilibrium distribution of ^{198}Au in K, finding the effective mass of the diffusing species to be essentially equal to that of a single Au atom. This result lends considerable support to the conclusion that Au both dissolves and diffuses interstitially in K.

F. Precipitation Experiments

Rossolimo and Turnbull (1973) have studied the kinetics and morphology of precipitation of $AuPb_3$ from Pb at temperatures between 20 and 130°C. The precipitate particles were typically thin plates of $AuPb_3$, about 1 μ across and 1000 Å thick, separated by distances of the order of 2–3 μ. Their growth was found to be diffusion controlled, with diffusivity $D_p \simeq 45 \exp(-20.8/RT)$ cm^2/sec, which is 2.5–4 orders of magnitude lower than the extrapolated high temperature diffusivity of Au in this temperature region. The authors proposed two models to account for this slower low temperature diffusion. In the first, diffusion proceeds by interstitial–vacancy pairs, provided vacancy sinks are at least as separated as the precipitate particles. In the second, the Au diffusion is supposed to be controlled by the equilibrium between interstitials and some slower diffusing defect, such as an associated pair or small cluster, which is energetically more stable but entropically disfavored. The specific model of disubstitutionals was developed and it predicts that Au diffusivity should decrease with increasing Au concentration at fixed temperature.

Rossolimo (1971) similarly measured the precipitation of Ag from Pb at room temperature but found its rate was described satisfactorily by the diffusivities from the extrapolated high temperature values. The implication

4. FAST DIFFUSION IN METALS

is that whatever defect is responsible for the slowing of diffusion in the Pb(Au) system is not stable in the Pb(Ag) system, at least in the temperature and concentration range studied.

G. Resistivity Measurements

Rossolimo and Turnbull (1973) have measured the resistivity of Pb(Au) alloys between -200 and $240°C$. After correcting for the resistivity of included precipitate particles, they were able to calculate the residual resistivity of Au in Pb as a function of temperature. These results are shown in Fig. 9. The sharp nonlinear departure from Mathiessen's Rule between 130 and $240°C$, with some saturation near $240°C$, led Rossolimo to postulate a change in state for the dissolved Au between these two temperatures. Using a statistical thermodynamic calculation for the concentration of defects in alloys, Warburton (1973b) was able to duplicate these curves theoretically, assuming the transition occurs between either Au–Au disubstitutionals (see Section IV.C) or diinterstitial vacancy triplets at low temperatures and i–v pairs at high temperatures. The model is insensitive to

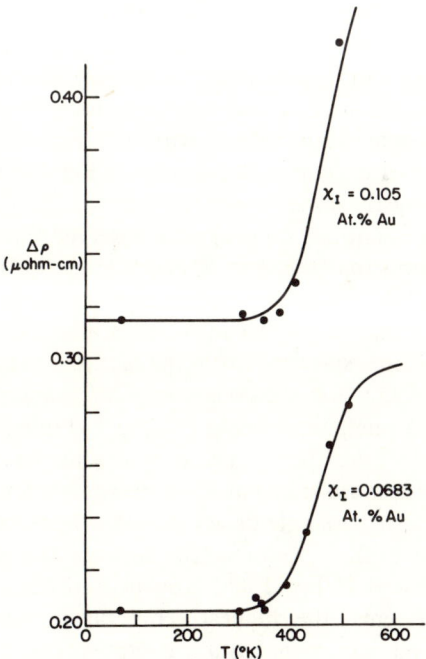

Fig. 9. Residual resistivity of gold in lead as a function of temperature at two concentrations. The data points ● have been fitted with a theoretical curve (—).

the exact defects used, so long as the low temperature defect contains two Au atoms and the high temperature defect contains just one Au atom and would work as well if the high temperature defect were a Au interstitial. The standard Gibbs free energy difference between the defects which gave the best fit to the experimental values of $\Delta\rho_i$ for the 0.0683 at. % Au alloy was

$$g^0_{2-1} = (9 \pm 2) \text{ kcal/mole} - T \text{ °K} (11.5 \pm 4.0) \text{ cal/mole °K} \quad (15)$$

This curve is superimposed on the data in Fig. 9 and may be seen to duplicate its features fairly accurately. As a test of both the model and the value of g^0_{2-1} reported in Eq. (15), it may be seen that the data for the 0.105 at. % Au alloy are also fit quite well, merely by using the appropriate concentration in the calculations. This theoretical work, then, also supports the existence of Au–Au disubstitutionals at low temperatures.

H. Internal Friction Measurements

Turner and associates (1972, 1974) studied the internal friction of the Pb(Au), Pb(Ag), and Pb(Cu) alloy systems as a function of temperature at 92 kHz. The samples were prepared by plating single-crystal Pb bars with the appropriate noble metal, allowing diffusion to occur at room temperature, and etching away the excess plate. They were then annealed at various temperatures, quenched to 77°K, and measured for internal friction between 77 and 320°K in quasi-equilibrium, with the temperature rising about 1°K/min. Internal friction curves are shown in Fig. 10, for $\langle 100 \rangle$ oriented crystals, for two anneal temperatures for each of the three alloy systems. Turner reports that essentially no peaks are observed for crystals with $\langle 111 \rangle$ orientation, indicating that the defects responsible for anelastic relaxation have some type of $\langle 100 \rangle$ symmetry. As the temperature at which the samples are equilibrated is changed, the various peaks also change in intensity, as may be seen for all three systems. From the fact that the peak height *ratios*, $Q^{-1}_{\tau_1'}/Q^{-1}_{\tau_1}$ and $Q^{-1}_{\tau_2'}/Q^{-1}_{\tau_2}$, remain constant in the course of these changes, Turner deduced that one pair of peaks (τ_1 and τ_1') must be associated with one defect while the other pair (τ_2 and τ_2') is associated with a second defect which becomes more stable at lower temperatures. The occurrence of two relaxation peaks for a single defect, coupled with the anisotropy noted above, implies that both types of defect have $\langle 100 \rangle$-orthorhombic point symmetry (Nowick and Heller, 1965; Nowick and Berry, 1972).

In all three systems, the highest temperature peak results from a relaxation process having an activation energy equal, within experimental error, to that of solute diffusion, as measured at high temperature and infinite dilution (see Table VI). Both processes may therefore be identified

TABLE VI

Activation Energies of Diffusion in Pb(Noble Metal) Systems and the Arrhenius Constants of their Anelastic Relaxation Times†

System	H_D (kcal/mole)	Q_{τ_1} (kcal/mole)	τ_1 (sec)	Q_{τ_2} (kcal/mole)	τ_2 (sec)
Pb(Cu)	8.0 ± 0.4	7.82	3.0×10^{-15}	7.2	1.8×10^{-16}
Pb(Ag)	14.4 ± 0.5	15.0	6.5×10^{-16}	14.1	1.8×10^{-17}
Pb(Au)	$10.0 \pm \sim 0.5$	10.0	1.1×10^{-15}	9.5	3.8×10^{-17}

† Turner et al. (1972).

with the same defect. Similarly, it is reasonable to identify the defect responsible for the lower temperature pair of relaxation peaks in Pb(Au) with the defect which produces the low temperature slowing of diffusion in Rossolimo's precipitation studies, the change of Au residual resistivity in his resistance measurements, or the slowing of diffusion at high concentrations in Warburton's Pb(Au) dehancement experiments. By extension, the Pb(Ag) internal friction data suggest that the equivalent Ag defect does not become stable until below room temperature, which would explain why Rossolimo saw no slowing of Ag diffusivity at temperatures above 20°C in his precipitation experiments.

Further indication that similar defects are occurring in all three systems may be found in the fact that the ratios of the temperatures of the peak maxima for τ_1- and τ_2-type defects are identical in all three systems, being given by

$$T_{\tau_1}/T_{\tau_1'} = 1.63 \pm 0.02 \quad \text{and} \quad T_{\tau_2}/T_{\tau_2'} = 1.89 \pm 0.02 \quad (16)$$

Very recently, it has come to our attention that Sagues and Nowick (1974), in an extended series of experiments, were unable to duplicate the results of Turner et al. reported above. Their samples were single crystals grown of nominal 5–9's Pb in reactor grade graphite, which were tested between 100 and 290°K. The samples were doped with Au, Ag, and Cu by either plating (both vacuum and electroplating were used) and annealing at 240°C for 16 hr, or by growing in the impurities. In no cases were peaks discovered and in most cases the background in the vicinity of the expected first peak was of order 10 times that given by Turner. A Pb(Cu) sample with 1% Bi was also run on the suspicion that impurity pinning might lower backgrounds. An exceptionally low background was obtained, but no peaks were found, though the detection limit was 0.5×10^{-5}.

The discrepancy between these sets of experiments is most hopefully viewed as a reflection of the difficulty in working with these particular

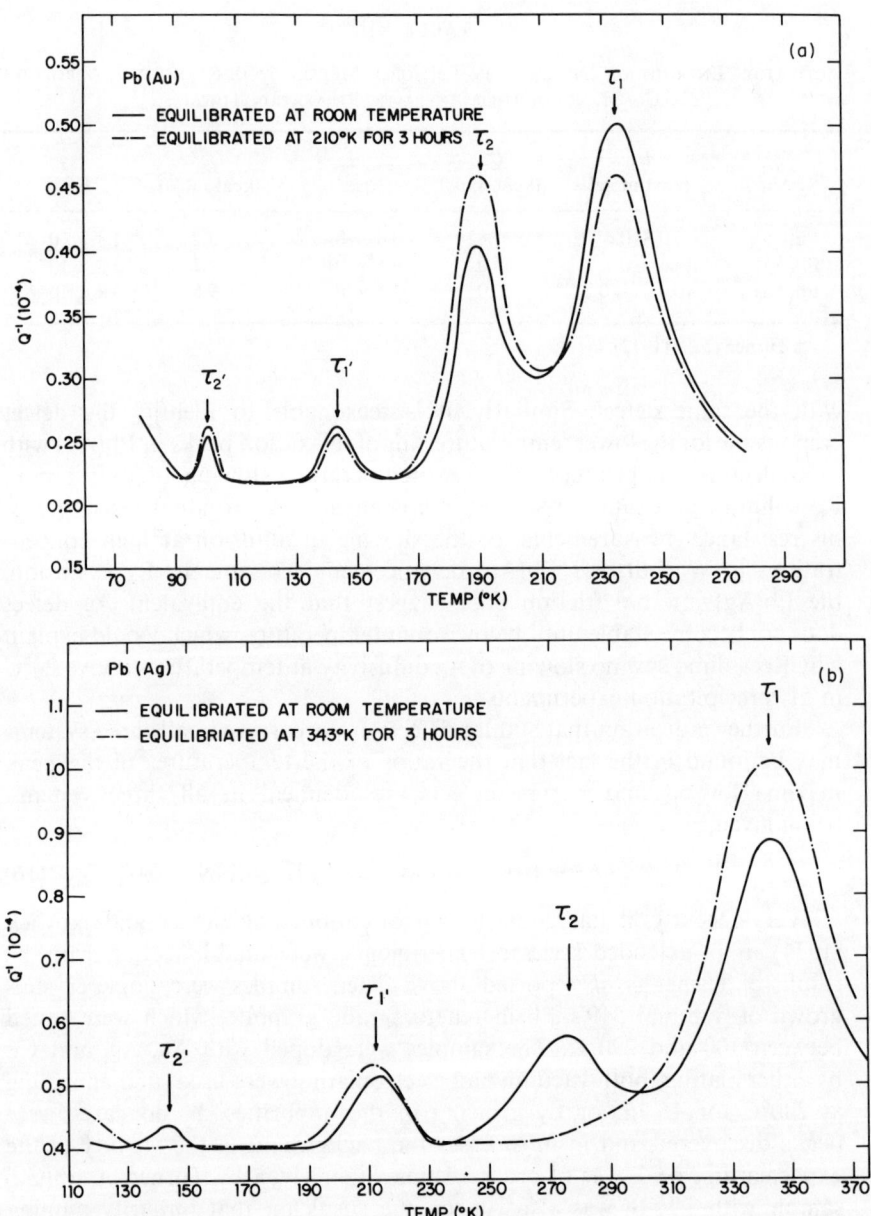

Fig. 10. The internal friction as a function of temperature for $\langle 100 \rangle$ single crystals of lead doped with equilibrium concentrations, at room temperature, of (a) gold, (b) silver, and (c) copper. The samples were equilibrated at the indicated temperatures and quenched to 77°K prior to the runs.

systems. It is our own experience that reliably plating Pb with Au for diffusion studies is no easy matter. Furthermore, the precipitation studies of Rossolimo (1971) on Pb(Au) indicate dislocations as the most probable nucleation sites for Au precipitation, at a supersaturation ratio of about 2 : 1. In this view, two explanations are possible. The first is that an unknown impurity in Turner's samples reduced the background by pinning dislocations and simultaneously poisoned them as precipitation sites, allowing large amounts of noble metal impurity to remain in solution. Alternatively, Sagues and Nowick, with their higher concentrations, may have exceeded the critical supersaturation ratio, allowing precipitation to occur. An explanation more damaging to our present exposition is that the unknown impurity in Turner's systems is also directly responsible for the internal friction peaks when combined with a noble metal. In this case it is difficult to see how the noble metal heats of diffusion are reproduced so well. The fact remains that the reliability of Turner's findings is now questionable and that further research will be necessary.

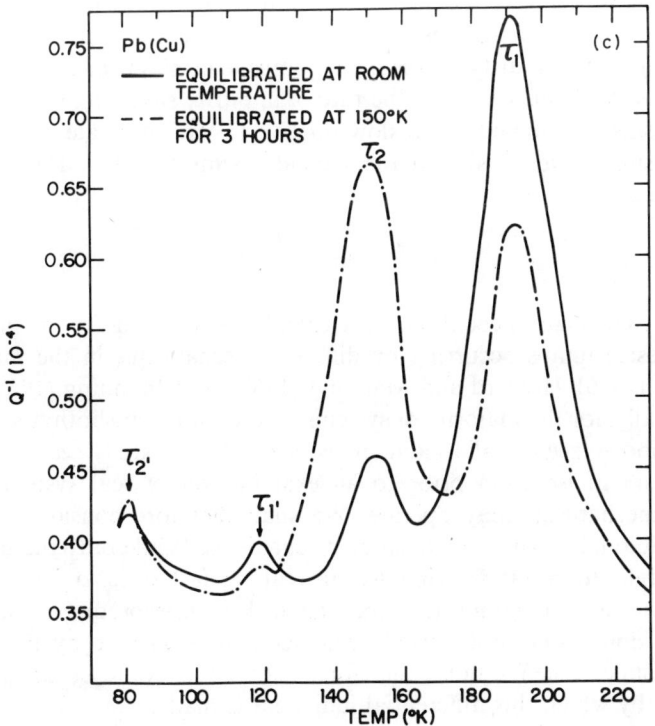

IV. Fast Diffusion Mechanisms

A. Introduction

All the models developed to describe fast diffusion behavior share certain common assumptions. First, the diffusion is assumed to proceed by a bulk mechanism rather than along surfaces or line defects for the reasons given in Section II.B.1. Second, the mechanisms operating in any given system are assumed to be separable allowing the total impurity diffusivity to be written as the sum

$$D_t = \sum_i c_i D_i \tag{17}$$

over all the mechanisms available to the impurity, c_i being the average fraction of time spent in the ith mechanism, and D_i the diffusivity of that mechanism being given by

$$D_i = \tfrac{1}{6} a_i^2 \Gamma_i f_i \tag{18}$$

where a_i^2, Γ_i, and f_i are the jump length, jump frequency, and correlation coefficient, which is calculated as if only that single mechanism were operating in the system. This treatment is equivalent to assuming that, on the average, the impurity makes many jumps via a single mechanism before it changes mechanisms. Thus, the rate of impurity sampling of the various mechanisms is assumed to be slow relative to its jump rate in any single mechanism. However, should the second assumption fail, Eq. (17) must be replaced by

$$D_t = \sum_i \tfrac{1}{6} a_i^2 \Gamma_i f_i \tag{19}$$

where the correlation coefficients are calculated, allowing for the possibility of successive jumps occurring by different mechanisms, in the manner of Howard (1966), Howard and Manning (1967), and Manning (1967).

Actually, for the majority of systems studied thus far, both assumptions stated above seem valid, and it is sensible to study each diffusion mechanism in isolation prior to an examination of real systems, where several mechanisms may operate. We shall therefore consider the interstitial, diplon, i–v-pair, and vacancy mechanisms. While only the first three are thought to result in what we are calling fast diffusion, the vacancy mechanism is so common that the first task in interpreting an unusually fast diffusion system is to decide whether it is explicable by the vacancy mechanism or not. We must, therefore, examine it too, in order to indicate the tests by which this differential judgment is made.

B. THE VACANCY MECHANISM

This mechanism has been treated in great detail in the review by Peterson (1968) and, for the most part, only results will be presented here. Briefly, impurity atoms are assumed to be dissolved substitutionally, to move only by exchange with nearest neighbor vacancies and to affect diffusion processes in the alloy only by local modifications of atom–vacancy-exchange rates. If "local" is taken to mean "nearest neighbor", the resulting model and jump frequencies are indicated in Fig. 11. The four frequency ratios w_1/w_0–w_4/w_0 are not independent since w_3 and w_4

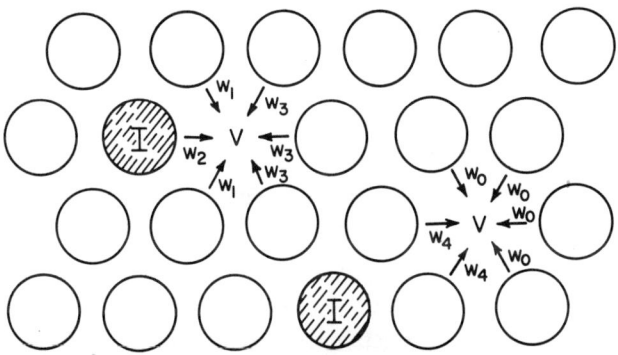

Fig. 11. Model of the vacancy mechanism of diffusion. Shown is a (111) plane of a fcc lattice, indicating the jump frequencies of atoms into both associated and dissociated vacancies. Key: open circle: solvent atom; filled circle: impurity atom; missing circle: vacancy.

are related so as to maintain the population of nearest neighbor impurity–vacancy pairs in equilibrium. Lidiard (1960) has shown this relationship to be

$$w_4/w_3 = \exp(-E_b/RT) \tag{20}$$

where E_b is the energy of association of the pair. The three independent frequency ratios are then usually taken to be w_2/w_1, w_3/w_1, and w_4/w_0.

The three major factors which determine whether an impurity diffuses faster than its host by the vacancy mechanism are (1) the binding between impurities and vacancies, (2) the impurity jump rate into vacancies, and (3) the correlation of these jumps. The first factor usually is determined by whether the impurity is electropositive or electronegative compared to the host and thus attracted or repelled by vacancies which have a net negative charge. Strains due to misfit of the impurity with the host (Swalin, 1957) also can bind vacancies to the impurities, but this effect, which was considered by Anthony (1970a) in the interpretation of fast diffusion, seems to be much less important than the coulombic effects. The third factor, correlation, which is important since it can effectively cancel the effects of a fast impurity–vacancy exchange rate, is determined by the relative jump rates of the impurity and host atoms bounding a vacancy. Thus both the second and third factors are fundamentally connected to the barriers to diffusion seen by host and impurity atoms in the various configurations encountered. There is currently no treatment available by which either these barriers or the impurity–vacancy association energies can be generally calculated. The state of the art, an electrostatic theory introduced by Lazarus (1954) and refined by LeClaire (1962), has fair success only with electropositive solutes in the noble metals. This theory represents a diffusing atom and its neighboring vacancy (or pair of partial vacancies at the saddle point of the jump) as point charges interacting through a screened potential. LeClaire thus obtains, for the difference in heats of diffusion for an impurity and host atom in the same lattice

$$\Delta Q_D = Z_v Z_i e^2 \alpha (16/11a) \exp(-11qa/16) \tag{21}$$

where Z_i, the effective charge of an impurity on a lattice site, is the difference between host and impurity valences; Z_v, the vacancy's effective charge, equals the host valence; e is one electronic charge, a the separation, q the screening constant, and α a constant dependent on the Z's. Such a treatment implies that monovalent or divalent impurities should diffuse more slowly in polyvalent hosts than the host itself, contrary to observation in systems such as Pb(Au) and Pb(Cd). The case against the vacancy mechanism in such systems is further strengthened by the calculation of March and Murray (1961) of the electrostatic potential surrounding a point charge in a polyvalent metal, which indicates that electron densities are so high as to screen defect charges in less than an atomic radius and so eliminate impurity–vacancy binding entirely. Similar considerations should also affect impurity energies at the saddle point in diffusive jumps. Coupled with the LeClaire treatment, this implies that all impurities in polyvalent hosts should diffuse at essentially the host self-diffusion rate. As reported above, such behavior is approximated by In and Hg in Sn, and by Sn and Tl in Pb. The diffusion of a variety of impurities in Al with essentially constant heat of diffusion has also been cited in this context (Miller, 1969a; Peterson and Rothman, 1970), but NMR work of Rowland and Fradin (1969) points to the existence of very high impurity jump frequencies in these systems, suggesting that they may be more complex than indicated by the above analysis. The general conclusion remains, however, that in terms of these two theories, fast diffusion should not occur in polyvalent metal hosts by a vacancy mechanism.

Several relationships between observable quantities may be derived for the vacancy mechanism in terms of the three independent jump frequency ratios and can provide a clear-cut test of the applicability of this model to a given diffusion system when sufficient experimental data are available. The first relation, between solvent and solute diffusivities, was derived by Peterson (1968) from Lidiard's now classic expressions for $D_2(0)$ and $D_1(0)$, the solute and solvent diffusivities at zero impurity concentration:

$$D_2(0) = (z/6)a^2 f_2 w_2 \exp(-E_{v_2}/RT) \tag{22}$$

$$D_1(0) = (z/6)a^2 f_0 w_0 \exp(-E_{v_1}/RT) \tag{23}$$

where the w's are the model jump frequencies, the f's are the correlation coefficients for these jumps, a is the jump distance, z is the coordination of the lattice, and the E's are the energies of forming a vacancy next to an atom about to jump. Dividing (22) by (23) and noting that $E_{v_2} - E_{v_1}$ is just E_b, the binding energy of a vacancy to the impurity gives, using Eq. (20) for E_b,

$$D_2(0)/D_1(0) = (f_2/f_0)(w_2/w_1)(w_4/w_0)(w_1/w_3) \tag{24}$$

The second relation is for f_2 itself, which Manning (1962, 1964) has evaluated as a function of the three jump frequency ratios, finding

$$f_2 = \frac{1.0 + 3.5 \, F(w_4/w_0)(w_3/w_1)}{1.0 + (w_2/w_1) + 3.5 \, F(w_4/w_0)(w_3/w_1)} \tag{25}$$

where

$$F(X) = 1.0 - \frac{1}{7} \frac{10X^4 + 180.5X^3 + 927X^2 + 1341}{2X^4 + 40.2X^3 + 254X^2 + 597X + 435} \tag{26}$$

and goes from 1.0, as w_4/w_0 goes to zero, to $\frac{2}{7}$ as w_4/w_0 becomes infinite. Experimentally, f_2 is measured by an isotope experiment, as has been fully discussed by Peterson (1968). The third expression is for the linear coefficient of enhancement of the self-diffusion b_{11} defined in Eq. (8). Hoffman et al. (1955, 1957) recognized the necessity for such an effect, but it remained for Howard and Manning (1967) to extend Lidiard's (1960) treatment of the effect to include solvent atom correlations and obtain

$$b_{11} = -18.0 + (4.0/f_0)(w_4/w_0)[\chi_1(w_1/w_3) + \tfrac{7}{2}\chi_2] \tag{27}$$

where the χ_1 and χ_2 are partial correlation coefficients for w_1 and w_2 jumps and are tabulated for a variety of values of the ratios w_2/w_1, w_4/w_0, and w_3/w_1. Starting with this expression, Miller (1969a) demonstrated that the smallest value of b_{11} compatible with a given finite value of $D_2(0)/D_1(0)$ could only be obtained by simultaneously letting $w_4/w_0 \to 0$

and $w_2/w_1 \to \infty$. Under these conditions, w_3/w_1 must also go to zero [see Eq. (20)] and the values for χ_1 and χ_2 are such that

$$b_{min} = -18.0 + 1.9448[D_2(0)/D_1(0)] \tag{28}$$

as previously noted. Miller was also able, by making several rather drastic simplifying assumptions, to extend Lidiard's treatment to the problem of impurity enhancement of its own diffusion through the formation of trimers consisting of a vacancy and two impurities (1969c). The expression he obtained, as corrected by Warburton (1973b), is

$$b_{21} = 10[\exp(-E_{b_2}/RT) - 1.0] \tag{29}$$

which, as a result of his assumptions, depends only on the energy of binding a second impurity to an impurity–vacancy dimer E_{b_2}.

Equations (24), (25), and (27) or (28) can be used to compare the predictions of the vacancy model to the observed behavior of a given diffusion system. The reliability of such comparisons will clearly depend upon the type and quality of data available. If only the ratio $D_2(0)/D_1(0)$ is known, then only the remarks made concerning the theory of LeClaire [Eq. (21)] are applicable. If b_{11} has been measured as well and is smaller than b_{min}, then the vacancy mechanism can be ruled out entirely. Systems where b_{11} is close to this boundary, as in the case of Pb(Hg), should be examined very carefully before being interpreted in terms of a vacancy mechanism, if only because of the physically extreme conditions required to achieve the minimum. When an isotope experiment has also been done and a value for f_2 obtained, then Eqs. (24), (25), and (27) form a set of three equations in the three unknowns w_2/w_1, w_3/w_1, and w_4/w_0. Thus Rothman and Peterson (1967) solved for the values of these ratios in the Ag(Zn) system. Should no solution exist, however, as in the case of Pb(Cd) (Miller and Edelstein, 1969), then the behavior is again incompatible with the vacancy model. Such calculations are easily done by hand, assuming a value of w_4/w_0 and iteratively calculating consistent values of w_2/w_1, w_3/w_1, and thus f_2, for comparison with experiment. Warburton (1973c) used this method to determine a maximum value of f_2 for the system Pb(Hg) if the diffusion were to go by a vacancy mechanism. In summary, our survey has indicated the following critical tests for the vacancy mechanism:

(1) A solute showing valence deficiency should also exhibit slower diffusivity than its host, except in polyvalent solvents, where the two should be nearly equal.

(2) The existence of b_{min} provides a strong experimental test of the applicability of the vacancy model.

(3) The experimental values of b_{11}, f_2, and $D_2(0)/D_1(0)$ must form a consistent set if diffusion occurs by the vacancy mechanism.

C. Interstitial-Type Mechanisms

In the general interstitial mechanism, transport is effected by a defect which is interstitial in the Simmons–Baluffi (1960, 1962) sense, i.e., the total number of lattice sites is decreased when the defect forms from atoms which are initially substitutional. The average local configuration around the defect will vary with the nature of the interaction between the interstitial and its neighboring solvent atoms. We will consider here two distinct types of interstitial configurations. In one, the interstitial atom is sited, on the average, at equal distances from several neighboring substitutional atoms (Fig. 12). We shall call this defect the "conventional" interstitial.

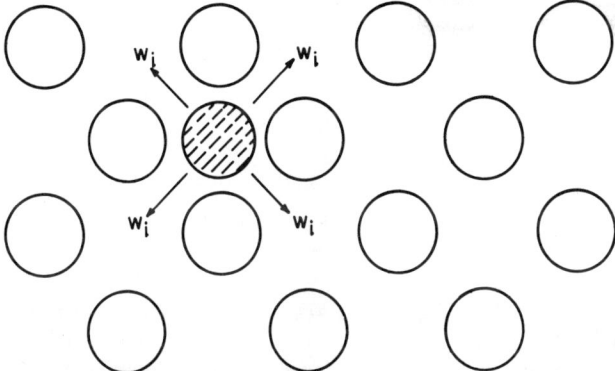

Fig. 12. Model of the interstitial mechanism of diffusion. Same view as in Fig. 11, except that the (100) plane is shown.

An example would be an atom at the center of an octahedral interstice in a close packed structure. In the second configuration, illustrated in Fig. 13a, two atoms are paired at a single substitutional position of the structure. The pair may be composed of host–host, host–impurity, or impurity–impurity atoms. In the literature, this type of defect has been called split-interstitial, off-center interstitial, disubstitutional or dumbbell, but not consistently. None of these terms seems quite satisfactory to us. For example, the term disubstitutional seems to suppress the sense that the defect is interstitial in the Simmons–Baluffi type experiment, while the term split-interstitial has been used traditionally for another type of defect (Peterson, 1968); moreover, when applied to impurity–impurity pairs, this term does not indicate that a host atom is missing (Ray et al., 1972). Therefore, we have coined and offer for consideration the designation "diplon" for this kind of defect.

Since a diplon has only rotational symmetry about its axis, the point symmetry of the lattice about the occupied lattice site will be modified,

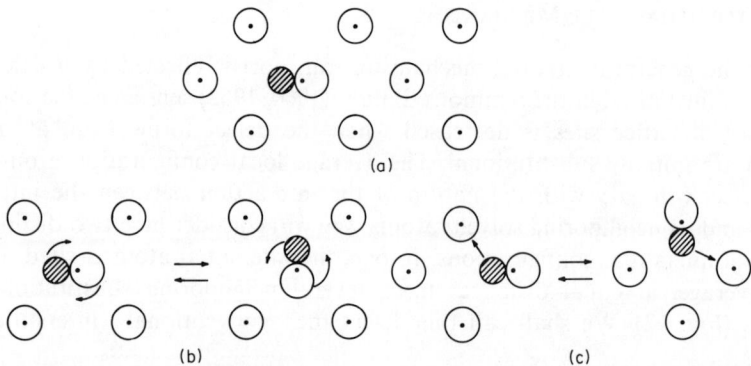

Fig. 13. Model of the diplon mechanism of diffusion. Same view as in Fig. 12. Lattice sites are also indicated. (a) shows a single diplon. (b) and (c) show the reorientation effects of two possible types of diplon motion.

allowing the diplon to interact elastically with applied strain fields. In cases where the orientation of the diplon axis is important, it can also be indicated. When the diplon axis lies along some axis of symmetry of the lattice, this symmetry may be retained or may degrade in Jahn–Teller fashion as the host structure relaxes about the defect. For a full discussion of symmetry effects, see Nowick and Heller (1965). Such modifications of the structure symmetry can often be deduced from internal friction experiments, which thus become a major tool for deciding whether impurity atoms in a solution exhibiting interstitial behavior are in a conventional (C) interstitial or diplon state.

Diffusive transport of impurity may be effected by a motion of the host–impurity (HI) diplon involving reorientation of the diplon and jumping of the impurity from one host atom to an adjacent one. The formal theory for diffusion by this mechanism, as well as for the equilibrium between the diplon and other impurity states, is the same as for the conventional interstitial.

Thus for either type of interstitial defect, the free energy of activation for diffusion will, provided the vacancies are in equilibrium with the external surface, be the sum of $g_1^0 + g_m$, where g_i^0 has already been defined as the standard free energy for forming an interstitial from a substitutional atom and g_m is the free energy required for a diffusive jump of the defect. g_i^0 may be positive or negative and we expect that g_m, although its magnitude may be small, will be positive.

In the following, we examine and compare some of the expected characteristics of diffusive motion by each of the two interstitial mechanisms.

4. FAST DIFFUSION IN METALS

1. *Conventional (C) Interstitial Mechanism*

As indicated in Fig. 12, this mechanism involves only a single jump frequency, w_i. No other defect is required for the jump step and it is uncorrelated. Thus $f = 1.0$ and

$$D_i = (Z/6)a^2 w_i \tag{30}$$

where Z is the number of accessible interstitial sites and a is the separation between them. It is difficult to make any further quantitative statements about this mechanism. Any attempt to relate w_i to more basic atomic parameters requires a solution of the sort of problems dealt with by LeClaire (1962) in the development of the vacancy mechanism. Similar problems arise in developing a theory for the ΔK term in the product $f \Delta K$ measured in isotope experiments. In the case of Si(Li) reported above, (Section II.A), $f \Delta K \simeq 0.94$, so that ΔK must be rather close to unity.

Qualitative inferences about crystalline anisotropy effects on diffusion can be drawn from examination of the connectivity between similar types of interstitial voids. Thus, in close packed structures, there are two types of voids, tetrahedral and octahedral, and twice as many of the former as the latter. In the hcp structure, the octahedral voids share common faces in the c direction but not in the a direction, while in the fcc structure, they share no common faces, only common edges. The tetrahedral voids share only common edges in both structures. Because the polyhedra edges represent nearest neighbor atoms; it seems most likely that transition paths between interstices which pass through edges will be energetically much less favorable than paths which pass through faces. An impurity diffusing as a C interstitial in a fcc host must then pass through both types of voids as must an impurity diffusing in the a direction in a hcp host. In the hcp c-axis direction, on the other hand, only octahedral sites need to be occupied. Anthony et al. (1968) have thus interpreted the faster c-axis diffusion of Au and Ag in Tl in terms of C-interstitial motion.

The prediction for tin, which is body-centered tetragonal with two atoms per lattice point, is that diffusion should be faster in the c direction, where interstitial voids form essentially continuous channels of almost constant diameter, than in the a direction, where passage is far more restricted. The experience on the diffusion of the noble metals in tin (Section II.B.4) accords remarkably well with this prediction and this is reflected by the activation energies which are much smaller for diffusion along the c than along the a axis.

It has been assumed that the enhancement coefficient, b_{11}, would be zero or very small for the interstitial mechanism (Peterson, 1968) but there has been no demonstration that this is a necessary feature of the model.

Interstitials could influence vacancy jump rates in their vicinity through strain fields or through other interactions. Thus, the existence of an enhancement coefficient cannot be taken *per se* as evidence that some mechanism other than interstitial diffusion is operating in an alloy.

In conclusion, fast diffusion values of $f \Delta K$ near unity and tiny enhancements relative to the impurity diffusivity have been typical of the systems in which the C-interstitial mechanism of diffusion has been considered to be predominant. When taken together, these characteristics form a convincing argument for the mechanism, yet none of them, taken singly, is strictly required by the model.

2. *Diffusion by Host–Impurity (HI) Diplons*

Mechanisms for diplon diffusion have not yet been generally elucidated but, from the constraints of spatial geometry, certain of their features are clear. Since either atom of a host–host diplon can remain at the lattice site when the diplon moves on, such a defect moves essentially like a self-interstitial. An impurity–impurity diplon, on the other hand, must perform some sort of exchange with neighboring host atoms if it is to move. If it is tightly bound as well, its mechanism of motion must be somewhat similar to that of a substitutional impurity. Notice that it can reorient itself in an applied strain field, and thus show an anelastic effect, but that these reorienting movements are not necessarily capable of producing diffusion. Finally, there is the host–impurity diplon, which might move by a vacancy mechanism, but will be most interesting when it does not. Then, in structures with closed interstitial voids, in the sense that the energy barriers to passage out of a given void through its various faces are all approximately equal, and as opposed to structures like white Sn, where the voids connect to form channels along the c axis, neither a diplon motion by translation of the impurity atom about the void w_t nor a rotation of the diplon about its lattice site w_r is sufficient to cause net diffusion of the impurity atom (see Figs. 13b and 13c). Both are required, or else a single motion which produces the effects of both (w_c) is required, if the impurity is to show any long range motion. Furthermore, if the barriers to the appropriate motions are not too high, the resulting diffusion can be very fast, as is the case of C interstitials. A comparison between diffusion and internal friction data will decide between the two possibilities. Thus, if two motions w_r and w_t are producing diffusive motion, then the one with the higher activation energy will be rate controlling. In internal friction experiments, on the other hand, it is the faster step which produces equilibrium after an applied strain and which is thus measured (Nowick and Berry, 1972). So the two measurements will produce different energies, the diffusion being sensitive to the high energy step. Should these two energies be identical, then either there has been an unlikely coincidence, or else only a single-defect motion exists. Such a correspondence would also imply that the formation energy of the defect is zero or negative.

Under certain circumstances the presence of host–impurity diplons may also lead to a reduced mass effect in a diffusion isotope experiment. In the case of two motions w_r and w_t producing its diffusion, and w_r being rate controlling, it is clearly the mass of the pair which is important in the frequency of the pair's rotation. The isotope effect in such a system will clearly be smaller than in a system where the diffusing defect is a single impurity atom. (This is shown in Table VII for several systems.)

TABLE VII

"Naive" Values of "$f \Delta K$"†

System	$n = 0$	$n = 1$	$n = 2$
Pb(Au)	1.00	0.484	0.320
Pb(Ag)	1.00	0.339	0.204
Pb(Cu)	1.00	0.238	0.135
Sn(Ag)	1.00	0.472	0.309
Sn(Cu)	1.00	0.352	0.214

† Expected from isotope measurement if moving defect is actually a noble metal atom plus n host atoms, assuming the true $f \Delta K$ is equal to unity. Calculated from $D_\beta/D_\alpha - 1 = f \Delta K \times ((M_\alpha/M_\beta)^{1/2} - 1)$.

Similarly, if some sort of compound motion is producing diplon motion, then the masses of the participating atoms will have to be appropriately included in the expression for the jump frequency.

It is easy to see that the host diffusivity might be enhanced considerably by dissolution of impurity as HI diplons. The host atom in this defect is displaced in a way which ought to greatly facilitate its exchange with a neighboring vacancy. This effect may account for the rather large enhancement coefficient for the Au impurity in Pb reported by Miller (1970).

In conclusion, diplons may be expected to be distinguished from conventional interstitials or substitutionals both by their response to applied stress fields in internal friction experiments and, under certain circumstances, by a reduced mass effect in an isotope experiment.

D. Interstitial–Vacancy Pairs

This mechanism was originally viewed by Miller (1969c) as a means of achieving essentially one-for-one exchanges of impurity and solvent atoms without invoking a direct-exchange mechanism. The model Miller developed (1969c, 1969b) is indicated schematically in Fig. 14a, where the various jump frequencies are labeled. The i–v pairs are created with frequency v_1, annihilated with frequency v_2, dissociated by interstitial jumps k_1 and by solvent jumps w_1, associated by solvent jumps w_3, and maintained in association by interstitial jumps k_2 and by solvent jumps w_2. Here and throughout we use the same label to designate both a jump type and its jump frequency. Since fixed concentrations of both i–v pairs and interstitials must be maintained in equilibrium, there will be relations similar to Eq. (20) between v_1 and v_2 and also among k_1, w_1, and w_3, so that there are only five independent frequencies in the model. The model

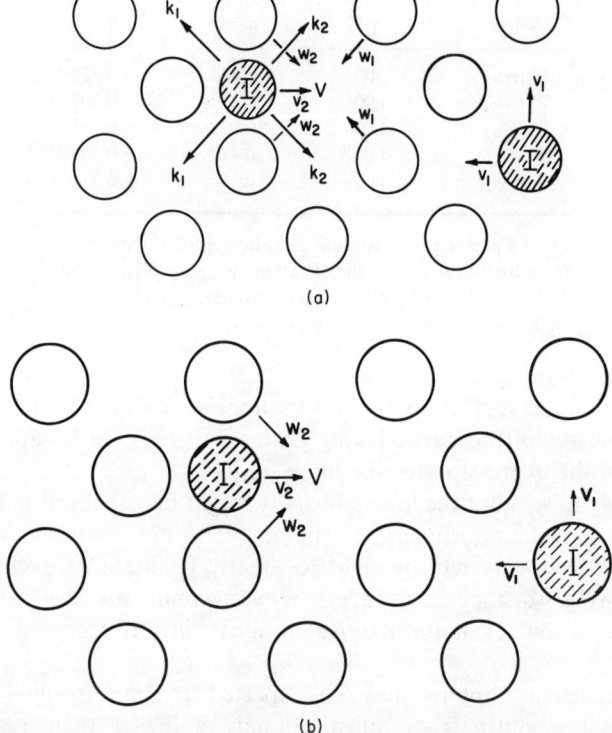

Fig. 14. Model of diffusion by interstitial–vacancy pairs. Same view as in Fig. 12. (a) shows the model of Miller, (b) the model of Warburton.

4. FAST DIFFUSION IN METALS

is of most interest in the regime where interstitial formation is energetically difficult, so that frequencies k_1 and w_1 are assumed to be small.

Miller developed the model particularly for the case where the frequency k_2 is much larger than any of the others. In this limit, the impurity essentially orbits about the vacancy in a series of k_2 jumps without occupying it. When a solvent atom makes a w_2 jump into the vacancy, the impurity proceeds to orbit about the new vacancy position, which leads to an effective exchange of the solvent and impurity atoms. The result of these exchanges is that

$$b_{11} \approx D_2(0)/D_1(0) \tag{31}$$

Since most of the jumps are of type k_2, which does not cause any effective diffusion, one should also expect to find that

$$f_2 \approx 0 \tag{32}$$

with the diffusion being limited by w_2 jumps. Miller therefore proceeds with an analysis of the type of Lidiard, assuming that the correlation coefficients for all types of solvent atom jumps are identical, and finds that

$$D_2(0) = a^2 \exp[-(E_{v_1} + I + B)/RT]$$
$$\times \left[\frac{(4k_2 + v_2)(4w_2 + 8w_1 + 12k_1)}{4w_2 + 8w_1 + 4k_2 + 8k_1 + v_2} + 8k_1 \right] \tag{33}$$

and

$$b_{11} = \frac{f_0 D_2(0)}{D_1(0)} \left[1 + \frac{5k_1}{w_2 + 2w_1} \right]^{-1} \tag{34}$$

where a is the jump distance, f_0 is the correlation coefficient in the host material, and E_{v_1}, I, and B, respectively, are the energies of forming an isolated vacancy, forming an isolated interstitial, and associating the two. In the case where the defects are tightly bound, we have $k_2, v_2 \gg w_2 \gg w_1, k_1$, so that dividing (33) by the host diffusivity $D_1(0)$ gives

$$D_2(0)/D_1(0) \cong (2w_2/w_0 f_0) \exp[-(I + B)/RT] \tag{35}$$

showing that the diffusion is controlled by solvent w_2 jumps and that it can become arbitrarily large for appropriate w_2/w_0 ratios. Also, in this limit,

$$b_{11} \cong f_0 D_2(0)/D_1(0) \tag{36}$$

so that the assumption of tight binding is in accord with the experimental result that

$$b_{11} \cong D_2(0)/D_1(0) \tag{37}$$

the data not being sufficiently accurate to distinguish between these two expressions. Furthermore, according to Eq. (35), the diffusivity would be essentially independent of the impurity jump; hence, in accord with experience, it should vary little with the isotopic mass.

Warburton (1973d) reexamined this model following his experiments on the Pb(Hg) system. He found that the diffusivity behavior of this system is very similar to that of the Pb(Cd) system, but that

$$b_{11} \cong 1.5 \, D_2(0)/D_1(0) \tag{38}$$

considerably exceeding the upper limiting value [Eq. (35)] permitted by the Miller model. He noted that, if the i–v pair were relatively more stable, so that $w_2 \gg v_2$, and also relatively immobile, so that k_2 would be small, then several solvent atoms might exchange with the vacancy before it was annihilated again. In such a process, the ratio of solvent atom jumps to solute atom jumps could easily exceed unity and b_{11} correspondingly could exceed the ratio $D_2(0)/D_1(0)$. Because this effect is strongly dependent on the correlations of the jumping solvent atoms, the model had to be greatly simplified in order to calculate these correlations explicitly. Thus frequencies k_2, k_1, and w_1 are set equal to zero, corresponding to a relatively immobile, tightly bound defect, as schematically shown in Fig. 14b. Since v_2/v_1 is such as to maintain the population of i–v pairs, there are only two independent parameters in the model: $w_2/v_2 = K$ and v_2/w_0, where w_0 is the intrinsic jump frequency for host atoms.

With these assumptions, Warburton (1973d) calculated

$$D_2(0)/D_1(0) = (12x_v f_0)^{-1} \, (x_p/x_2)(v_2/w_0)[1 + (4K)^{-1}]^{-1} \tag{39}$$

so that

$$f = [1 + (4K)^{-1}]^{-1} \tag{40}$$

while

$$b_{11} = f_2'(4K + 1) \, D_2(0)/D_1(0) \tag{41}$$

In these relations, x_v, x_2, and x_p are, respectively, the concentrations of vacancies, solute atoms, and i–v pairs in the alloy; f_2 is the correlation coefficient for impurity diffusion; and f_2' is that for solvent atoms making w_2 jumps. The diffusivity ratio may be made arbitrarily large by increasing the ratio v_2/w_0, providing only that K remains large enough to keep the motion from becoming excessively correlated. Values of f_2' were calculated numerically, using the method of Howard (1966), Howard and Manning (1967), and Manning (1967) to obtain values of $f_2'(4K + 1)$, which correspond to α in Eq. (12), as a function of K. These results are indicated in Fig. 15. It should be noted that the

4. FAST DIFFUSION IN METALS

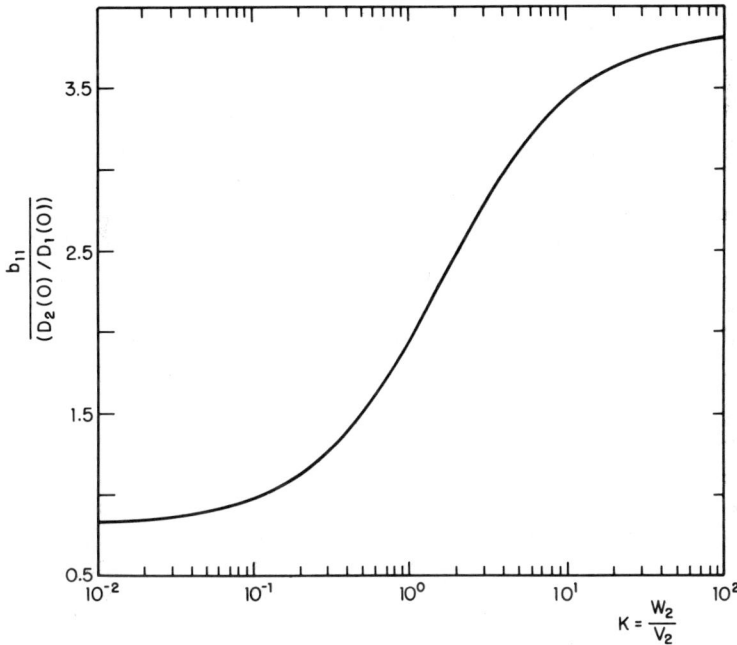

Fig. 15. Results of numerical calculation of $b_{11}/[D_2(0)/D_1(0)]$ as a function of the jump frequency ratio $K = w_2/v_2$, for Warburton's model of the i–v-pair mechanism.

values of α allowed in this version of the i–v-pair model range from approximately f_0 to ~ 3.80, as K goes from zero to infinity, while f_2 goes from zero to unity over the same range. Since both α and f_2 depend only on the single parameter K, a measurement of both provides a reasonable test of the consistency of the model with the data from any particular system.

In conclusion, the i–v-pair model has been developed in two approximations for comparison with experimental data. Both assume tight binding of the defect $(w_1, k_1 \approx 0)$ but the model of Miller assumes that the impurity is very mobile (large k_2) while that of Warburton assumes it is not $(k_2 \approx 0)$. A priori, neither set of assumptions is preferable. Rather, determining which model provides the best explanation of a particular system will tell something about the nature of the i–v-pair defect in that system.

E. Choosing among the Mechanisms

Having outlined the chief features of the various models currently developed, we now group them together in abbreviated form for their utility both in deciding which model offers the best description of a particular system and for suggesting critical tests on those systems which have not yet been fully investigated. This is done in Table VIII. The limits indicated in this table, of course, are not absolute, but are rather meant only to be typical of the systems which have been clearly associated with the indicated mechanisms. Thus, for example, while the diffusivity ratio in Ag(Ge) is nearly 60, the diffusion still seems to occur predominantly by the vacancy mechanism (Dyson et al., 1967). Similarly,

TABLE VIII

CHARACTERISTIC PROPERTIES OF DIFFUSION BY VARIOUS MODELS

Model	$D_2(0)/D_1(0)$	b_{11}	f
Vacancy	Order of 10 or less	$b_{11} \geq b_{min}$	Consistent with both b_{11} and $D_2(0)/D_1(0)$
Interstitial	Order of 100 or more	$b_{11} \ll b_{min}$	Unity expected
Diplon	Order of 100 or more	$b_{11} \ll b_{min}$	Unity expected if correct mass used
Interstitial-vacancy pair (Miller)	Order of 10 to 100	$b_{11} \sim b_{min}$ $b_{11} \leq f_0 \dfrac{D_2(0)}{D_1(0)}$	Small
Interstitial-vacancy pair (Warburton)	Order of 10 to 100	$b_{11} \sim b_{min}$ $f_0 \dfrac{D_2(0)}{D_1(0)} \leq b_{11} \leq 3.8 \dfrac{D_2(0)}{D_1(0)}$	$0 < f < 1$, but consistent with value of b_{11}

while there is no theoretical reason why interstitial diffusion need be fast, the experimental situation is such that, if it is not, it cannot be distinguished readily from other mechanisms.

Final grounds for choosing one diffusion mechanism over another may come from nondiffusion measurements. Thus, a major difference between conventional interstitials and diplons lies in their responses to applied strains in an internal friction experiment. We must realize that the mechanisms presented here are by no means a closed set and that other mechanisms and models may provide better explanations of existing facts or become necessary to explain future observations. In this vein, a further use for Table VIII is the isolation of systems which are not well described by the models developed to date.

V. Interpretation of Fast Diffusion Behavior of Particular Systems

Studies of certain systems, for example Pb(Au), have been extensive enough to provide important tests of the models for fast diffusion described in the preceding section as well as a stimulus for further model development. Here we discuss these tests with some of the results surveyed in Section II.

A. Pb(Au)

The small enhancement, b_{11}, coupled with Au's extremely rapid diffusion, rules out the vacancy, direct-exchange, and i–v-pair mechanisms as the dominant ones for diffusion in this system (Miller, 1970). This leaves the interstitial-type mechanisms, either conventional or diplon, as possibilities. The internal friction data of Turner et al. (1972) would indicate that the defect responsible for diffusion has $\langle 100 \rangle$-orthorhombic symmetry, favoring the diplon mechanism. In this case, since the enthalpy for diffusion is the sum of the enthalpies of defect formation and motion, its equivalence to the activation energy for anelastic relaxation, which is the barrier to defect motion, would further suggest that the enthalpy of forming Au–Pb diplons from substitutional Au is negligible and that the mode of motion is unique. These conclusions are subject to the provisos mentioned in Section III.H concerning the work of Sagues and Nowick (1974).

The other observations on this system can best be understood in terms of the existence of a thermodynamic equilibrium between two types of Au defects which, since the equilibrium shows concentration dependence, must contain different numbers of Au atoms. Again, if the data of Turner et al. (1972) are correct, the internal friction behavior would indicate that this defect also has $\langle 100 \rangle$-orthorhombic symmetry. In this case, the simplest assumption is that the high temperature defect is a Au–Pb diplon while the low temperature defect is a Au–Au diplon. It should be emphasized that the existence of an equilibrium between two defect types in Pb(Au) is principally indicated by the dehancement and precipitation experiments, while the internal friction results are of primary utility in speculation concerning their geometric configurations. Lacking the constraint of the internal friction measurements, many other defects can be considered.

Rossolimo and Turnbull (1973), for example, suggested diinterstitial–vacancy triplets in explaining the Pb(Au) precipitation results. The Au–Au diplons (or another low temperature defect) must be assumed to diffuse much more slowly than the Au–Pb (or another high temperature defect) in order to explain both the dehancement and precipitation results. This behavior appears to be inconsistent with the anelastic data, which show the Au–Au diplon's relaxation motions to occur at lower energies than those of the Au–Pb diplon. As indicated in Section IV.C.2, this inconsistency may be resolved if the relaxation motions are incapable of producing effective diffusive motion. For example, the Au–Au diplons might be so strongly bound that they would require some second defect, such as a vacancy, in order to diffuse.

In these terms, the Pb(Au) system's behavior may be explained as follows. At low concentration, entropic effects, both configurational and vibrational, assure that the Au is entirely dissolved as Au–Pb diplons. At constant concentration, as the temperature drops, these effects decrease in importance and more Au–Au diplons are formed until, at sufficiently low temperature, all the Au is so dissolved. Essentially, it is the resultant decrease in number of scattering centers which accounts for the changes in Au residual resistivity observed by Rossolimo (1971). Similarly, at a fixed temperature, increases in Au concentration will drive the equilibrium toward increased Au–Au diplon concentrations since energy of formation terms will increase faster than entropy terms. This increased fraction of Au–Au diplons is accompanied by a resultant decrease in Au diffusivity, which accounts for the negative enhancement results. Decreasing temperature decreases the entropic stabilization of the Au–Pb diplons so that the dehancement effect becomes even more pronounced and the apparent Au diffusivity observed in precipitation from supersaturated solutions falls further from the value extrapolated from high temperature measurements.

Thus, we have an apparently consistent view of the behavior of the Pb(Au) system, subject to the following provisos. First, as mentioned in Section III.H, the ambiguities concerning internal friction peaks must be resolved. Second, some theoretical work needs to be done on the mechanisms of diplon diffusion in order to explain the relation between the internal friction peaks and the observed diffusion behavior. Third, it is possible, by measuring dehancement coefficients carefully (Warburton, 1973a) and also by studying the anneal temperature dependence of the internal friction peak heights, to obtain two independent measurements of the difference in formation energy between the two diplons. These experiments should be done and checked for consistency with the value obtained from the resistivity measurements. Fourth, the precise distribution

4. FAST DIFFUSION IN METALS

of Au impurity between diplon and other, particularly substitutional, defects may require determination.

B. POLYVALENT SOLVENT (NOBLE METAL) ALLOYS

1. Pb(Ag) *and* Pb(Cu)

Judging from internal friction and diffusivity studies, the nature and behavior of the defects in these systems are quite similar to those in the Pb(Au) system. However, the internal friction results indicate that Ag at the room temperature equilibrium concentration does not associate appreciably into Ag–Ag diplons until temperatures well below 300°K are reached. This result indicates that the binding energy of the Ag–Ag diplon is substantially lower than that of the Au–Au diplon and it is consistent with Rossolimo's (1971) observations that the precipitation rates of Ag from Pb at 300°K and higher are as predicted by the high temperature $D = f(T)$ relation.

For further illumination of the action and relative stabilities of solute–solvent and solute–solute diplons in the Pb(Ag) and Pb(Cu) systems, precipitation, diffusivity enhancement, and resistivity studies over the temperature range of interest, as indicated by the internal friction results, would be especially useful. Because of the relatively small Ag diffusivity below room temperature, enhancement experiments on the Pb(Ag) system may be difficult, but the diffusivity of Cu seems sufficiently high to make such experiments on the Pb(Cu) system relatively easy.

Since no detailed description of diplon motions is available, neither is a prediction of what sort of isotope effect should be expected for solutes diffusing by this mechanism. As an approximation, however, if we consider the diffusive motion to be something like a rotation, in the case of impurity–Pb diplons, and substitute the mass of the pair into the expression used to obtain $f \Delta K$ from the experimental data, then the values of $f \Delta K$ obtained from Miller *et al.* (1972) on Pg(Ag) become

$$f \Delta K = 0.74 \pm 0.18 \quad \text{at} \quad 178°C \tag{42}$$

$$f \Delta K = 0.94 \pm 0.06 \quad \text{at} \quad 300°C \tag{43}$$

which would be in good agreement with the value $f = 1.0$ expected for an uncorrelated mechanism. Other diffusive motions are possible, however, and the isotope effect of Cu in Pb would be especially helpful in this connection, since it is particularly sensitive to whether the diffusive motion is a one-, two-, or three-body process. Miller (1972) reports that this work has been done and should be published shortly.

There remains the question of the equilibrium distribution of the noble metal atoms (N) between the N–Pb diplon and substitutional N states at high temperatures. This question might be resolved by precise measurements of lattice parameter and specific volume, in the manner of Simmons and Baluffi (1960).

On the basis of various lines of indirect evidence already presented, a plausible case can be made that the majority of the noble atoms are dissolved as N–Pb diplons at high temperatures. In particular, there is the equivalence, already alluded to, between the activation energy for anelastic relaxation, attributed to movement of N–Pb diplons (N = Cu, Ag, or Au) and for N diffusion. That Ag precipitates from Pb at room temperature at the rate calculated from the Ag diffusivity given by the high temperature $D_{Ag} = f(T)$ relation, supports the view that most of the Ag also was dissolved as N–Pb diplons. If it had been mostly in a substitutional state, then, as Rossolimo and Turnbull (1973) noted, the vacancies generated by the precipitation should have trapped the interstitials frequently enough to lower the effective Ag diffusivity to levels far below that given by the high temperature relation $D_{Ag} = f(T)$.

2. Tin (Noble Metals)

The noble metals appear to diffuse in tin by some type of interstitial mechanism, as evidenced by their extreme rapidity of motion and by their lack of enhancement of the host diffusivity. As yet there is not sufficient data to indicate whether the predominant defect is a conventional or some other type of interstitial. Steric considerations offer a partial explanation of the anisotropy of diffusion observed, if diffusion proceeds by conventional interstitials, since the paths connecting adjacent sites along the channels running in the c direction are much more open than the paths connecting sites in the a direction. On the other hand, following the analysis used for $f\Delta K$ in Pb(Ag), the diplon mechanism seems to account nicely for $f\Delta K = 0.40$ for diffusion of Ag in the a direction of Sn. No explanation for why $f\Delta K$ (≈ 0.01) is so small for diffusion of Ag in the c direction has been deduced from either the conventional interstitial or diplon models. Further illumination on the operation of these or other mechanisms in Sn might be gained from internal friction and enhancement experiments and from isotope effect studies on Sn(Cu) and Sn(Au). Also, it would be interesting to determine whether new defects form in these systems at low temperatures, as they do in Pb, or whether there is something special about the Pb host in this connection. Here again, the problem of the equilibrium distribution of impurity between the interstitial-type and purely substitutional states has not been resolved definitively.

3. Indium and Thallium (Noble Metals)

In these metals, as in Sn, transport of the noble metals apparently is effected primarily by an interstitial-type mechanism, but information is insufficient to indicate which of the types is the principal one.

The small observed anisotropy of diffusion between the a and c directions in hcp Tl seems to be consistent with either interstitial mechanism. Therefore, internal friction, isotope effect, and enhancement studies would be helpful for further delimiting the mechanism.

C. Pb(Cd) AND Pb(Hg)

Solute transport in these two systems seems adequately explained at present by the interstitial–vacancy pair mechanisms, as developed by Miller (1969b, 1969c) and Warburton (1973d). The model developed by Miller, in which the jump rate of the vacancy is negligible compared with the creation jump rate of the interstitial, would account adequately for the diffusivity behavior of the Pb(Cd) system, where $b_{11} \approx D_2(0)/D_1(0)$ (i.e., $\alpha \approx 1$). However, the results for the Pb(Hg) system, where b_{11} exceeds $D_2(0)/D_1(0)$ ($\alpha > 1$) substantially, are explained adequately only by Warburton's model which allows arbitrary ratios for these jump frequencies. Of course, Warburton's model is consistent with the results for Pb(Cd) as well. Warburton (1973c) notes that sufficiently accurate measurements of the solute enhancement in Pb(Hg) and possibly Pb(Cd) as well, will give direct information about the kinetic barriers to defect motion for the i–v pairs. Other experiments on these systems which might prove especially illuminating include internal friction measurements and the isotope effect measurement for Pb(Hg). Although the mass difference between ^{197}Hg and ^{203}Hg is only 3%, and the experiment therefore difficult, this result would provide a direct test of Warburton's model, which predicts a large value of f, in contrast to Pb(Cd) where f is small. The particular configuration of the i–v pairs would be of considerable theoretical interest. An i–v pair was originally thought to be a vacancy with the solute atom occupying a nearest neighbor interstitial site, but it was not clear how such a pair would be stabilized. In view of the Pb(Au) results, it now seems possible that the i–v-pair defect may actually consist of a vacancy and a nearest neighbor impurity–lead diplon, the energy for the creating of the vacancy being compensated to some extent by the diplon formation. Again, there is no evidence concerning the stabilities of the i–v pairs relative to other defects. Judging from the relatively large solubilities of impurity in these systems, it appears that substitutionals may be the predominant species. Warburton (1973c) has suggested various experiments which should help to resolve this problem. Experiments in which x-ray and dilatometric densities are simultaneously measured (Simmons and Baluffi, 1960) would be of no use for distinguishing between the i–v and substitutional states.

The diffusivity of the system Pb(Zn) has not yet been thoroughly investigated, probably because of the difficulties of plating the tracer. In the aqueous plating of Zn, it is extremely difficult to avoid oxidation at the plated surface, which leads to severe hold-up of tracer during the diffusion anneal. This problem might be circumvented by vapor deposition of tracer or by the cold welding technique which was applied successfully to diffusivity studies of the Pb(Na) system (Owens and Turnbull, 1972). The diffusivity behavior of Pb(Zn) would be especially interesting, in view of the very fast diffusion of Zn in Sn and of the results of the Pb(Cd) and Pb(Hg) studies.

Acknowledgments

We thank Drs. T. J. Turner, J. W. Miller, and H. B. Vanfleet for informing us, in advance of publication, of their recent results. Our research at Harvard in the field of fast diffusion has been supported in part by grants from the Advanced Research Projects Agency, Office of Naval Research, and National Science Foundation.

References

ANTHONY, T. R. (1967). Ph.D. Thesis, Harvard University, Cambridge, Massachusetts.
ANTHONY, T. R. (1970a). Interstitial Metal Impurity Diffusion in Metals. In "Vacancies and Interstitials in Metals" (A. Seeger et al., eds.). North-Holland Publ., Amsterdam.
ANTHONY, T. R. (1970b). Acta Met. **18**, 877.
ANTHONY, T. R., AND TURNBULL, D. (1966). Phys. Rev. **151**, 495.
ANTHONY, T. R., DYSON, B. F., AND TURNBULL, D. (1968). J. Appl. Phys. **39**, 1391.
ASCOLI, A. (1960). J. Inst. Metals **89**, 218.
BARR, L. W., AND LECLAIRE, A. D. (1969). Phil. Mag. **20**, 1289.
BARR, L. W., AND SMITH, F. A. (1969). Phil. Mag. **20**, 1293.
BARR, L. W., MUNDY, J. N., AND SMITH, F. A. (1969). Phil. Mag. **20**, 389.
BERGNER, D., AND LANGE, W. (1966). Phys. Status Solidi **18**, 67.
CANDLAND, C. T., AND VANFLEET, H. B. (1973). Phys. Rev. **B7**, 575.
CANDLAND, C. T., DECKER, D. L., AND VANFLEET, H. B. (1972). Phys. Rev. **B5**, 2085.
COSTON, C., AND NACHTRIEB, N. H. (1964). J. Phys. Chem. **68**, 2219.
DARIEL, M. P. (1970). Phil. Mag. **22**, 563.
DARIEL, M. P. (1971). J. Appl. Phys. **42**, 2251.
DARIEL, M. P., EREZ, G., AND SCHMIDT, G. M. J. (1969a). Phil. Mag. **19**, 1053.
DARIEL, M. P., EREZ, G., AND SCHMIDT, G. M. J. (1969b). J. Appl. Phys. **40**, 2746.
DARIEL, M. P., EREZ, G., AND SCHMIDT, G. M. J. (1969c). Phil. Mag. **19**, 1045.
DEVRIES, K. L., BAKER, G. S. AND GIBBS, P. (1963). J. Appl. Phys. **34**, 2254, 2258.
DICKEY, J. E. (1959). Acta Met. **7**, 350.

Dyson, B. F. (1966). *J. Appl. Phys.* **37**, 2375.
Dyson, B. F., Anthony, T. R., and Turnbull, D. (1966). *J. Appl. Phys.* **37**, 2370.
Dyson, B. F., Anthony, T. R., and Turnbull, D. (1967). *J. Appl. Phys.* **38**, 3408.
Enderby, J. E., North, D. M., and Egelstaff, P. A. (1966). *Phil. Mag.* **14**, 961.
Flinn, P. A., Gonser, U., Grant, R. W., and Housely, R. M. (1967). *Phys. Rev.* **157**, 530.
Frank, F. C., and Turnbull, D. (1956). *Phys. Rev.* **104**, 617.
Giessen, B. C., Ray, R., and Hahn, S. (1971). *Phys. Rev. Lett.* **26**, 509.
Hägg, G. (1929). *Z. Phys. Chem.* **6B**, 221.
Hägg, G. (1930). *Z. Phys. Chem.* **7B**, 339; (1930). **8B**, 445.
Halder, N. C., and Wagner, C. N. J. (1967). *J. Chem. Phys.* **47**, 4385.
Hall, R. N., and Racette, J. H. (1964). *J. Appl. Phys.* **35**, 379.
Hansen, M. (1958). "Constitution of Binary Alloys." McGraw-Hill, New York.
Hoffman, R. E., Turnbull, D., and Hart, E. W. (1955). *Acta Met.* **3**, 417.
Hoffman, R. E., Turnbull, D., and Hart, E. W. (1957). *Acta Met.* **5**, 74.
Howard, R. E. (1966). *Phys. Rev.* **144**, 650.
Howard, R. E., and Manning, J. R. (1967). *Phys. Rev.* **154**, 561.
Huang, F. H., and Huntington, H. B. (1972). *Bull. Amer. Phys. Soc.* **17**, 244.
Hudson, J. B., and Hoffman, R. E. (1961). *Trans. Met. Soc. AIME* **221**, 761.
Hultsch, R. A., and Barnes, R. G. (1961). *Phy. Rev.* **125**, 1832.
Kidson, G. V. (1966). *Phil. Mag.* **13**, 247.
Kohler, C. R., and Ruoff, A. L. (1965). *J. Appl. Phys.* **36**, 2444.
Lazarus, D. (1954). *Phys. Rev.* **93**, 973.
LeClaire, A. D. (1962). *Phil. Mag.* 8, **7**, 141.
Letaw, H., Portnoy, W. M., and Slifkin, L. (1956). *Phys. Rev.* **102**, 636.
Lidiard, A. B. (1960). *Phil. Mag.* **5**, 1171.
Lodding, A., and Ott, A. (1971). *Z. Naturforsch.* **26a**, 81.
Lodding, A., Mundy, J. N., and Ott, A. (1970). *Phys. Status Solidi* **38**, 559.
Longini, R. L., and Greene, R. F. (1956). *Phys. Rev.* **102**, 992.
Manning, J. R. (1962). *Phys. Rev.* **128**, 2169.
Manning, J. R. (1964). *Phys. Rev.* **136**, A1758.
Manning, J. R. (1967). "Diffusion Kinetics for Atoms in Crystals," 87–104. Van Nostrand-Reinhold, Princeton, New Jersey.
March, N. H., and Murray, A. M. (1961). *Proc. Roy. Soc.* **A261**, 119.
Meakin, J. D., and Klokholm, D. (1960). *Trans. AIME* **218**, 463.
Millea, M. F. (1966). *J. Phys. Chem. Solids* **27**, 309.
Miller, J. W. (1969a). *Phys. Rev.* **181**, 1095.
Miller, J. W. (1969b). *Phys. Rev.* **188**, 1074.
Miller, J. W. (1969c). Ph.D. Thesis, Harvard University, Cambridge, Massachusetts.
Miller, J. W. (1970). *Phys. Rev.* **B2**, 1624.
Miller, J. W. (1971). Interstitial Solute-Vacancy Pairs in Lead Alloys. *In* "Diffusion Processes" (J. N. Sherwood *et al.*, eds.), Vol. 1. Gordon & Breach, New York.
Miller, J. W. (1972). Private communication of work in progress.
Miller, J. W., and Edelstein, W. A. (1969). *Phys. Rev.* **188**, 1081.
Miller, J. W., Rothman, S. J., and Mundy, J. N. (1972). *Bull. Amer. Phys. Soc. Ser. II* **17**, 244.
Mundy, J. N. (1971). *Phys. Rev.* **B3**, 2431.
Mundy, J. N., Barr, L. W., and Smith, F. A. (1966). *Phil. Mag.* **14**, 785.
Mundy, J. N., Ott, A., and Löwenberg, L. (1967). *Z. Naturforsch.* **22a**, 2113.
Mundy, J. N., Ott, A., Löwenberg, L., and Lodding, A. (1969). *Phys. Status Solidi* **35**, 359.

MUNDY, J. N., MILLER, T. E., AND PORTE, R. J. (1971). *Phys. Rev.* **B3**, 2445.
NACHTRIEB, N. H., RESING, H. A., AND RICE, S. A. (1959). *J. Chem. Phys.* **31**, 135.
NORTH, D. M., AND WAGNER, C. N. J. (1970). *Phys. Chem. Liquids* **2**, 87.
NOWICK, A. S., AND BERRY, B. S. (1972). "Anelastic Relaxation in Crystalline Solids." Academic Press, New York.
NOWICK, A. S., AND HELLER, W. R. (1965). *Advan. Phys.* **14**, 101.
OTT, A. (1968). *Z. Naturforsch.* **23a**, 1683.
OTT, A. (1969). *J. Appl. Phys.* **40**, 2395.
OTT, A. (1970). *Z. Naturforsch.* **25a**, 1477.
OTT, A. (1971a). *Phys. Status Solidi* **B43**, 213.
OTT, A. (1971b). *J. Appl. Phys.* **42**, 2999.
OTT, A., AND LODDING, A. (1970). *Z. Naturforsch.* **25a**, 1765.
OTT, A., AND NORDÉN-OTT, A. (1968). *Z. Naturforsch.* **23a**, 473.
OTT, A., AND NORDÉN-OTT, A. (1971). *Phys. Rev.* **42**, 3745.
OTT, A., MUNDY, J. N., LÖWENBERG, L., AND LODDING, A. (1968). *Z. Naturforsch.* **23a**, 771.
OTT, A., LODDING, A., AND LAZARUS, D. (1969). *Phys. Rev.* **188**, 1088.
OWENS, C. W., AND TURNBULL, D. (1972). *J. Appl. Phys.* **43**, 3933.
PELL, E. M. (1960). *Phys. Rev.* **119**, 1014.
PETERSON, N. L. (1968). *Solid State Phys.* **22**, 409.
PETERSON, N. L., AND ROTHMAN, S. J. (1964). *Phys. Rev.* **136A**, 842.
PETERSON, N. L., AND ROTHMAN, S. J. (1970). *Phys. Rev.* **B1**, 3264.
POWELL, G. W., AND BRAUN, J. D. (1964). *Trans. AIME* **230**, 694.
RAY, R., SEGNINI, M., AND GIESSEN, B. C. (1972). *Solid State Commun.* **10**, 163.
REISS, H., AND FULLER, C. S. (1959). *In* "Semiconductors" (N. B. Hannay, ed.), Chapter VI. Van Nostrand-Reinhold, Princeton, New Jersey.
RESING, H. A., AND NACHTRIEB, N. H. (1961). *J. Phys. Chem. Solids* **21**, 40.
ROBERTS-AUSTEN, W. C. (1896). *Proc. Roy. Soc.* **59**, 281.
ROBERTS-AUSTEN, W. C. (1900). *Proc. Roy. Soc.* **67**, 101.
ROSS, R. A., VANFLEET, H. B., AND DECKER, D. L. (1974). *Phys. Rev.* **B9**, 4026.
ROSSOLIMO, A. N. (1971). Ph.D. Thesis, Harvard University, Cambridge, Massachusetts.
ROSSOLIMO, A. N., AND TURNBULL, D. (1973). *Acta Met.* **21**, 21.
ROTHMAN, S. J., AND PETERSON, N. L. (1967). *Phys. Rev.* **154**, 552.
ROWLAND, T. J., AND FRADIN, F. Y. (1969). *Phys. Rev.* **182**, 760.
SAGUES, A. A., AND NOWICK, A. S. (1974). *Solid State Commun.* **15**, 239.
SAWATZKY, A. (1958). *J. Appl. Phys.* **29**, 1303.
SEEGER, A., AND CHIK, K. P. (1968). *Phys. Status Solidi* **29**, 455.
SEITH, W., AND ETZOLD, H. (1934). *Z. Elektrochem.* **40**, 829.
SEITH, W., AND KEIL, A. (1933). *Z. Phys. Chem.* **B22**, 350.
SEITH, W., AND KEIL, A. (1935). *Z. Phys. Chem.* **27**, 213.
SEITH, W., AND LAIRD, J. G. (1932). *Z. Metallkunde* **24**, 193.
SHIRN, G. A. (1955). *Acta Met.* **3**, 87.
SHOCKLEY, W., AND MOLL, J. L. (1960). *Phys. Rev.* **119**, 1480.
SIMMONS, R. O., AND BALUFFI, R. W. (1960). *Phys. Rev.* **117**, 52; (1960). **119**, 600.
SIMMONS, R. O., AND BALUFFI, R. W. (1962). *Phys. Rev.* **125**, 862.
SMITH, F. A., AND BARR, L. W. (1970). *Phil. Mag.* **21**, 633.
SWALIN, R. A. (1957). *Acta Met.* **5**, 443.
TOMLINSON, P. N., AND HOWIE, A. (1968). *Phys. Lett.* **27A**, 8, 491.
TURNBULL, D. (1966). *Proc. Mem. Lecture Meeting, 10th Ann. of Nat. Res. Inst. Metals, Tokyo,* pp. 1–7.

TURNER, T. J. (1973). Private communication.
TURNER, T. J., AND NIELSENN, C. H. (1974). *Solid State Commun.* **15**, 243.
TURNER, T. J., PAINTER, S., AND NIELSENN, C. H. (1972). *Solid State Commun.* **11**, 577.
VAN DER MAESEN, F., AND BRENKMAN, J. A. (1954). *Philips Res. Rep.* **9**, 225.
VANFLEET, H. B., DECKER, D. L., AND CANDLAND, C. T. (1973). *Bull. Am. Phys. Soc.* **18**, 428.
WAGNER, C. (1938). *Z. Phys. Chem.* **B38**, 325.
WAGNER, C. N. J., HALDER, N. C., AND NORTH, D. M. (1969). *Z. Naturforsch.* **24A**, 432.
WARBURTON, W. K. (1972). *Phys. Rev.* **B6**, 2161.
WARBURTON, W. K. (1973a). *Scr. Met.* **7**, 105.
WARBURTON, W. K. (1973b). *J. Phys. Chem. Solids* **34**, 451.
WARBURTON, W. K. (1973c). *Phys. Rev.* **B7**, 1330.
WARBURTON, W. K. (1973d). *Phys. Rev.* **B7**, 1341.
WEYLAND, J. A., DECKER, D. L., AND VANFLEET, H. B. (1971). *Phys. Rev.* **B4**, 4225.
WOODBURY, H. H., AND TYLER, W. W. (1957). *Phys. Rev.* **105**, 84.

5

Hydrogen Diffusion in Metals

J. VÖLKL AND G. ALEFELD

PHYSIK-DEPARTMENT DER TECHNISCHEN UNIVERSITÄT MÜNCHEN
GARCHING, WEST GERMANY

I. Introduction	232
II. Site Location, Phase Diagrams, and Solubility	233
III. Experimental Methods	240
A. Permeation Methods	241
B. Electrochemical Methods	242
C. Mechanical Relaxation Methods	242
D. Magnetic Disaccommodation	243
E. Resistivity Relaxation	244
F. Tracer Method	244
G. Other Methods	244
H. Nuclear Magnetic Resonance (NMR)	245
I. Quasi-Elastic Neutron Scattering (QNS)	245
IV. Values for the Diffusion Coefficients at Small Concentrations (α Phases)	246
A. Palladium	247
B. Nickel	248
C. Iron (α Phase)	250
D. Niobium	252
E. Tantalum	255
F. Vanadium	257
V. High Hydrogen Concentrations	258
VI. Isotope Dependence	260
VII. Deviations of the Diffusion Coefficient from the Arrhenius Relation	264
VIII. Dependence on Alloying	266
IX. Influence of Traps	267
X. Influence of Structure	268
XI. Conclusions	269
Tables	271
References	295

I. Introduction

The diffusion of hydrogen in metals has attracted special attention for several reasons. These have been motivated from points of view of both basic as well as applied research.

(1) The diffusivity of hydrogen in metals is extremely high [e.g., 2×10^{12} jumps/sec at room temperature for H in vanadium (Schaumann et al., 1970)] and exceeds that of heavy interstitials such as oxygen and nitrogen at this temperature by 15–20 orders of magnitude. In Fig. 1, a comparison is shown

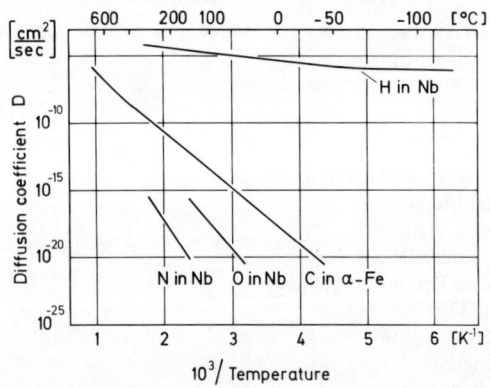

Fig. 1. Diffusion coefficients of H (Schaumann et al., 1970), N, and O (Powers and Doyle, 1959) in Nb and C in α-Fe (Lord and Beshers, 1966) after Völkl et al. (1970).

for Nb. Phenomenologically, this high diffusivity is a consequence of the low activation energy for hydrogen diffusion [e.g., 0.05 eV for hydrogen in vanadium (Schaumann et al., 1970)].

(2) Hydrogen diffusion can be observed at low temperatures so that quantum effects in the diffusion of hydrogen can be expected. It may be possible that the hydrogen atoms remain mobile even at zero temperature. In palladium, e.g., Fritz et al. (1961) have found heat production below 1°K, possibly resulting from a rearrangement of hydrogen atoms. Recent quenching experiments by Hanada (1973) on H in Ta have shown hydrogen mobility at 11°K. By inelastic neutron scattering, the energy difference between vibrational states of the hydrogen atoms (local modes) can be determined. For hydrogen in niobium, one finds 0.11 eV (Verdan et al., 1968; Brand, 1971) which corresponds to a temperature of 1370°K. So for this system, room temperature can already be considered as low.

(3) For hydrogen, three isotopes are available with the largest possible relative mass difference.

(4) Hydrogen–metal alloys can be prepared relatively easily by loading either from the gas phase or electrolytically. Also recently, the method of ion implantation, which allows loading even at He temperature, has been applied (Stritzker and Buckel, 1972). For some metals such as niobium, vanadium, tantalum, and palladium, in certain temperature regions, the interstitial concentration can be varied from 0 to $\sim 100\%$ without passing a phase boundary. It is therefore possible to study regions of low concentrations in which the interstitials diffuse more or less independently (e.g., Fig. 17) and also regions of high concentrations in which cooperative effects become significant (e.g., Fig. 20).

(5) Many new experimental techniques have been developed and are now available as a result of extended interest in the special properties of these systems (see Section III).

(6) The technical interest in hydrogen–metal systems extends into many areas: hydrogen embrittlement of technical alloys especially steels, hydrogen–metal alloys as moderators in reactors, fuel cells, purification of hydrogen for chemical application, permeation of hydrogen through vessels or pipelines, and storage of hydrogen as metal–hydrogen alloys, for instance. Many technical problems concerning the behavior of hydrogen in metals must be solved in connection with fusion-reactor designs. Moreover, the recent discovery of superconductivity of PdH (Skośkiewicz, 1972), especially the high transition temperature for $PdAg_{0.2}H_x$ ($T_c \approx 16°K$) (Stritzker and Buckel, 1972; Buckel and Stritzker, 1973), will stimulate further activity in the field of hydrogen–metal alloys.

In Section II, some information about hydrogen–metal phase diagrams and solubility of hydrogen in metals will be provided, and in Section III, a short review of experimental techniques will be presented. This latter section is important since the reliability of data depends strongly on the techniques used.

II. Site Location, Phase Diagrams, and Solubility

Hydrogen is in general dissolved interstitially into metals. In palladium, hydrogen, or deuterium, atoms occupy the octahedral sites, as has been found by neutron diffraction experiments (Worsham et al., 1957; Bergsma and Goedkoop, 1960; Ferguson et al., 1965; Nelin, 1971) and also from the results of quasi-elastic neutron scattering (Sköld and Nelin, 1967; Rowe et al., 1972). In Fig. 2, the tetrahedral and octahedral site for a bcc lattice is shown. According to neutron-diffraction experiments on deuterided niobium (Somenkov et al., 1968) and deuterided tantalum (Somenkov et al., 1969; Petrunin et al., 1970), the tetrahedral sites are occupied by deuterium,

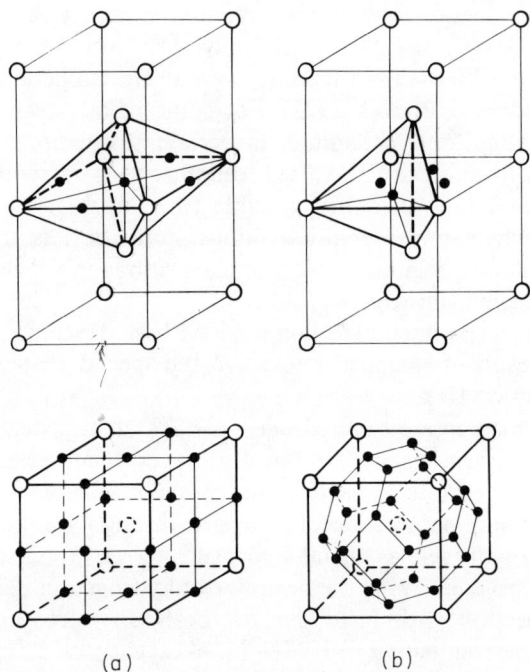

Fig. 2. (a) Octahedral sites and (b) tetrahedral sites for a bcc lattice.

whereas for vanadium (Chervyakov et al., 1972; Somenkov et al., 1972) the authors suggest that both sites are taken by deuterium. For Nb, additional experiments on hydrided samples exist (Somenkov et al., 1968) with the same result as for deuterium. Also, channeling experiments by Carstanjen and Sizmann (1972) strongly favor the tetrahedral sites for D in Nb. For H in Fe, Ni, or Cu, no position determination is known to us.

Figures 3–6 show the phase diagrams for hydrogen in palladium, niobium, tantalum, and vanadium. Figures 7–10 show pressure-composition isotherms for the same four systems. From these latter diagrams, it can be seen that the nonideality which for decreasing temperature leads to a miscibility gap is a common feature to all four metals, although in the phase diagrams for vanadium and tantalum (Figs. 4 and 6) the miscibility gap is hidden behind other phase transitions. Extrapolating from high temperatures yields $T_c = 47.1°C$ (H in V) and $T_c = -58.5°C$ (H in Ta) for the miscibility gap (Veleckis, 1960). The miscibility gap has also been found in a series of Pd alloys, especially $PdAg_x$ (Brodowsky and Poeschel, 1965; Buck and G. Alefeld, 1972). Furthermore, solubility measurements for Ni (Baranowski, 1972) suggest a similar phase diagram as for Pd.

5. HYDROGEN DIFFUSION IN METALS

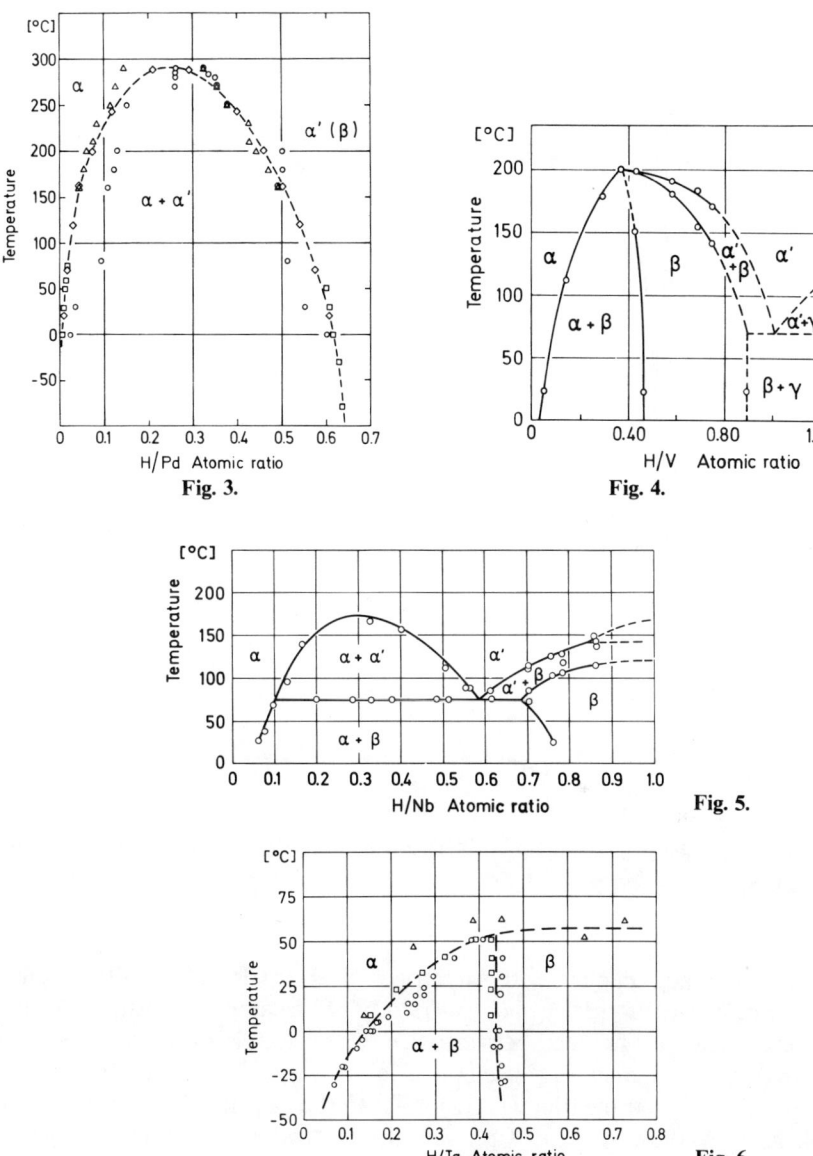

Fig. 3. Phase diagram for H in Pd (○ Gillespie and Hall, 1926; Gillespie and Galstaun, 1936; △ Brüning and Sieverts, 1933; □ Wicke and Nernst, 1964; ◇ Frieske and Wicke, 1973).

Fig. 4. Phase diagram for H in V (Maeland, 1964).

Fig. 5. Phase diagram for H in Nb (Walter and Chandler, 1965).

Fig. 6. Phase diagram for H in Ta (△ Kofstad and Butera, 1963, as evaluated by Cannelli and Mazzolai, 1969; ○ Zierath, 1969; □ Pryde and Tsong, 1971a).

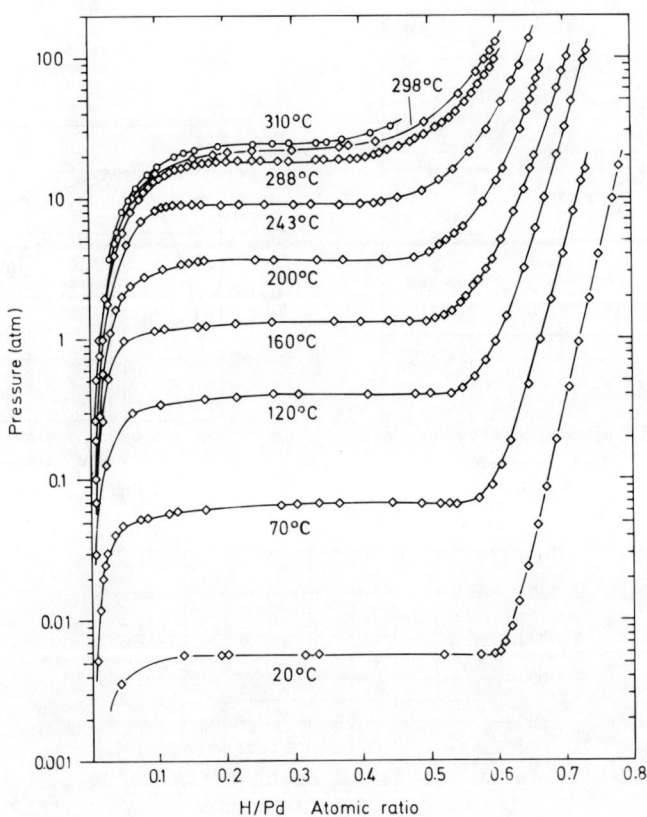

Fig. 7. Solubility isotherms for H in Pd (○ Gillespie and Galstaun, 1936; ◇ Frieske and Wicke, 1973).

In the α and α' phases, the hydrogen interstitials occupy the possible sites randomly, in the α' phase, certainly, with some short range order. So the essential difference between the two phases is only the density of hydrogen which manifests itself in a larger lattice parameter for the α' phase. The phase transition α–α' is similar to the gas–liquid phase transition (G. Alefeld, 1969). In the β phases, the hydrogen orders in two ways: (a) of the three tetrahedral sites, only one or maybe two are occupied; and (b) in these particular sites, the hydrogen may assume a periodic arrangement and thus a long range order which is different from the trivial long range order as imposed by the host metal. As a consequence, superstructure reflections have been found experimentally (e.g., Somenkov et al., 1968).

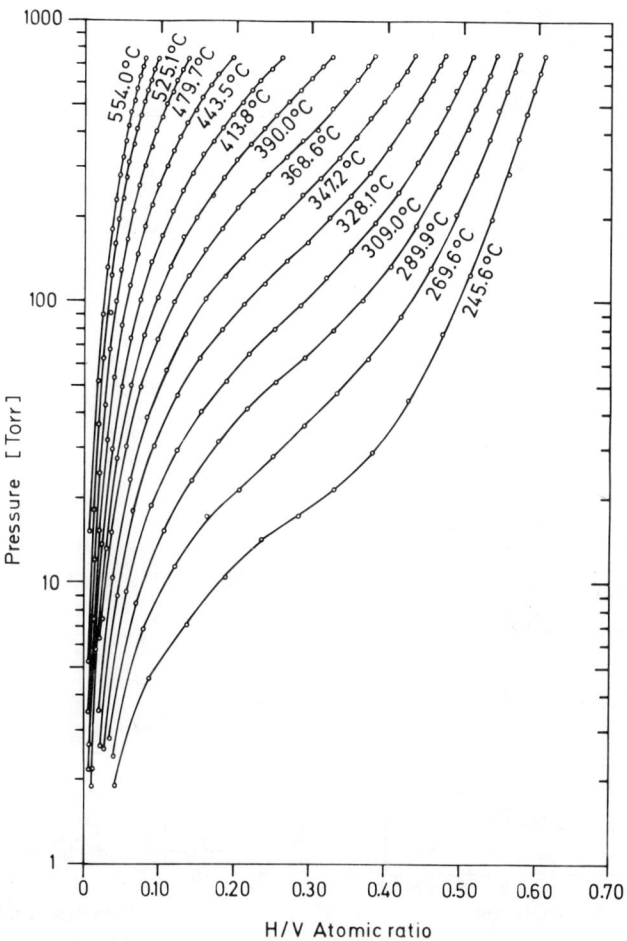

Fig. 8. Solubility isotherms for H in V (Veleckis, 1960).

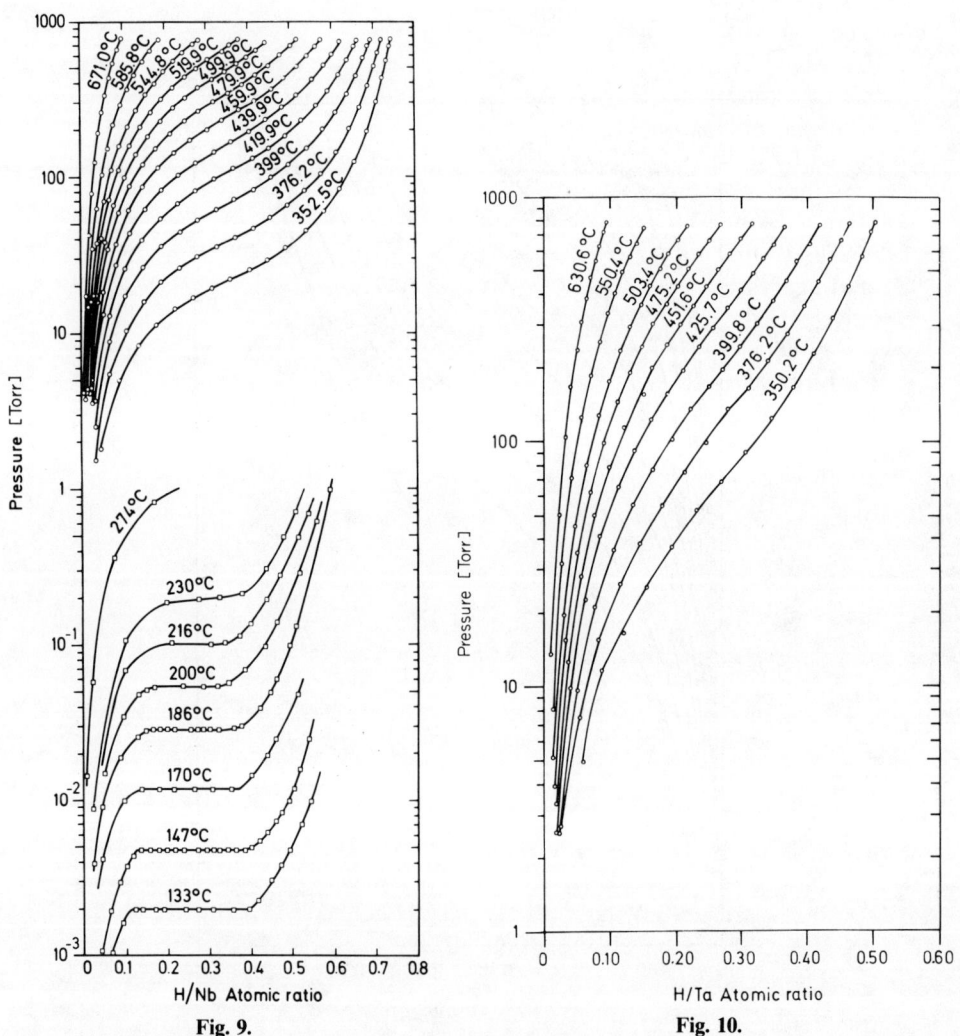

Fig. 9. Solubility isotherms for H in Nb (○ Veleckis, 1960; □ Pryde and Titcomb, 1969).
Fig. 10. Solubility isotherms for H in Ta (Veleckis, 1960).

5. HYDROGEN DIFFUSION IN METALS

The solubility data in Figs. 7–10 are not only important for questions of sample preparation but also for evaluation of diffusion data. From these results, thermodynamic properties of the hydrogen–metal systems can be derived: e.g., the derivative of the chemical potential with respect to concentration which, depending on the experimental technique, enters into the measured diffusion coefficient.

The relevant parameter for the question of how much hydrogen is absorbed by a metal at a given temperature and pressure is the sign and size of the enthalpy of solution. Figure 11, in which the enthalpy of solution is plotted as a function of atomic number (McLellan and Oats, 1973), shows that there is a maximum in exothermicity at Groups II and III and again at the elements palladium and nickel of Group VIIIC. For negative enthalpy of solution, the solubility of hydrogen is high and further increases with decreasing temperature. The sign and magnitude of the enthalpy of solution is determined by the following considerations (Ebisuzaki and O'Keeffe, 1967): The binding energy of the hydrogen molecule plus ionization energy of the hydrogen atom is balanced against the work function of an electron plus work function of a proton in the metal considered. The latter quantity consists of the energy involved in screening the proton, the interaction energy of the screened proton with the metal ions, and the energy associated with the resulting strain field. A small

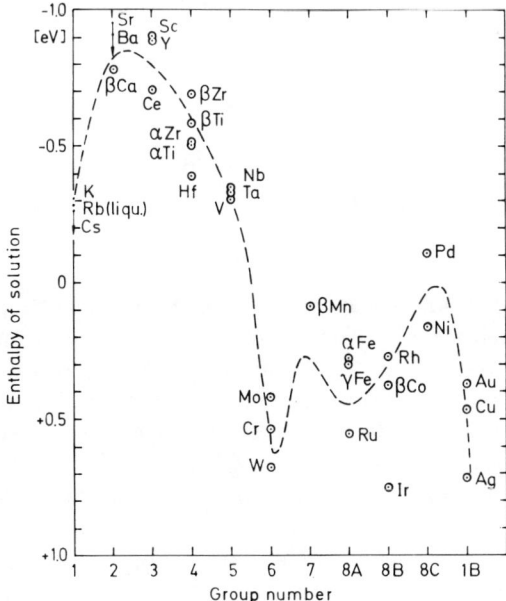

Fig. 11. Enthalpy of solution for H in metals. [Reprinted with permission from McLellan and Oates, *Acta Met.* (1973), Pergamon Press.]

change in this balance between large energy quantities causes large changes in the solubility. For example, the difference of the enthalpy of solution between Pd and Ni of 0.3 eV (Fig. 11) amounts to only $\sim 2\%$ of the above balance but, because of the exponential factor, $\exp(0.3/kT)$, the solubility is increased by 5 orders of magnitude at room temperature.

III. Experimental Methods

In general, hydrogen diffusion is studied by setting up a nonequilibrium distribution and measuring the time to reach equilibrium under some given conditions. There are also two methods by which hydrogen diffusion is studied at equilibrium, namely, nuclear magnetic resonance (NMR) and quasi-elastic neutron scattering (QNS). The nonequilibrium experiments have the advantage that besides a time constant from which the diffusion coefficient is derived, an equilibrium property can be measured as well [e.g., the derivative of the chemical potential μ with respect to the hydrogen density ρ (atoms/cm^3)]. The same quantity appears as a factor in the measured diffusion coefficient which can easily be seen as follows:

$$s = M\rho \text{ grad } \mu(\rho) = M\rho(\partial\mu/\partial\rho) \text{ grad } \rho \tag{1}$$

where M is the mobility and s the current. The diffusion coefficient, measured by studying the decay of a concentration gradient, is therefore given by

$$D^* = M\rho \, \partial\mu/\partial\rho \tag{2}$$

For low concentrations, $\partial\mu/\partial\rho$ is given by

$$\partial\mu/\partial\rho = kT/\rho \tag{3}$$

Thus, Eq. (2) can be written as

$$D = MkT \tag{4}$$

Equation (4) is known as Einstein relation, connecting the diffusion coefficient with the mobility. The measured diffusion coefficient D^* is directly proportional to the mobility of the individual particles only in the limit of low concentration. Therefore, in general, the measurement of the equilibrium property, $\partial\mu/\partial\rho$, is also necessary to eliminate temperature and concentration dependencies not characteristic of the individual jump process.

With NMR and QNS, only information about the diffusion coefficient or the mean time-of-stay can be derived. Yet this information is already free of the factor $\partial\mu/\partial\rho$ which results from cooperative behavior. In both methods, the motion of individual particles which are marked by their spin is being studied. (For QNS, this statement holds only for incoherent scattering which is dominant for hydrogen, but not for deuterium.) It may be mentioned that the quantity $\partial\mu/\partial\rho$ can also be determined by neutron scattering, namely, from the diffuse intensity for small scattering angles (Conrad et al., 1974).

The experimental techniques involving nonequilibrium distributions can be subdivided either according to the method of producing the nonequilibrium state or according to the method by which the relaxation to equilibrium is measured. We will follow the second criterion but also make statements about the first. It should also be mentioned that the results of the techniques described in Sections A and B may depend on surface conditions. This is not the case for the techniques described in Sections III.C–III.I.

A. PERMEATION METHODS

The stationary rate of permeation of hydrogen through a metal foil yields only the product of the diffusion coefficient and solubility. Therefore, the latter quantity and especially its temperature dependence must be known very accurately to determine the diffusion coefficients by this method. Since it has been shown that the solubility can be influenced strongly by lattice defects (e.g., Carnuth, 1963), the solubility should be measured individually for each sample.

There are more direct methods in which permeation under nonstationary conditions is measured, e.g., the time dependence of permeation after a sudden drop or rise of the pressure. This group of experiments utilizes so called "time-lag" methods (Daynes, 1920; Barrer, 1939, 1940). The diffusion coefficient can be deduced from the relaxation time to the new stationary state and the change in solubility can be determined from the change in permeation. Wagner and Sizmann (1964) have developed a method in which the pressure is applied periodically, e.g., in addition to a stationary pressure. The diffusion coefficient can be derived from the phase difference between the pressure and hydrogen current, and the solubility is determined from the amplitudes. A special case of nonstationary permeation is the widely used method of absorption and desorption developed by Dünwald and Wagner (1934).

In general, all permeation methods suffer from the problem that the surface reaction should be very fast compared with the rate of bulk diffusion. Hydrogen absorption is easily poisoned by many substances which are adsorbed at the surface of the metal. An important test that bulk diffusion is being measured is the check that the measured relaxation time, from which D is deduced, is proportional to the square of the diameter of the sample.

B. ELECTROCHEMICAL METHODS

These techniques are variations of the permeation methods, using electrochemical techniques to measure and often, also, to produce the hydrogen current. The time constants are measured for the redistribution of hydrogen in a sample separating two electrolytic cells. The various methods differ by the boundary conditions imposed experimentally on the sample. The electrochemical "time-lag" method (Devanathan and Stachurski, 1962) is analogous to the technique in which the hydrogen pressure on one side is raised suddenly. The time dependence of the hydrogen current through the sample or the electric current into the sample, respectively, is measured. The "discharge" methods (Küssner, 1963) correspond to the gas volumetric desorption method. In both the time-lag and discharge methods, instead of measuring the current through the sample, it is also possible to measure the change of the electrochemical potential caused by the hydrogen arriving at the opposite side of the sample (Holleck and Wicke, 1967). Also, an electrochemical oscillation method has been applied recently (Züchner and Boes, 1972). From the frequency, the diffusion coefficient can be derived. The information contained in the amplitude has not been evaluated so far. As with the permeation techniques, the electrochemical methods can suffer greatly from problems of surface barriers. A general review of electrochemical methods has recently been given by Züchner and Boes (1972).

C. MECHANICAL RELAXATION METHODS

Experimentally, these methods include stress and strain relaxation, internal friction, and modulus change as well as ultrasonic absorption and changes in sound velocity. With these methods, the *Snoek effect*, local redistribution of atoms over equivalent sites under stress (Snoek, 1939b, 1941; Nowick and Berry, 1972), or the *Gorsky effect*, long range diffusion of atoms under the action of a gradient in dilatation (Schaumann et al., 1968), is being studied. With the second effect, the diffusion coefficient is measured directly without any adjustable parameter. Using the Snoek effect requires knowledge of the mean jump distance; i.e., the interstitial position must be known in order to convert the relaxation time into the diffusion coefficient. Furthermore, it must also be established that the measured relaxation process has something to do with the individual steps leading to long range diffusion.

1. *Snoek Effect*

Although the Snoek effect has been used to study the diffusion of heavy interstitials such as oxygen and nitrogen or carbon in iron, niobium, vanadium, and tantalum (e.g., Powers and Doyle, 1959), its application to hydrogen diffusion has not been very successful despite the fact that many attempts have been made (for discussion, see Section IV.D). Several hydrogen-induced internal friction peaks have been observed, but in no case has it been established conclusively that any of the peaks result from the usual Snoek effect. The peaks may be caused by the relaxation of hydrogen in combination with impurities or other imperfections, or by hydrogen pairs. In order to prove that one of these peaks arises from the diffusive jump of hydrogen, it would be necessary to show that the strength of this peak is proportional to the hydrogen concentration, independent of impurity concentration, properly orientation dependent, etc. It is therefore not safe to relate the relaxation time of these peaks to the mean time-of-stay as required to calculate the diffusion coefficient. Since the Snoek effect yields information about a relaxation time for a process which occurs on an atomistic scale, more work along this direction would be desirable. Since it has been shown recently that the relaxation strength for the Snoek effect of H or D in Nb or Ta is at least 2 orders of magnitude lower than for heavy interstitials (Buchholz et al., 1973), the observation of the Snoek effect will not be easy.

2. Gorsky Effect

A review of the Gorsky effect has recently been written by Völkl (1972). With the Gorsky-effect technique, one measures the time required for diffusion across the sample diameter. This is a macroscopic rather than an atomistic relaxation time. If the sample diameter is known, the absolute value of D can be determined without any adjustable parameter. The measured diffusion coefficient is therefore the macroscopic diffusion coefficient which, in the worst case, could be the result of several processes. The Gorsky effect arises because hydrogen expands the lattice; therefore, a gradient in dilatation (produced by bending a sample) may be used to set up a gradient in concentration. From the relaxation strength, the lattice expansion and $\partial\mu/\partial\rho$ can be deduced (G. Alefeld et al., 1970). By determining $\partial\mu/\partial\rho$ simultaneously with the relaxation time, the diffusion coefficient can be corrected for cooperative effects (see, e.g., Fig. 20). One of the main advantages of the Gorsky effect is the fact that bulk diffusion is studied without any influence of surface effects.

D. Magnetic Disaccommodation

With this technique (Lord Rayleigh, 1887) the pinning of Bloch walls due to defect motion is studied. As with mechanical relaxation, one can study an orientation (Snoek, 1938, 1939a–c) and a long range diffusion effect (Dietze, 1959). The advantage of this method is its extreme sensitivity, although there is the disadvantage that it is limited to ferromagnetic substances. A summary paper has been written by Kronmüller (1970).

E. RESISTIVITY RELAXATION

The residual resistivity per atomic percent for hydrogen has been reported to be 0.65–0.75 $\mu\Omega$-cm in niobium (Borgucci and Verdini, 1965; Westlake, 1969; Sasaki and Amano, 1969; Pryde and Titcomb, 1969) and 0.7–0.9 $\mu\Omega$-cm in tantalum (Borgucci and Verdini, 1965; Ducastelle et al., 1970; Pryde and Tsong, 1971b). Since resistivity can be measured very accurately, this property can be used to measure the time dependence of hydrogen distribution in a sample. Several methods to produce a concentration gradient have been used in this connection:

(a) loading locally and following the diffusion by measuring locally along the sample (e.g., Merisov et al., 1966, 1971; Piper, 1966);

(b) electromigration: a current through the sample can produce a gradient in concentration due to electrotransport [The time constant of this process yields the mobility, and the concentration gradient yields the effective charge (Coehn and Specht, 1930; Coehn and Jürgens, 1931).]‡;

(c) strain: if a dumbbell-shaped specimen is put under axial stress, the hydrogen diffuses into the narrow section which causes a resistivity increase [The information derived from this method (Schnabel, 1972) is the same as with the Gorsky effect.]; and

(d) temperature: a gradient in temperature causes a gradient in the hydrogen distribution. From the magnitudes of the gradients, the heat of transport can be determined; the diffusion coefficient is deduced from the redistribution of the hydrogen at constant temperature (Wipf, 1972).

F. TRACER METHOD

Since the isotope tritium is radioactive, the tracer method is applicable (Sicking and Buchold, 1971; Buchold and Sicking, 1972; Katz et al., 1971).

G. OTHER METHODS

Instead of using the Gorsky effect or resistivity relaxation, the lattice expansion of hydrogen can be measured locally directly with X rays (Züchner and Wicke, 1969). In general, it is difficult to convert the local lattice parameter into the local concentration, since, for instance, long

‡ This topic is dealt with extensively by Huntington in Chapter 6.

range strain fields can induce lattice expansion in areas which are free of hydrogen.

The gravimetric method makes use of the fact that as a result of diffusion, the center of gravity of an inhomogeneously loaded specimen changes as a function of time (Wicke and Obermann, 1972).

H. Nuclear Magnetic Resonance (NMR)

Recent reviews on the application of nuclear magnetic resonance to studies of hydrogen diffusion have been written by Pedersen (1965) and Cotts (1972). The NMR technique is a well-established method for studying atomic motion in solids. Data usually consist of measurements of the longitudinal spin–lattice relaxation time T_1 or the transverse spin–spin relaxation time T_2. From the absolute values or the frequency and temperature dependence of these quantities, the rate of atomic motion can be calculated. The analysis also yields the activation energy. The best way to measure T_1 and T_2 is with the pulsed NMR technique. Nevertheless, NMR line widths have been measured with steady state techniques especially in the temperature regime in which motional narrowing is observed. Two relatively new NMR techniques, the so called rotating-frame technique and the direct measurement of the diffusion coefficient by observation of the spin-echo decay in the presence of an applied magnetic field are beginning to be utilized in solid state diffusion experiments (Cotts, 1972). Along with diffusion parameters, NMR techniques also yield valuable information on phase diagrams, changes of electronic properties caused by hydrogen, and the symmetry of the position if quadrupole splitting of deuterium can be observed (Pedersen and Slotfeldt-Ellingsen, 1971; Lütgemeier et al., 1972b).

I. Quasi-Elastic Neutron Scattering (QNS)

Recent reviews on this technique have been written by Springer (1972) and Gissler (1972) who emphasized results on metal–hydrogen systems especially. The method first applied by Sköld and Nelin (1967) to a solid state system is based on the following physical phenomenon: a monoenergetic neutron beam is scattered incoherently by the protons in the metal. As a result of the diffusive motion of the protons, the beam will be energetically broadened and the width of the line will depend on the rate of diffusion. Small and large scattering angles contain different information. For small momentum transfer, i.e., for small scattering angles (since the scattering is nearly elastic), the width at half maximum ΔE is given by

$$\Delta E = 2\hbar D\kappa^2 \tag{5}$$

where $\hbar$ is Planck's constant, κ the momentum transfer [$\kappa = 4\pi \sin(\vartheta/\lambda)$, where ϑ is the angle between the scattered and incident beams and λ the wavelength], and D the diffusion coefficient. D describes the diffusive motion of one particular proton (marked by its spin). D certainly can be changed by interaction with other protons, yet it does not contain the factor $\partial\mu/\partial\rho$ which enters in nonequilibrium measurements. From the line width at large scattering angles, more atomistic details of the jump diffusion process can be obtained, namely, the mean jump distance or the mean jump rate. This technique therefore allows one to determine which of the two jump models, octahedral–octahedral or tetrahedral–tetrahedral, describes the physical process (see Fig. 12).

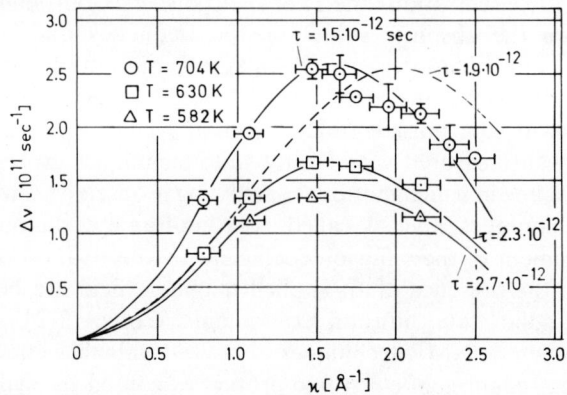

Fig. 12. Observed linewidth ($\Delta v = \Delta E/h$) for neutrons, scattered quasi-elasticly on H in Pd, versus momentum transfer κ for three temperatures (Sköld and Nelin, 1967), (–) are fitted to the octahedral model, (- - -) to the tetrahedral model.

With conventional spectrometers, the quasi-elastic n-scattering technique is limited to $D > 10^{-5}$–10^{-6} cm²/sec. With the recent development of the high resolution back scattering spectrometer (B. Alefeld, 1972) diffusion coefficients which are 1–2 orders of magnitude smaller can still be measured (see Section V, Fig. 21).

IV. Values for the Diffusion Coefficients at Small Concentrations (α Phases)

In the subsequent sections, the values and the temperature dependence of some representative metals will be discussed. The experimental data are collected in figures as well as in tables; the authors are cited in the tables. The relation between symbols in figures and authors can also be found in the appropriate tables. Furthermore, Table XIII contains a collec-

5. HYDROGEN DIFFUSION IN METALS

tion of diffusion data for hydrogen in a variety of additional metals without claim for completeness. In all tables, the activation energies are given in units of electron volts. In the conversion of published activation energies to electron volts, no attempt was made to ascertain whether the third decimal place had any experimental significance.

A. PALLADIUM

The palladium–hydrogen system, along with H in Ni and Fe, has attracted the greatest attention. Practically all experimental methods except the Snoek effect have been applied. In Fig. 13, the data of 23 authors are collected (see also Table I). The consistency of the data is remarkably good. Because of relatively well-defined surface conditions, the different permeation techniques give reliable results in this case. We regard the

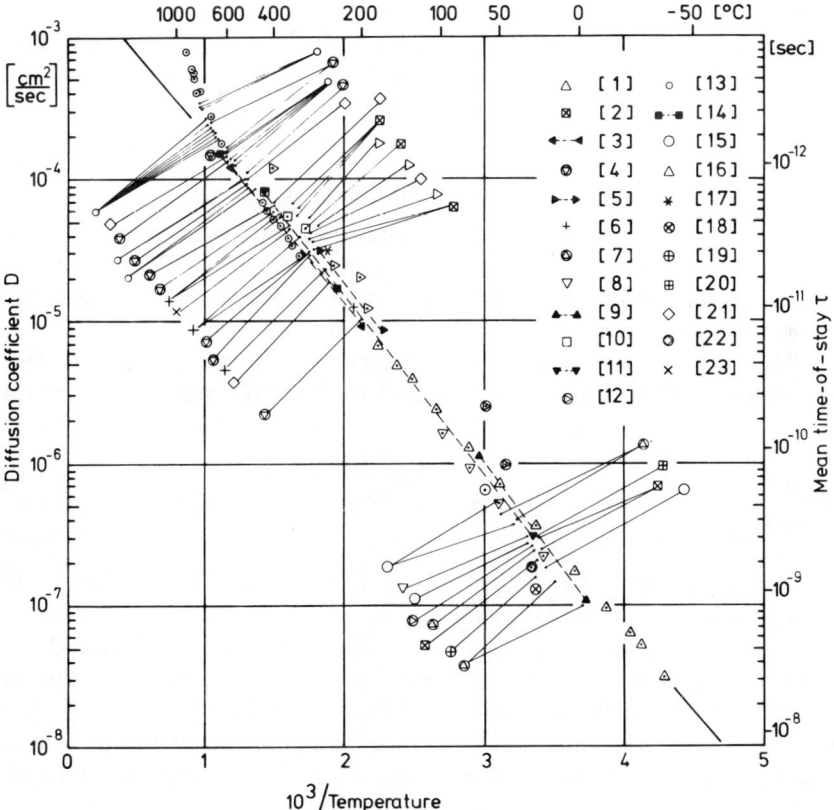

Fig. 13. Diffusion coefficient of H in Pd. For references, see Table I. The right hand scale for the mean time-of-stay refers to octahedral–octahedral jumps.

solid line drawn in the figure as the best fit to the available data. This line yields the following values for the frequency factor D_0 and the activation energy U:

$$D_0 = 2.90 \cdot 10^{-3} \text{ cm}^2/\text{sec} \quad \text{and} \quad U = 0.230 \text{ eV}$$

D_0 is about an order of magnitude larger than those values found for H in V, Nb, or Ta. In contrast to this, the vibrational frequency of the local mode associated with the hydrogen atom (Bergsma and Goedkoop, 1960) is a factor of 2 smaller than those of the bcc group mentioned. The energy level spacing associated with the local mode, $\hbar\omega = 0.056$ eV, is also small compared to the activation energy, 0.23 eV; therefore, a classical model for diffusion appears to be a good approximation. The predictions of the jump model have been very well verified by quasi-elastic neutron scattering. Figure 12 shows the width of the quasi-elastic line as measured for polycrystalline Pd (Sköld and Nelin, 1967). As has been indicated in Section III.I, the absolute value of D as well as the mean time-of-stay τ or the mean jump distance can be extracted from these measurements, or, even more accurately, from those done recently on single crystals of Pd (Rowe et al., 1972). The jump distance is consistent with $a/\sqrt{2}$, which would apply to jumps between octahedral sites, instead of $a/2$ for jumps between tetrahedral sites. Since in the phase diagram for H in Pd no phase transition line between the α and α' phase exists above the critical point, it is safe to assume that the octahedral position, which has been established by neutron scattering only for the α' phase (Worsham et al., 1957; Bergsma and Goedkoop, 1960; Ferguson et al., 1965; Nelin, 1971) is also occupied in the α phase (Ebisuzaki and O'Keeffe, 1967).

The apparent deviation from the Arrhenius behavior in Fig. 13 at high temperature will be discussed in Section VII. The small although systematic scatter around room temperature toward lower values may be attributed to surface problems (Wicke and Meyer, 1969).

B. NICKEL

The consistency of the data on nickel (Fig. 14; see also Table IV) is as good as in Pd for the region above 100°C; below this temperature, the scatter increases. Since with one exception (Cermak and Kufudakis, 1966, 1968) all measurements have been performed with surface-sensitive methods, this consistency is quite remarkable considering the large scatter for other metals. Apparently, as with palladium, the influence of surface contamination is negligible. Because of the smaller solubility of H in Ni, other experimental methods have limited applicability. At the Curie point of nickel, Belyakov and Ionov (1961) have reported a small step in the

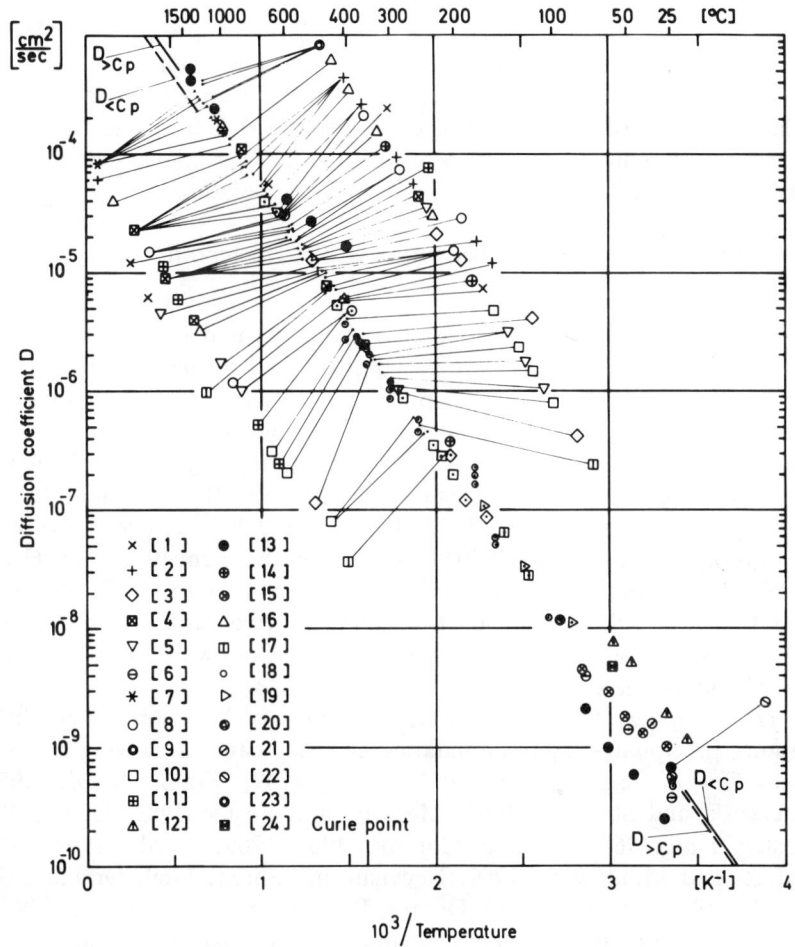

Fig. 14. Diffusion coefficient of H in Ni. For references, see Table IV. ($D_{<C_p}$ and $D_{>C_p}$ are best values below and above the Curie point.)

diffusion coefficient. The best values indicated in Fig. 14 can be described by

$$D_0 = 6.87 \cdot 10^{-3} \text{ cm}^2/\text{sec} \quad \text{and} \quad U = 0.42 \text{ eV}, \quad T > T_c$$
$$D_0 = 4.76 \cdot 10^{-3} \text{ cm}^2/\text{sec} \quad \text{and} \quad U = 0.41 \text{ eV}, \quad T < T_c$$

The activation energy is about twice as large as that in Pd, but D_0 is also larger. As a consequence, the absolute value of D for nickel at room temperature is 3 orders of magnitude lower than D in palladium, but at 600°C, this ratio has decreased to below 1 order of magnitude.

C. Iron (α Phase)

In contrast to Figs. 13 and 14, the measurements for iron in Fig. 15 and Table V show quite unsatisfactory scatter dispite the endeavors of a large number of investigations. Only above 200°C is the scatter below 1 order of magnitude and it decreases slightly with higher temperature. The absolute value of D for iron in this temperature region is higher than for Pd.

Several possibilities have been discussed in the literature to explain the large scatter.

(1) No surface independent method (such as is described in Sections III.C–III.I) has been applied to iron until now. (For internal friction and magnetic disaccommodation, see p. 252.) This is probably due to the low solubility of H in Fe. The influence of different surface conditions has been studied experimentally as well as theoretically, e.g., by Smithells and Ransley (1935), Baukloh and Zimmerman (1936), Baukloh and Retzlaff (1937), Baukloh and Gehlen (1937), Baukloh and Wenzel (1937), Baukloh and Müller (1938), Fast (1942), P. L. Chang and Bennett (1952), Heath (1952), Ewing and Ubbelohde (1955), Bastien and Amiot (1958); Schenck and Taxhet (1959), Carmichael *et al.* (1960), Gorman and Nardella (1962), Radhakrishnan and Shreir (1967), Wach and Miodownik (1968), Joppien (1968) and Gonzalez (1969).

(2) Trapping of H at lattice imperfections such as impurities, dislocations, precipitates, grain boundaries, etc. may cause a slowing down of the diffusion (e.g., Darken and Smith, 1949; Frank *et al.*, 1958; Raczyński and Stelmach, 1961; McNabb and Foster, 1963; Gaus, 1965; Foster *et al.*, 1965; Coe and Moreton, 1967, 1969; Gibala, 1967, 1970; Sturges and Miodownik, 1969; Newman and Shreir, 1969; Oriani, 1970; Miodownik and Sturges, 1970; Domke, 1971; Heumann and Domke, 1972).

(3) The formation of immobile diinterstitials has been discussed by Ono and Rosales (1968).

(4) The formation of molecular hydrogen in micro- or macropores, either already existing in the material or produced by excess loading, may cause slowing down or also enhancement of the apparent diffusion process (e.g., Smialowski, 1957; Hill and Johnson, 1959; Vibrans, 1961; Harhai *et al.*, 1965; Naeser and Dautzenberg, 1965; Evans, 1966; Boniszewski and Moreton, 1967; Lange, 1969; Evans and Rollason, 1969a,b; Ellerbrock *et al.*, 1972).

Considering the relative significance of the effects discussed, a comparison of the Fe measurements with those for Ni seems useful. In both metals the solubility is low and for both metals surface-sensitive methods have

5. HYDROGEN DIFFUSION IN METALS

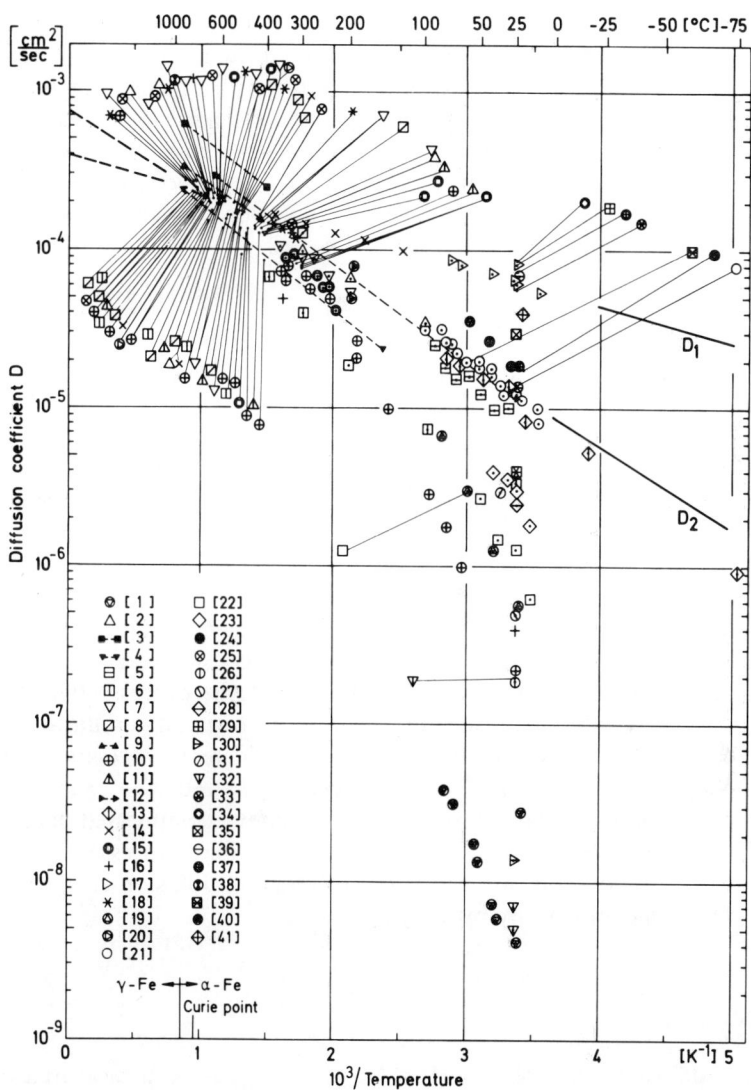

Fig. 15. Diffusion coefficient of H in α-Fe. For references, see Table V.

been applied almost exclusively. If imperfections in the metals play an important role, the Ni data should show comparable scatter. Furthermore, it has been shown for H in Nb (see Section IX) that impurities such as nitrogen did not significantly change the diffusion constant. Therefore, we feel that surface effects are the cause of error. Thus, surface independent measurements on Fe are necessary before the influence of imperfections can be inferred. Assuming that surface effects and possibly trapping are responsible for slowing down the diffusion, it is reasonable to identify the highest values measured with the bulk-diffusion coefficient (D_1 in Fig. 15). Then the diffusion coefficient can be described by

$$D_0 = 4.0 \cdot 10^{-4} \text{ cm}^2/\text{sec} \quad \text{and} \quad U = 0.047 \text{ eV}$$

On the other hand, if one were to believe the large amount of data at $D \approx 10^{-5}$ cm²/sec for room temperature, one may draw a line as indicated by D_2 with the values

$$D_0 = 7.5 \cdot 10^{-4} \text{ cm}^2/\text{sec} \quad \text{and} \quad U = 0.088 \text{ eV}$$

For H in Fe, internal friction and magnetic disaccommodation measurements have been performed at lower temperatures and have been interpreted in terms of the motion of individual H atoms. In Fig. 16, these data, converted to diffusion coefficients, are shown together with the extrapolated lines D_1 and D_2 of Fig. 15. Although the experimental data fall between the range of D_1 and D_2, the consistency must be considered as being poor. In no case has it been shown conclusively that the process studied has anything to do with the elementary process of long range diffusion. The large scatter may be due to the influences of different processes. Indeed, magnetic disaccommodation measurements by Steeb and Kronmüller (1973) and Au and Birnbaum (1973) show quite a complicated spectrum for H in α-Fe at low temperatures with at least four processes involved.

Finally, it may be mentioned that the apparent diffusion coefficient of iron can change drastically by alloying.

D. NIOBIUM

The diffusion coefficient of H in Nb (Fig. 17) has been measured over the largest $1/T$ region known to us, even larger than for C in iron (see Fig. 1). Because of the relatively small changes in D, it was possible to cover nearly the whole temperature region with one experimental method, the Gorsky effect (Schaumann et al., 1968, 1970; Schaumann, 1969).

5. HYDROGEN DIFFUSION IN METALS

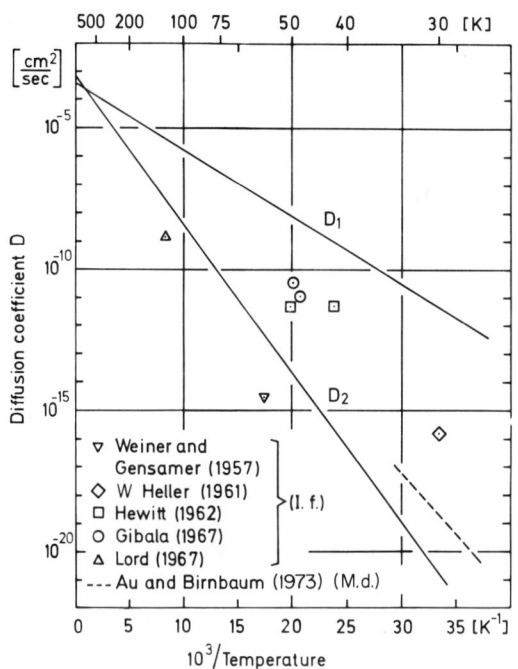

Fig. 16. Extrapolated diffusion coefficient of H in Fe (see Fig. 15), compared with internal friction (I.f.) and magnetic disaccommodation (M.d.) data. D_1 and D_2 represent the lines of Fig. 15.

The good consistency of the data shown in Fig. 17 exists only if just the results of surface independent methods are compared. For Nb, the surface is so difficult to control that even at high temperatures surface dependent methods as applied by Albrecht *et al.* (1959), Ryabchikov (1964a), and Oğurtani (1971) are not very reliable (see Table VII). The systematic deviation of the results of Cantelli *et al.* (1969) below $-50°C$ are very likely caused by precipitation of hydrogen.

The results of applying the Gorsky-effect technique all refer to hydrogen concentration $c \to 0$. Since with the Gorsky effect the quantity $\partial \mu / \partial \rho$ can be measured independently, possible corrections of D as a result of this factor are easily made (Schaumann, 1969; Schaumann *et al.*, 1970). The concentration and isotope dependence as well as the break in temperature dependence will be discussed later. The absolute values of D at room temperature are larger than for Pd.

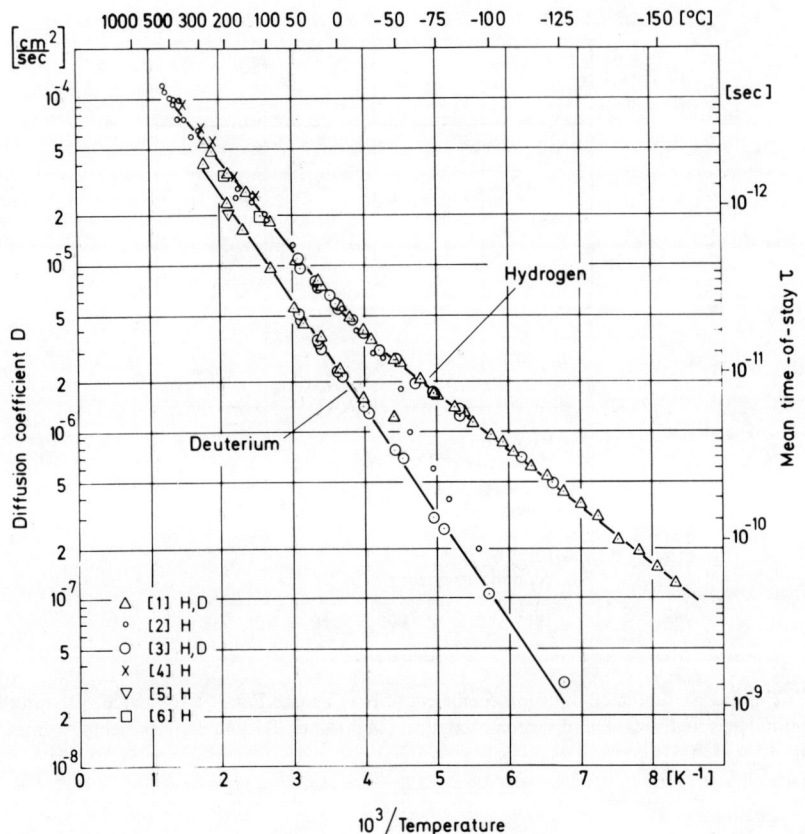

Fig. 17. Diffusion coefficients of H and D in Nb. For references, see Table VII. The mean time-of-stay τ is calculated for tetrahedral–tetrahedral jumps.

Suggestions for best values are the following.

Hydrogen in Nb:

$D_0 = 5.0 \cdot 10^{-4}$ cm^2/sec   and $U = 0.106$ eV, $T > 0°C$

$D_0 = 0.9 \cdot 10^{-4}$ cm^2/sec   and $U = 0.068$ eV, $T < -50°C$

Deuterium in Nb:

$D_0 = 5.4 \cdot 10^{-4}$ cm^2/sec   and $U = 0.129$ eV

The low temperature activation energy for H in Nb has recently been confirmed by Faber and Schultz (1972) at about 60°K by studying quenched-in hydrogen.

The mean time-of-stay τ on the right hand scale in Fig. 17 is calculated for tetrahedral–tetrahedral jumps; the assumption of this position is justified by the results of Somenkov et al. (1968).

It should be pointed out that the high temperature activation energy is about equal to the local-mode energy 0.11 eV (Verdan et al., 1968; Brand, 1971), whereas the low temperature value is considerably lower. It has been shown by Brand (1971) and Conrad et al. (1974) that the local-mode frequency of D in Nb is about a factor of $\sqrt{2}$ lower than for H. Therefore, in this case, the activation energy has a value between the first and second excited level of the deuterium atom.

Internal friction measurements by Cannelli and Verdini (1966a) on H in Nb and by Schiller and Schneiders (1968) on H and D in Nb show for kilocycle frequencies internal friction peaks at about 100°K which have been discussed as Snoek peaks for H or D. Buchholz et al. (1973) have shown experimentally that the Snoek-relaxation strength for H or D in Nb and Ta is at least 2–3 orders of magnitude smaller than for heavy interstitials. Their conclusion, combined with the low concentration in solution at 100°K (Westlake, 1969) makes it appear unlikely that the Snoek effect has been observed. Furthermore, the presence of the peaks depends on the impurity concentration (Schiller, 1968; Cannelli and Mazzolai, 1973). It is therefore unlikely that the process leading to internal friction is connected with the diffusion process. Cannelli and Mazzolai (1973) have recently adopted this view themselves.

The apparent discrepancy between NMR data by Zamir and Cotts (1964) and the diffusion data in Fig. 17 has recently been removed by Lütgemeier et al. (1972a) who have shown that, besides the diffusion of the protons, the interaction with the conduction electrons contributes to the spin–lattice relaxation. By removing this part, the latter authors have determined an activation energy of 0.11 eV.

E. TANTALUM

In Fig. 18, only the resulting data for surface independent methods are included. Nevertheless, discrepancies in the numerical values of D (though not so much in the slopes) can be noted. The quasi-elastic neutron-scattering results have not been performed at very low concentration but at 15% H/Ta. Since, in Nb, the diffusion coefficient decreases with increasing H concentration (see Section V, Fig. 20), this may also explain the somewhat lower values of D observed in Ta.

The systematically lower values of Cantelli et al. (1971, 1973) may have several explanations. The large number of data points are based on one internal friction peak. The authors used a method in which the

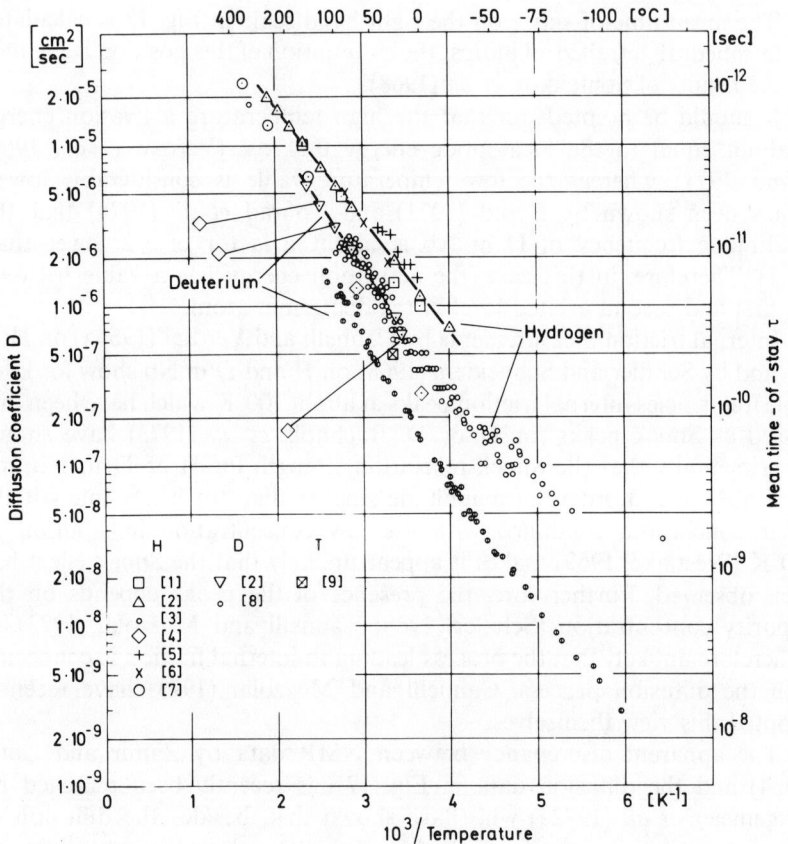

Fig. 18. Diffusion coefficients of H and D in Ta. For references, see Table IX. The mean time-of-stay τ is calculated for tetrahedral–tetrahedral jumps.

magnitude of D is deduced from the shape of the internal friction peak. This method is not free of ambiguities, since corrections for the temperature dependence of the relaxation strength as well as similar corrections of D for $\partial\mu/\partial\rho$ must be applied (except at very low concentrations). The latter correction has not been applied to the deuterium data of Cantelli *et al.* (1973) in Fig. 18. Furthermore, the impurity content was not controlled very carefully. Impurities, as shown for Nb, can lower the measured value of the diffusion constant (see Section IX, Fig. 28). Also, the data points at low temperatures should have large error bars as a result of uncertainties in background corrections and the possibility of precipitation.

5. HYDROGEN DIFFUSION IN METALS

The lines shown in Fig. 18 correspond to the following values.

Hydrogen in Ta: $D_0 = 4.4 \cdot 10^{-4}$ cm^2/sec and $U = 0.140$ eV

Deuterium in Ta: $D_0 = 4.9 \cdot 10^{-4}$ cm^2/sec and $U = 0.163$ eV

Again, the activation energy for H in Ta is about equal to the local-mode energy (Sakamoto, 1964; Brand, 1971). For the internal friction peaks observed at lower temperatures (Cannelli and Verdini, 1966a,b), the same statements hold as given for Nb in the preceding section.

F. Vanadium

Figure 19 shows only data of surface independent methods. The absolute value at room temperature is the largest value so far measured [except perhaps for iron and CuPd (Piper, 1966)]. At $-100°$C, the diffusion coefficient for H in V is still as high as for H in Pd at 300°C.

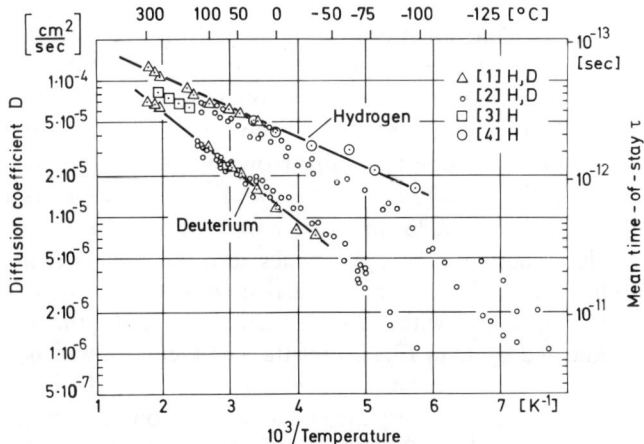

Fig. 19. Diffusion coefficients for H and D in V. For references, see Tables XI and XII. The mean time-of-stay τ is calculated for tetrahedral–tetrahedral jumps.

The scatter of the data by Cantelli et al. (1970) at low temperatures should not be taken too seriously. The same arguments as given for Ta apply, especially at low temperatures.

The quasi-elastic neutron-scattering data by Rowe et al. (1971) have been obtained from a sample with $c = 20\%$ H/V. From the data of Rowe et al. (1971) on samples with different concentrations, it can be deduced that D decreases with increasing H content. Therefore, an extrapolation of the 20% H data to $c = 0$ would bring them closer to the Gorsky-effect data.

The lines drawn in Fig. 19 correspond to the following values:

Hydrogen in V: $D_0 = 2.9 \cdot 10^{-4}$ cm^2/sec   and $U = 0.043$ eV

Deuterium in V: $D_0 = 3.7 \cdot 10^{-4}$ cm^2/sec   and $U = 0.080$ eV

The activation energy for H in V is the lowest value for solid state diffusion so far published. Since the local-mode energy for H in V has been determined as 0.12 eV (Verdan et al., 1968; Rush and Flotow, 1968; Brand, 1971), the "activation energy" is far below the local-mode energy, similarly, as is the case for the low temperature value of H in Nb. Rowe (1972) has shown that the local-mode frequencies for H and D scale as $\sqrt{2}$. Therefore, in this case, the activation energy for deuterium is close to the local-mode frequency.

V. High Hydrogen Concentrations

As shown in the introduction, for high hydrogen concentrations, the measured diffusion coefficient depends on the experimental method used. Except for NMR and quasi-elastic neutron scattering, the experimental diffusion coefficient is modified by the concentration and temperature dependent equilibrium property $\partial\mu/\partial\rho$. How drastic this influence can be is shown in Fig. 20. D^* represents the diffusion coefficient as measured with the Gorsky effect for 33 and 40% H/Nb. For $T \to T_c$, the relation $\partial\mu/\partial\rho \to 0$ holds. Therefore, D^* approaches zero (G. Alefeld et al., 1969) as in Fig. 20. This effect is known as "critical slowing down." If D^* is divided by $\partial\mu/\partial\rho$ and multiplied with kT/ρ, which is $\partial\mu/\partial\rho$ in the limit $\rho \to 0$, the curves indicated by D in Fig. 20 result. These curves, which now show the temperature and concentration dependence for individual jumps of the protons, can be compared with quasi-elastic neutron-scattering measurements by Gissler et al. (1970). The agreement of the results of both methods is very good. The upper curve shows, for comparison, the results of both methods in the limit of small concentrations (Schaumann et al., 1970; Gissler et al., 1970) with similarly good agreement. The remaining concentration dependence of D is due to partly to blocking of sites and partly to a real concentration dependence for individual jumps.

Wicke and Bohmholdt (1964), Bohmholdt and Wicke (1967), and Holleck and Wicke (1967) have measured the diffusion coefficient of the α' phase of Pd (formerly called the β phase) by a permeation method and an electrochemical method. Solubility data of Wicke and Nernst (1964) have been used to correct for $\partial\mu/\partial\rho$. These authors also correct for blocking by dividing the diffusion coefficient by $(1 - c)$. After these corrections,

5. HYDROGEN DIFFUSION IN METALS

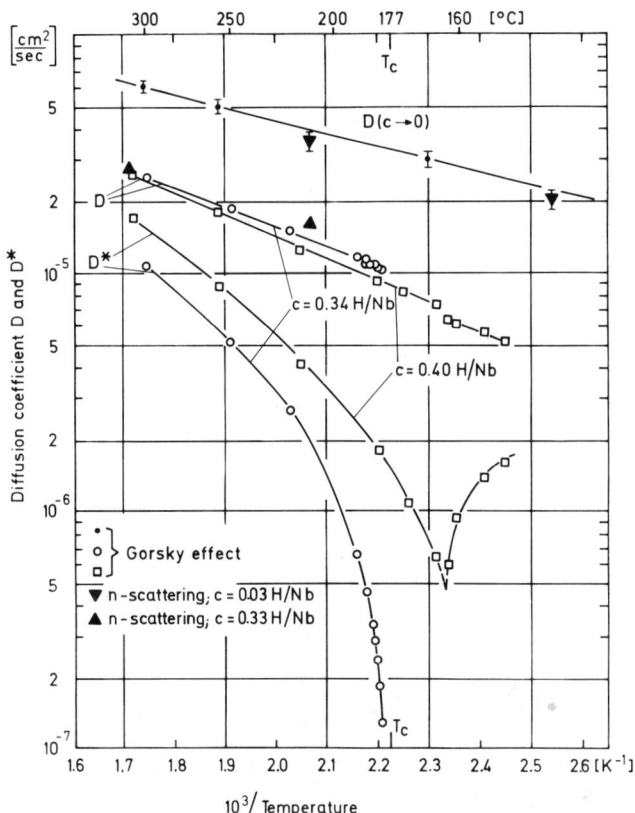

Fig. 20. Diffusion coefficient of H in Nb in the region of the critical concentration compared with $D(c \to 0)$. The Gorsky-effect measurements are taken from Tretkowski (1969) and Völkl (1972); the quasi-elastic neutron-scattering data are taken from Gissler et al. (1970). [D^* and D are defined in Eqs. (2) and (4).]

only a weak concentration dependence of D remains. Yet these data are not in agreement with measurements by Beg and Ross (1970) (quasi-elastic neutron scattering, see Table III). The numerical values differ by an order of magnitude and this difference becomes even worse if the data of Beg and Ross (1970) are corrected for blocking.

A comparison of the diffusion data listed in Table III with elastic relaxation measurements by Arons et al. (1970) is difficult since it is not clear what effect is studied and therefore how the measured relaxation time is affected by $\partial \mu / \partial \rho$.

In the transition from the α or α' to the β phases of the bcc metals, more drastic changes in D are reported (see Tables VIII, X and XII). The

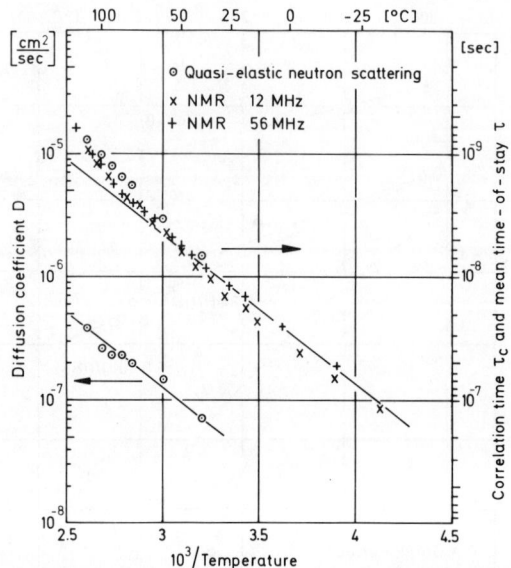

Fig. 21. Diffusion coefficient and mean time-of-stay of H in the β phase of Nb (0.9 H/Nb) (B. Alefeld et al., 1972).

absolute values decrease by 1–2 orders of magnitude. As an example, in Fig. 21 the diffusion coefficient of H in Nb for the β phase (90% H/Nb) is shown. Comparing with Fig. 17, we find a reduction of 2 orders of magnitude. The reduction is not as pronounced, however, when compared with the α' phase of equal concentration. The upper curves in Fig. 21 represent measurements of the mean time-of-stay with NMR and quasi-elastic neutron scattering (B. Alefeld et al., 1972). The agreement is remarkably good. From these data and from the measured diffusion coefficient, the mean jump distance can be calculated with the result $\Delta l = 4.8 \pm 0.8$ Å. This value is surprisingly large compared with 1.2 Å for tetrahedral–tetrahedral jumps and 1.65 Å for octahedral–octahedral jumps, yet it is consistent with the distance of occupied sites in the ordered β structure as suggested by Somenkov et al. (1968).

VI. Isotope Dependence

For all metal–hydrogen systems, deviations from the predictions of classical rate theory (Wert and Zener, 1949; Vineyard, 1957) $D_1/D_2 = (m_1/m_2)^{1/2}$ have been observed. This is not too unexpected in the light of the small masses and the high Einstein temperatures of the hydrogen

5. HYDROGEN DIFFUSION IN METALS

isotopes [see Section I(2)]. Surprisingly, the characteristic features of the isotope dependence seem to be correlated with the structure of the host metal, as demonstrated for the bcc metals V (Fig. 19), Nb (Fig. 17), and Ta (Fig. 18) and the fcc metals Pd (Fig. 22), Ni (Fig. 23), and Cu (Fig. 24).

For the three bcc metals (Figs. 17–19), hydrogen diffuses faster than deuterium in the complete temperature region investigated. The pre-exponentials are almost isotope independent (for Nb above 300°K), whereas for the activation energies, $U_H < U_D$. As a consequence, the ratio D_H/D_D shown in Fig. 25 for the three metals is temperature dependent. Only at a definite temperature which is about the same for all three metals does this ratio happen to be $\sqrt{2}$, whereas at lower temperatures, ratios up to ~ 20 have been observed (Wipf, 1972). For tritium diffusion, only one value (for Ta at room temperature, see Fig. 18) has been published (Sicking and Buchold, 1971).

For the fcc metals, the preexponential factors are mass dependent and scale as $(m_1/m_2)^{1/2}$ for H and D in all three metals within experimental error. The activation energies are mass dependent as well, but in contrast to the three bcc metals, one finds: $U_H > U_D$. This fact leads to a reversed isotope dependence below a certain temperature ($\approx 500°C$ for Pd), i.e., deuterium diffuses faster than hydrogen. This was first observed experimentally by Bohmholdt and Wicke (1967) for Pd and is shown in Fig. 22 for data obtained with the Gorsky effect (Völkl et al., 1971). For Ni and Cu, an extrapolation of high temperature diffusion data by Katz et al. (1971) would yield similar results. In Fig. 26, the ratio D_H/D_D for diffusion in Pd is shown. In contrast to bcc metals, all values are below $\sqrt{2}$ and furthermore, for $T < 500°C$, below 1, indicating the reversed isotope dependence. For tritium, the experimental situation is as follows: Katz et al. (1971) have found that for Ni and Cu, the tritium-diffusion coefficient follows the same mass dependence as observed for H and D, i.e., $D_{0H} : D_{0D} : D_{0T} \approx 1 : \sqrt{2} : \sqrt{3}$ and $U_H > U_D > U_T$. Therefore, extrapolating to low temperatures, tritium would diffuse fastest. The same result has been found by Salmon and Randall (1954) for H and T in Pd, using the permeation technique. Since solubility data for T in Pd were not available, the authors calculated the diffusion coefficient of T by using solubility data of hydrogen (Favreau et al., 1954) and correcting for the higher binding energy of the tritium molecule. In contrast to this result for the temperature region 200–600°C, Sicking and Buchold (1971) found for tritium in Pd in the room temperature region a normal isotope dependence (see Figs. 22 and 26).

In comparing the influence of structure on the isotope effect, a result of Piper (1966) is of interest: Pd with 58% Cu changes to a bcc structure.

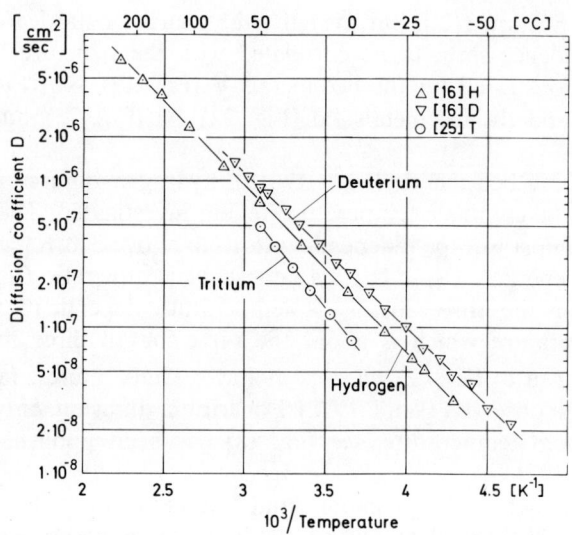

Fig. 22. Diffusion coefficients of isotopes of hydrogen in Pd in the room temperature region. For references, see Tables I and II.

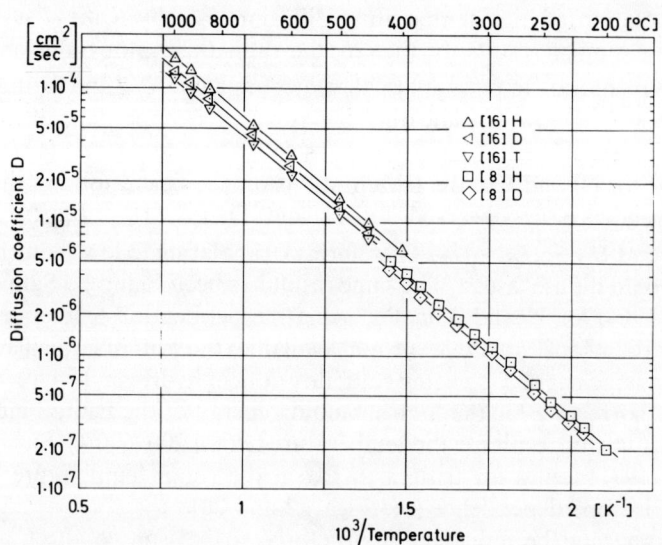

Fig. 23. Diffusion coefficients of isotopes of hydrogen in Ni. For references, see Table IV.

5. HYDROGEN DIFFUSION IN METALS

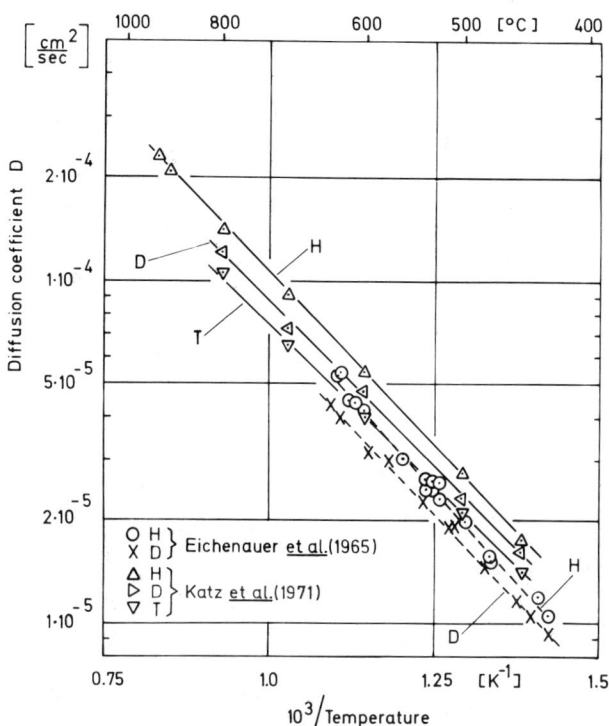

Fig. 24. Diffusion coefficients of isotopes of hydrogen in Cu.

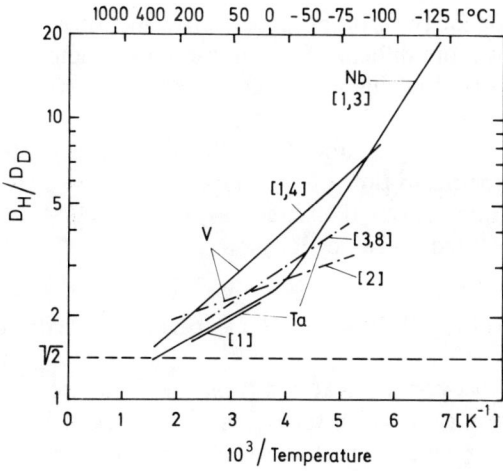

Fig. 25. Ratios of diffusion coefficients of H and D in V, Nb and Ta. For references see Tables VII (Nb), IX (Ta), and XI (V).

Piper finds for this structure an isotope ratio between 1.5 and 1.6, which is more characteristic of bcc metals (Fig. 25) than of fcc metals (Fig. 26).

The observation of Heumann and Primas (1966) of $D_H/D_D = 1.2$ for bcc and fcc iron has been made in a temperature regime ($\approx 1000°C$), in which, according to Figs. 25 and 26, the influence of structure is small.

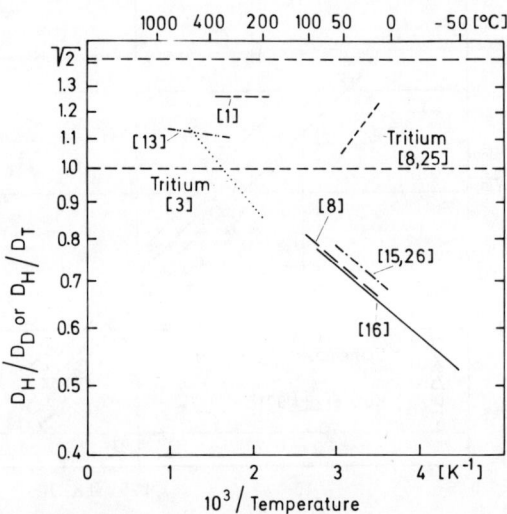

Fig. 26. Ratios of diffusion coefficients of isotopes of hydrogen in Pd. (The indicated curves refer to tritium.) For references, see Tables I and II.

A comparison of the isotope dependence measured at high hydrogen concentrations is more difficult since samples with identical concentrations are hard to prepare. Yet this is a necessary requirement since the observed diffusion coefficients depend upon the concentration. As an example, Bohn (1972) reports, for 78% H and D in the β phase of Nb, the following parameters for the mean time-of-stay. Hydrogen: $\tau_0 = 1 \cdot 10^{-12}$ sec; $U_H = 0.22$ eV. Deuterium: $\tau_0 = 2 \cdot 10^{-13}$ sec; $U_D = 0.31$ eV. More data on the α' phase in Pd are listed in Table III.

VII. Deviations of the Diffusion Coefficient from the Arrhenius Relation

Because of the nonclassical diffusion behavior of H and the large temperature region over which the diffusion coefficient has been measured, it is by no means evident that the diffusion coefficient of H in metals should obey an Arrhenius behavior over the complete temperature region. (See Chapter 1.) Nevertheless, empirically, an exponential temperature dependence seems to be a good fit, although it must be admitted that, for H in V, a power dependence on T would have fitted the data equally well.

In α-Pd, the deviation at high temperatures shown in Fig. 13 have been observed by Gol'tsov et al. (1970) by a permeation technique. Since these authors report hysteresis effects for heating and cooling in the temperature range of the deviation, a confirmation of their results by other techniques would be desirable.

For the α' phase of Pd, Arons et al. (1970) reported marked deviations for H as well as for D by combining diffusion, NMR, internal friction, and elastic after-effect measurements. For both isotopes, with decreasing temperatures, smaller activation energies as well as smaller preexponential factors are observed (see Table III). Since the process causing the internal friction and the elastic after-effect is not known in detail, it is not evident that the elementary process responsible for long range diffusion is being studied. A similar decrease of the activation energy for the diffusion of H in Ni with decreasing temperature has been postulated by Combette and Grilhé (1970) and Combette et al. (1972a,b) (see Table IV). Unfortunately, the low temperature values have not been determined by direct diffusion experiments, but by studying the time dependence of internal friction, modulus defect, or yield point. These properties are influenced by hydrogen arranging in the strain field of dislocations.

For Fe, often the opposite behavior, namely, a drastic increase of the activation energy as well as the preexponential factor with decreasing temperature, has been discussed quite frequently (for summaries, see e.g. Birnbaum and Wert, 1972; Heumann and Domke, 1972). Inspecting the data in Fig. 15, such a deviation is difficult to justify. In some experiments (e.g., Johnson and Hill, 1960), a marked break in the temperature dependence of the permeation rate has indeed been observed. However, since this break is not reported by other authors using different samples, it is possible that a change from bulk-diffusion controlled permeation to surface controlled permeation has been observed.

The break in the temperature dependence of D, as shown in Fig. 17 for H in Nb, has been observed with the Gorsky-effect method and has been confirmed by Wipf (1972) using a completely different method, which was based on resistivity changes caused by bulk diffusion of H. No model for this marked break in the temperature dependence exists. It should be mentioned that for impure samples, the process causing this enhanced diffusion at low temperatures disappears (see Section IX, Fig. 28). No break has been observed for deuterium in the temperature region so far investigated.

For H in Ta, a break in the temperature dependence similar to that for H in Nb may be inferred from the data of Cantelli et al. (1971). Hanada *et al.* (1972) and Hanada (1973) have studied the annealing of quenched-in hydrogen in Ta. If the resistivity changes observed at 15°K can be attributed to long range diffusion of H (which still has to be shown), the diffusion coefficient for H in Ta at this temperature must be far greater than that which would be predicted by extrapolation of the data in Fig. 18.

VIII. Dependence on Alloying

From the many investigations on alloys, a representative result is shown in Fig. 27. The diffusion coefficient at 30°C for Pd is rather insensitive to

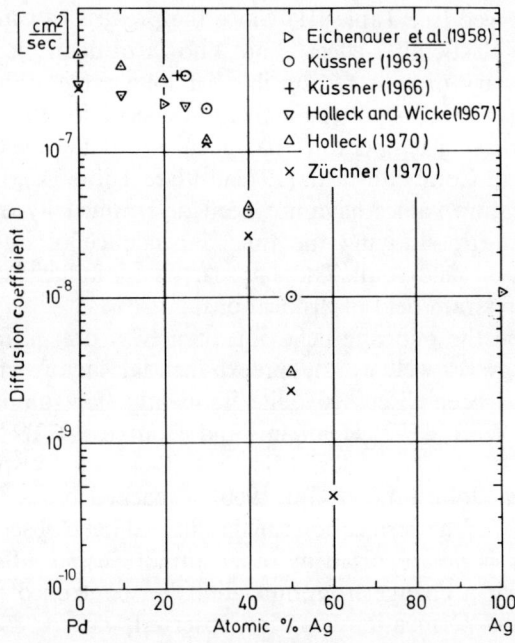

Fig. 27. Diffusion coefficient of H in PdAg alloys at 30°C (Züchner, 1970).

small Ag concentrations, but falls off strongly in the region of 40 at. % Ag to a value which is 3 orders of magnitude smaller than for pure Ag. Whether this apparent minimum in D occurs for homogeneous solid solution of PdAg, which would be a rather interesting effect, or whether it is caused by the formation of intermediate phases (see Section X) cannot be decided. Ordered phases in the AgPd phase diagrams have been suggested but not confirmed by other authors (see Elliot, 1965; Shunk, 1969).

IX. Influence of Traps

Because of the very high mobility of hydrogen still existing at and below room temperature, one is inclined to expect that traps significantly influence the diffusion constant and its temperature dependence. For instance, the strong slowing down of H diffusion in α iron observed by some authors in the region of room temperature has been attributed to trapping (e.g., McNabb and Foster, 1963; Oriani, 1970).

According to Baker and Birnbaum (1973), H–interstitial pairs like H–N or H–O with 0.09 eV binding energy are formed in Nb. The existence of internal friction peaks (Cannelli and Verdini, 1966a; Schiller and Schneiders, 1968) which depend on impurity content as well as on H concentration also suggests the existence of such clusters. From these observations, one would expect that for Nb interstitial traps strongly influence the macroscopic diffusion constant. In Fig. 28, the diffusion constant for H in Nb is shown

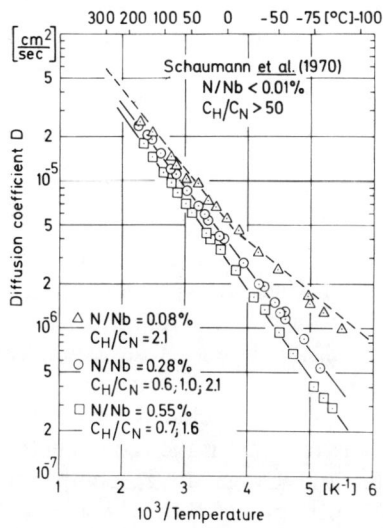

Fig. 28. Diffusion coefficient of H in Nb for varying nitrogen concentration (Münzing, 1972). (---) refers to measurements by Schaumann et al. (1970) on outgassed Nb (see Fig. 17).

with the nitrogen content varying between $c_N < 0.01$ and 0.55% N/Nb (Münzing, 1972). The ratio of hydrogen to nitrogen traps c_H/c_N has been varied between 50 and 0.6, i.e., twice as many traps as hydrogen in the latter case. The nitrogen was kept in solution by quenching from above 1000°C and afterward H was loaded electrolytically at room temperature. Figure 28 shows that the macroscopic diffusion coefficient decreases surprisingly by only about a factor of 2 above room temperature. The absolute value of D seems to depend only on the absolute nitrogen content and not on the H/N ratio. Below room temperature, the influence of N on the diffusion coefficient is stronger. The break in the activation energy, as discussed in Section VII, disappears with increasing N content or is shifted to lower temperatures. This observation is consistent with results of Sasaki and Amano (1969) (see Table VII), who found for relatively impure Nb the high temperature activation energy of 0.1 eV at about 80°K instead of the low temperature value by studying the precipitation of the α phase into the β phase. The recent observations of Hanada et al. (1972) and Hanada (1973) on quenched-in H and D in Ta are very likely attributable to the interaction of H with traps. Hanada (1973) observes marked steps in the isochronal annealing curves of the quenched-in resistivity with the steps essentially independent of the H isotope. If H were to migrate freely, such steps are inexplainable and, furthermore, a strong isotope dependence should exist at 10°K. The steps may be interpreted as a release of H from shallow traps with different binding energies. This interpretation implies that H is freely migrating at helium temperature in a trap-free lattice. Hanada's results (1973) may also be interpreted as stress assisted migration in the strain field of traps into positions of lower energy. Such an explanation is possible only if trapping of H manifests itself as a resistivity change.

X. Influence of Structure

Structure dependent properties of H diffusion have been noted in connection with isotope effects (see Section VI). There are several examples in which the transition from a close packed to a bcc structure leads to a drastic decrease of the activation energy; Fig. 29 shows, as an example, the $\alpha-\gamma$ transition for iron. Only data of those authors who have used the same measuring technique in both phases have been included. For the γ phase, an activation energy of $U_\gamma \approx 0.5$ eV is reported rather consistently (see Table VI and the summary by Bester and Lange, 1972), compared to U_α between 0.05 and 0.1 eV. For Ti, Wasilewski and Kehl (1954) have found: U (bcc) $= 0.29$ eV and U (hcp) $= 0.54$ eV. Piper (1966) reports for PdCu alloys a drop of the activation energy to 0.1 eV, correlated with the transition to a bcc structure at about 45 at. % Cu. This value is considerably smaller than for the pure substances (Figs. 13 and 24) which are both face-centered.

It is also worth noticing that the smallest values for U have been found for the four bcc metals V, Nb, Ta, α-Fe.

Figure 29 shows that also the absolute value of D changes; it increases with the transition to the bcc phase. Similar observations have been made for PdCu (Piper, 1966) and NiFe (Dresler, 1971; Dresler and Frohberg,

5. HYDROGEN DIFFUSION IN METALS

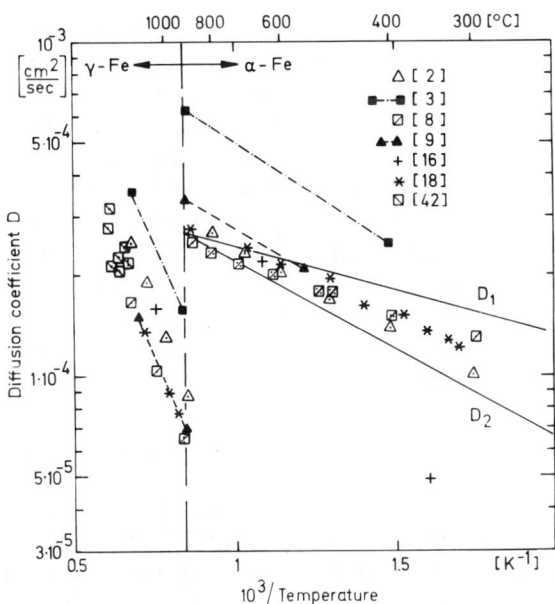

Fig. 29. Diffusion coefficients of H in α- and γ-Fe. For references, see Tables V and VI. D_1 and D_2 represent the lines of Fig. 15.

1972) when changing from fcc to bcc by varying the composition. One would be inclined to attribute the higher diffusivity to the less closely packed structure of the bcc metals. Yet, as Piper (1966) has pointed out, for Ti as well as PdCu, the close packed structures for these systems are less dense than the bcc phases.

Correlations between the H-diffusion coefficient and the degree of order have also been noted. Thus, Dresler (1971) and Dresler and Frohberg (1972) have found for Ni_3Fe an increase of D (measured at 58°C) with long range order, whereas the data of Dus and Smialowski (1967) evaluated by Dresler (1971) show a decrease of D (measured at 400°C) with long range order. Also, the drastic change in D for PdCu (2–3 orders of magnitude) (Piper, 1966) is not only connected with the fcc–bcc transition but also with disorder–order, as has been pointed out by Dresler and Frohberg (1972).

XI. Conclusions

In this article, an attempt was made to summarize the essential experimental observations concerning hydrogen diffusion in metals. The observations are exemplified on those systems which have received the greatest attention in the past.

We have intentionally omitted thus far interpretation and theoretical discussion of observations. A theory for hydrogen diffusion must explain the following facts. In some metals (for example, see Figs. 17–19) the absolute jump rate of hydrogen reaches values which are equal to the highest lattice frequencies of the host metal [e.g., Haas et al., 1963 (V); Nakagawa and Woods, 1965 (Nb); Woods, 1964 (Ta)]. The jump rate even comes close to the local-mode frequency of the hydrogen atom. For some bcc metals (V and Nb), "activation energies" have been found which are smaller than the first excited state of the hydrogen atom. It is evident that, under these conditions, the concepts of classical rate theory even with additional semiclassical quantum corrections cannot be applicable. For reviews of these theories, see Chapter 1 and Stoneham (1972).

Of special importance is the fact that the observed isotope dependence is still unexplained. In the bcc metals (V, Nb, and Ta), the preexponential factors D_0 are isotope independent, whereas the "activation energies" show strong isotope dependence. Therefore, a temperature dependent ratio D_H/D_D is observed, reaching a value of about 20 for Nb at 150°K. In this system also a remarkably strong change in the temperature dependence of the diffusion coefficient of hydrogen has been observed, which is absent for deuterium.

In fcc metals, D_0 shows a mass dependence as expected from classical rate theory. Yet again, in contrast to bcc metals for which the relation $U_H < U_D$ holds, one finds for fcc metals, $U_H > U_D$. Thus, for these metals, deuterium diffuses faster than hydrogen in the room temperature region and below.

A comparison of the diffusion coefficients of the light interstitials H and D with those of heavy interstitials O, N, or C for the metals V, Nb, Ta yields the following facts tabulated by Schaumann et al. (1970). The activation energies of the light interstitials are a factor 10–20 smaller, yet surprisingly also the preexponential factors D_0 are smaller by 1–2 orders of magnitude. A similar correlation between a small U and a small D_0 is shown in Fig. 29 at the α–γ transition of iron.

In our opinion, the existing theories for hydrogen diffusion, including band theories, tunnelling theories, hopping theories, etc., are not yet able to explain the experimental results.

ACKNOWLEDGMENT

The financial support by the Bundesministerium für Forschung und Technologie is gratefully acknowledged.

Tables I-XIII

TABLE I

Diffusion Coefficients, Activation Energies, and D_0 for H in Pd (α Phase)

U (eV)	D_0 (cm²/sec)	Temp. range (°C)	Method	Author(s)	Symbol no. Figs. 13, 22, 26
0.248	5.95×10^{-3}	192–302.5	Absorption	Jost and Widmann (1935)	[1]
0.248	8.43×10^{-3}	120–402	Absorption	Jost and Widmann (1940)	[2]
0.447	0.31	294–343	Permeation, time-lag	Barrer (1940)	[3]
0.241	3.55×10^{-3}	200–600	Permeation, using solubility data of Favreau et al. (1954)	Salmon and Randall (1954)	
0.244	4.3×10^{-3}	200–700	Permeation (using solubility data of Sieverts, 1929) and permeation, time-lag	Davis (1954)	[4]
0.291^a	$1.9 \times 10^{-2\,a}$				
0.245	5.18×10^{-3}	170–290	Permeation, using solubility data of Sieverts and Danz (1938)	Toda (1958)	[5]
0.236^a	$8.5 \times 10^{-3\,a}$				
0.217	1.3×10^{-4}	231–334	Permeation, time-lag	van Swaay and Birchenall (1960)	[6]
0.243	4.3×10^{-3}	213–379	Permeation, using solubility data of Moon (1956)	Katz and Gulbransen (1960)	[7]
0.260	6.1×10^{-3}	0–50	Absorption, measured electrochemically and by resistivity change	Simons and Flanagan (1965a)	
0.243	$3.65 \times 10^{-3\,b}$	20–100	Permeation, using solubility data of Wicke and Nernst (1964)	Bohmholdt and Wicke (1967)	[8]
0.252	4.5×10^{-3}	0–80	Electrochemical	Holleck and Wicke (1967)	[9]
0.161	—	309–431	Quasi-elastic neutron scattering	Sköld and Nelin (1967)	[10]
0.249	4.94×10^{-3}	27–436	Permeation, using solubility data of Favreau et al. (1954)	Koffler et al. (1969)	[11]
0.182	—	27–60	Electrochemical	Bockris et al. (1970)	[12]
0.270	5.25×10^{-3}	320–650	Permeation	Gol'tsov et al. (1970)	[13]

5. HYDROGEN DIFFUSION IN METALS

	D (cm²/sec)	Temp. (°C)	Method	Reference	
0.228	2.94×10^{-3}	260–640	Permeation, time-lag	Holleck (1970)	[14]
0.250	4×10^{-3}	20–60	Electrochemical	Züchner (1970)	[15]
0.226	2.5×10^{-3}	−40–220	Gorsky effect	Völkl et al. (1971)	[16]
	4.5×10^{-6}	≈25	Local electrolytic loading, and outgassing sections of the sample after a certain time	Duhm (1935)	
	3.3×10^{-5}	260	Permeation, time-lag	Toda (1958)	[17]
	1.3×10^{-7}	≈25	Electrochemical	Devanathan and Stachurski (1962)	[18]
	1.6×10^{-7}	≈25	Permeation, time-lag	von Stackelberg and Ludwig (1964)	[19]
	$(0.87–1.3) \times 10^{-7}$	23	Permeation	La Pietra and Castellan (1964)	
	3.1×10^{-7} [c]	25	Permeation	Makrides and Jewett (1966)	[20]
	4.6×10^{-5}	231	Quasi-elastic neutron scattering	Sköld and Nelin (1967)	[10]
	5.5×10^{-5}	357	Quasi-elastic neutron scattering	Sköld and Nelin (1967)	[10]
	8.4×10^{-5}	431	Quasi-elastic neutron scattering	Sköld and Nelin (1967)	[10]
	1.9×10^{-5}	260	Desorption	Knaak and Eichenauer (1968)	[21]
	4.2×10^{-5}	300	Desorption	Knaak and Eichenauer (1968)	[21]
	6.9×10^{-5}	400	Desorption	Knaak and Eichenauer (1968)	[21]
	1.0×10^{-4}	500	Desorption	Knaak and Eichenauer (1968)	[21]
	1.43×10^{-4}	600	Desorption	Knaak and Eichenauer (1968)	[21]
	1.9×10^{-7}	26	Electrochemical	Breger and Gileadi (1971)	[22]
	4.6×10^{-5}	350	Quasi-elastic neutron scattering	Rowe et al. (1972)	[23]

[a] Reevaluated by Bohmholdt and Wicke (1967), using solubility data compiled by Brodowsky (1965).
[b] Value for D_0 probably too low by about 20%, see Wicke and Meyer (1969).
[c] Result of Jewett and Makrides (1965) reevaluated using solubility data of Wicke and Nernst (1964) and Simons and Flanagan (1965b).

TABLE II

DIFFUSION COEFFICIENTS, ACTIVATION ENERGIES, AND D_0 FOR D AND T IN Pd (α PHASE)

Deuterium

U (eV)	D_0 (cm²/sec)	Temp. range (°C)	Method	Author(s)	Symbol no. Figs. 22, 26
0.248	6.7×10^{-3}	200–300	Absorption	Jost and Widman (1940)	[1]
0.224	2.5×10^{-3}	20–100	Permeation, using solubility data of Wicke and Nernst (1964)	Bomholdt and Wicke (1967)	[8]
0.267	4.46×10^{-3}	320–650	Permeation	Gol'tsov et al. (1970)	[13]
0.206	1.7×10^{-3}	−55–60	Gorsky effect	Völkl et al. (1971)	[16]
0.230	2.6×10^{-3}	0–60	Electrochemical	Boes (1971)	[26]

	D (cm²/sec)	Temp. (°C)			
	3.01×10^{-5}	302.5	Absorption	Jost and Widmann (1935)	
	0.97×10^{-5}	192.5	Absorption	Jost and Widmann (1935)	

Tritium

U (eV)	D_0 (cm²/sec)	Temp. range (°C)	Method	Author(s)	Symbol no. Figs. 22, 26
0.213	2.1×10^{-3}	200–600	Permeation, using H-solubility data of Favreau et al. (1954), the theory of solubility of Lacher (1937) and the zero-point energy for the tritium molecule given by W. M. Jones (1948)	Salmon and Randall (1954)	[3]
0.276	1.05×10^{-2}	16.5–50	Time-lag (tracer)	Sicking and Buchold (1971)	[25]

TABLE III

DIFFUSION COEFFICIENTS, ACTIVATION ENERGIES, AND D_0 FOR H AND D IN Pd (α' PHASE)

Hydrogen

U^* (eV)	D_0^* (cm²/sec)	Temp. range (°C)	Method	Author(s)
0.161	1×10^{-5}	−55–75	NMR ($c = 0.53$ H/Pd)	Norberg (1952)
0.24	—	> −53	NMR ($c = 0.64$ H/Pd)	Torrey (1958)
0.08	—	< −53	NMR ($c = 0.64$ H/Pd)	Torrey (1958)
0.104	5×10^{-7}	−175–0	NMR ($c = 0.63$ H/Pd)	Spalthoff (1961)
	$\tau_0 = 3.1 \times 10^{-10}$ sec			
0.21	—	> −43	NMR [$c = (0.68–0.73)$ H/Pd]	Burger et al. (1961)
0.06	—	< −43	NMR [$c = (0.68–0.73)$ H/Pd]	Burger et al. (1961)
0.126	—	−123 to −73	Time dependence of heat evolution ($c = 0.125$, 0.25 H/Pd)	Mitacek and Aston (1963)
0.278	—	−73 to −23	Time dependence of heat evolution ($c = 0.125$, 0.25 H/Pd)	Mitacek and Aston (1963)
0.239	—	−73 to −23	Time dependence of heat evolution ($c = 0.75$ H/Pd)	Mitacek and Aston (1963)
0.248	1.9×10^{-2}	30–100	Permeation, using solubility data of Wicke and Nernst (1964) ($0.6 < c < 0.7$ H/Pd)	Wicke and Bohmholdt (1964)
0.295	2.6×10^{-2}	3.6–15.3	Permeation ($c = ?$)	Damour and Castellan (1964)
0.127	3.8×10^{-4}	30–70	Permeation, using solubility data of Gillespie and Hall (1926) ($c \approx 0.6$ H/Pd)	Cleary and Greene (1965)
0.250[a]	5.7×10^{-3} [a]	20–100	Permeation, using solubility data of Wicke and Nernst (1964) ($c = 0.74$ H/Pd)	Bohmholdt and Wicke (1967)
0.265[a]	1.1×10^{-2} [a]	0–60	Electrochemical ($c = 0.69$ H/Pd)	Holleck and Wicke (1967)

TABLE III—*Continued*

U^* (eV)	D_0^* (cm²/sec)	Temp. range (°C)	Method	Author(s)
0.216	$\tau_0 = 5 \times 10^{-13}$ sec	−183 to −148	Internal friction and elastic after-effect [$c = (0.60–0.77)$ H/Pd]	Arons (1969) and Arons et al. (1970)
0.152	$\tau_0 = 4 \times 10^{-9}$ sec	−223 to −203	Internal friction and elastic after-effect [$c = (0.60–0.77)$ H/Pd]	Arons (1969) and Arons et al. (1970)
0.147	1.1×10^{-3}	20–200	Quasi-elastic neutron scattering ($c = ?$)	Beg and Ross (1970)

	D^* (cm²/sec)	Temp. (°C)	Method	Author(s)
	5.5×10^{-6}	≈25	Local electrolytic loading and outgassing sections of the sample after a certain time ($c = 0.64$ H/Pd)	Duhm (1935)
	10^{-5}–10^{-6}	80	Absorption ($c = 0.54$ H/Pd)	Jost and Widmann (1940)
	4×10^{-6}	≈25	Permeation ($c \approx 0.7$ H/Pd)	von Stackelberg and Ludwig (1964)
	4×10^{-7}	25	Permeation, electrochemical, using solubility data from literature ($c \approx 0.7$ H/Pd)	von Sturm and Kohlmüller (1965)
	1.5×10^{-6}	27	Permeation, using solubility data of Gillespie and Hall (1926) ($c = 0.6$–0.8 H/Pd)	Jewett and Makrides (1965)
	2.7×10^{-6}	20	Permeation, electrochemical, using solubility data of Nernst (1961, 1963) ($c = 0.72$ H/Pd)	Kahrig et al. (1966)
	1.45×10^{-6}	25	Permeation, using solubility data of Wicke and Nernst (1964) ($c = 0.74$ H/Pd)	Bohmholdt and Wicke (1967)
	1.7×10^{-6}	25	Electrochemical ($c = 0.69$ H/Pd)	Holleck and Wicke (1967)

Deuterium

U^* (eV)	D_0^* (cm²/sec)	Temp. range (°C)		
0.230[a]	3.45×10^{-3} [a]	20–100	Permeation, using solubility data of Wicke and Nernst (1964) ($c = 0.71$ D/Pd)	Bohmholdt and Wicke (1967)
0.248[a]	8×10^{-3} [a]	0–60	Electrochemical ($c = 0.66$ D/Pd)	Holleck and Wicke (1967)
0.202	$\tau_0 = 2.5 \times 10^{-13}$ sec	−183 to −148	Internal friction and elastic aftereffect [$c = (0.61$–$0.73)$ D/Pd]	Arons (1969) and Arons et al. (1970)
0.125	$\tau_0 = 7.2 \times 10^{-8}$ sec	−223 to −203	Internal friction and elastic aftereffect [$c = (0.61$–$0.73)$ D/Pd]	Arons (1969) and Arons et al. (1970)

	D^* (cm²/sec)	Temp. (°C)		
	1.93×10^{-6}	25	Permeation, using solubility data of Wicke and Nernst (1964) ($c = 0.71$ D/Pd)	Bohmholdt and Wicke (1967)
	2.5×10^{-6}	25	Electrochemical ($c = 0.66$ D/Pd)	Holleck and Wicke (1967)

[a] Values corrected for the dependence of D^* on $\partial \mu / \partial \rho$ (using solubility data of Wicke and Nernst, 1964) and $(1 - c)$.

TABLE IV Diffusion Coefficients, Activation Energies, and D_0 for H, D, and T in Ni (α Phase)

Hydrogen

U (eV)	D_0 (cm²/sec)	Temp. range (°C)	Method	Author(s)	Symbol no. Figs. 14, 23
0.378	2.04×10^{-3}	?	?	Barrer (1942)	
0.429	7.6×10^{-3}	400–700	Desorption	Ransley and Talbot (1955)	[1]
0.373	4.47×10^{-3}	380–986	Desorption	Hill and Johnson (1955)	[2]
0.642[a]	0.2[a]	458–693	Permeation	Lombard (1923)	
0.469[a]	1.7×10^{-2} [a]	200–550	Permeation	Borelius and Lindblom (1927)	
0.365[a]	1.1×10^{-3} [a]	478–798	Permeation	Hendricks and Ralston (1929)	
0.460[a]	1.5×10^{-2} [a]	376–600	Permeation	Ham (1932)	
0.469[a]	2.3×10^{-2} [a]	248–400	Permeation	Smithells and Ransley (1936)	
0.391[b]	3.7×10^{-3} [b]	85–165	Desorption	Euringer (1935)	
0.238[b]	1×10^{-3} [b]	400–600	Absorption	Lieser and Witte (1954)	
0.438	1.07×10^{-2}	162.5–496	Desorption	A. G. Edwards (1957)	[3]
0.447	9.5×10^{-3}	430–850	Permeation, using solubility data of Armbruster (1943)	Grimes (1959)	[4]
0.365[c]	4.22×10^{-3} [c]	303–694	Desorption	Eichenauer (1960)	
0.416	5.73×10^{-3}	250–350 (below Cp)	Permeation, time-lag	Belyakov and Ionov (1961)	[5]
0.386	4.21×10^{-3}	350–600 (above Cp)	Permeation, time-lag	Belyakov and Ionov (1961)	[5]
0.23, 0.39	—	31–122	Desorption	Marchand (1962)	
0.356	—	−45 to −17	Temperature dependence of strain rate	Boniszewski and Smith (1963)	
0.404	—	−140 to −130	Temperature dependence of strain aging time	Boniszewski and Smith (1963)	
0.412	3.8×10^{-3}	25–75	Desorption, followed by resistivity	Olsen and Larkin (1963)	[6]
0.387	5.5×10^{-3}	970–1313	Desorption	Ryabchikov (1964b)	[7]

	D (cm^2/sec)	Temp. (°C)	Method	Reference	
0.295	—	−60 to −20	Temperature dependence of crack propagation	Wilcox and Smith (1965)	[8]
0.411	6.73×10^{-3}	385–620	Desorption	Eichenauer et al. (1965)	[9]
0.331	4.68×10^{-3}	1080–1396	Absorption	Schenck and Lange (1966)	[10]
0.415	5.22×10^{-3}	200–420	Permeation, time-lag	Ebisuzaki et al. (1967b)	[11]
0.417	5.39×10^{-3}	350–700	Permeation, time-lag, and permeation, using solubility data of Sieverts and Hagenacker (1909) and Armbruster (1943)	Fischer (1967)	
0.377	4.38×10^{-3}	400–600	Desorption	Dus and Smialowski (1967)	[12]
0.44	3.5×10^{-3}	0–55	Desorption, followed by resistivity	Scherrer et al. (1967)	
0.41	5.4×10^{-3}	206–606	Desorption	Combette and Azou (1969)	
0.382	$(4.65–5.0) \times 10^3$	27–75	Electrochemical	Bockris et al. (1970)	[13]
0.42–0.43		200–650	Desorption	Combette and Azou (1970)	[14]
0.395–0.40	$(5.4–6.5) \times 10^{-3}$	200–650	Absorption	Combette and Azou (1970)	
0.356	—	27–70	Electrochemical	Beck et al. (1971)	[15]
0.409	7.04×10^{-3}	400–1000	Desorption	Katz et al. (1971)	[16]
0.425	8.1×10^{-3}	121–249	Permeation, time-lag	Louthan et al. (1971)	[17]
0.407	4.02×10^{-3}	25–400	Permeation, time-lag	Robertson (1972)	[18]
0.19–0.26	$\tau_0 = 10^{-6}$ sec	−160 to −20	Dislocation pinning (internal friction, modulus change, yield points)	Combette et al. (1972a,b) and Combette and Grilhé (1970)	
	1.16×10^{-8}	85	Desorption	Euringer (1935)	[19]
	3.4×10^{-8}	125	Desorption	Euringer (1935)	[19]
	10.5×10^{-8}	165	Desorption	Euringer (1935)	[19]
	1.63×10^{-5}	400	Absorption	Lieser and Witte (1954)	[20]
	2.79×10^{-5}	500	Absorption	Lieser and Witte (1954)	[20]
	4.16×10^{-5}	600	Absorption	Lieser and Witte (1954)	[20]
	4.4×10^{-7}	850	Desorption	Doolittle et al. (1963)	
	$(4–16) \times 10^{-12}$	20	Absorption, followed by measuring the saturation magnetization	von Aufschnaiter and Bauer (1964)	

TABLE IV—Continued

U (eV)	D_0 (cm²/sec)	Temp. range (°C)	Method	Author(s)	Symbol no. Figs. 14, 23

	D (cm²/sec)	Temp. (°C)			
	1.6×10^{-9}	35	Elastic after-effect	Cermak and Kufudakis (1966)	[21]
	6.91×10^{-10}	≈25	Elastic after-effect	Cermak and Kufudakis (1968)	[22]
	5.83×10^{-10}	≈25	Permeation, time-lag	Cermak and Kufudakis (1968)	
	1.19×10^{-8}	94	Permeation, time-lag	Ebisuzaki and O'Keeffe (1968)	[23]
	4.95×10^{-9}	58	Permeation, nonsteady state	Dresler (1971)	[24]

Deuterium

U (eV)	D_0 (cm²/sec)	Temp. range (°C)	Method	Author(s)	Symbol no. Figs. 14, 23
0.417	4.15×10^{-3}	250–350 (below Cp)	Permeation, time-lag	Belyakov and Ionov (1961)	
0.386	2.99×10^{-3}	350–600 (above Cp)	Permeation, time-lag	Belyakov and Ionov (1961)	
0.399	4.76×10^{-3}	410–650	Desorption	Eichenauer et al. (1965)	[8]
0.404[d]	2.5×10^{-3d}	217–441	Permeation, time-lag	Ebisuzaki et al. (1967a)	
0.401	5.27×10^{-3}	450–1000	Desorption	Katz et al. (1971)	[16]

Tritium

| 0.395 | 4.32×10^{-3} | 450–1000 | Desorption (tracer) | Katz et al. (1971) | [16] |

[a] Calculated by Hill and Johnson (1955) using solubility data of Sieverts (1911) and Luckemeyer-Hasse and Schenck (1932/33).
[b] Evaluated by Hill and Johnson (1955). [c] Value influenced by impurities, see Eichenauer (1967).
[d] Estimated by Birnbaum and Wert (1972).

TABLE V Diffusion Coefficients, Activation Energies, and D_0 for H and D in α-Fe

Hydrogen

U (eV)	D_0 (cm²/sec)	Temp. range (°C)	Method	Material	Author(s)	Symbol no. Figs. 15,29
0.379	1.1×10^{-2}	23–80	Permeation, time-lag	Iron	Barrer (1940)	[1]
0.40	1.65×10^{-2}	?	?	Iron	Barrer (1942)	
0.099[a]	7.6×10^{-4} [a]	20–800	Permeation	Iron	Sykes et al. (1947)	[2]
0.126[b]	2.17×10^{-3} [b]	400–900	Permeation	Iron	Geller and Sun (1950)	[3]
0.132	8.85×10^{-4}	150–900	Desorption	Iron	Stross and Tomkins (1956)	[4]
0.143	2.2×10^{-3}	30–90	Permeation, time-lag	Iron	Raczyński (1958) (See also Smialowski, 1967)	[5]
0.167	1.2×10^{-3}	25–900	Desorption	Iron	Zitter and Krainer (1958)	[6]
0.117	9.3×10^{-4}	200–774	Desorption	Armco iron	Eichenauer et al. (1958)	[7]
0.068	4.95×10^{-4}	298–878	Desorption	Iron "pure"	Maas (1958)	[8]
0.112	1.02×10^{-3}	550–900	Permeation, using solubility data of Sieverts et al. (1939)	Iron	Schenck and Taxhet (1959)	[9]
0.139	1.4×10^{-3}	200–780	Desorption	Iron	Johnson and Hill (1960)	[10]
0.340	0.12	25–200	Desorption	Iron	Johnson and Hill (1960)	[10]
0.095	6.7×10^{-4}	430–725	Absorption and desorption	Iron	Carmichael et al. (1960)	[11]
0.135	—	25–650	Desorption	Iron	Lee et al. (1961)	[12]
0.130	—	−75–80	Permeation, time-lag	Iron	Raczyński and Stelmach (1961)	[13]
0.220	1.7×10^{-3}	?	Desorption	Iron	Plusquellec et al. (1962)	
0.047	3.87×10^{-4}	126–693	Permeation, time-lag	Ferrovac E	Bryan and Dodge (1963)	[14]
0.142	1.42×10^{-3}	209–600	Permeation, nonsteady state	Iron	Wagner and Sizmann (1964)	[15]
0.232	3.65×10^{-3}	25–650	Desorption	Iron	Schwarz and Zitter (1965)	[16]
0.058	6×10^{-4}	10–75	Electrochemical	Armco iron	Beck et al. (1966)	[17]
0.083	6.42×10^{-4}	300–900	Desorption	Iron	Heumann and Primas (1966)	[18]

TABLE V—*Continued*

U (eV)	D_0 (cm²/sec)	Temp. range (°C)	Method	Material	Author(s)	Symbol no. Figs. 15, 29
0.369	1.14	23–84	Desorption	Iron	Evans and Rollason (1969a)	[19]
0.08	—	200–700	Desorption	Iron	Erdmann-Jesnitzer and Hieber (1969)	[20]
0.135	2.2×10^{-3}	10–100	Permeation, nonsteady state	Iron	Choi (1970)	[21]
0.282	6.5×10^{-2}	15–60	Permeation, nonsteady state	Iron A	Reiermann (1970)	[22]
0.258	6.1×10^{-2}	15–60	Permeation, nonsteady state	Iron B	Reiermann (1970)	[23]
0.195	—	27–60	Electrochemical	Armco iron	Bockris et al. (1970)	[24]
0.061	4.74×10^{-4}	326–816	Desorption	Iron	Domke (1971) and Heumann and Domke (1972)	[25]
0.108	2.2×10^{-3}		Permeation, time-lag	Iron "pure"	Donovan et al. (1971)	
	D (cm²/sec)	Temp. (°C)				
	$(0.14–1.1) \times 10^{-4}$	31	Desorption	Iron	Güntherschulze et al. (1939)	[26]
	1.9×10^{-7}	25	Absorption	Armco iron	Schuetz and Robertson (1957)	[27]
	5.0×10^{-7}	20	Desorption	Iron	Baranowski et al. (1957)	[28]
	2.5×10^{-6}	≈ 25	Electrochemical	Armco iron	Bastien and Amiot (1958)	[29]
	8.3×10^{-5}	25	Electrochemical	Armco iron	Devanathan et al. (1960) and Devanathan and Stachurski (1964)	
	1.39×10^{-8}	≈ 25	Electrochemical	Armco iron	Alikin (1960)	[30]
	3×10^{-6}	30–40	Permeation, time-lag	Armco iron	Palczewska and Ratajczyk (1961)	[31]
	$(0.5, 0.7, 2.0) \times 10^{-7}$	≈ 25	Permeation, nonsteady state	Iron	Kuznietsov and Subbotina (1965)	[32]
	4.9×10^{-5} ᶜ					
	6.05×10^{-5}	25	Electrochemical	Iron, zone-refined	Beck et al. (1966)	[17]

5. HYDROGEN DIFFUSION IN METALS

8.25×10^{-5}		25	Electrochemical	Armco iron, single crystal	Beck et al. (1966)	[17]
1.4×10^{-5}		25	Electrochemical	Iron	Radhakrishnan and Shreir (1967)	[33]
$(7.6–9.7) \times 10^{-5}$		25	Electrochemical	"Swedish iron"	Radhakrishnan and Shreir (1967)	[33]
4.5×10^{-5}		25	Permeation, time-lag	Iron	Wach and Miodownik (1968)	[34]
3×10^{-5}		25	Electrochemical	Iron	Joppien (1968)	[35]
8×10^{-5}		24	Electrochemical	Iron, zone-refined	Dillard and Talbot-Besnard (1969)	
7×10^{-5}		23	Permeation, time-lag	Iron, zone-refined	Raczyński and Talbot-Besnard (1969)	[36]
$(7.0–7.8) \times 10^{-5}$		24	Electrochemical	Iron, zone-refined	Dillard (1970)	[37]
2.45×10^{-4}		700	Desorption	Iron, zone-refined	Domke (1971) and Heumann and Domke (1972)	[38]
4.1×10^{-6}		25	Permeation, nonsteady state	Iron, "pure"	Dresler (1971)	[39]
6×10^{-5}		27	Electrochemical	Iron	Beck et al. (1971)	[40]
1.8×10^{-5} [d]		20	Desorption	Iron	Ellerbrock et al. (1972)	[40]
1.9×10^{-5} [d]		23	Desorption	Iron	Ellerbrock et al. (1972)	[40]
4.05×10^{-5}		22	Electrochemical	Armco iron	Govindan Namboodhiri and Nanis (1973)	[41]

Deuterium

0.083	$5.55 \times 10^{-4} D_0$	300–900	Desorption	Iron "pure"	Heumann and Primas (1966)
	1.1×10^{-4}	31	Desorption	Iron	Güntherschulze et al. (1939)

[a] Calculated using permeability data of Smithells and Ransley (1935) and solubility data of Sieverts (1911).
[b] Calculated using permeability data of Bennek and Klotzbach (1941) and solubility data of Sieverts et al. (1939), Martin (1929/30), and Luckemeyer-Hasse and Schenck (1932/33).
[c] Reevaluated by Wach and Miodownik (1968).
[d] Calculated using desorption data of Lange (1969) and Evans (1966).

TABLE VI

Diffusion Coefficients, Activation Energies, and D_0 for H and D in γ-Fe

Hydrogen

U (eV)	D_0 (cm²/sec)	Temp. range (°C)	Method	Author(s)	Symbol no. Fig. 29
0.516[a]	1.5×10^{-2} [a]	900–1200	Permeation	Sykes et al. (1947)	[2]
0.432[b]	1.07×10^{-2} [b]	900–1200	Permeation	Geller and Sun (1950)	[3]
0.486	8.2×10^{-3}	915–1205	Desorption	Maas (1958)	[8]
0.467	6.73×10^{-3}	900–1160	Permeation, using solubility data of Sieverts et al. (1939)	Schenck and Taxhet (1959)	[9]
0.465	6.63×10^{-3}	930–1120	Desorption	Heumann and Primas (1966)	[18]
0.334	2.9×10^{-3}	1222–1369	Absorption	Schenck and Lange (1966)	[42]
	D (cm²/sec)	Temp. (°C)			
	1.6×10^{-4}	1050	Desorption	Schwartz and Zitter (1965)	[16]

Deuterium

U (eV)	D_0 (cm²/sec)	Temp. range (°C)	Method	Author(s)	
0.465	4.85×10^{-3}	930–1120	Desorption	Heumann and Primas (1966)	

[a] Calculated using permeability data of Bennek and Klotzbach (1941) and solubility data of Sieverts (1911).
[b] Calculated using permeability data of Bennek and Klotzbach (1941) and solubility data of Sieverts et al. (1939), Martin (1929/30), and Luckemeyer-Hasse and Schenck (1932/33).

TABLE VII

Diffusion Coefficients, Activation Energies, and D_0 for H and D in Nb (α Phase)

Hydrogen

U (eV)	D_0 (cm²/sec)	Temp. range (°C)	Method	Author(s)	Symbol no. Figs. 17, 25
0.41	2.15×10^{-2}	600–700	Absorption	Albrecht et al. (1959)	
0.83	5.6×10^{-2}	1112–1500	Desorption	Ryabchikov (1964a)	
0.16[a]	$\tau_0 = 5.3 \times 10^{-13}$ sec	>110	NMR ($c = 0.7$ H/Nb)(α' phase)	Zamir and Cotts (1964)	[1]
0.106	5.0×10^{-4}	20–300	Gorsky effect	Schaumann et al. (1968, 1970) and Schaumann (1969)	
0.068	0.9×10^{-4}	−153–20	Gorsky effect	Cantelli et al. (1969)	[2]
0.109	5.4×10^{-4}	−40–560	Gorsky effect		
0.10	—	−196 to −183	Kinetic of precipitation	Sasaki and Amano (1969)	
0.08–0.1	—	100–250	Quasi-elastic neutron scattering	Rubin and Claessen (1970)	
0.43	1.8×10^{-2}	200–300, 600–700	Absorption, followed by measuring the hardness	Oğurtani (1971)	
0.093	2.9×10^{-4} [b]	−50–50	Resistivity relaxation	Wipf (1972)	[3]
0.061	5.6×10^{-5} [b]	−120 to −50	Resistivity relaxation	Wipf (1972)	[3]
0.11	6.5×10^{-4}	130–400	Quasi-elastic neutron scattering ($c = 0.14$ H/Nb)	Birchall and Ross (1972)	[4]
0.11	—	−50–203	NMR	Bohn (1972) and Lütgemeier et al. (1972a)	
0.072	—	−269 to −170	Kinetic of precipitation	Faber and Schultz (1972)	

TABLE VII—Continued

U (eV)	D_0 (cm²/sec)	Temp. range (°C)	Method	Author(s)	Symbol no. Figs. 17, 25
	D (cm²/sec)	Temp. (°C)			
	2×10^{-5}	200	Quasi-elastic neutron scattering ($c = 0.09$ H/Nb)	Verdan et al. (1968)	[5]
	2×10^{-5}	120	Quasi-elastic neutron scattering	Gissler et al. (1970)	[6]
	3.5×10^{-5}	210	Quasi-elastic neutron scattering	Gissler et al. (1970)	[6]
Deuterium					
U (eV)	D_0 (cm²/sec)	Temp. range (°C)			
0.129	5.4×10^{-4}	−50–300	Gorsky effect	Schaumann et al. (1968, 1970) and Schaumann (1969)	[1]
0.123	4.1×10^{-4} [b]	−120–50	Resistivity relaxation	Wipf (1972)	
0.16	—	0–200	NMR ($c = 0.1$ D/Nb)	Bohn (1972)	[3]

[a] Value has been corrected by Bohn (1972) and Lütgemeier et al. (1972a).
[b] Adjusted to the value of Schaumann et al. (1970) at 300°K.

TABLE VIII

Activation Energies and Mean Time-of-Stay τ_0 of H and D in Nb (β Phase)

Hydrogen

U (eV)	τ_0 (sec)	Temp. range (°C)	Method	Author(s)
0.15 (U^*)	—	600	Kinetic of hydride formation ($c = 1.0$ H/Nb)	Samsonov and Antonova (1961)
0.22	2×10^{-12}	−50–60	NMR ($c = 0.7$ H/Nb)	Zamir and Cotts (1964)
0.18	2.5×10^{-11}	−80 to −110	NMR ($c = 0.709$ H/Nb)	Staliński and Zogał (1965)
0.23	1.7×10^{-12}	−80 to −110	NMR ($c = 0.784$ H/Nb)	Staliński and Zogał (1965)
0.21	7.7×10^{-12}	−80 to −110	NMR ($c = 0.830$ H/Nb)	Staliński and Zogał (1965)
0.18	4.5×10^{-11}	−80 to −110	NMR ($c = 0.833$ H/Nb)	Staliński and Zogał (1965)
0.222	10^{-12}	−269–27	NMR ($c = 0.78$ H/Nb)	Bohn (1972) and Lütgemeier et al. (1972a)
0.24	10^{-12}	−30–120	Quasi-elastic neutron scattering and NMR ($c = 0.9$ H/Nb)	B. Alefeld et al. (1972)

Deuterium

U (eV)	τ_0 (sec)	Temp. range (°C)	Method	Author(s)
0.31	2×10^{-13}	−269–27	NMR ($c = 0.78$ D/Nb)	Bohn (1972)

288 J. VÖLKL AND G. ALEFELD

TABLE IX

DIFFUSION COEFFICIENTS, ACTIVATION ENERGIES, AND D_0 FOR H AND D IN Ta (α PHASE)

Hydrogen

U (eV)	D_0 (cm²/sec)	Temp. range (°C)	Method	Author(s)	Symbol no. Figs. 18, 25
1.4	1560	500–700	Absorption ($c = 0.05$ H/Ta)	Mallett and Koehl (1962)	
1.46	13,960	500–700	Absorption ($c = 0.1$ H/Ta)	Mallett and Koehl (1962)	
0.07	—	−20–130	NMR	Pedersen et al. (1965)	
0.63	7.5×10^{-2}	675–800	Permeation	Makrides et al. (1965)	[1]
0.15	6.1×10^{-4}	0–160	Diffusion, followed by resistivity	Merisov et al. (1966)	[2]
0.14	4.4×10^{-4}	−20–300	Gorsky effect	Schaumann et al. (1970)	[3]
0.15	3×10^{-4}	−40–250	Gorsky effect	Cantelli et al. (1971)	[4]
0.21	1.7×10^{-3}	0–150	Diffusion, followed by resistivity	Merisov et al. (1971)	[5]
0.16	—	0–50	Diffusion, followed by weight changes	Wicke and Obermann (1972)	
0.15	25°C: 2.4×10^{-6}	30–85	Diffusion, followed by lattice parameter changes	Züchner (1972)	[6]
0.11	25°C: 1.8×10^{-6}	148–340	Quasi-elastic neutron scattering	de Graaf et al. (1972)	[7]

	D (cm²/sec)	Temp. (°C)			
	1.3×10^{-6}	25	Diffusion, followed by lattice parameter changes	Züchner and Wicke (1969)	
	6.5×10^{-6} [a]	148	Quasi-elastic neutron scattering	de Graaf et al. (1972)	[7]
	1.35×10^{-5} [a]	248	Quasi-elastic neutron scattering	de Graaf et al. (1972)	[7]
	2.5×10^{-5}	340	Quasi-elastic neutron scattering	de Graaf et al. (1972)	[7]

5. HYDROGEN DIFFUSION IN METALS

Deuterium

U (eV)	D_0 (cm^2/sec)	Temp. range (°C)			
0.163	4.9×10^{-4}	20–150	Gorsky effect	Schaumann et al. (1970)	[2]
0.176	3.3×10^{-4}	−70–120	Gorsky effect	Cantelli et al. (1973)	[8]

	D (cm^2/sec)	Temp. (°C)			
	1.4×10^{-6}	25	Diffusion, followed by lattice parameter changes	Züchner and Wicke (1969)	

Tritium

| | 0.5×10^{-6} | 25 | Time-lag (tracer) | Sicking and Buchold (1971) | [9] |

[a] Value deduced from data contained in Fig. 3 of de Graaf et al. (1972).

TABLE X
Activation Energies for H in Ta (β Phase)

U (eV)	τ_0 (sec)	Temp. range (°C)	Method	Author(s)
0.25	—	> −30	NMR ($c = 0.75$ H/Ta)	Torrey (1958)
0.11	—	< −30	NMR ($c = 0.75$ H/Ta)	Torrey (1958)
0.11	4.4×10^{-10}	−150 to −5	NMR ($c = 0.66$ H/Ta)	Spalthoff (1961)
0.16–0.11	2×10^{-13}–10^{-11}	−50–10	NMR [$c = (0.1$–$0.66)$ H/Ta]	Pedersen et al. (1965)

TABLE XI
Diffusion Coefficients, Activation Energies, and D_0 for H and D in V (α Phase)

U (eV)	D_0 (cm²/sec)	Temp. range (°C)	Method	Author(s)	Symbol no. Figs. 19, 25
Hydrogen					
0.05	3.5×10^{-4}	20–300	Gorsky effect	Schaumann (1969) and Schaumann et al. (1970)	[1]
0.059	4.4×10^{-4}	−110–130	Gorsky effect	Cantelli et al. (1970)	[2]
0.039	3.5×10^{-4} [a]	−100–25	Resistivity relaxation	R. Heller (1973)	[4]
Deuterium					
0.08	3.7×10^{-4}	−40–300	Gorsky effect	Schaumann (1969) and Schaumann et al. (1970)	[1]
0.073	3.1×10^{-4}	−70–130	Gorsky effect	Cantelli et al. (1970)	[2]

[a] Adjusted to the value of Schaumann et al. (1970) at 300°K.

TABLE XII

Diffusion Coefficients, Activation Energies, and D_0 for H in V (High Concentrations)

U (eV)	D_0 (cm²/sec)	Temp. range (°C)	Method	Author(s)	Symbol no. Fig. 19
0.17	—	−200–27	NMR ($c = 0.2$ H/V)	von Meerwall and Schreiber (1968)	
0.07	—	−200–27	NMR ($c = 0.45$ H/V)	von Meerwall and Schreiber (1968)	
0.11	—	−200–27	NMR ($c = 1.8$ H/V)	von Meerwall and Schreiber (1968)	
0.047	—	145–250	Quasi-elastic neutron scattering ($c = 0.2$ H/V)	Rowe et al. (1971)	[3]

	D (cm²/sec)	Temp. (°C)	Method	Author(s)	Symbol no. Fig. 19
	6.4×10^{-5} [a]	145	Quasi-elastic neutron scattering ($c = 0.2$ H/V)	Rowe et al. (1971)	[3]
	7.0×10^{-5} [a]	175	Quasi-elastic neutron scattering ($c = 0.2$ H/V)	Rowe et al. (1971)	[3]
	7.5×10^{-5} [a] $[D(c \to 0) = 8 \times 10^{-5}]$	210	Quasi-elastic neutron scattering ($c = 0.2$ H/V)	Rowe et al. (1971)	[3]
	8.3×10^{-5} [a]	250	Quasi-elastic neutron scattering ($c = 0.2$ H/V)	Rowe et al. (1971)	[3]

[a] Value deduced from data contained in Fig. 5 of Rowe et al. (1971).

TABLE XIII

ACTIVATION ENERGIES AND D_0 FOR DIFFUSION OF H (AND D, T) IN SOME SELECTED METALS

Material	U (eV)	D_0 (cm²/sec)	Temp. range (°C)	Method	Author(s)
Ag–H	0.326	2.82×10^{-3}	388–600	Desorption	Eichenauer et al. (1958)
Al–H	1.45	1.2×10^{5}	400–600	Desorption	Ransley and Talbot (1955)
Al–H	0.47	0.21	455–590	Desorption	Eichenauer and Pebler (1957)
Al–H	0.42	0.11	400–600	Desorption	Eichenauer (1960) and Eichenauer et al. (1961)
Al–H	0.52	2×10^{-2}	570–630	Desorption	Matsuo and Hirata (1967)
Au–H	0.245	5.6×10^{-4}	500–940	Desorption	Eichenauer and Liebscher (1962)
Ba–H	0.196	4.0×10^{-3}	200–620	Concentration profile, determined by outgassing sections of the sample	Peterson and Hammerberg (1968)
Co–H	0.27	2.49×10^{-3}	1090–1416	Absorption	Schenck and Lange (1966)
Co–H	0.52	10.0	600–1150	?	Primas and Heumann (1972)
Co–D	0.55	11.3	600–1150	?	Primas and Heumann (1972)
Cu–H	0.49	6.8×10^{-2}	250–500	Desorption	Ransley and Talbot (1955)
Cu–H	0.40	1.1×10^{-2}	432–651	Desorption	Eichenauer and Pebler (1957)
Cu–D	0.12	8.9×10^{-8}	−46–20	Increase of D concentration, followed by detection of neutrons from the $D(d,n)^3$He reaction	Robinson et al. (1960)
Cu–H	0.42	1.15×10^{-2}	430–640	Desorption	Eichenauer et al. (1965)
Cu–D	0.39	6.2×10^{-3}	430–640	Desorption	Eichenauer et al. (1965)
Cu–H	0.403	11.3×10^{-5}	450–925	Desorption	Katz et al. (1971)
Cu–D	0.382	7.3×10^{-3}	450–800	Desorption	Katz et al. (1971)
Cu–T	0.378	6.1×10^{-3}	450–800	Desorption	Katz et al. (1971)
Hf–H	0.6–0.8	$\tau_0 = (3-1.2) \times 10^{-14}$ sec	220–400	NMR ($c = 1.35$–1.98 H/Hf)	Khodosov and Shepilov (1971)
La–H	1.0–0.13	$\tau_0 = (10^{-14}$–$10^{-11})$ sec	−197–400	NMR ($c = 1.4$–2.85 H/La)	Schreiber and Cotts (1963)
Mo–H	0.64	5.9×10^{-2}	575–980	Desorption	Hill (1960)

5. HYDROGEN DIFFUSION IN METALS

System			Method	Temp	Reference
Mo–H	0.40	—	Kinetic of annealing of yield point	25–130	Lawley et al. (1961)
Mo–H	0.96	0.158	Desorption	1023–1643	Ryabchikov (1964a)
Mo–H	—	$D(1712°C) = 2.7 \times 10^{-7}$	Kinetic of thermal dissociation	1712	Moore and Unterwald (1964)
Mo–H	0.36	7.6×10^{-5}	Permeation	525–625	P. M. S. Jones et al. (1966)
Pt–H	0.42	$D(70°C) = 3.4 \times 10^{-9}$	Electrochemical	50–80	Gileadi et al. (1966)
Pt–H	0.26	6×10^{-3}	Permeation	340–630	Ebisuzaki et al. (1968)
Sc–H	0.55	—	NMR ($c = 1.7$ H/Sc)	280–630	Weaver (1972)
Sc–H	0.73	—	NMR ($c = 1.8$ H/Sc)	390–780	Weaver (1972)
Sc–H	0.42	—	NMR ($c = 1.9$ H/Sc)	320–630	Weaver (1972)
Sc–H	1.13	—	NMR ($c = 1.9$ H/Sc)	630–780	Weaver (1972)
Sc–H	0.37	—	NMR ($c = 1.98$ H/Sc)	280–500	Weaver (1972)
Sc–H	0.79	—	NMR ($c = 1.98$ H/Sc)	500–780	Weaver (1972)
Th–H	0.42	2.92×10^{-3}	Desorption, absorption	300–900	Peterson and Westlake (1960)
Th–H	0.23	$\tau_0 = 4.8 \times 10^{-11}$ sec	NMR ($c = 3.5$ H/Th)	−65–230	Spalthoff (1961)
β-Ti–H	0.29	1.95×10^{-3}	Absorption	650–1000	Wasilewski and Kehl (1954)
α-Ti–H	0.54	1.8×10^{-2}	Absorption	500–824	Wasilewski and Kehl (1954)
Ti–H	0.38	5.7×10^{-3}	?	650–1000	Someno et al. (1960)
γ-Ti–H	0.41–0.44	$\tau_0 = (2.5–0.8) \times 10^{-12}$ sec	NMR ($c = 1.61–1.97$ H/Ti)	−196–200	Staliński et al. (1961)
Ti–H	0.25	$\tau_0 = 1.5 \times 10^{-9}$ sec	NMR ($c = 1.98$ H/Ti)	25–250	Spalthoff (1961)
α-Ti–H	0.64	3×10^{-2}	Absorption	610–900	Papazoglou and Hepworth (1968)
γ-Ti–H	0.72–0.76	$\tau_0 = 10^{-13}$ sec	NMR ($c = 1.94$ H/Ti)	150–190	Roberts and Merideth (1969)
γ-Ti–H	0.51	$\tau_0 = (1.2–5.0) \times 10^{-13}$ sec	NMR ($c = 1.3–1.9$ H/Ti)	25–500	Korn and Zamir (1970)
α-Ti–H	0.56		NMR ($c = 0.08$ H/Ti)	170–430	Korn and Zamir (1970)
γ-Ti–T	0.50	$\tau_0(T)/\tau_0(H) = 1.3^a$	NMR ($c = 1.5$ T/Ti)	190–470	Weaver (1971)
(α + γ)-Ti–H	0.49–0.47	$\tau_0 = (8–7) \times 10^{-13}$ sec	NMR ($c = 1.25–1.62$ H/Ti)	180–350	Khodosov and Shepilov (1971)
γ-Ti–H	0.47–0.49	$\tau_0 = (6–9) \times 10^{-13}$ sec	NMR ($c = 1.7–1.79$ H/Ti)	180–350	Khodosov and Shepilov (1971)
α-U–H	0.28	1.4×10^{-3}	Permeability, desorption	450–650	Davis (1956)
γ-U–H		$D(800°C) = 14.2 \times 10^{-5}$	Desorption	800	Oblinger and Dube (1961)
β-U–H	0.26	$\tau_0 = 7.7 \times 10^{-10}$ sec	NMR ($c = 3$ H/U)	50–300	Spalthoff (1961)
β-U–H	0.36	$\tau_0 = 2.2 \times 10^{-10}$ sec	NMR ($c = 3$ H/U)	80–240	Grunzweig-Genossar et al. (1970)

TABLE XIII—*Continued*

Material	U (eV)	D_0 (cm²/sec)	Temp. range (°C)	Method	Author(s)
β-U-D	0.39	$\tau_0 = 2.7 \times 10^{-10}$ sec	80–180	NMR ($c = 3$ D/U)	Grunzweig-Genossar et al. (1970)
W-H	0.86	8.1×10^{-2}	1055–1570	Desorption	Ryabchikov (1964a)
W-H	1.8	7.25×10^{-4}	1510–1900	Kinetic of thermal dissociation	Moore and Unterwald (1964)
W-H	0.39	4.1×10^{-3}	830–2130	Desorption	Frauenfelder (1969)
Zr-H	0.5	1.09×10^{-3}	60–250	Absorption ($c = 0$–1.7 H/Zr)	Gulbransen and Andrew (1954)
Zr-D	0.5	0.73×10^{-3}	60–250	Absorption	Gulbransen and Andrew (1954)
Zr-H	0.53	$\tau_0 = 2.6 \times 10^{-11}$ sec ($D_0 = 9 \times 10^{-6}$)	200–300	NMR ($c = 1.4$ H/Zr)	Spalthoff (1961)
α-Zr-H	0.31	7×10^{-4}	305–610	Concentration profile, determined by outgassing sections of the sample	Mallett and Albrecht (1957)
β-Zr-H	2.0	6.14×10^4	650–850	Permeation (using solubility data of Ells and McQuillan, 1956, and R. K. Edwards et al., 1955) and permeation, time-lag [$c = (9–33)^b$]	Albrecht and Goode (1959)
α-Zr-H	0.41	4.15×10^{-3}	450–700	Analysis of the diffusion gradient	Someno (1960)
β-Zr-H	0.37	7.35×10^{-3}	870–1100	Analysis of the diffusion gradient	Someno (1960)
Zr-T	0.31	1.53×10^{-3}	149–240	Concentration profile, determined by autoradiography	Cupp and Flubacher (1962)
Zr-H	0.54	0.6×10^{-4}	40–120	NMR ($c = 1.54$ H/Zr)	Hon (1962)
β-Zr-H	0.36	5.3×10^{-3}	760–1010	Absorption [$c = (0–41)^b$]	Gelezunas et al. (1963)
δ-Zr-H	0.77	0.25	650–800	Absorption [$c = (61–63)^b$]	Paetz and Lücke (1971)

^a τ_0(H) taken from Korn and Zamir (1970).   ^b At. % H.

5. HYDROGEN DIFFUSION IN METALS

References

Albrecht, W. M., and Goode, W. D., Jr. (1959). Battelle Mem. Inst. Rep. BMI-1373.
Albrecht, W. M., Goode, W. D., and Mallett, M. W. (1959). *J. Electrochem. Soc.* **106**, 981.
Alefeld, B. (1972). *Kerntechnik* **14**, 15.
Alefeld, B., Bohn, H. G., and Stump, N. (1972). *Jül-Ber. Jül-Conf.*, 6th **1**, 286.
Alefeld, G. (1969). *Phys. Status Solidi* **32**, 67.
Alefeld, G., Schaumann, G., Tretkowski, J., and Völkl, J. (1969). *Phys. Rev. Lett.* **22**, 697.
Alefeld, G., Völkl, J., and Schaumann, G. (1970). *Phys. Status Solidi* **37**, 337.
Alikin, V. P. (1960). *Isv. Estestvennonauchn. Inst. pri Permsk. Univ.* **14**, 19. (Abstract in: *Chem. Abstr.* **57** (1962), col. 4468c.)
Armbruster, M. H. (1943). *J. Amer. Soc.* **65**, 1043.
Arons, R. R. (1969). Thesis, Univ. Amsterdam, The Netherlands.
Arons, R. R., Tamminga, Y., and deVries, G. (1970). *Phys. Status Solidi* **40**, 107.
Au, J. J., and Birnbaum, H. K. (1973). *Scr. Met.* **7**, 595.
Baker, C., and Birnbaum, H. K. (1973). *Acta Met.* **21**, 865.
Baranowski, B. (1972). *Ber. Bunsenges. Phys. Chem.* **76**, 714.
Baranowski, B., Śmiałowski, M., and Szkłarska-Śmiałowska, Z. (1957). *Bull. Acad. Pol. Sci., Sér. Sci. Chim.* **5**, 191.
Barrer, R. M. (1939). *Trans. Faraday Soc.* **35**, 628.
Barrer, R. M. (1940). *Trans. Faraday Soc.* **36**, 1235.
Barrer, R. M. (1942). *Trans. Faraday Soc.* **38**, 78.
Bastien, P., and Amiot, P. (1958). *Rev. Met.* **55**, 24.
Baukloh, W., and Gehlen, K. (1937). *Arch. Eisenhüttenw.* **11**, 253.
Baukloh, W., and Müller, R. (1938). *Arch. Eisenhüttenw.* **11**, 509.
Baukloh, W., and Retzlaff, W. (1937). *Arch. Eisenhüttenw.* **11**, 97.
Baukloh, W., and Wenzel, W. (1937). *Arch. Eisenhüttenw.* **11**, 273.
Baukloh, W., and Zimmermann, G. (1936). *Arch. Eisenhüttenw.* **9**, 459.
Beck, W., Bockris, J. O'M., McBreen, J., and Nanis, L. (1966). *Proc. Roy. Soc. Ser. A* **290**, 220.
Beck, W., Bockris, J. O'M., Genshaw, M. A., and Subramanyan, P. K. (1971). *Met. Trans.* **2**, 883.
Beg, M. M., and Ross, D. K. (1970). *J. Phys. C* **3**, 2487.
Belyakov, Yu. I., and Ionov, N. I. (1961). *Sov. Phys.-Tech. Phys.* **6**, 146.
Bennek, H., and Klotzbach, G. (1941). *Stahl Eisen* **61**, 597, 624.
Bergsma, J., and Goedkoop, J. A. (1960). *Physica* **26**, 744.
Bester, H., and Lange, K. W. (1972). *Arch. Eisenhüttenw.* **43**, 207.
Birchall, J. H. L., and Ross, D. K. (1972). *Jül-Ber. Jül-Conf.*, 6th **1**, 313.
Birnbaum, H. K., and Wert, C. A. (1972). *Ber. Bunsenges. Phys. Chem.* **76**, 806.
Bockris, J. O'M., Genshaw, M. A., and Fullenwider, M. (1970). *Electrochim. Acta* **15**, 47.
Boes, N. (1971). Master's thesis, Univ. Münster, Germany. [See also Züchner, H., and Boes, N. (1972). *Ber. Bunsenges. Phys. Chem.* **76**, 783.]
Bohmholdt, G., and Wicke, E. (1967). *Z. Phys. Chem. N.F.* **56**, 133.
Bohn, H. G. (1972). Jül-Ber. Jül-853-FF.
Boniszewski, T., and Smith, G. C. (1963). *Acta Met.* **11**, 165.
Boniszewski, T., and Moreton, J. (1967). *Brit. Weld. J.* **14**, 321.
Borelius, G., and Lindblom, S. (1927). *Ann. Phys.* **82**, 201.
Borgucci, M. V., and Verdini, L. (1965). *Phys. Status Solidi* **9**, 243.
Brand, K. (1971). *Atomkernenergie* **17**, 113.
Breger, V., and Gileadi, E. (1971). *Electrochim. Acta* **16**, 177.

BRODOWSKY, H. (1965). Z. Phys. Chem. N.F. **44**, 129.
BRODOWSKY, H., AND POESCHEL, E. (1965). Z. Phys. Chem. N.F. **44**, 143.
BRÜNING, H., AND SIEVERTS, A. (1933). Z. Phys. Chem. A **163**, 409.
BRYAN, W. L., AND DODGE, B. F. (1963). AIChE J. **9**, 223.
BUCHHOLZ, J., VÖLKL, J., AND ALEFELD, G. (1973). Phys. Rev. Lett. **30**, 318.
BUCHOLD, H., AND SICKING, G. (1972). Jül-Ber. Jül-Conf., 6th **2**, 391.
BUCK, H., AND ALEFELD, G. (1972). Phys. Status Solidi **49**, 317.
BUCKEL, W., AND STRITZKER, B. (1973). Phys. Lett. A **43**, 403.
BURGER, J. P., POULIS, N. J., AND HASS, W. P. A. (1961). Physica **27**, 514.
CANNELLI, G., AND MAZZOLAI, F. M. (1969). Nuovo Cimento B **64**, 171.
CANNELLI, G., AND MAZZOLAI, F. M. (1973). Appl. Phys. **1**, 111.
CANNELLI, G., AND VERDINI, L. (1966a). Ric. Sci. **36**, 98.
CANNELLI, G., AND VERDINI, L. (1966b). Ric. Sci. **36**, 246.
CANTELLI, R., MAZZOLAI, F. M., AND NUOVO, M. (1969). Phys. Status Solidi **34**, 597.
CANTELLI, R., MAZZOLAI, F. M., AND NUOVO, M. (1970). J. Phys. Chem. Solids **31**, 1811.
CANTELLI, R., MAZZOLAI, F. M., AND NUOVO, M. (1971). J. Phys. (Paris) **32**, C2–59.
CANTELLI, R., MAZZOLAI, F. M., AND NUOVO, M. (1973). Appl. Phys. **1**, 27. [See also CANTELLI, R., MAZZOLAI, F. M., AND NUOVO, M. (1972). Jül-Ber. Jül-Conf., 6th **2**, 770.]
CARNUTH, W. (1963). Z. Angew. Phys. **15**, 291.
CARMICHAEL, D. C., HORNADAY, J. R., MORRIS, A. E., AND PARLEE, N. A. (1960). Trans. Met. Soc. AIME **218**, 826.
CARSTANJEN, H. D., AND SIZMANN, R. (1972). Ber. Bunsenges. Phys. Chem. **76**, 1223.
CERMAK, J., AND KUFUDAKIS, A. (1966). Mém. Sci. Rev. Met. **63**, 767.
CERMAK, J., AND KUFUDAKIS, A. (1968). Mém. Sci. Rev. Met. **65**, 375.
CHANG, P. L., AND BENNET, W. D. G. (1952). J. Iron Steel Inst. **170**, 205.
CHERVYAKOV, A. YU., ÉUTIN, I. R., SOMENKOV, V. A., SHIL'SHTEIN, S. SH., AND CHERTKOV, A. A. (1972). Sov. Phys.-Solid State **13**, 2172.
CHOI, J. Y. (1970). Met. Trans. **1**, 911.
CLEARY, H. J., AND GREENE, N. D. (1965). Electrochim. Acta **10**, 1107.
COE, F. R., AND MORETON, J. (1967). Brit. Weld. J. **14**, 313.
COE, F. R., AND MORETON, J. (1969). Met. Sci. J. **3**, 209.
COEHN, A., AND JÜRGENS, H. (1931). Z. Phys. **71**, 179.
COEHN, A., AND SPECHT, W. (1930). Z. Phys. **62**, 1.
COMBETTE, P., AND AZOU, P. (1969). C. R. Acad. Sci. Sér. C **267**, 677.
COMBETTE, P., AND AZOU, P. (1970). Mém. Sci. Rev. Met. **67**, 17.
COMBETTE, P., AND GRILHÉ, J. (1970). Mém. Sci. Rev. Met. **67**, 491.
COMBETTE, P., RENARD, M., AND GRILHÉ, J. (1972a). Phys. Status Solidi **11**, 677.
COMBETTE, P., RENARD, M., AND GRILHÉ, J. (1972b). Jül-Ber. Jül-Conf., 6th **2**, 821.
CONRAD, H., BAUER, G., ALEFELD, G., SPRINGER, T., AND SCHMATZ, W. (1974). Z. Physik **266**, 239.
COTTS, R. M. (1972). Ber. Bunsenges. Phys. Chem. **76**, 760.
CUPP, C. R., AND FLUBACHER, P. (1962). J. Nucl. Mater. **6**, 213.
DAMOUR, P. L., AND CASTELLAN, G. W. (1964). J. Electrochem. Soc. **111**, 1280.
DARKEN, L. S., AND SMITH, P. P. (1949). Corrosion **5**, 1.
DAVIS, W. D. (1954). USAEC Rep. KAPL-1227.
DAVIS, W. D. (1956). USAEC Rep. KAPL-1548.
DAYNES, H. A. (1920). Proc. Roy. Soc. Ser. A **97**, 286.
DE GRAAF, L. A., RUSH, J. J., LIVINGSTON, R. C., FLOTOW, H. E., AND ROWE, J. M. (1972). Jül-Ber. Jül-Conf., 6th **1**, 301.
DEVANATHAN, M. A. V., AND STACHURSKI, Z. (1962). Proc. Roy. Soc. Ser. A **270**, 90.

DEVANATHAN, M. A. V., AND STACHURSKI, Z. (1964). *J. Electrochem. Soc.* **111**, 619.
DEVANATHAN, M. A. V., STACHURSKI, Z., AND BECK, W. (1960). *J. Electrochem. Soc.* **110**, 886.
DIETZE, H. D. (1959). *Tech. Mitt. Krupp* **17**, 67.
DILLARD, J.-L., AND TALBOT-BESNARD, S. (1969). *C. R. Acad. Sci. Sér. C* **269**, 1173.
DILLARD, J.-L. (1970). *C. R. Acad. Sci. Sér. C* **270**, 669.
DOMKE, E. (1971). Thesis, Univ. Münster, Germany.
DONOVAN, J. A., DERRICK, R. G., DEXTER, A. H., AND LOUTHAN, M. R., JR. (1971). Rep. DPST (NASA)-71-2.
DOOLITTLE, H. D., SINGER, B., AND VÁRADI, P. F. (1963). *Nuovo Cimento Suppl.* **I**, 593.
DRESLER, W. (1971). Thesis, Tech. Univ. Berlin (D83), Germany.
DRESLER, W., AND FROHBERG, M. G. (1972). *Jül-Ber. Jül-Conf.*, 6th **2**, 516.
DUCASTELLE, F., CAUDRON, R., AND COSTA, P. (1970). *J. Phys. Chem. Solids* **31**, 1247.
DÜNWALD, H., AND WAGNER, C. (1934). *Z. Phys. Chem. B* **24**, 53.
DUHM, B. (1935). *Z. Phys.* **94**, 434; **95**, 801.
DUS, R., AND SMIALOWSKI, M. (1967). *Acta Met.* **15**, 1611.
EBISUZAKI, Y., AND O'KEEFFE, M. (1967). *Prog. Solid State Chem.* **4**, 187-211.
EBISUZAKI, Y., AND O'KEEFFE, M. (1968). *J. Chem. Phys.* **48**, 1867.
EBISUZAKI, Y., KASS, W. J., AND O'KEEFFE, M. (1967a). *J. Chem. Phys.* **46**, 1373.
EBISUZAKI, Y., KASS, W. J., AND O'KEEFFE, M. (1967b). *J. Chem. Phys.* **46**, 1378.
EBISUZAKI, Y., KASS, W. J., AND O'KEEFFE, M. (1968). *J. Chem. Phys.* **49**, 3329.
EDWARDS, A. G. (1957). *Brit. J. Appl. Phys.* **8**, 406.
EDWARDS, R. K., LEVESQUE, P., AND CUBICCIOTTI, D. (1955). *J. Amer. Chem. Soc.* **77**, 1307.
EICHENAUER, W. (1960). *Mém. Sci. Rev. Met.* **57**, 943.
EICHENAUER, W. (1967). *Z. Naturforsch.* **22a**, 2115.
EICHENAUER, W., AND LIEBSCHER, D. (1962). *Z. Naturforsch.* **17a**, 355.
EICHENAUER, W., AND PEBLER, A. (1957). *Z. Metallk.* **48**, 373.
EICHENAUER, W., KÜNZIG, H., AND PEBLER, A. (1958). *Z. Metallk.* **49**, 220.
EICHENAUER, W., HATTENBACH, K., AND PEBLER, A. (1961). *Z. Metallk.* **52**, 682.
EICHENAUER, W., LÖSER, W., AND WITTE, H. (1965). *Z. Metallk.* **56**, 287.
ELLERBROCK, H.-G., VIBRANS, G., AND STÜWE, H.-P. (1972). *Acta Met.* **20**, 53.
ELLIOT, R. P. (1965). "Constitution of Binary Alloys," First Suppl., p. 16. McGraw-Hill, New York.
ELLS, C. E., AND MCQUILLAN, A. D. (1956). *J. Inst. Metals* **85**, 89.
ERDMANN-JESNITZER, F., AND HIEBER, H. (1969). *Arch. Eisenhüttenw.* **40**, 73.
EURINGER, G. (1935). *Z. Phys.* **96**, 37.
EVANS, G. M. (1966). Thesis, Univ. Birmingham, England.
EVANS, G. M., AND ROLLASON, E. C. (1969a). *J. Iron Steel Inst.* **207**, 1484.
EVANS, G. M., AND ROLLASON, E. C. (1969b). *J. Iron Steel Inst.* **207**, 1591.
EWING, V. C., AND UBBELOHDE, A. R. (1955). *Proc. Roy. Soc. Ser. A* **230**, 301.
FABER, K., AND SCHULTZ, H. (1972). *Scr. Met.* **6**, 1065.
FAST, J. D. (1942). *Philips Tech. Rev.* **7**, 74.
FAVREAU, R. L., PATTERSON, R. E., RANDALL, D., AND SALMON, O. N. (1954). USAEC Rep. KAPL-1036.
FERGUSON, G. A., JR., SCHINDLER, A. I., TANAKA, T., AND MORITA, T. (1965). *Phys. Rev. A* **137**, 483.
FISCHER, W. (1967). *Z. Naturforsch.* **22a**, 1581.
FOSTER, P. K., MCNABB, A., AND PAYNE, C. M. (1965). *Trans. Met. Soc. AIME* **233**, 1022.
FRANK, R. C., SWETS, D. E., AND FRY, D. L. (1958). *J. Appl. Phys.* **29**, 892.
FRAUENFELDER, R. (1969). *J. Vac. Sci. Technol.* **6**, 388.
FRIESKE, H., AND WICKE, E. (1973). *Ber. Bunsenges. Phys. Chem.* **77**, 48.

FRITZ, J. J., MARIA, H. J., AND ASTON, J. G. (1961). *J. Chem. Phys.* **34**, 2185.
GAUS, H. (1965). *Z. Naturforsch.* **20a**, 1298.
GELEZUNAS, V. L., CONN, P. K., AND PRICE, R. H. (1963). *J. Electrochem. Soc.* **110**, 799.
GELLER, W., AND SUN, T.-H. (1950). *Arch. Eisenhüttenw.* **21**, 423.
GIBALA, R. (1967). *Trans. Met. Soc. AIME* **239**, 1574.
GIBALA, R. (1970). *Scr. Met.* **4**, 77.
GILEADI, E., FULLENWIDER, M. A., AND BOCKRIS, J. O'M. (1966). *J. Electrochem. Soc.* **113**, 926.
GILLESPIE, L. J., AND HALL, F. P. (1926). *J. Amer. Chem. Soc.* **48**, 1207.
GILLESPIE, L. J., AND GALSTAUN, L. S. (1936). *J. Amer. Chem. Soc.* **58**, 2565.
GISSLER, W. (1972). *Ber. Bunsenges. Phys. Chem.* **76**, 770.
GISSLER, W., ALEFELD, G., AND SPRINGER, T. (1970). *J. Phys. Chem. Solids* **31**, 2361.
GOL'TSOV, V. A., DEMIN, V. B., VYKHODETS, V. B., KAGAN, G. YE., AND GEL'D, P. V. (1970). *Phys. Metals Metallogr.* **29**, 195.
GONZALEZ, O. D. (1969). *Trans. Met. Soc. AIME.* **245**, 607.
GORMAN, W. R., AND NARDELLA, W. R. (1962). *Vacuum* **12**, 19.
GOVINDAN NAMBOODHIRI, T. K., AND NANIS, L. (1973). *Acta Met.* **21**, 663.
GRIMES, H. H. (1959). *Acta Met.* **7**, 783.
GRUNZWEIG-GENOSSAR, J., KUZNIETZ, M., AND MEEROVICI, B. (1970). *Phys. Rev.* **B1**, 1958.
GÜNTHERSCHULZE, A., BETZ, H., AND KLEINWÄCHTER, H. (1939). *Z. Phys.* **111**, 657.
GULBRANSEN, E. A., AND ANDREW, K. F. (1954). *J. Electrochem. Soc.* **101**, 560.
HAAS, R., KLEY, W., KREBS, K. H., AND RUBIN, R. (1963). In "Inelastic Scattering of Neutrons in Solids and Liquids," Vol. II, p. 145. Int. At. Energy Ag., Vienna.
HAM, W. R. (1932). *J. Chem. Phys.* **1**, 476.
HANADA, R. (1973). *Scr. Met.* **7**, 681.
HANADA, R., SUGANUMA, T., AND KIMURA, H. (1972). *Scr. Met.* **6**, 483.
HARHAI, J. G., VISWANATHAN, T. S., AND DAVIS, H. M. (1965). *Trans. ASM* **58**, 210.
HEATH, H. R. (1952). *Brit. J. Appl. Phys.* **3**, 13.
HELLER, W. R. (1961). *Acta Met.* **9**, 600.
HELLER, R. (1973). Master's thesis, Techn. Univ. München, Germany.
HENDRICKS, B. C., AND RALSTON, R. R. (1929). *J. Amer. Chem. Soc.* **51**, 3278.
HEUMANN, TH., AND DOMKE, E. (1972). *Jül-Ber, Jül-Conf.*, 6th **2**, 492.
HEUMANN, TH., AND PRIMAS, D. (1966). *Z. Naturforsch.* **21a**, 260.
HEWITT, J. (1962). Hydrogen in Steel, BISI Rep. 73, 83.
HILL, M. L. (1960). *J. Metals* **12**, 725.
HILL, M. L., AND JOHNSON, E. W. (1955). *Acta Met.* **3**, 566.
HILL, M. L., AND JOHNSON, E. W. (1959). *Trans. Met. Soc. AIME* **215**, 717.
HOLLECK, G., AND WICKE, E. (1967). *Z. Phys. Chem. N. F.* **56**, 155.
HOLLECK, G. L. (1970). *J. Phys. Chem.* **74**, 503.
HON, J. F. (1962). *J. Chem. Phys.* **36**, 759.
JEWETT, D. N., AND MAKRIDES, A. C. (1965). *Trans. Faraday Soc.* **61**, 932.
JOHNSON, E. W., AND HILL, M. L. (1960). *Trans. Met. Soc. AIME* **218**, 1104.
JONES, W. M. (1948). *J. Chem. Phys.* **16**, 1077.
JONES, P. M. S., GIBSON, R., AND EVANS, J. A. (1966). AWRE Rep. 0-16/66 Aldermaston, England.
JOPPIEN, K.-D. (1968). Thesis, Tech. Univ. Clausthal, Germany.
JOST, W., AND WIDMANN, A. (1935). *Z. Phys. Chem. B* **29**, 247.
JOST, W., AND WIDMANN, A. (1940). *Z. Phys. Chem. B* **45**, 285.
KAHRIG, E., KIRSTEIN, D., AND LANGE, FR. (1966). *Ber. Bunsenges. Phys. Chem.* **70**, 592.
KATZ, L., GUINAN, M., AND BORG, R. L. (1971). *Phys. Rev.* **B4**, 330.

KATZ, O. M., AND GULBRANSEN, E. A. (1960). *Rev. Sci. Instrum.* **31**, 615.
KHODOSOV, E. F., AND SHEPILOV, N. A. (1971). *Phys. Status Solidi B* **47**, 693.
KNAAK, J., AND EICHENAUER, W. (1968). *Z. Naturforsch.* **23a**, 1783.
KOFFLER, S. A., HUDSON, J. B., AND ANSELL, G. S. (1969). *Trans. Met. Soc. AIME* **245**, 1735.
KOFSTAD, P., AND BUTERA, R. A. (1963). *J. Appl. Phys.* **34**, 1517.
KORN, CH., AND ZAMIR, D. (1970). *J. Phys. Chem. Solids* **31**, 489.
KRONMÜLLER, H. (1970). In "Vacancies and Interstitials in Metals" (A. Seeger, D. Schumacher, W. Schilling, and J. Diehl, eds.), pp. 667. North-Holland Publ., Amsterdam.
KÜSSNER, A. (1963). *Z. Phys. Chem. N.F.* **36**, 383.
KÜSSNER, A. (1966). *Z. Naturforsch.* **21a**, 515.
KUZNIETSOV, V. V., AND SUBBOTINA, N. I. (1965). *Elektrokhimija* **1**, 1096.
LACHER, J. R. (1937). *Proc. Roy. Soc. Ser. A* **161**, 525.
LANGE, G. (1969). *Arch. Eisenhüttenw.* **40**, 635.
LA PIETRA, R. A., AND CASTELLAN, G. W. (1964). *J. Electrochem. Soc.* **111**, 1276.
LAWLEY, A., LIEBMANN, W., AND MADDIN, R. (1961). *Acta Met.* **9**, 841.
LEE, R. W., SWETS, D. E., AND FRANK, R. C. (1961). *Mém. Sci. Rev. Met.* **58**, 36.
LIESER, K. H., AND WITTE, H. (1954). *Z. Phys. Chem.* **202**, 321.
LOMBARD, M. V. (1923). *C. R. Acad. Sci.* **177**, 116.
LORD, A. E., JR., (1967). *Acta Met.* **15**, 1241.
LORD, A. E., JR., AND BESHERS, D. N. (1966). *Acta Met.* **14**, 1659.
LOUTHAN, M. R., JR., DERRICK, R. G., AND DEXTER, A. H. (1971). Rep. DPST (NASA)-71-4.
LUCKEMEYER-HASSE, L., AND SCHENCK, H. (1932/33). *Arch. Eisenhüttenw,* **6**, 209.
LÜTGEMEIER, H., ARONS, R. R., AND BOHN, H. G. (1972a). *J. Magn. Resonance* **8**, 74.
LÜTGEMEIER, H., BOHN, H. G., AND ARONS, R. R. (1972b). *J. Magn. Resonance* **8**, 80.
MAAS, H. (1958). Thesis, Univ. Münster, Germany.
MAELAND, A. J. (1964). *J. Phys. Chem.* **68**, 2197.
MAKRIDES, A. C., AND JEWETT, D. N. (1966). *Engelhard Ind. Tech. Bull.* **7**, 51.
MAKRIDES, A. C., WRIGHT, M., AND MCNEILL, R. (1965). Final Rep. Contract DA-49-189-AMC-136(d), Tyco Lab., Waltham, Massachussetts.
MALLETT, M. W., AND ALBRECHT, W. M. (1957). *J. Electrochem. Soc.* **104**, 142.
MALLETT, M. W., AND KOEHL, B. G. (1962). *J. Electrochem. Soc.* **109**, 968.
MARCHAND, A. (1962). *C. R. Acad. Sci.* **254**, 4284.
MARTIN, E. (1929/30). *Arch. Eisenhüttenw.* **3**, 407.
MATSUO, S., AND HIRATA, T. (1967). *Nippon Kinzoku Gakkaishi* **31**, 590.
MCLELLAN, R. B., AND OATES, W. A. (1973). *Acta Met.* **21**, 181.
MCNABB, A., AND FOSTER, P. K. (1963). *Trans. Met. Soc. AIME* **227**, 618.
MERISOV, B. A., KHOTKEVICH, V. I., AND KARNUS, A. I. (1966). *Phys. Metals Metallogr.* **22**, 163.
MERISOV, B. A., SERDYUK, A. D., FAL'KO, I. I., KHADZHAY, G. YA., AND KHOTKEVICH, V. I. (1971). *Phys. Metals Metallogr.* **32**, 154.
MIODOWNIK, A. P., AND STURGES, C. (1970). *Scr. Met.* **4**, 143.
MITACEK, P., JR., AND ASTON, J. G. (1963). *J. Amer. Chem. Soc.* **85**, 137.
MOON, K. A. (1956). *J. Phys. Chem.* **60**, 502.
MOORE, G. E., AND UNTERWALD, F. C. (1964). *J. Chem. Phys.* **40**, 2639.
MÜNZING, W. (1972). Master's thesis, Tech. Univ. München, Germany.
NAESER, G., AND DAUTZENBERG, N. (1965). *Arch. Eisenhüttenw.* **36**, 175.
NAKAGAWA, Y., AND WOODS, A. D. B. (1965). *J. Phys. Chem. Solids Suppl.* **1**, 39.
NELIN, G. (1971). *Phys. Status Solidi B* **45**, 527.
NERNST, G. H. (1961). Master's thesis, Univ. Hamburg, Germany.

NERNST, G. H. (1963). Thesis, Univ. Münster, Germany. (See also WICKE, E., AND NERNST, G. H. (1964). *Ber. Bunsenges. Phys. Chem.* **68**, 224.)
NEWMAN, J. F., AND SHREIR, L. L. (1969). *J. Iron Steel Inst.* **207**, 1369.
NORBERG, R. E. (1952). *Phys. Rev.* **86**, 745.
NOWICK, A. S., AND BERRY, B. S. (1972). "Anelastic Relaxation in Crystalline Solids," pp. 226. Academic Press, New York.
OBLINGER, C. J., AND DUBE, H. A. (1961). *Nucl. Sci. Eng.* **11**, 263.
OĞURTANI, T. Ö. (1971). *Met. Trans.* **2**, 3035.
OLSEN, K. M., AND LARKIN, C. F. (1963). *J. Electrochem. Soc.* **110**, 86.
ONO, K., AND ROSALES, L. A. (1968). *Trans. Met. Soc. AIME* **242**, 244.
ORIANI, R. A. (1970). *Acta Met.* **18**, 147.
PAETZ, P., AND LÜCKE, K. (1971). *Z. Metallk.* **62**, 657.
PALCZEWSKA, W., AND RATAJCZYK, I. (1961). *Bull. Acad. Pol. Sci., Sér. Sci. Chim.* **9**, 267.
PAPAZOGLOU, T. P., AND HEPWORTH, M. T. (1968). *Trans. Met. Soc. AIME* **242**, 682.
PEDERSEN, B. (1965). *Tidsskr. Kjemi, Bergv. Met.* **25**, 63.
PEDERSEN, B., AND SLOTFELDT-ELLINGSEN, D. (1971). *J. Less-Common Metals* **23**, 223.
PEDERSEN, B., KROGDAHL, T., AND STOKKELAND, O. E. (1965). *J. Chem. Phys.* **42**, 72.
PETERSON, D. T., AND HAMMERBERG, C. C. (1968). *J. Less-Common Metals* **16**, 457.
PETERSON, D. T., AND WESTLAKE, D. G. (1960). *J. Phys. Chem.* **64**, 649.
PETRUNIN, V. F., SOMENKOV, V. A., SHIL'SHTEIN, S. SH., AND CHERTKOV, A. A. (1970). *Sov. Phys.-Crystallogr.* **15**, 137.
PIPER, J. (1966). *J. Appl. Phys.* **37**, 715.
PLUSQUELLEC, J., VEYSSEYRE, H., AZOU, P., AND BASTIEN, P. (1962). *C. R. Acad. Sci.* **255**, 518.
POWERS, R. W., AND DOYLE, M. V. (1959). *J. Appl. Phys.* **30**, 514.
PRIMAS, D., AND HEUMANN, TH. (1972). Private communication.
PRYDE, J. A., AND TITCOMB, C. G. (1969). *Trans. Faraday Soc.* **65**, 2758.
PRYDE, J. A., AND TSONG, I. S. T. (1971a). *Trans. Faraday Soc.* **67**, 297.
PRYDE, J. A., AND TSONG, I. S. T. (1971b). *Acta Met.* **19**, 1333.
RACZYŃSKI, W. (1958). *Arch. Hutn.* **3**, 59.
RACZYŃSKI, W., AND STELMACH, S. (1961). *Bull. Acad. Pol. Sci., Sér. Sci. Chim.* **9**, 633.
RACZYŃSKI, W., AND TALBOT-BESNARD, S. (1969). *C. R. Acad. Sci. Sér. C* **269**, 1253.
RADHAKRISHNAN, T. P., AND SHREIR, L. L. (1967). *Electrochim. Acta* **12**, 889.
RANSLEY, C. E., AND TALBOT, D. E. J. (1955). *Z. Metallk.* **46**, 328.
RAYLEIGH, LORD (1887). *Phil. Mag.* **23**, 225.
REIERMANN, B. K. (1970). Thesis, Tech. Univ. Berlin (D83), Germany.
ROBERTS, E. M., AND MERIDETH, CH. W. (1969). *Phys. Rev.* **179**, 381.
ROBERTSON, W. M. (1972). *Jül-Ber. Jül-Conf.*, 6th **2**, 449; *Z. Metallk.* **64**, 436 (1973).
ROBINSON, M. T., SOUTHERN, A. L., AND WILLIS, W. R. (1960). *J. Appl. Phys.* **31**, 1474.
ROWE, J. M. (1972). *Solid State Commun.* **11**, 1299.
ROWE, J. M., SKÖLD, K., FLOTOW, H. E., AND RUSH, J. J. (1971). *J. Phys. Chem. Solids* **32**, 41.
ROWE, J. M., RUSH, J. J., DE GRAAF, L. A., AND FERGUSON, G. A. (1972). *Phys. Rev. Lett.* **29**, 1250.
RUBIN, R., AND CLAESSEN, Y. (1970). *Solid State Commun.* **8**, 1321.
RUSH, J. J., AND FLOTOW, H. E. (1968). *J. Chem. Phys.* **48**, 3795.
RYABCHIKOV, L. N. (1964a). *Ukr. Fiz. Zh.* **9**, 293.
RYABCHIKOV, L. N. (1964b). *Ukr. Fiz. Zh.* **9**, 303.
SAKAMOTO, M. (1964). *J. Phys. Soc. Japan* **19**, 1862.
SALMON, O. N., AND RANDALL, D. (1954). USAEC Rep. KAPL-984.
SAMSONOV, G. V., AND ANTONOVA, M. M. (1961). *Zh. Fiz. Khim.* **35**, 900.
SASAKI, Y., AND AMANO, M. (1969). *Trans. Japan Inst. Met.* **10**, 29.

5. HYDROGEN DIFFUSION IN METALS
301

Schaumann, G. (1969). Thesis, Tech. Hochschule Aachen, Germany; Jül-Ber. Jül-606-FN.
Schaumann, G., Völkl, J., and Alefeld, G. (1968). *Phys. Rev. Lett.* **21**, 891.
Schaumann, G., Völkl, J., and Alefeld, G. (1970). *Phys. Status Solidi* **42**, 401. [See also Völkl, J., Schaumann, G., and Alefeld, G. (1970). *J. Phys. Chem. Solids* **31**, 1805.]
Schenck, H., and Lange, K. W. (1966). *Arch. Eisenhüttenw.* **37**, 809.
Schenck, H., and Taxhet, H. (1959). *Arch. Eisenhüttenw.* **30**, 661.
Scherrer, S., Lozes, G., and Deviot, B. (1967). *C. R. Acad. Sci. Sér. B* **264**, 1499.
Schiller, P. (1968). Private communication.
Schiller, P., and Schneiders, A. (1968). *Jül-Ber. Jül-Conf.*, 2nd **2**, 871.
Schnabel, D. (1972). Thesis, Tech. Hochschule Aachen, Germany; Jül-Ber. Jül-878-FF.
Schreiber, D. S., and Cotts, R. M. (1963). *Phys. Rev.* **131**, 1118.
Schuetz, A. E., and Robertson, W. D. (1957). *Corrosion* **13**, 437t.
Schwarz, W., and Zitter, H. (1965). *Arch. Eisenhüttenw.* **36**, 343.
Shunk, F. A. (1969). "Constitution of Binary Alloys," Second Suppl., p. 7. McGraw-Hill, New York.
Sicking, G., and Buchold, H. (1971). *Z. Naturforsch.* **26a**, 1973.
Sieverts, A. (1911). *Z. Phys. Chem.* **77**, 591.
Sieverts, A. (1929). *Z. Metallk.* **21**, 44.
Sieverts, A., and Danz, W. (1938). *Z. Phys. Chem. B* **38**, 46.
Sieverts, A., and Hagenacker, J. (1909). *Ber. Deut. Chem. Ges.* **42**, 338.
Sieverts, A., Zapf, G., and Moritz, H. (1939). *Z. Phys. Chem.* **183**, 19.
Simons, J. W., and Flanagan, T. B. (1965a). *J. Phys. Chem.* **69**, 3581.
Simons, J. W., and Flanagan, T. B. (1965b). *J. Phys. Chem.* **69**, 3773.
Sköld, K., and Nelin, G. (1967). *J. Phys. Chem. Solids* **28**, 2369.
Skośkiewicz, T. (1972). *Phys. Status Solidi A* **11**, K123.
Smialowski, M. (1957). *Neue Hütte* **2**, 611.
Smithells, C. J., and Ransley, C. E. (1935). *Proc. Roy. Soc. Ser. A* **150**, 172.
Smithells, C. J., and Ransley, C. E. (1936). *Proc. Roy. Soc. Ser. A* **157**, 292.
Snoek, J. L. (1938). *Physica* **5**, 663.
Snoek, J. L. (1939a). *Physica* **6**, 161.
Snoek, J. L. (1939b). *Physica* **6**, 591.
Snoek, J. L. (1939c). *Physica* **6**, 797.
Snoek, J. L. (1941). *Physica* **8**, 711.
Somenkov, V. A., Gurskaya, A. V., Zemlyanov, M. G., Kost, M. E., Chernoplekov, N. A., and Chertkov, A. A. (1968). *Sov. Phys.-Solid State* **10**, 1076.
Somenkov, V. A., Gurskaya, A. V., Zemlyanov, M. G., Kost, M. E., Chernoplekov, N. A., and Chertkov, A. A. (1969). *Sov. Phys.-Solid State* **10**, 2123.
Somenkov, V. A., Eutin, I. R., Chervyakov, A. Yu., Shil'shtein, S. Sh., and Chertkov, A. A. (1972). *Sov. Phys.-Solid State* **13**, 2178.
Someno, M. (1960). *Nippon Kinzoku Gakkaishi* **24**, 249.
Someno, M., Nagasaki, K., and Kagaku, S. (1960). *Vak. Chem.* **8**, 145.
Spalthoff, W. (1961). *Z. Phys. Chem. N. F.* **29**, 258.
Springer, T. (1972). "Springer Tracts in Modern Physics," Vol. 64. Springer-Verlag, Berlin and New York.
Staliński, B., Coogan, C. K., and Gutowsky, H. S. (1961). *J. Chem. Phys.* **34**, 1191.
Staliński, B., and Zogał, O. J. (1965). *Bull. Acad. Pol. Sci., Sér. Sci. Chim.* **13**, 397.
Steeb, H., and Kronmüller, H. (1973). *Phys. Status Solidi A* **16**, K 175.
Stoneham, A. M. (1972). *Ber. Bunsenges. Phys. Chem.* **76**, 816.
Stritzker, B., and Buckel, W. (1972). *Z. Phys.* **257**, 1.
Stross, T. M., and Tomkins, F. C. (1956). *J. Chem. Soc.* **159**, 230.

STURGES, C. M., AND MIODOWNIK, A. P. (1969). *Acta Met.* **17**, 1197.
SYKES, C., BURTON, H. H., AND GEGG, C. G. (1947). *J. Iron Steel Inst.* **156**, 155.
TODA, G. (1958). *Hokkaido Univ. Res. Inst. Catalysis J.* **6**, 13.
TORREY, H. C. (1958). *Nuovo Cimento Suppl.* **9**, 95.
TRETKOWSKI, J. (1969). Jül-Ber. Jül-626-FN.
VAN SWAAY, M., AND BIRCHENALL, C. E. (1960). *Trans. Met. Soc. AIME* **218**, 285.
VELECKIS, E. (1960). Thesis, Illinois Inst. Technol. Chicago. [See also VELECKIS, E., and EDWARDS, R. K. (1969). *J. Phys. Chem.* **73**, 683.]
VELECKIS, E., AND EDWARDS, R. K. (1969). *J. Phys. Chem.* **73**, 683.
VERDAN, G., RUBIN, R., AND KLEY, W. (1968). *Proc. IAEA Symp. Inelastic Scattering of Neutrons, Copenhagen* **1**, 223.
VIBRANS, G. (1961). *Arch. Eisenhüttenw.* **32**, 667.
VINEYARD, C. H. (1957). *J. Phys. Chem. Solids* **3**, 121.
VÖLKL, J. (1972). *Ber. Bunsenges. Phys. Chem.* **76**, 797.
VÖLKL, J., SCHAUMANN, G., AND ALEFELD, G. (1970). *J. Phys. Chem. Solids*, **31**, 1805.
VÖLKL, J., WOLLENWEBER, G., KLATT, K.-H., AND ALEFELD, G. (1971). *Z. Naturforsch.* **26a**, 922.
VON AUFSCHNAITER, ST., AND BAUER, H. J. (1964). *Z. Angew. Phys.* **17**, 209.
VON MEERWALL, E., AND SCHREIBER, D. S. (1968). *Phys. Lett. A* **27**, 574.
VON STACKELBERG, M., AND LUDWIG, P. (1964). *Z. Naturforsch.* **19a**, 93.
VON STURM, F., AND KOHLMÜLLER, H. (1965). *Naturwissenschaften* **52**, 31.
WACH, S., AND MIODOWNIK, A. P. (1968). *Corros. Sci.* **8**, 271.
WAGNER, R., AND SIZMANN, R. (1964). *Z. Angew. Phys.* **18**, 193.
WALTER, R. J., AND CHANDLER, W. T. (1965). *Trans. Met. Soc. AIME* **233**, 762.
WASILEWSKI, R. J., AND KEHL, G. L. (1954). *Metallurgica* **50**, 225.
WEAVER, H. T. (1971). *Phys. Lett. A* **35**, 417.
WEAVER, H. T. (1972). *Phys. Rev.* **B 5**, 1663.
WEINER, L. C., AND GENSAMER, M. (1957). *Acta Met.* **5**, 692.
WERT, C., AND ZENER, C. (1949). *Phys. Rev.* **76**, 1169.
WESTLAKE, D. G. (1969). *Trans. Met. Soc. AIME* **245**, 287.
WICKE, E., AND BOHMHOLDT, G. (1964). *Z. Phys. Chem. N. F.* **42**, 119.
WICKE, E., AND MEYER, K. (1969). *Z. Phys. Chem. N. F.* **64**, 225.
WICKE, E., AND NERNST, G. H. (1964). *Ber. Bunsenges. Phys. Chem.* **68**, 224.
WICKE, E., AND OBERMANN, A. (1972). *Z. Phys. Chem. N. F.* **77**, 163.
WILCOX, B. A., AND SMITH, G. C. (1965). *Acta Met.* **13**, 331.
WIPF, H. (1972). Thesis, Tech. Univ. München, Germany; Jül-Ber. Jül-876-FF. [See also WIPF, H., AND ALEFELD, G. (1974). *Phys. Status Solidi* **23**, 175.]
WOODS, A. D. B. (1964). *Phys. Rev.* **A 136**, 781.
WORSHAM, J. E., JR., WILKINSON, M. K., AND SHULL, C. G. (1957). *J. Phys. Chem. Solids* **3**, 303.
ZAMIR, D., AND COTTS, R. M. (1964). *Phys. Rev. A* **134**, 666.
ZIERATH, J. (1969). Thesis, Univ. Münster, Germany.
ZITTER, H., AND KRAINER, H. (1958). *Arch. Eisenhüttenw.* **29**, 401.
ZÜCHNER, H. (1970). *Z. Naturforsch.* **25a**, 1490.
ZÜCHNER, H. (1972). *Z. Phys. Chem. N. F.* **82**, 240.
ZÜCHNER, H., AND BOES, N. (1972). *Ber. Bunsenges. Phys. Chem.* **76**, 783.
ZÜCHNER, H., AND WICKE, E. (1969). *Z. Phys. Chem. N. F.* **67**, 154.

6

Electromigration in Metals

H. B. HUNTINGTON

DEPARTMENT OF PHYSICS
RENSSELAER POLYTECHNIC INSTITUTE
TROY, NEW YORK

I. Introduction	303
II. Formal Background	306
A. Irreversible Thermodynamics	307
B. Some Formal Solutions of the Diffusion Equation with Forcing Term	313
III. Techniques for Measurement	314
IV. The Nature of the Driving Force	321
A. The Electron Wind Force	321
B. Electrostatic Force	324
V. Interstitial Electromigration	325
VI. Monovalent Metals and Their Alloys	328
A. Self-Electromigration of the Noble Metals	328
B. Electromigration in Noble Metal Alloys	330
C. Electromigration of the Alkali Metals	334
VII. Divalent Metals—Anisotropy in Single Crystals	336
VIII. Electromigration in Trivalent Metals	338
IX. Electromigration in Metals of More Complex Electronic Structure	340
A. Transition Metals	340
B. Anomalous BCC Metals	341
C. Quadrivalent Metals	344
D. Concentrated Alloys	345
X. Electromigration in Thin Films: Problem for Integrated Circuitry	345
References	349

I. Introduction

From an etymological point of view, diffusion is primarily the process of pouring out, hence, spreading or homogenizing. For solids, such processes occur fundamentally by random atom motions whereby various species may move relative to the crystal lattice. By common usage, the term diffusion has become almost synonymous with the study of such motions, in the solid state at least. Hence, it is quite appropriate to include in a book on diffusion a discussion of nonrandom atomic motions—those that owe their origin to directed driving forces—even though in some cases the net effect of such processes may be segregation rather than homogenization. Examples of such "uphill diffusion" on the small scale are the growth of precipitates around nuclei or the collecting of impurity atmospheres about dislocations. In this chapter, we shall be concerned with flows more characteristic of the specimen's dimensions. Electric fields and thermal gradients can give rise to easily observed mass transport and it is the case of the former with which we are primarily concerned here.

The term "electromigration," as used here, is primarily applied to those cases of mass transport where the charge transport number is small. As a consequence, the term electromigration will be limited to metals and semiconductors. For the corresponding transport due to thermal gradient, thermomigration is used. Formerly, in some quarters electro- and thermodiffusion were used but have now fallen into disfavor. On the other hand, the words "electrotransport" and "thermotransport" are close competitors to the "-migration" nomenclature. At a recent international conference, it was apparent that the division was pretty evenly matched.

The observation of mass motion in condensed phases caused by the electric fields goes back many years. The first recorded example is the study of Gerardin (1861) of motion in molten alloys of lead–tin and mercury–sodium. Because of the smallness of the effect in solids, systematic studies came much later. There was considerable work done during the late twenties and thirties of this century in Germany and the literature of that period has been reviewed first by Schwarz (1940). Subsequent reviews include those by Jost (1952) and Seith (1955). In much of the early work, the strength of the electromigration effect was measured in terms of the transport number, gram moles transported times valence, per Faraday flowing through the specimen. These transport numbers are always small for metals, roughly of the order of 10^{-7}, depending strongly of course on the diffusivity which is a factor in the ionic mobility.

The modern period in the study of electromigration really started in the early 1950's with the work of Seith and Wever (1953) and Wever and Seith (1955). First, they introduced the technique of markers on the specimen surface to display the net motion with respect to the crystal lattice. For example, in this way it was possible not only to observe in a binary alloy the relative motion of A with respect to B by chemical or radioactive analysis after the run, but also to determine the absolute velocity of both A and B with respect to the crystal lattice during the run. Their second major contribution was a conceptual one arising out of their study of electromigration as a function of composition across the phase diagrams of Hume–Rothery alloys. In the α phases of these alloy systems, simple electronic conduction prevails and the mass transport net flow is toward the anode. The same general situation holds also for the β phases, but in the complex γ phases, the electron bands are more nearly full and holes conduction prevails. In these phases, the net mass transport is toward the cathode. These observations set Seith to thinking that momentum exchange with the moving charges might provide an important driving force for the electromigration. He thereby resurrected (and modernized) an old suggestion of Skaupy (1914) who proposed that "the electron wind" could be important in inducing mass motion. As we shall see, this suggestion has proved most fruitful, particularly in understanding self-electromigration whose observations the moving markers now made possible. Wever (1956) made the first self-electromigration measurements of mass motion of the matrix atoms in copper and followed up with similar studies for iron and nickel (Wever, 1959). For a comprehensive summary of this period and thereafter, the reader is referred to the review by Verhoeven (1963).

In the succeeding decade, self-electromigration has been systematically studied in a large number of pure metals, as such exploration was a natural avenue toward understanding the basic nature of the phenomenon. There has also been considerable study of the electromigration of dilute impurities and a few treatments of concentrated binary alloys as a function of composition. The fundamental scientific interest in all this centers on electromigration as a probe of electron-defect interaction in the high temperature regime. Because the effect appears to depend on the sign of the charge carriers, the electronic band structure is intimately involved and there is a strong interest in learning how to predict the electromigration drive for those metals with both electron and holes surfaces. There is also a basic connection with the usual transport theory at high temperatures, since the anisotropy of the electron relaxation time, as limited by phonon scattering, has a bearing on the effect as well.

The practical aspects of electromigration seemed for a long time rather minor. It has proved to have some usefulness as a technique for the purification of metal specimens. In the same metal, all interstitials tend to move in the same direction under current flow and this fact facilitates the simultaneous elimination of many trace impurities in the same electromigration treatment. Somewhat more technologically important are the deleterious effects of this phenomenon. It plays an important role in failure of incandescent filaments run under dc current (Johnson, 1938). It has been shown to have a role in the wear of tool steel (Hehenkamp, 1962). Its biggest impact by far, however, has been in that section of the electronic industry producing integrated circuitry. Somewhat over a half dozen years ago, it was discovered that a high rate of failure of the units packaged for computer usage could be traced to the electromigration taking place in the thin aluminum stripes which are used as connecting elements in the circuitry. Because of the large investment tied up in the associated production, several companies launched intensive "crash" programs to understand better the basic mechanism of the failures and how to mitigate their frequency. It was apparent from the start that bulk mass transport in aluminum could not be important at the operating temperatures of the solid state circuitry—seldom above 200°C. Electromigration in thin films was then a really new area where grain boundary migration and, just possibly, migration at the free surface played the dominant role. Section X reviews briefly the salient findings in this area.

II. Formal Background

As indicated in Section I, it is now more customary to present the results of electromigration experiments in terms of the effective driving force rather than in terms of the transport number. The usual nomenclature is to write this force in the rest frame of the lattice as

$$F_{\text{eff}} = |e|Z^*\mathscr{E} \tag{1}$$

where $|e|$ is the absolute value of the electronic charge and $\mathscr{E}$ stands for the electric field. The quantity Z^* is a parameter with which to measure the strength of the electromigration effect, and its use in this capacity is almost universal. Z^* is dimensionless and it has been called "the effective charge number." It ranges in value between $\sim 10^{-1}$–10^2. It can be thought of as the number of charges the moving species has to possess in order that it be subject to an electrostatic force equal to the driving force for electromigration. In view of what has already been remarked, it is natural to divide Z^* into two parts,

$$Z^* = Z_{\text{el}} + Z_{\text{wd}} \tag{2}$$

where Z_{el} represents the direct electrostatic force on the moving ion and, hence, its value might be expected to be that of the ion's nominal valence (see Section IV). The second term Z_{wd} accounts for the electron wind force which is, in general, the larger. In Section II.A, the formalism of irreversible thermodynamics will be applied to show that this intuitive division follows in a completely general treatment.

A. Irreversible Thermodynamics

The active development of the study of irreversible processes (Denbigh, 1958; Prigogine, 1967) has been widely applied to the field of forced motion in solids. Although the formalism is rather abstract and cumbersome, we feel it worthwhile to demonstrate the application here because of the generality of the method. In particular, the foundation for Eq. (2) will be established and the nature of "correlation" considerations for forced motion will be developed.

According to the usual procedure, one begins with an expression for the density of entropy generation as a sum of terms. Each term is evolved from two factors, a driving force and a flux. For the condition close to equilibrium, the flux can all be expressed as linear functions of the driving forces. The matrix of the linear coefficients so introduced is, according to the Onsager relations, symmetric—barring the inclusion of magnetic fields. It is in the application of these relations that the power of the method lies. For a quite detailed treatment of atom motions in the solid state, the reader is referred to Adda and Philibert (1966, Chapter V).

The various relevant terms that contribute to σ, the increase in entropy density per unit time, are

$$\rho = -T^{-1} \sum_1^n J_i \cdot \nabla \mu_i - T^{-1} J_e \cdot \nabla \mu_e - \sum_1^n (q_i/T) J_i \cdot \nabla \phi - (q_e/T) J_e \cdot \nabla \phi \quad (3)$$

Here J_i and μ_i are, respectively, the mass current density and chemical potential of the ith component of the system, and J_e and μ_e are the number current density and chemical potential of the electrons. The latter can be set at the Fermi level or equal to zero. The q_i's represent the respective charges and ϕ is the electrostatic potential. From (3), the phenomenological equations between the currents and driving forces near equilibrium can be established.

$$J_i = -T^{-1} \sum_j L_{ij} \nabla(\mu_j + q_j \phi) - T^{-1} L_{ie} q_e \nabla \phi \quad (4a)$$

$$J_e = -T^{-1} \sum_j L_{ej} \nabla(\mu_j + q_j \phi) - T^{-1} L_{ee} q_e \nabla \phi \quad (4b)$$

Here the L's are the usual phenomenological coefficients of transport theory. The diagonal members are related to the diffusion coefficients, except for L_{ee} which involves the conductivity. Nondiagonal members embody the correlation effects and introduce such cross terms as the "electron wind."

The Onsager relations are not applied at this point because these coefficients are not uniquely determined since there is a linear relationship between the J_i and another between the driving forces, $-\nabla(\mu + q\phi)_i$. This is most readily seen if the J's are taken to be mass current densities and the μ_i to refer to unit mass. Then at mechanical equilibrium in the center of mass system, $\sum J_i = 0$ and likewise for the sum of the driving forces, $\sum \nabla(\mu + e\phi)_i = 0$. While transformation to another reference frame changes the form of these equations, they still remain linear relations.

To remove the indeterminacy of the L_{ij} one of the matter components can be eliminated from the transport equations by use of the equilibrium condition. This is best done starting with the equation for σ. We write μ_i' for $(\mu_i + q_i\phi)$ and eliminate the nth component through $J_n = -\sum_1^{n-1} J_i$ to get

$$\sigma = -T^{-1} \sum_1^{n-1} J_i(\nabla\mu_i' - \nabla\mu_n') - (q_e/T) J_e \cdot \nabla\phi \qquad (3')$$

This expression suggests a new set of phenomenological transport equations with coefficients $\mathscr{L}_{ij}$:

$$J_i = -T^{-1} \sum_{j=1}^{n-1} \mathscr{L}_{ij}(\nabla\mu_j' - \nabla\mu_n') - T^{-1} \mathscr{L}_{ie} q_e \nabla\phi \qquad (4'a)$$

$$J_e = -T^{-1} \sum_{j=1}^{n-1} \mathscr{L}_{ej}(\nabla\mu_j' - \nabla\mu_n') - T^{-1} \mathscr{L}_{ee} q_e \nabla\phi \qquad (4'b)$$

Since J_i and $\nabla\mu_i'$ are now all independent, the $\mathscr{L}_{ij}$ are uniquely determined. They should obey the Onsager relations as established by the usual argument of detailed balance, $\mathscr{L}_{ij} = \mathscr{L}_{ji}$. It is of interest to look back at the original L_{ij} and see what restrictions on them would bring Eqs. (4) to Eqs. (4'). For each Eq. (4a), we add and subtract $\sum_j L_{ij} \nabla\mu_n'$ to get

$$J_i = -T^{-1}\left\{\sum_{j=1}^{n-1} L_{ij} \nabla(\mu_j' - \mu_n') + \left(L_{in} + \sum_{j=1}^{n-1} L_{ij}\right) \nabla\mu_n'\right\} - T^{-1} L_{ie} q_e \nabla\phi \qquad (5)$$

These equations are equivalent to (4'a) only if $L_{ij} = \mathscr{L}_{ij}$ $(i, j \neq n)$, $L_{ie} = \mathscr{L}_{ie}$, and

$$\sum_{j=1}^{n} L_{ij} = 0 \qquad (6)$$

Adding together all of the Eqs. (4a) and noting that $\sum J_i = 0$, we obtain

$$\sum_{i=1}^{n} L_{ij} = 0 \qquad (7)$$

Comparison with Eq. (6) and the symmetry for the first $n-1$ terms leads to

$$L_{in} = L_{ni} \qquad (8)$$

and hence the Onsager relations hold for the L_{ij} as well as for the uniquely determined $\mathscr{L}_{ij}$. There are then $\frac{1}{2}n(n+1)$ different L_{ij} for which there are n restricting conditions, Eq. (6) or (7). It follows then that the number of independent L_{ij} is $\frac{1}{2}n(n-1)$ which checks with the number of $\mathscr{L}_{ij}$.

When we make applications of the transport theory to crystalline solids, there is some advantage in changing over the J_i to number rather than mass flux and in making the corresponding change in μ_i. The crystalline lattice then plays the role of the preferred reference frame for which $\sum J_i = 0$, instead of the center of mass system. We shall make application to several different situations: (a) motion of interstitial atoms, (b) motion of vacancies in a pure metal, (c) motion of marked isotopes in a pure metal, and (d) motion of a dilute impurity in an otherwise pure metal, treated as a special case of a concentrated alloy. As we proceed, we shall note that the conventional diffusion constant which appears in the homogenizing term may, as in the case for vacancy diffusion, be reduced by a correlation factor f as developed by LeClaire and Lidiard (1955). However, the diffusion coefficient which appears in the mobility for the forcing term is uncorrelated for a self-isotope and altered in a rather complicated way for the electromigration of an impurity.

1. Motion of Interstitial Atoms

This is the simplest situation to treat. Intuitively, one senses that the problem is equivalent to a single species moving over an empty lattice. However, for didactic purposes, we shall treat it more elaborately. Index the lattice atoms by A and the interstitials by I. Then the transport equations are

$$\begin{aligned}
J_I &= -T^{-1}L_{II}\nabla\mu_I' - T^{-1}L_{IA}\nabla\mu_A' - T^{-1}L_{Ie}q_e\nabla\phi \\
J_A &= -T^{-1}L_{AI}\nabla\mu_I' - T^{-1}L_{AA}\nabla\mu_A' - T^{-1}L_{Ae}q_e\nabla\phi \\
J_e &= -T^{-1}L_{eI}\nabla\mu_I' - T^{-1}L_{eA}\nabla\mu_A' - T^{-1}L_{ee}q_e\nabla\phi
\end{aligned} \qquad (9)$$

Since the J_A and J_I occur on different lattices, they are essentially independent. We therefore neglect cross terms and write, for the diffusivity of the interstitial (Adda and Philibert, 1966, p. 570),

$$D_I = T^{-1}L_{II}\,\partial\mu_I/\partial n_I = (kL_{II}/n_I)(1 + \partial \ln \gamma_I/\partial \ln n_I) \qquad (10)$$

Now $\mu_I = \mu_{0I} + kT \ln a_I$ where a_I the activity is $\gamma_I \times n_I$, γ_I is the activity coefficient, and n_I the interstitial concentration. From the expression for the part of J_I in Eq. (9) caused by the electric field

$$-T^{-1}(L_{II} q_I + L_{Ie} q_e)\nabla\phi \tag{11}$$

it is easy to pick out the first term as due to the electrostatic force. The identification, $Z_{el} = q_I/|e|$, follows. In the absence of any cross coupling, one recognizes the Nernst–Einstein relation for the mobility B in an ideal solution ($\gamma_I = 1$)

$$B_I = -J_I/n_I \nabla\phi = q_I D_I/kT \tag{12}$$

In the case of general cross coupling, the contributions to J will include additional terms and the simple Nernst–Einstein relation will not hold. This situation is familiar mainly from cases where the correlation corrections have been included. Here the only cross coupling term arises from the electron wind current embodied in $-T^{-1} L_{Ie} q_e \nabla\phi$ in Eq. (11). On this basis, one concludes

$$Z_{\text{wd}} = (L_{Ie} q_e/L_{II}|e|) \tag{13}$$

2. Motion of Vacancies in a Pure Metal

The treatment of the vacancies as a kind of atomic species is not trivial since they have the special property that they can be created or annihilated in the bulk of the material. Strictly speaking, the entropy generation associated with this process should be included in Eq. (3).

For vacancies, the transport equations take the form

$$\begin{aligned} J_v &= -T^{-1} L_{vv} \nabla(\mu_v - \mu_A') - T^{-1} L_{ve} q_e \nabla\phi \\ J_e &= -T^{-1} L_{ev} \nabla(\mu_v - \mu_A') - T^{-1} L_{ee} q_e \nabla\phi \end{aligned} \tag{14}$$

where the subscript v denotes vacancies. Because the sum of the atom and vacancy flow must add to zero in the lattice frame, it is convenient to eliminate the atom terms [see Eq. (4')]. Next, we note that

$$\mu_v - \mu_A = kT \ln (n_v/n_A) + \mu_0 \tag{15}$$

Since $n_A \gg n_v$, one can write the diffusion coefficient for the vacancies, with disregard for $\nabla \ln n_A$, as

$$D_v = kL_{vv}/n_v \tag{16}$$

and, by replacing μ_A' by $\mu_A + q_A \phi$ as before,

$$J_v = -D_v \nabla n_v + (n_v D_v/kT)[q_A - (L_{ve}/L_{vv})q_e]\nabla\phi \tag{17}$$

Here as before, the electrostatic and wind forces appear clearly separated,

6. ELECTROMIGRATION IN METALS

but of course their influence is not exerted directly on the vacancies but on the atoms which exchange with the vacancies. If we had chosen to write an equation for J_A instead of J_v as above, then the resulting nomenclature would have born this out more clearly, since L_{Ae} would have replaced L_{ve}, etc. It should be emphasized that the D_v appearing in the second term of Eq. (17) does not include any cross linkage, i.e., it is an uncorrelated diffusion coefficient. If there is a vacancy flow, $J_v = n_v V_v$, established uniformly in the lattice reference frame where V_v is the drift velocity of the vacancies, then in the laboratory frame the atom flow will be transformed to zero and marks fixed to the lattice, such as fiducial scratches or inert inclusions, will appear to move with velocity V. It is assumed that there is a uniform vacancy flow transversing the specimen. The condition that the molar flux vanishes determines V

$$nV + (-n_v V_v) = 0 \quad \text{or} \quad V = +N_v V_v \tag{18}$$

where N_v is the fraction of vacant lattice sites.

3. Motion of Marked Isotopes in a Pure Metal

We consider first a simplified system where there are only two constituents besides the vacancies, both isotopes of the same element. Let the principal isotope be A with B a dilute radioactive species, initially inhomogeneously distributed. The appropriate equations follow:

$$J_A = -T^{-1}[L_{AA}\nabla(\mu_A - \mu_v) + L_{AB}\nabla(\mu_B - \mu_v)]$$
$$\qquad - T^{-1}[L_{AA} q_A + L_{AB} q_B + L_{Ae} q_e]\nabla\phi \tag{19a}$$

$$J_B = -T^{-1}[L_{BB}\nabla(\mu_B - \mu_v) + L_{BA}\nabla(\mu_A - \mu_v)]$$
$$\qquad - T^{-1}[L_{BA} q_A + L_{BB} q_B + L_{Be} q_e]\nabla\phi \tag{19b}$$

$$J_A + J_B + J_v = 0 \tag{20}$$

We next assume that the vacancy sources and sinks are sufficiently effective to keep μ_v constant in the region of interest. One can then use the Gibbs–Duhem relation in the general case to combine the first two terms in each equation. Since this system is ideal (and dilute), the activity coefficient is unity, hence,

$$J_A = -k[L_{AA}/n_A - L_{AB}/n_B]\nabla n_A - T^{-1}[L_{AA} q_A + L_{AB} q_B + L_{Ae} q_e]\nabla\phi \tag{21a}$$

The diffusion coefficient in the first term, $k[L_{AA}/n_A - L_{AB}/n_B]$ differs from kL_{AA}/n_A because of correlation effects introduced by the cross term L_{AB}. Since $n_A \gg n_B$ and L_{AA}/n_A, L_{AB}/n_B, and L_{BB}/n_B are all roughly comparable, it follows that L_{BA}/n_A which appears in the expression for J_B is negligible by virtue of the Onsager relation. By use of this simplification and the fact that $q_A = q_B$, the equation for J_B becomes

$$J_B = -kL_{BB}/n_B \nabla n_B - T^{-1}[(L_{BB} + L_{AB})q_A + L_{Be}q_e]\nabla\phi \qquad (21b)$$

In the event that there is no electric field, there is no driving force to push a vacancy current and $J_v = 0$. As a consequence, Eq. (20) becomes $J_A + J_B = 0$ and $n_A + n_B = \text{const}$. It follows then from Eqs. (21a–b) that

$$L_{BB}/n_B = L_{AA}/n_A - L_{AB}/n_B \qquad (22)$$

Such a relation, however, holds independent of presence of the electric field. It shows the extent to which the uncorrelated coefficient of diffusion kL_{AA}/n_A is reduced by correlation considerations arising from the cross term. The correlated coefficient is, of course, that appropriate to the isotope, L_{BB}/n_B. On substitution, we get for J_B,

$$J_B = -D^* \nabla n_B - Dn_B Z_B^*|e|\nabla\phi(kT)^{-1} \qquad (21'b)$$

Here $D^* = kL_{BB}/n_B$ is the correlated diffusivity and the uncorrelated diffusivity D is as above, kL_{AA}/n_A. Also $Z_B^*|e| = q_B + L_{Be}(L_{BB} + L_{AB})^{-1}q_e$, where the second term is $Z_{wd}|e|$. The correlation coefficient f equals D^*/D.

4. The Case of the Dilute Alloy

The preceding analysis is altered by the fact that q_A no longer equals q_B when the marked impurity is not an isotope of the solvent atom. Moreover the presence of two chemically different atoms introduces the complication that the vacancies may have more of an attraction for one species that the other. Because of this binding energy, the vacancy free energy varies with microscopic position and its gradient can not be dropped from Eqs. (19). As a result, the line of development leading to Eq. (22) can not be followed if there is a vacancy binding energy.

It is found generally convenient to express the L_{Ae} and L_{Be} in terms of two new independent quantities,

$$L_{Ae}q_e = L_{AA}q_A' + L_{AB}q_B' \qquad (23a)$$
$$L_{Be}q_e = L_{AB}q_A' + L_{BB}q_B' \qquad (23b)$$

We now identify the q's with $|e|Z_{wd}$ for each element so that they can readily be combined with q_A and q_B to give the respective $Z^*|e|$. The expressions for the current densities are then simplified. For example,

$$J_B = -kL_{BB}\nabla\ln n_B - T^{-1}(L_{BB}Z_B^* + L_{BA}Z_A^*)|e|\nabla\phi \qquad (24)$$

The first term in the forced flow of J_B is just a result of the direct drive on species B and it involves the *correlated* diffusion constant only. If the A constituent also feels a drive ($Z_A^* \neq 0$), J_B will also be affected through the cross coupling term L_{AB}. This quantity has been called by Manning (1968) the "vacancy flow term" and he has stressed that it is incorrect to consider it simply as a correlation correction. We have seen in Section II.A.3 that for the tracer of a matrix atom the effect is just to increase the effective diffusivity from the correlated to uncorrelated value, $D_B^* \to D_B$. If the tracer is chemically different but there is no effective vacancy binding, then Eq. (22) shows that $L_{AB}/L_{BB} = (1 - f)/f$ and hence is always positive. The vacancy flow term may in general, however, have either sign as shown by Manning (1968) who evaluates L_{AB}/L_{BB} as $2\langle n_p \rangle$ where $\langle n_p \rangle$ is a complex function of the vacancy jump probabilities. Physically, one sees how the vacancy flux might either increase or decrease the solute flow depending on circumstance.‡ In the simplest case, the flow of A, say, to the right will cause a counter flow of vacancies to the left which, meeting the B atoms on the right side, will induce them to flow to the right more than they otherwise would. However, the vacancies can also draw the B atoms to the left along with them under the circumstances that there is considerable binding between B atoms and vacancies. This effect will be augmented if $Z_A^* > Z_B^*$ so that a vacancy which contacts a B atom will always, in due course, work its way around to the left side of the atom before the pair breaks up. Not only can the ratio L_{AB}/L_{BB} go negative but for Z_A^*/Z_B^* large enough J_B can become oppositely directed from J_A even though both Z^*'s have the same sign.

B. Some Formal Solutions of the Diffusion Equation with Forcing Term

In this section, some formal solutions of the diffusion equation with forcing term are briefly presented.

For those situations which do not involve a boundary condition fixed in a particular reference frame nor a variation of D or T with position, the electromigration–diffusion equation can be reduced by a shift of reference frame to the simple diffusion equation. In one dimension

$$\partial n_A/\partial T = (\partial/\partial x)\{D^* \partial n_A/\partial x - n_A F\mu\} \tag{25}$$

where $F = -Z^*|e|\nabla\phi$ and $\mu = D/kT$. Let $n_A(x, t) \to n_A'(x - vt, t)$ so that

$$\partial n_A'/\partial t = D^* \partial^2 n_A'/\partial x^2 \tag{26}$$

if $v = F\mu$, the drift velocity. The standard solutions can now be applied for an initial step or thin planar source in the concentration $n(x)$. Boundary

---

‡ This topic is considered in detail by Anthony in Chapter 7.

conditions fixed in the initial reference frame are troublesome but a solution does exist for the case of the thin film surface deposit. Suppose that initially $\int n(x)\,dx = N$ and $n(x) = 0$ for $x < 0$. For all values of t, $\partial n/\partial x = 0$ at $x = 0$. The solution for this problem is (Jost, 1960)

$$n(x, t) = N\{(\pi Dt)^{-1/2} \exp[-(x - F\mu t)^2/4Dt]$$
$$- (F\mu/2D) \exp(F\mu x/D) \operatorname{erfc}[(x + F\mu t)/(4Dt)^{1/2}]\} \qquad (27)$$

The availability of this solution makes possible quantitative experiments with a thin tracer layer backed on one side by an impenetrable barrier.

III. Techniques for Measurement

There are two basic types of measurements in electromigration work and any particular experiment may involve either or both. The two types are the determination of the changes in composition brought about in the specimen by the passage of the electric current and the measurement of the net matter transport with respect to a fixed reference frame. For work in those solids which crystallize, it is natural to use the lattice as the reference frame.

The determinations of the current-caused changes in composition usually start with an homogeneous specimen and the composition is then determined at the end of the run by whatever technique the experimenter favors. If the specimen is subsequently sectioned, chemical analysis may be applied to the determination of the concentration profile but radioisotope counting will certainly afford higher sensitivity (Adda and Philibert, 1966, Chapter IV). In some cases, optical absorption observations (Jousset and Huntington, 1969) have sufficed. Of those techniques which are nondestructive, the use of the electron probe (Hehenkamp, 1968) has proved to be most effective and sensitive in skilled hands. In addition, autoradiographic techniques could be used. For certain systems such as carbon in iron, hardness measurements have been widely used to map out composition profiles (Okabe and Guy, 1970).

Continuing an electromigration run until steady state conditions are established requires that $(D^*t)^{1/2}$ be of the order of several times the specimen thickness, i.e., the dimension in the direction of the electric field (Okabe and Guy, 1970). Where this is possible, the concentration profile becomes independent of time. A more important advantage can be appreciated from setting the left hand side of Eq. (24) equal to zero. The concentration profile is then given by

$$\nabla \ln n_B = (f'/fkT)Z_B^* e\mathscr{E} \qquad (28)$$

The f' stands for $(1 + Z_A^* L_{BA}/Z_B^* L_{BB})$, a sort of correlation factor for forced motion, while $\mathscr{E}$ is the electric field. Because of the rapid variation of D with temperature for most systems, it is difficult to unravel the temperature dependence of Z^* with confidence by other techniques. The advantage here is that the cancellation of D by D^* to give f^{-1} eliminates this effect and allows the temperature dependence of Z^* to be determined directly.

Such composition measurements suffice to give the whole picture in a system where only interstitial migration is taking place. Since these problems frequently involve gases in metals, an alternative technique (Oriani and Gonzales, 1967) is the use of a steady state flow method where the specimen in rod form is gripped at the ends with hollow electrodes that connect through a glass system. Here an oil drop in a glass capillary can be used to indicate with its motion the pressure imbalance brought about by the electromigration at the electrodes.

Besides the determination of how the electromigration process has changed the chemical concentrations in the specimen, a complete specification of the Z^*'s in a multicomponent system requires that a measurement be made of the net motion of the atoms with respect to the lattice if vacancy flow is involved. In fact, for self-electromigration this is the only measurement that can be made. Basically there are two rather different techniques for doing this.

(a) The first of these, which we shall call the "isothermal, isotope method," involves inserting a thin layer of radioisotope perpendicular to the direction of current flow in the specimen (see Fig. 1). The usual technique is to weld together the two specimen halves after one of them has been plated with the isotope. The primary measurement is that of the displacement of the maximum of the radioactive distribution with respect to the weld interface. Usually it is possible to fix the position of the interface to ± 1 or 2 μm from occluded material at the weld. In the event that the welding technique is so perfect as to eliminate such markers, it may be necessary to include a second radioisotope at the interface (Gilder and Lazarus, 1966), this time of a completely inert material. In the past, a layer

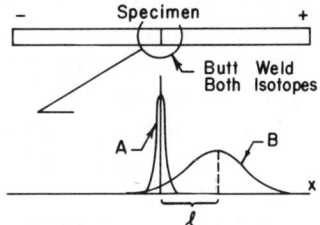

Fig. 1. The isothermal isotope method: A, inert marker concentrations, B, concentration of matrix isotope. $v_a(T) = l/t$ where t is duration of experiment. Temperature determined from the diffusional spreading B.

of Hf has been laid down and allowed to oxidize. The hafnium oxide has proved to be completely inert and immobile at temperatures below 1100°C. The method calls for isothermal conditions at the position of the weld and the absence of thermal gradient. The Joule heating naturally causes a variation in temperature for any rod-like specimen but the gradient at the weld can be minimized by shaping the specimen (thicker in the middle) or, as is more usually done, by inserting high resistance elements, sections of graphite (Hehenkamp, 1968) or stainless steel, placed symmetrically on either side of the specimen. For a self-electromigration experiment the Z^* is determined by

$$Z^* = -(d/t)(kT/|e|D\mathscr{E}) \qquad (29)$$

where d is the displacement between the center of the isotopic activity and the original interface, t the duration of the run, $|e|$ the charge of the electron, and D the uncorrelated diffusivity.

Since the determination of Z^* depends on d/tD where D is strongly temperature dependent, it is worthwhile to know the temperature of the active region as accurately as possible. A great advantage of this method is that it naturally offers a built-in thermometer of great accuracy, namely the diffusional spreading of the isotopic concentration. The shape as well as the position of the displaced distribution is usually determined accurately by the standard sequential sectioning technique. By measuring the diffusion in the critical region, one determines the temperature of the specimen with high precision since the diffusivity is highly temperature dependent. Moreover, if the specimen temperature fluctuates somewhat during the course of the run, the diffusional spreading of the peak averages the temperature fluctuations in the same way as does the electromigration displacement.

The really restrictive requirement for this method is the obtaining of a sound weld between the two parts of the specimen. In some systems, this requirement is a considerable stumbling block. Various methods have been tried to circumvent this obstacle with results of varying reliability. In one approach (Kuz'menko et al., 1960) a weld of sorts was obtained and after the experiment the specimen was broken apart at the original interface and sectioned. The uncertain conditions at the interface and the probable extra joule heating involved makes this method open to question. A variation of this method (Kuz'menko, 1962) is to use plated but unwelded specimens on the assumption that electric but no mass current transverses the interface. The insertion of an impermeable foil would insure this boundary condition and the application of Eq. (27) would make quantitative analysis feasible. The difficulty with temperature control at the interface would remain. A third method (Kalinovich et al., 1962), which apparently gives consistent results, consists in preparing a cylindrical specimen with a band of radio-

6. ELECTROMIGRATION IN METALS 317

isotope around the central part of the lateral area. After passing a high current during the electromigration anneal, the specimen is measured for motion and diffusional smearing of the edges of the radioactive band. Although the surface of the specimen is lightly etched after the anneal to remove the transport by surface diffusion, there is still some question as to what effect the surface electromigration, followed by diffusion into the bulk, might have.

One final comment on this method is to point out that it is indeed the uncorrelated D that appears in Eq. (29). Our first reaction might be to use the correlated D^* as determined by the standard diffusion measurement which this arrangement so closely resembles. It has been pointed out by Manning (1968) that the current induces a counter vacancy flow which enhances the isotope flow by a factor of f^{-1} so that D^* rather than D enters into the mobility. This result is also implicit in the irreversible thermodynamics treatment presented in Section II.A. On intuitive grounds, one can also argue that in self-electromigration the currents of both active and inactive isotopes are proportional to their concentrations and their sum must balance the counterflowing vacancies. Since the motion of these is uncorrelated, then so also must be that of the marked atoms. However, for a radioimpurity the effective diffusivity is a more complex quantity, as we have already seen.

(b) The second method for determining atom flow with respect to the lattice, which we shall call the "vacancy flux method," depends on measuring the dimensional changes of the specimen during the course of electromigration to map out vacancy creation and annihilation as a function of position. Initially, this was done by observing the motion of surface markers (Huntington and Grone, 1961) scratches, indentations, or inert inclusions, but it soon became apparent (Penney, 1964) that for most materials it was important to monitor changes in the dimensions at right angles to the current flow as well. If $y(x, t)$ is taken to be the marker displacement function, as established by the history of a finite number (10–30) of markers, and $d(x, t)$ is specimen diameter as a function of position and time, then the local dilatation Δ can be constructed

$$\Delta = \nabla \cdot y_m + 2(\delta d/d_0), \quad \text{where} \quad \delta d = d(x, t) - d_0 \quad \text{and} \quad d_0 = d(x, 0)$$

Next, one can determine the vacancy flow $J_v(x)$ by integrating

$$n_0^{-1} \nabla \cdot J_v(x) = \partial \Delta/\partial t = \nabla \cdot v_m + 2\, \partial(\delta d/d_0)/\partial t, \quad \text{where} \quad v_m = \partial y_m/\partial t \quad (30)$$

and n_0 is the number of atoms per unit volume. It follows that

$$J_v(x) = n_0 \left[v_m(x) + \int_{x_0}^{x} (\delta \dot{d}/d_0)\, dx \right] \quad (31)$$

Since in a pure material the vacancy flux is simply the reverse of the atom flow by the vacancy mechanism, one can determine Z^* directly from $Z^* = -(J_v/n_v)(kT/|e|D_v)\mathscr{E}^{-1}$, where D_v is the vacancy diffusivity.

$$Z^* = -(kT/|e|D\mathscr{E})[v_m + \int_{x_0}^{x} (\delta \dot{d}/d_0)\, dx], \quad \text{since} \quad D = (n_v/n_0)D_v \quad (32)$$

Throughout, x_0 refers to some reference point cold enough so that the dilation is negligible there.

A typical situation is represented schematically in Fig. 2 where an electric field to the left drives electrons to the right, which in turn push the moving atoms in the same direction. This means a counterflowing vacancy

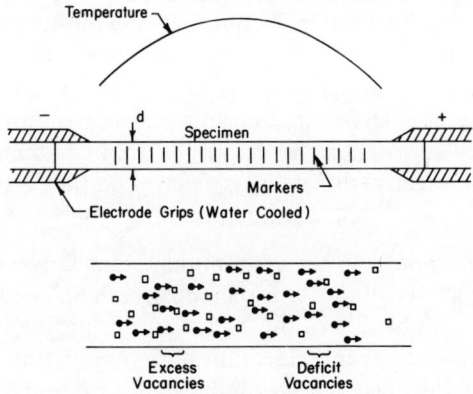

Fig. 2. Vacancy flux method—schematic experiment: ● electron, □ vacancy.

flux to the left. Because of the nearly parabolic temperature variation, vacancies must be generated on the right and annihilated on the left. This means positive dilatation on the right and contraction on the left. The change may show up in all three directions by expanding or shrinking the diameter and by the increase or decrease of the spacing between the marks. These effects are illustrated in the typical curves of Fig. 3 for experimental measurements of marker velocity and rate of change in transverse diameter as a function of position. The directions of the motions are consistent with the experiment shown in Fig. 2. Note that on the right the diameter is swelling and the distance between scratches grows with time.

The use of this method requires some precautions to maintain the specimen in a stress-free condition to prevent continued creep throughout the run. Differential thermal expansion makes it infeasible to hold the specimen in rigid clamps. In some instances, the specimen is pressed between spring electrodes of weak force constant. Other investigators have used a liquid metal bath to float one electrode, perhaps in mercury at low temperature or in a gallium eutectic where low vapor pressure is required.

The temperature variation in the specimen plays an essential role in the method and therein lies one of its greatest weaknesses or strengths, depending on the point of view of the investigator. The temperature variation necessarily reduces the accuracy with which the temperature can be known at a particular locality. Intrinsically, the method is less accurate than the isothermal isotope method but it does have certain advantages for rapid search. One avoids the very delicate and time consuming process of obtaining a good internal radioisotope layer which becomes increasingly

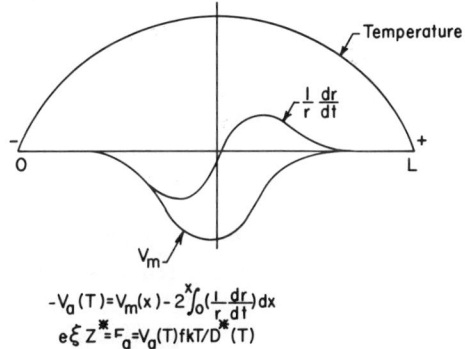

$$-V_a(T) = V_m(x) - 2\int_0^x (\tfrac{1}{r}\tfrac{dr}{dt})dx$$
$$e\xi Z^* = F_a = V_a(T) f kT/D^*(T)$$

Fig. 3. Vacancy flux method—typical curves of marker displacement and transverse diameter change.

difficult with the more readily oxydizable metals that are hard to weld satisfactorily. In addition, a single run with this method gives Z^* as a function of position [Eq. (29)] and hence over a range of temperature so that two or three runs with this method may suffice to give a good idea of the temperature variation of J_v and the degree to which Z^* may be a constant.

Further details of the method will come up later for discussion in connection with particular experiments. Some comment should be made here on the technique of temperature measurement which, for this method, is limited to a precision of 2 or 3° at best. Many procedures have been tried but that which seems to be most universally applicable and satisfactory for the primary measurement is to paint on small aquadag spots and observe these with an infrared pyrometer. The emissivity of the aquadag is constant and well known as are the corrections for absorption and reflectivity of the glass port. Running the specimen to the melting point gives a check for the accuracy and a chance to correct for any temperature difference between aquadag and specimen because of the greater emissivity of the former.

In the past, there has been some skepticism on the validity of relying on surface markers to monitor the internal bulk motions. The primary question rests on whether the markers, whatever their nature, act as traps, sources, or sinks for the vacancies. If none of these, then the vacancy flow will be unimpeded and the markers will truly remain fixed in the lattice frame of reference. There is some evidence that is not always so for the inert markers. One would expect difficulty particularly with conducting inclusions since the electric current lines might flow up to and through the inclusion leaving the vacancies stranded at the interface. The same difficulty would not occur at scratches or indentations, since here the current flow lines would flow around the mark, drawing the mass flow along the same path. One naturally worries about the effect of surface diffusion and surface electromigration. Of course, the former does limit the time that the surface markers will hold the sharpness of their definition. There may be other unfortunate effects but the evidence accumulated from a wide body of measurements seems to indicate that such difficulties do not play a major role. For one thing, the information from the markers appears to be quite consistent with that obtained from measurements of transverse diameter, as read with a bifilar eyepiece. Moreover, the temperature dependence of the measured J_v is in general more characteristic of bulk, rather than the surface, mobility.

For this method, too, it is the uncorrelated diffusion coefficient that enters into the analysis for Z^*.

To summarize, then, what has been just said about these two methods, the "isothermal isotope method" and the "vacancy flux method": the former relies on following the electrotransport of marked atoms, presumably by sequential sectioning, starting with a thin tracer layer. The method requires skill and patience but has the potential for high accuracy, mainly because the effective temperature of the run is measured very sensitively by the diffusional spreading of the tracer atoms. The "vacancy flux method," on the other hand, gives results of limited accuracy primarily because of the difficulty in knowing the effective temperature accurately at each marker and, secondarily, because of some uncertainty about the possible influence of surface transport. It is, however, well suited for search, getting quick results with a minimum of careful specimen preparation with an internal accuracy of perhaps $\sim 10\%$ in Z^* under favorable circumstances.

6. ELECTROMIGRATION IN METALS

IV. The Nature of the Driving Force

A. The Electron Wind Force

As indicated in Section II, Eq. (2), the drive for electromigration can be resolved into a purely electrostatic force and the interaction with the moving charge carriers. There have been two quite different methods for treating the latter effect, which, incidentally, is usually the dominant one. The first of these might be called the "ballistic method" and emphasizes the actual collisions between the current carrying charges and the moving atoms of the metal. It was presented originally by Fiks (1959) and Huntington and Grone (1961). The idea has been put very simply by Fiks. The number of collisions per unit time between the electrons and a moving atom is the product of the electron density n_e, their average velocity v_e, and the atom's intrinsic cross section for collision with the electrons σ_e. For each collision, the electron confers on the average the momentum that it has acquired during one relaxation time τ_e or $e\mathscr{E}\tau_e$. On the basis, the electron wind force is

$$|e|Z^*_{\text{wd}}\mathscr{E} = -|e|n_e\mathscr{E}\lambda_e\sigma_e \tag{33}$$

where $\lambda_e = v_e\tau_e$. This expression has been extended by Glinchuk (1959) to cover the situation where both electrons and holes are involved in the transport process

$$Z_{\text{wd}} = -n_e\lambda_e\sigma_e + n_h\lambda_h\sigma_h \tag{34}$$

The treatment by Huntington and Grone (1961) postulates a transition probability per unit time between free electron states induced by the moving atoms acting as lattice defects. One can then write a formal expression for the momentum transfer per unit time, i.e., the force transmitted from the electrons to a moving atom. Actually it turns out to be important (Huntington and Ho, 1963) to treat the electron pseudomomentum (or the momentum of electron plus lattice) rather than $m_0 v_g$, the momentum of the electron as a free particle. The calculation of the force on the ion then involves a summation over the initial and final states of the scattered electrons but this expression can be simplified by substituting the quantity, the contribution to the electron reciprocal relaxation time arising from collisions with the moving atoms—assumed to be independent of electron states. The authors emphasize that this quantity is not a constant during the diffusion jump, particularly for a noninterstitial type of motion, but will increase as the atom moves from its equilibrium position into a saddle-point configuration where its scattering power is enhanced. The final result can be put into a quite simple form in terms of specific resistivities,

$$Z_{\text{wd}} = -z(\rho_d/N_d)(N/\rho)|m^*|/m^* \tag{35}$$

where z is the electron–atom ratio and the first parenthesis contains the space-averaged specific resistivity per moving defect. The second parenthesis holds the specific resistivity per normal atom of the lattice. The m^* is an average effective mass for the electrons. The final factor shows in a crude way that the sign of the charge carriers determines the sign of the direction of mass transport. It is quite easy to demonstrate that Eq. (35) reduces to Eq. (33) since $zN = n$ and $(\rho_d/\rho) = \tau\sigma_d N_d v$. If the scattering power of the moving atom varies sinusoidally on the way from the lattice point to saddle point to lattice point, then it is appropriate to take ρ_d to be $\frac{1}{2}$ its saddle-point value. For impurity electromigration, however, there will be appreciable ρ_d when the impurity is at a normal site.

In the treatment of Bosvieux and Friedel (1962), no explicit mention is made of the collision concept but the nature of the defect shielding by the free electron charge is explored. Part of this shielding which is permanent and static has a symmetric distribution and part, which exists only when there is an unbalanced flow of charge carriers, is antisymmetric about the defect in the direction of the current. For example, one can imagine how the trajectories of an initially uniform sheet of electrons might be deflected by an attractive defect to give a negatively charged concentration in back of the defect, which would draw the defect backward. The potential field from the asymmetric distribution always draws the defect in the direction of the charge carrier motion. Interestingly enough, the end result of the Bosvieux–Friedel calculation is in essential agreement with the predictions of the ballistic approaches. To relate the theoretical prediction Eq. (35) to experiments on the electromigration of impurities, the (ρ_d/N_d) has been set equal to the average of its values for the (substitutional) impurity at a lattice position and at a saddle point. For the former, one takes the incremental resistivity of the impurity as determined by experiment or by Linde's rule (1932), which implies $\rho_d/N_d \sim Z_\Delta{}^2$ where Z_Δ is the difference between the Z' of the valence of the impurity, and Z the valence of the matrix atoms. At the saddle point, the conduction electrons are scattered by the full charge Z' rather than Z_Δ and the $\rho_d/N_d \sim (Z')^2$.

Bosvieux and Friedel also considered the force exerted by the asymmetric distribution in the neighborhood of a vacancy on a marked atom at a next neighbor site. They found it to be small but not negligible and quite dependent on the structure of the matrix atoms. This particular result illustrates one of the advantages of this method over the ballistic approaches which always left rather unresolved the question of how the scattered momentum of the charge carriers was to be distributed between the moving atom and its immediate neighbors. For the jumping atom at the saddle point, there seems to be little ambiguity as to where the recoil would be mainly felt but for the atom near its equilibrium position, the situation is not so clear.

6. ELECTROMIGRATION IN METALS   323

All of these methods incorporated two basic assumptions: (1) they relied on first-order perturbation theory or Born approximation and (2) they employed an essentially free electron model for the metal. As a result of the latter shortcoming, it was not easy to see what the specific influence of band structure might be on electromigration. While the correct result for the limiting case of a pure hole band seems moderately apparent for the ballistic models, there is only an intuitive statement on the part of the polarization treatment that the passage of holes should drive in the opposite direction from the electrons. The experimental evidence in general seems to indicate that pure metals with complex band structure do indeed have smaller Z^* than those with simple structures. The question arises how to extend the theories to treat these intermediate situations.

A promising development of the polarization model involving a more general type of band structure has been carried by Sorbello (1970, 1973). By adapting the pseudopotential approach, he showed how to incorporate the real metal properties of the lattice, thereby going one step beyond the free electron model used previously. He was able to establish formally the application of the pseudowave function and pseudopotential to the problem of the current-caused force felt by a defect atom or by the atoms near a defect. The wave functions so obtained could be used, together with the change in electron state distribution caused by the electric field, to give the current-associated polarization, from which the forces on the moving atom and all others could be calculated. The final result could be expressed succinctly in terms of a simple potential, one term of which gave just the force to be expected from the scattering of the electrons by the moving atom considered as an isolated entity. The remaining terms of the potential consisted of two-body interaction, proportional to the current, which accounted for the interference scattering effects. The numerical calculations built on this model show quite satisfactory agreement with experiment.

There is also real metal variation of the ballistic approach initiated by Feit (1969) and Huntington et al. (1971) which appears promising. The Bloch wave functions for the regular lattice are constructed of linear combinations of plane waves in accord with electron–lattice coupling implicit in an appropriate pseudopotential. In the usual way, the transition rate formula based on the second-order, time dependent, perturbation theory of quantum mechanics is used to give the changes of the Bloch wave functions with time because of scattering by the defect structure: moving atom, associated vacancies, and elastic displacements. As in Sorbello's work, the force on the moving atom is calculated from the rigorous quantum mechanical formulation involving the commutator of the complete Hamiltonian with the conjugate momentum operator for the moving atom. The resulting expression for the driving force involves the square of the scattering form factor but the structure factor of the defect complex appears only to the first power. While this approach resembles Sorbello's in many particulars, the emphasis of the latter is mainly on the various forces exerted by the electron wind on all the atoms whereas this development concentrates on the influence of the band gap energy discontinuities on the sign and anisotropy of Z^* in simple metals where Brillouin zone cutting occurs. Numerical results from this approach are still in progress.

In all these theoretical treatments, it has been assumed almost tacitly that first-order scattering theory (Born approximation) should suffice. Quite recently, Landauer and Woo (1974) have pointed out that higher orders may well be important since the strength of the scattering by the even orders must be responsible for the intensification of the electric field in the regions of higher specific resistivity. The work grew out of earlier considerations by the senior author (Landauer, 1957) wherein he introduced the concept of residual resistivity dipoles localized at impurity scattering sites. It is a formidable task to estimate quantitatively how large an effect the second-order term might have but it appears that the contribution to Z^* should be in the same direction as Z_{el}.

B. Electrostatic Force

The second part of Z^* [see Eq. (3)], called Z_{el}, represents that part of the force on the moving ion due to the direct electrostatic action of the electric field. The evaluation of this quantity has been subject to some controversy (Frohberg, 1971) because of the complications arising from the shielding by the conduction electrons. Moreover, this controversy has taken on a rather academic flavor because Z_{el} is usually much smaller than Z_{wd} and it requires extremely sensitive experiments to reveal its influence. In this matter, the point of view of the author has been to rely on the intuitive picture one gets from the classical model of a continuum electronic fluid containing fixed compensating positive charges. A rather laborious calculation (Huntington, 1969) shows quite generally that the shielding of the electronic fluid does not affect the electrostatic force felt by the moving ion since the ion's motion does not cause any net displacement of the electronic fluid. Unlike the situation for an electron in a bound state, the shielding charge from conduction electrons is not carried along by the moving charge but reforms locally at every point along its path. Although this argument is admittedly nonquantum mechanical, Sorbello (1970), using more sophisticated arguments, has come to the same conclusion. Of course, a situation in which the moving ion changed its valence character as the number of nearest neighbors changed during the jump process would require a different treatment.

V. Interstitial Electromigration

In this and the subsequent sections, electromigration in various contexts will be discussed. Here we shall be concerned with the motion of interstitials of the small atoms: hydrogen, nitrogen, oxygen, and carbon. This mode of mass transport affords some simplifications for theoretical treatment as compared to vacancy motion: (1) the geometry of the saddle-point configuration more nearly resembles that of the equilibrium position and (2) the complication of the associated vacancy and its mitosis at the saddle point is absent. As a consequence, the driving force is more nearly constant throughout the jump and, in the case of electromigration, more nearly what one woud expect from the interstitial's incremental resistivity. In addition, the complication of the correlation consideration is avoided.

A considerable number of interstitial systems have been studied by electromigration up to this time, about 45 in all. Table I gives a compilation of these experiments and the direction of motion in each case. Two general observations are that in the same matrix the interstitial impurities go in the same direction—with two rather questionable exceptions—and that the direction of migration correlates closely with the position of the column of the solvent in the periodic table (R. E. Einziger, 1973). The sign of Z^* is positive for columns V, VI, and VIII and negative for columns III and IV and for the two rare earths. The fact that all solutes tend to go in the same direction in the same solvent bears out the electron wind picture since it is reasonable to expect that the direction at least of its force should be the same for all interstitials in the same matrix. The positive Z^*'s for those columns of the periodic table with unfilled d shell suggest that it may be the holes in high density bands which are dominating the electromigration of the interstitials. A few comments about work on particular systems follow.

Because of the high mobility of H in Pd, it was natural that this should be one of the first interstitial systems to be studied (Wagner and Heller, 1940). More recently, the measurements of deuterium in palladium have also been made (Knack and Eichenauer, 1968) but over a rather narrow temperature range within which the temperature dependence appears to be at variance with that observed for hydrogen.

Naturally, the carbon-in-iron system as been extensively studied with a rather wide range of results. The most convincing study is that of Okabe and Guy (1970) who used the steady state technique and thereby could study the behavior independent of D. They find Z^* to be around 4 and quite constant with temperature.

The work on Th, V, Nb, Ta, Ga, Y, and Zr, using several impurities in each (Carlson *et al.*, 1966; Schmidt *et al.*, 1970; Peterson *et al.*, 1966;

TABLE I

Electromigration of Interstitial Impurities

Solvent	Column in periodic table	Impurity + to Cathode	Impurity − to Anode	Ref.
Silver	IB		H, D	a
Yttrium	II		C, O, N, H	b
Titanium	IV	C		c
	IV		O	d
Zirconium	IV		O, N, C	e
Thorium	Actinide		C, N	f
Vanadium	V	C, O, N		g
	V	H, D		h
Niobium	V	C		i
	V	O		j
	V	H, D		h
Tantalum	VI	C		c, i
	VI	H, D		h
Tungsten	VI	C		c
Iron α	VIII	H, D		k
	VIII	C		l
Iron γ	VIII	C		m
	VIII	B	N	n–s
Cobalt	VIII	C		c
Nickel	VIII	H, D		k
	VIII	C		c
Palladium	VIII	H		o
	VIII	H, D		p
Gadolinium	Rare earth		C, N, O	q
Lutetium	Rare earth		C, N, O	r

[a] Einziger and Huntington (1973).
[b] Carlson et al. (1966).
[c] Kovenskii (1963).
[d] Claisse and Koenig (1956).
[e] Schmidt et al. (1970).
[f] Peterson et al. (1966).
[g] Schmidt and Warner (1967).
[h] Herold et al. (1971).
[i] Schmidt and Carlson (1972).
[j] Rudman (1965).
[k] Oriani and Gonzales (1967).
[l] Okabe and Guy (1970).
[m] Dayal and Darken (1950).
[n] Seith and Daur (1938).
[o] Wagner and Heller (1940).
[p] Knack and Eichenauer (1968).
[q] Peterson and Schmidt (1972).
[r] Peterson and Schmidt (1969).
[s] Bibby et al. (1966).

Schmidt and Warner, 1967; Peterson and Schmidt, 1969, 1972) was carried by the group in the metallurgy department of Iowa State University. Their motivation centered primarily on the use of electromigration as a tool for purification. In certain cases, it proved to be more effective than zone solidification. The usefulness of the method is greatly enhanced by the fact that all the interstitial impurities tend to flow in the same direction in any given matrix.

Perhaps the most interesting study of interstitial electromigration is that of Oriani and Gonzales (1967) who made careful measurements of the flow of the hydrogen isotopes, H and D, in Fe and Ni over about 200°C. The gas flow was measured by the velocity of an oil drop in a capillary system connected to the two (hollow) electrodes. The specimen was always immersed in the appropriate hydrogen gas and heated by the passage of current. No effect was observed for ac and the motion of the drop reversed when dc current was reversed. The results showed rather small Z^*, about $\frac{1}{4}$ for H in Fe and $\frac{1}{2}$ for H in Ni, with roughly the same *flow* for both isotopes under the same conditions. However, this means that Z^* was some 40–50% higher for the heavy isotope because of its lower mobility. There seems to be no simple explanation for this effect, although there is promise that it can be understood in terms of the new quantum mechanical approaches that are now developing to treat the motion of light atoms.

The theoretical situation has been reviewed by Sussman (1967, 1971). Perhaps the most useful and specific treatment is that of Flynn and Stoneham (1970). In regard to the electromigration aspect, Stoneham (1972) has emphasized the point that any isotope dependence for Z^* almost surely must be a quantum effect and has hazarded the prediction that such an effect must be due to displacements which are symmetric with respect to the plane bisecting the jump. The electron drag would be bigger for the proton than the deuteron because presumably its larger zero-point energy would cause the larger symmetric displacements (Stoneham and Flynn, 1973). If one takes Z_{el} as 1 for hydrogen, then Z_{wd} is $(Z^* - 1)$, which is in the direction to explain the results of Oriani. However, the recent results by Einziger and Huntington (1973) who also used a flow method patterned after Oriani and Gonzales (1967), have shown that the hydrogen isotopes in silver flow strongly toward the anode but again the effect is greater for deuterium than for normal hydrogen, which, because of the change of sign of Z^*, runs counter to Stoneham's prediction. Because the literature values for the permeability of hydrogen in silver scatter widely and none exist for deuterium in silver, it was necessary to make these measurements to deduce values for Z^*. This was done by a modification of the apparatus used to measure the electromigration flows. The result for the permeability v_i gave $v_D/v_H = 0.8$ and for the effective charge numbers, $Z_H^* \simeq 6$ and

$Z_D^* \simeq 15$. The largeness of the isotope effect in this system is difficult to explain but emphasizes the importance of the quantum mechanical aspects. The fact that the magnitude of the Z^*'s are about 10 times those found for the hydrogen isotopes in Ni and Fe suggests that there may be a real qualitative difference between the two cases. We propose that in the silver the hydrogen may function as a proton and so scatters with an effectively larger cross section than when it is accompanied by an electron in a bound state. Eastman *et al.* (1971) have emphasized the presence of the bound electron for hydrogen in Pd (and probably most other transition metals). The situation may, however, be different for silver because of the depth of the d bands below the Fermi surface.‡

In Sections VI–IX, the work on the electromigration of pure metals and some of the alloys will be reviewed, grouped as they appear in the columns of the periodic table. As this material covers quite a large field, it is not possible to be all-inclusive. Considerable selection has been involved, admittedly biased by the author's personal interest. In Table II, the data for the self-electromigration of several metals have been compiled. For each, the Z^*/f, the temperature range of the measurements, and the references, chosen to be as quantitative and recent as available, are given. The reason for chosing to give Z^*/f rather than just Z^* is that this is the quantity that the electromigration plus diffusion measurements give. While in most cases the correlation coefficient for pure metals is a well-known quantity depending only on crystal geometry, there can be exceptions where two or more simultaneous mechanisms are involved in the operation of self-diffusion.

VI. Monovalent Metals and Their Alloys

A. Self-Electromigration of the Noble Metals

From the standpoint of ease of operation, gold is in many ways the simplest of the metals to study. The first work on the electromigration of this metal (Huntington and Grone, 1961) was done initially on polycrystalline wires by a rudimentary form of the vacancy flux method. The results indicated a Z^*/f of ~ -9 but at lower temperatures, below 850°C, the Z^* increased quite rapidly. Later Gilder and Lazarus (1966) used the isothermal isotope method on single-crystal gold specimens to obtain Z^*/f equal to -8. The possibility that grain boundary transport may be responsible for the high Z^* at low temperatures is suggested by some recent studies (Bryant, 1971) of electromigration in 1 mil gold foil where the effect is even more

‡ We are indebted to Dr. D. Eastman for this suggestion (1972).

TABLE II
Z^*/f for Several Pure Metals

Metal	Z^*/f	Temp. range (C°)	Reference
Noble metals			
Gold	−9.5 to −7.5	850–1000	Huntington and Grone (1961)
	−8.0		Gilder and Lazarus (1966)
Silver	−21 ± 5	830–890	Doan (1971)
	−8.3 ± 1.8	795–900	Patil and Huntington (1970)
Copper	−5.5 ± 1.5	845–1030	Sullivan (1967a)
	−4.8 ± 1.5	870–1005	Grimme (1971)
Alkali metals			
Lithium	−2.5 to −1.6	90–160	Thernquist and Lodding (1968)
Sodium	−3.3 ± 0.7	45–80	Sullivan (1967b)
Divalent metals			
Zinc			
∥ to c axis	−2.5 ± 0.2	366–400	Routbort (1968)
⊥ to c axis	−5.5 ± 0.6		
Cadmium ∥	−2.0 ± 0.2	215–290	Alexander (1971)
⊥	−4.1 ± 0.4		
Magnesium	2.0 ± 0.3	500–580	Wohlgemuth (1973)
Trivalent metals			
Aluminum	−30 to −12	480–640	Penney (1964)
	Comparable	450–610	Heumann and Meiners (1966)
Indium	−11.5	115–150	Lodding (1965)
Gallium			
(liquid)	−1.3	18–312	Lodding (1967)
Thallium	−4.0 ± 0.5	233–303	Lodding et al. (1972)
Transition metals			
Nickel	−3.5	1000–1400	Hering and Wever (1967b)
Iron	+2 ± 1	700–1300	Hering and Wever (1967a)
Cobalt	+1.6 ± 0.3	1260–1360	Ho (1966)
Platinum	+0.28 ± 0.04	1480–1670	Huntington and Ho (1963)
Refractory metals			
Zirconium	+0.3	930–1730	Campbell and Huntington (1969)
Uranium	−1.6 ± 0.1	830–1100	D'Amico and Huntington (1969)
Quadrivalent metals			
Lead	$-47/f$	250	Kuz'menko (1962)
Tin	$-80/f$	190	Khar'kov and Kuz'menko (1960)
Tin	−18	180–213	Khosla (1973)

pronounced, presumably because grain size remains limited by the foil thickness during electromigration even at high temperatures.

The electromigration studies of silver show a rather large range of values for Z^*, with the recent measurements by the vacancy flux method (Patil and Huntington, 1970) well below the numbers obtained from the isotope studies (Khar'kov and Kuz'menko, 1960) (Doan and Brebec, 1970). Some of the early work on silver (Ho and Huntington, 1966) was done on a material of questionable purity. During these runs bubbles developed, mostly on the cathode side, and their number and size increased if the specimen were run in a hydrogen atmosphere. When the work was extended later (Patil and Huntington, 1970) with specimens of higher purity, it became necessary to impregnate the silver with oxygen to form visible bubbles. It turned out that this bubble formation occurred most readily during the recrystallization process when presumably the moving grain boundaries sweep the incipient steam bubbles together. For specimens of the highest purity (1 or 2 ppm impurities) bubbles did not form, even with oxygen impregnation, showing that a minimal number of impurity (presumably metal) atoms are necessary to act as nuclei for forming the bubbles. It was not possible to tell whether the greater density of bubbles on the cathode side was the result of bubble migration under the action of the electric field or whether densities of the interacting gases were sufficiently increased by electromigration (see Section V) that more bubbles were generated in this region.

Electromigration of copper demands excellent atmosphere control if the tendency for rapid oxidation of the surface is not to force a premature termination of the runs. The first two investigations of this metal (Wever, 1956; Grone, 1961) both reported that the sign of the effect changes in the high temperature region. Later investigations (Sullivan, 1967a; Grimme, 1971) have shown no evidence for this effect and it is a bit of a mystery what might have been the origin of the spurious findings. Incidently, the raw data (v_m/j vs T^{-1}) of Grimme is double that of Sullivan but they differ on assumptions on effect of transverse dimensional change on Z^*. Grimme (1971) went to the extent of exploring the influence of hydrogen and oxygen impurities in the copper. In neither case was any indication of a reversal observed although the presence of the oxygen did increase the activation energy for the self-electromigration.

B. Electromigration in Noble Metal Alloys

Over the years, the study of dilute impurities diffusing in matrices of the noble metals has been extensively and systematically pursued. The information so obtained has aided greatly to the understanding of interatomic relations in these systems. Naturally, a similar effort has been made to explore the systematic effect of impurity valence on electromigration. Here too the results have been illuminating. The work of Doan (1970) on the electromigration of impurities in silver is of particular importance.

Doan used the isothermal isotope method to measure the electromigration of the elements in the same row as silver in the silver solvent. His data for the Z^* are shown as functions of $z(z + 1)$, where z is the difference in nominal valence between the solute and silver (see Fig. 4). The wind force on a solute atom is taken to be roughly proportional to z^2 in accord with Mott's theory for Born scattering for impurities in a metal and with the experimental observations of Linde (1932). Correspondingly, the wind force for the atom at the saddle point is proportional to the square of its nominal valence or $(z + 1)^2$ since the valence of silver is taken to be one. For motion by the vacancy method, Z^* for the impurity depends on the average of the wind forces for the substitutional and saddle-point positions, hence the choice of $z(z + 1)$ for the abscissa. In Fig. 5, Doan shows the two contributions separately, the part for the substitutional position from Linde's data and the estimate for the saddle-point contribution by subtraction from his results. While his data tend to bear out this model well, for similar studies by Guilmin et al. (1973) on impurity electromigration in copper, the agreement is less convincing.

In Section II, it was stressed that the considerations which give rise to the correlation factor in diffusion become more complex for the case of a impurity under a driving force in that one must bring into the picture the

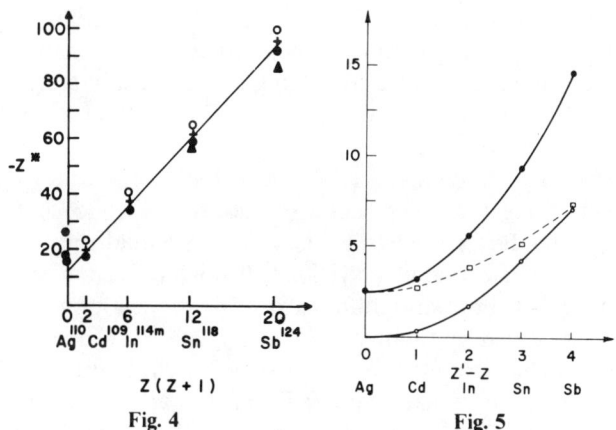

Fig. 4. Z^* as a function of $z(z + 1)$. (Courtesy of Doan, 1970.)

Fig. 5. Equilibrium and saddle-point contribution to Z^*: □ saddle-point contribution, ○ equilibrium configuration contribution, ● total Z^*. (Courtesy of Doan, 1970.)

effect of the vacancy flow induced by the Z^* of the solvent atoms. Only for self-electromigration does the influence of the vacancy flux exactly compensate the standard correlation factor and the tracer motion becomes strictly uncorrelated. For the electromigration of an impurity, the calculation of the corresponding flux follows from

$$J_B = -D_B^* \nabla n_B - |e| \nabla \phi\, D_B^*(Z_B^* + (L_{AB}/L_{BB})Z_A^*)n_B/kT$$

where $L_{AB}/L_{BB} = 2\langle n_p \rangle$. To define $\langle n_p \rangle$ as introduced by Manning (1968) one imagines an impurity on plane 0 and a vacancy ready to receive it on plane $+1$. The planes are normal to the electric field. If the vacancy moves away, it will cease to have any directed influence on the impurity when it (1) has moved far enough so that it has negligible chance to return, (2) has been annihilated at a sink, or (3) has reached the neutral plane 0. The n_p is defined as the average number of planes to the right of plane $+1$ where one of these three things happens. For the case of the fcc lattice, the expression for n_p is (Manning, 1965, 1968)

$$\langle n_p \rangle = [3\omega_3 - 2\omega_1 + (\omega_3/\omega_1)(\omega_0 - \omega_4)7(1 - F)]/(2\omega_1 + 7\omega_3 F)$$

where ω_1, ω_2, ω_3, and ω_4 represent, according to the standard notation, the relative jumping frequencies of the vacancy, respectively, for moving around the impurity, exchanging, dissociating, or recombining with the solute atom. The quantity F gives the fraction of the dissociating vacancies that never return. The value of $\langle n_p \rangle$ varies from -1 to $(1-f)/2f$ and, as explained in Section II, the negative value occurs when the vacancy tends to circumnavigate the solute, i.e., large ω_1/ω_3. An alternate attack on the influence of the vacancy wind (Huntington, 1969), which avoids the $\langle n_p \rangle$ concept, proceeds by writing out the kinetic equations for changing vacancy populations around the solute and solving for the steady state conditions. The J_B can then be directly determined. The method is conceptually simple but is somewhat cumbersome to apply.

Doan (1971) has made a special effort to look for vacancy flow effects in experiments using transition metal solutes, Mn, Fe, Co, and Ni in Ag. (The high Z^* and D^* of the earlier used solutes would make such efforts quite negligible.) The experiments yielded effective Z^*'s smaller than would have been expected on a straightforward basis in view of the incremental resistivities determined by Linde (1932) for impurities in silver. Perhaps the discrepancies can be attributed to the vacancy flux effect but the information on the ω_i is scarce and what changes Doan was able to compute seemed to be relatively minor. Doan (1972) has also pointed out that electromigration studies may also be useful in determining ω_i.

Studies of electromigration in the noble metals can also serve another purpose, namely to explore the effect of Z_{el}, the direct electrostatic coupling to the impurity motion. Because the Z_{wd} contribution varies as ρ^{-1}, one hopes to obtain Z_{el} as the zero intercept of a plot of Z^* vs ρ^{-1}. The most convincing work of this sort was carried out by Hehenkamp (1970, 1971) for Sb in Cu. The experiments were performed on welded tripartite specimens in which the central section contained the impurity up to 2–3% Sb. The concentration profile of the composite specimen was measured after the run with dc by microprobe with high accuracy. Figure 6 shows the overall results, with Z^* plotted against the reciprocal of the resistivity. The large scatter for the points at low resistivity (low impurity) is due to neglecting the change of diffusivity in this system. Taking this consideration into account and solving the ensuing nonlinear equation numerically for the specimens with higher concentration dramatically reduced the scatter as shown and allowed one to extrapolate to infinite resistivity. Presumably, the positive intercept so obtained can be identified with Z_{el}. In this case, the number is close to 4 whereas the nominal valence for Sb is of course 5. Similar experiments (Hehenkamp, 1971) for Sb and Sn in Ag give smaller intercepts but the extrapolation is also less convincing for these cases.

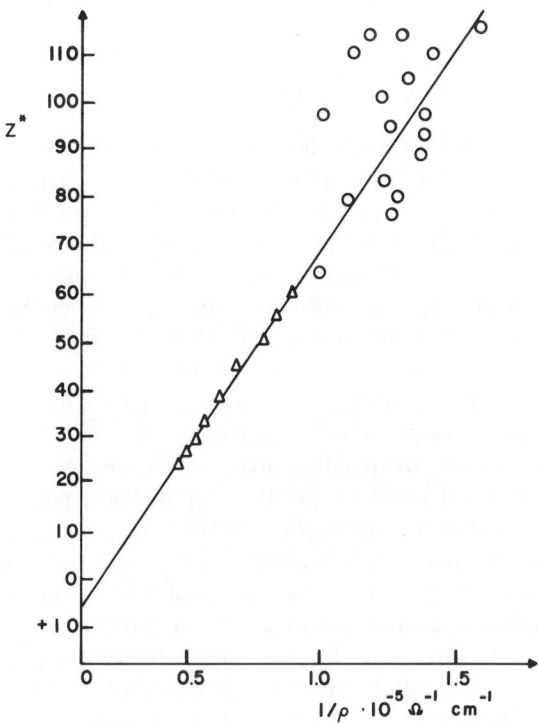

Fig. 6. Z^* versus reciprocal resistivity for Sb in Cu. (Courtesy of Hehenkamp, 1971.)

Hehenkamp also points out that there are several approximations in the application of theory which could cause small systematic errors that might nevertheless affect the extrapolation appreciably.

In regard to concentrated alloys involving the noble metals, there is one particularly complete investigation of the gold–silver system performed by Hofman and Guy (1972) which covers the whole range of composition. Since the D^*'s are also well known for this system, there is no trouble in determining the Z^*'s over the whole range and for both components. Although no effort is made to take the cross coupling transport coefficients into account in analyzing the data, this omission is not as serious a one as it would be in a system with more dissilimar constituents. The Z^*'s for both gold and silver turn out to be independent of composition within experimental accuracy and the Z^*_{Au} is only $\sim 10\%$ greater than Z^*_{Ag}. The value for Z^*_{Ag} for pure silver lies close to that of Patil and Huntington (1970). There has also been an interesting study of an intermetallic compound, Cu_3Sb, by Heumann et al. (1970). This is a system which is strongly defected off the stoichiometric ratio. A high density of vacancies is available and very rapid motion of the copper toward the cathode is reported.

C. Electromigration of the Alkali Metals

Studies of electromigration in the solid alkali metals have been practically limited to sodium and lithium. The first work on the former was done by Sullivan (1967b), using the vacancy flux method. A sample of the raw data is shown in Fig. 7. The mass motion is toward the anode. The slight asymmetry in the plot of the velocity versus marker position is evidence of an interesting effect that sometimes accompanies vacancy flux studies conducted in a substantial thermal gradient. It is the conjugate effect of the thermomigration or mass motion in a thermal gradient (Huntington, 1973). This mass flow may be either with or against the thermal gradient, depending on the sign of the appropriate off-diagonal coupling coefficient that relates the irreversible flows to the driving forces. In the case of sodium, the mass flow is up the thermal gradient and the asymmetrical part of the marker velocity curve is in the direction to indicate thickening of the specimen at the center where the temperature is highest. The fact that the thermal gradient is an asymmetric function about the specimen center whereas the electric field is symmetric, makes it simple to separate the two effects. The geometry of this experiment is not ideal for measuring thermomigration and it has been only in a limited number of cases (see Section IX) that it has been possible to get data on this effect as sort of bonus in the course of measuring electromigration by the vacancy flux method. The theory of

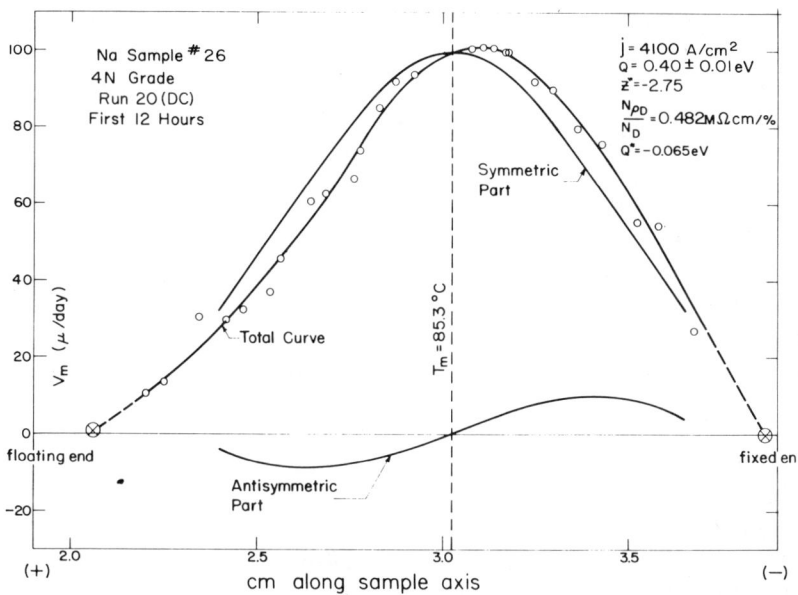

Fig. 7. Plots of the average velocities of fiducial scratches as a function of position along a sodium sample subject to electromigration. ○ represent data points. The best curve through the points has been resolved into its symmetric and antisymmetric parts as labeled. The antisymmetric part is caused by thermomigration. (Courtesy of Sullivan, 1967b.)

thermomigration is quite complex and, without going into detail, it can be remarked that the critical parameter is the heat of transport (less the heat of solution of the moving species). In the case of sodium, presumably in part at least a vacancy diffuser, (Mundy, 1971) one would expect that the flow of energy intrinsic in the vacancy formation would dominate the heat transport. Since this flow runs counter to the mass flow, its direction is at least qualitatively explained.

In passing, it might be mentioned that Sullivan's experience working with this metal provided a dramatic example of the importance of incorporating transverse measurements in a vacancy flux technique. Frequently, the early practice was to measure only marker motion and assume isotropic expansion. Initially, Sullivan cast his specimen about a ceramic tube containing a thermocouple. Taking the precaution of making one run without the benefit of the thermocouple assembly, he was amazed to find the marker motion enhanced by a factor of 4–6. Apparently, the inert ceramic greatly inhibited the axial expansion of the soft sodium. Later measurements, including those of transverse expansion, turned out to give much better reproducibility for Z^* independent of specimen constraints.

A good deal of work has been done on mass motion in lithium, both in the pure state and with dilute alloying, by the group at Chalmers Institute‡ under Lodding. The availability of two isotopes with large separation in relative mass and the capability of obtaining specimens with high purity in either ^{6}Li or ^{7}Li has made possible several interesting experiments, for example, in self-diffusion and thermomigration. The study of electromigration in pure lithium was performed by Thernquist and Lodding (1968) using the vacancy flux technique in vacuum rather than an inert atmosphere. The concomitant effect of thermomigration was even more pronounced in this investigation than it was for sodium, this time entering with a positive heat of transport, i.e., mass flow down the thermal gradient. The results for Z^* were particularly reproducible and showed a slow decrease with increasing temperature. On the assumption that $Z_{el} = 1$ in this case, substitution into Eqs. (8) and (11) gives a (ρ_d/N_d) equal to a 0.52 Ω-cm/% defect and independent of T within experimental error, as one might expect. It should be noted that this gratifying result depends rather crucially on taking the normal valence for Z_{el}.

There has been one interesting study of impurity electromigration in lithium, namely that of ^{22}Na as performed by Thermquist et al. (1972). The method involved analyzing the steady state profile of the radioisotope in a thin specimen and gave Z^* directly, since the diffusivity cancels out from both the force term and the back diffusion term (Section III). However, two corrections are involved: (1) transformation from barycentric coordinate system to the lattice frame which involves adding in the effect of the flowing solvent and (2) the usual correlation–vacancy flow factor which, in this case, is estimated to be small. The end result gave $Z^* = -4.7 \pm 1.9$ or about three times the Z^* for the solvent.

VII. Divalent Metals—Anisotropy in Single Crystals

Studies of electromigration in divalent metals have been limited to zinc, cadmium, and magnesium. For all three, the anisotropy of the effect has been investigated with single crystals.

For zinc, Routbort (1968) used the vacancy flux technique to measure electromigration in single crystals with orientations of 24°, 71°, and 83° for the angle between the c axis and the specimen axis. From flow data for these crystals, he could extrapolate quite accurately to the flow for orientations directly parallel or perpendicular to the hexagonal axis. The anisotropy ratio turned out to be about 1.35 in favor of flow in the basal plane. The result was somewhat unexpected, (1) because the conductivity

‡ Gothenburg, Sweden.

in zinc is only slightly anisotropic, being greater along the a axis and (2) because the diffusivity for zinc is some 40–50% greater along the a axis, a fact which implies a still greater anisotropy for the electromigration driving force. Following the usual procedure, one seeks to formulate the ratio of this driving force to the electric field in terms of a Z^* which now has tensor properties. However, the complications of crystalline tensors become trivial if one holds to an orthogonal system with one principal axis along the hexagonal direction, thereby putting all tensors in diagonal form. The ratio of mass to electron flow can then be expressed as the product of four tensors D^*, f^{-1}, Z^*e, and ρ^{-1}. The principal components of the diffusivity D^* and conductivity ρ^{-1} are known from experiment and, from the analysis of Mullen (1961), it appears that reasonable values for the correlation coefficients are $f_{\parallel} = 0.77$ and $f_{\perp} = 0.79$. On this basis, one finds $Z_{\parallel}^* = -1.95 \pm 0.16$ and $Z_{\perp}^* = -4.4 \pm 0.4$ or a quite large anisotropy ratio of 2.25 ± 0.3. The sign and magnitude of the anisotropy is established well outside of experimental accuracy. A measurement of a polycrystalline sample gave a value almost exactly two thirds the way between the parallel and perpendicular data—as might be expected.

The experiment gave another piece of information this time on the mechanism for formation and destruction of the vacancies. This information was obtained from the ratio of transverse to longitudinal dimensional change as a function of crystalline orientation. It was observed that these changes were always greater in the basal plane than along the c axis independent of the specimen orientation. Measurements by Gilder and Walmark (1969) on the thermal expansion of zinc and certain on-going x-ray measurements of lattice constant as a function of temperature‡ bear out the fact that the volume expansion from vacancy creation in zinc primarily affects the dimensions in the basal plane. It is postulated that this is a result of preponderance of active edge dislocations in the basal plane. The climb of these dislocations with consequent creation or destruction of vacancies would change only the dimensions at right angles to the hexagonal axis.

The electromigration of cadmium was investigated by Alexander (1971) with the results: $Z_{\parallel}^* = -1.5_4 \pm 0.15$ and $Z_{\perp}^* = -3.2_0 \pm 0.3$. Although the investigation paralleled that of zinc in most respects, there was one important difference. This was in regard to the influence of crystal orientation on the ratio of transverse to longitudinal dimensional change. For cadmium, the situation is the reverse of that for zinc in this respect, i.e., regardless of specimen orientation the dimensional changes were larger along

‡ Unpublished research initially by Apostolou and H. M. Gilder, continued by M. D. Current.

the c axis. The result is quite interesting in view of the rather markedly different plastic behavior shown by cadmium for which basal cleavage is difficult and basal slip is much less clearly indicated. Again the indications from the electromigration experiments are in agreement with the evidence from the thermal expansion–lattice constant measurements of Feder and Nowick (1972). The whole subject as to how such thermal expansion type measurements in noncubic crystals can be used to draw conclusions about the detailed mechanism of vacancy formation at the dislocations has been treated by Nowick and Feder (1972).

Work on magnesium‡ shows that the Z^* is negative and isotropic, but the experimental problem is complicated by the high reactivity of the metal and the apparently constraining influence of its oxide skin. For all three divalent metals, the Z^* are consistently negative even though the high temperature Hall coefficients may be of either sign.

The magnitude and direction of the electromigration anisotropy in zinc and cadmium is quite puzzling, especially in comparison with the behavior of the crystalline conductivity. Some attempt has been made to develop the theory (Huntington et al., 1971) to a point it could throw some light on the situation. Apparently the anisotropy of the phonon spectrum is the crucial consideration.

VIII. Electromigration in Trivalent Metals

The measurement of electromigration on aluminum was first made by Penney (1964) and later repeated by Heumann and Meiners (1966). The results subsequently had some indirect technological value since their existence in the literature enabled the engineers worried by failures of aluminum stripes in the units of integrated circuitry to eliminate motion in the bulk as the transport mechanism and to concentrate directly on intergranular and surface mechanisms.

The work done in our laboratory by the vacancy flux method (Penney, 1964) developed an interesting side study. We were just becoming aware of the importance of making measurements of changes in specimen diameter, as well as using marker positions to observe the longitudinal strains. In earlier work, we had assumed dilatational isotropy which meant that the ratio of the axial relative expansion to the total dilatation, a quantity which we have called α the anisotropy factor, would be approximately one third. With the aluminum, however, it was found that this was true only for long thin specimens and that, as the specimens were shortened, the α value tended to increase in the central portion. In interpreting these results, it was

‡ Wohlgemuth (1973).

assumed that the total dilatation depended only on the vacancy flux condition. How it was divided between longitudinal and transverse expansion could be strongly influenced by specimen geometry. In particular, in a short thick specimen the transverse expansion at a position along the rod where the dilatation was a maximum would be constrained by the mechanical stresses imposed by either side in which the dilatation was less. Measurements were made on several specimens of different diameter to length ratios (Fig. 8). The figure shows m, defined as the ratio of axial to transverse deformation, as a function of specimen radius divided by length [in the notation of Eq. (30)]:

$$m = (\delta \dot{d}/d_0) \div \nabla v_m = 2\alpha(1-\alpha)^{-1} \tag{36}$$

Although the errors on the individual measurement of m from the six specimens have wide limits, one can see that the points fall quite close to a parabolic relation, as predicted by the theory given in an appendix in Penney's paper. The proportionality constant for the parabolic relation of the theory involves both elastic and plastic parameters, so that the softer the material the more nearly isotropic will be the deformation.

The subsequent electromigration measurements by Heumann and Meiners (1966) on single crystals fall very closely on an Arrhenius plot of slope 1.16 eV. The Penney values are high by $\sim 20\%$ near 500°C but agree well at higher temperatures. Polycrystalline measurement by Heumann and Meiners also run higher at low temperatures and show more scatter than the single crystal data.

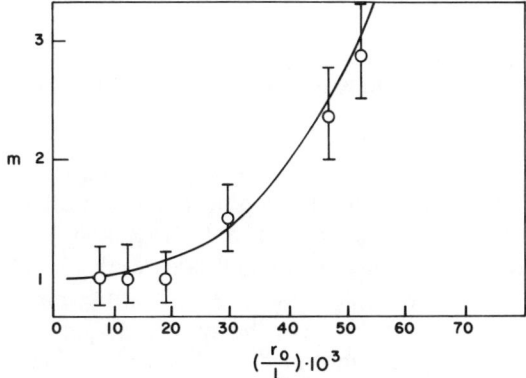

Fig. 8. The anisotropy index m is plotted for several aluminum specimens as a function of diameter to length ratio. Each specimen is represented by ○, a single experimental point. L is length of cylinder; r_0 radius of cylinder and difference of radii for hollow specimen. (—) gives the prediction of the theory from an appendix in the original paper. (Courtesy of Penney, 1964.)

As shown in Table II, electromigration has also been measured for pure indium and gallium by Lodding (1965, 1967). The indium work was on polycrystalline specimens and used only marker motion measurements for longitudinal deformation. The gallium study was for the liquid state and involved the separation of the isotopes ^{69}Ga and ^{71}Ga. Of course, in the liquid state there is no lattice frame available for reference nor is the vacancy flux a useful concept. Nevertheless it is known that high currents will give rise to the separation of isotopes. [The separation of isotopes in a pure metal by electromigration is called the Haeffner effect (Haeffner, 1953). It is observed generally that the lighter isotope is driven to the anode. For a while there was puzzlement as to how the centers of gravity of two different isotopes could be separated by a force which was presumably the same for both. The paradox was eventually explained by Klemm (1954). He pointed out that in a liquid the mobility of all the atoms is not the same, with those in the regions of lower density presumably being the greater. During such fluctuations, the atoms of lighter isotope will be more effectively influenced by the force field because of their higher intrinsic mobility arising from their higher vibration frequency.]

Gallium was a particularly attractive metal to study because of its low vapor pressure and extensive workable temperature range for the liquid state. Use of isotope separation to determine Z^* involves an estimate of the influence of isotopic mass on mobility, what is called E in the solid state diffusion field. Because of the highly cooperative nature of the diffusion step in liquids, this quantity is quite small. For the Z^* for liquid gallium obtained in this investigation, the value chosen for E was 0.2.

IX. Electromigration in Metals of More Complex Electronic Structure

A. TRANSITION METALS

Forced motion effects in the transition metals are often quite large. In the case of thermomigration for γ iron (Hering and Wever, 1967a) the effect is so large as to exceed the concomitant results for electromigration using the vacancy flux with dc power, i.e., the antisymmetric part of the marker displacement is greater than the symmetric part. The temperature dependence of these forced motions fitted an activation energy of 30.6 kcal/mol rather than the value of bulk-diffusion activation which was over twice as great. Therefore, in deducing the value for Z^* from these data the experimenters chose to take Q from measurements of grain boundary diffusion in γ iron and to quote as primary data only values for Z^*D_0. They assumed that the forced motion was primarily by short-circuit paths,

6. ELECTROMIGRATION IN METALS

perhaps in the grain boundaries, perhaps down isolated dislocations. Since the dislocation offers a nearly one-dimensional path, its correlation factor may be very small. If so, then its effect will be much more apparent in forced motion—which is uncorrelated—than in diffusion, where the correlation coefficient enters directly. This plausible hypothesis is interesting, not only because it points up the importance of dislocations in forced motion but also because it gives an indication that dislocation-augmented diffusion in γ iron is at least closely confined to the core region. The studies of forced motion in nickel, also reported by Hering and Wever (1967b) showed many of the same features: large concomitant thermomigration and abnormally small activation energy. However, they differed in one important respect in that the mass motion was for nickel in the direction of the anode in agreement with its negative sign for the (anomalous) Hall effect. The situation with nickel is rather puzzling since, as was pointed out earlier, the impurity interstitials move to the cathode in this metal.

Platinum (Huntington and Ho, 1963), on the other hand, shows a rather small positive Z^* although its Hall effect is negative. Evidently the correlation between the sign of Z^* and that of the Hall coefficient is weak whenever the latter is small. As a test case whether such correlation really did exist at all, cobalt was selected as a metal with one of the largest available Hall coefficients. The prediction was verified (Ho, 1966); its Z^* did turn out to be positive (1.6 ± 0.3).

B. ANOMALOUS BCC METALS

There has been a good deal of interest in atom movements with those so-called anomalous bcc metals which exhibit markedly nonlinear Arrhenius plots and unusually small activation energies.‡ A number of electromigration measurements has been carried out on zirconium (Campbell and Huntington, 1969), titanium (Dübler and Wever, 1968), and uranium (D'Amico and Huntington, 1969). In all cases, the bcc phase is reached only after passing through a low temperature phase of different structure.

The measurements on zirconium were of particular interest. It turned out that the Z^* was very small ($\sim +0.3$) so that it could only be detected unequivocally by long and carefully executed runs. On the other hand, the thermomigration effects were large and indicated a negative Q^* (material flowing to the hot center). Figure 9 shows this effect from the thermal gradient clearly. The existence of the phase transition to α-Zr at 860°C causes an additional interesting feature. The plot of the change in radial

‡ ASM Seminar Volume on "Diffusion in Body-Centered Cubic Metals," 1965. Amer. Soc. Metals, Metals Park, Ohio.

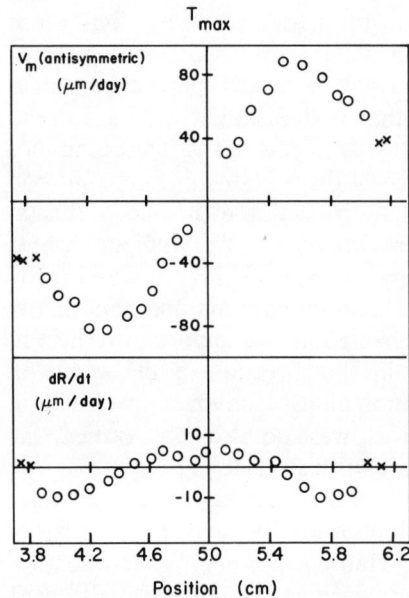

Fig. 9. Data taken on a zirconium specimen subject to a high dc current. Motion of fiducial scratches shown in upper plot, ○ in the β phase, x in the α phase. Lower plot shows change of specimen diameter with time. Both are plotted as a function of position along the specimen. Vertical axis indicate position of maximum temperature, hence position of thermal symmetry (Campbell and Huntington, 1969).

dimension (lower part of the figure) goes abruptly to zero at the transition temperature, indicating negligible mobility in the low temperature phase. The plot of marker motion versus position tells the same story and shows, in addition, an increase in specimen length. The latter effect is reversed for runs at lower maximum temperature where pronounced shortening is observed. Here the lengthening specimen (Fig. 9) shows marked discontinuities in the lateral displacement function at the phase boundaries. Presumably, the radial shrinkage of the bar just inside these boundaries supplies the material for the net elongation.

As one of the "anomalous bcc" metals, zirconium has a markedly nonlinear Arrhenius curve for the log of the diffusivity versus T^{-1}, as shown in Fig. 10. It turns out that this curve can be quite accurately and uniquely resolved into the sum of two exponential expressions (Kidson, 1965) but it is by no means clear as to what two distinct processes these lines may represent. We had some hope that the forced motion measurements might shed light on the mechanisms involved. In Fig. 10 are shown the data for DQ^*/f vs T^{-1}. Because of the nature of the experiment, the scatter is much greater than in the diffusion experiment. Nevertheless, it seems quite clear that the electromigration data come much closer to fitting a single exponential line and that this line is nearly parallel to the low

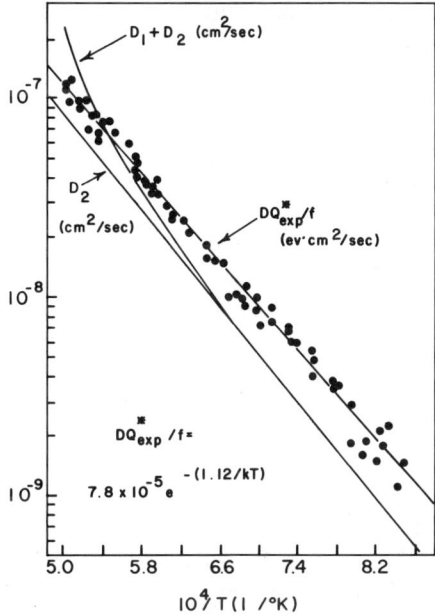

Fig. 10. Arrhenius plots for diffusion and thermomigration in Zr. The curved line represents a best fit to the diffusion data. It can be resolved into the sum of two Arrhenius terms, one of which (D_2) is shown. ● represent the electromigration results (arbitrary scale) fitted also with an Arrhenius line (Campbell and Huntington, 1969).

temperature component of the diffusion curve. Taking this at face value, one is led to surmise that the low temperature mechanism for diffusion dominates the forced motion at all temperatures. This line of reasoning immediately suggests short-circuit paths down dislocations as the appropriate mechanism which would be relatively more important for forced motion than for the random diffusion (see discussion of Fe and Ni in Section IX.A). This comes about because the correlation factor may be small indeed for diffusion down a dislocation (the motion of a vacancy down a strictly one-dimensional path will not cause any mixing action). The forced motion is undiminished by such considerations because in a pure material the motion of the vacancies is uncorrelated. Several competing mechanisms have been suggested but this suggestion of extensive dislocation line motion is a natural one to invoke for a crystal which has just come through a phase change on heating up to temperature. The suggestion was strengthened by a recent finding (Graham and Haines, 1970) that it is possible to decrease the diffusivity of the bcc phase substantially (by a factor of 3) with a prolonged anneal near the melting point (~1730°C) which presumably reduces the dislocation density. Because of the high temperature involved, it is not feasible to observe the dislocations in the bcc phase directly.

C. Quadrivalent Metals

Self-electromigration has been reported for both lead and tin (see Table II) using the technique of radioisotopes at the unwelded interface, a method which usually gives Z^* on the high side. Recently, the vacancy flux method has been applied to crystalline tin and a Z^*/f of 18 has been obtained.‡

Considerable interest has been attracted recently to the very rapid diffusion rates that have been observed for monovalent, and some divalent, impurities in these polyvalent solvents. It has been suggested (Dyson, 1966; Dyson et al., 1967) that these solutes move interstitially when exhibiting fast diffusion.§ However, this simple hypothesis has been complicated by subsequent suggestions of more complex mechanisms (Miller, 1969; Ascoli and Poletti, 1972; Warburton, 1972) devised to explain certain puzzling aspects uncovered by later experiments such as isotope effect studies, anisotropy of activation energy, and anomalous pressure dependence. Here the determination of the Z^* from electromigration may well be able to distinguish between a mechanism where a monovalent impurity moves solely interstitially and one where the impurity spends at least part of the time of the diffusion at or near a regular lattice site. The thinking is that, while the impurity acts as an interstitial, it has essentially a scattering charge of $z = +1$. While it occupies a regular lattice site, it behaves as a scattering center with charge of $z = -3$ and therefore has roughly nine times the scattering power. This difference should not be too difficult to observe.

Up to now there have been hardly any electromigration studies on such systems. The case of electromigration of radiosilver in lead has been explored by Kuz'menko et al. (1970), using an inert foil between two radioactively plated specimens. The Z^* did indeed turn out to be small, being negative at low temperatures (~280°C) but turning positive rather rapidly at higher temperature. Qualitative agreement with these results have been obtained by an investigator at the Chalmers Institute (Lodding, 1972). The rapid diffusivities of these systems make them ideal subjects for the steady state method of studying electromigration. This technique is being used to study the electromigration of gold into lead by Jeffery (1973).

‡ Khosla (1973).
§ See also Chapter 4.

6. ELECTROMIGRATION IN METALS

Preliminary results indicate a small negative Z^*, ~ -1 and not particularly temperature dependent. There is a wealth of similar systems for which this electromigration technique should prove an easy and valuable probe.

D. CONCENTRATED ALLOYS

Some quite extensive studies of concentrated alloys of complicated electronic structure have been carried out by a group at Kiev. They have investigated the completely soluble molybdenum–tungsten system and the iron–nickel system up to 2% nickel (Frantsevich et al., 1969) and the nickel–molybdenum system up to 23% molybdenum (Frantsevich et al., 1971). The measurements used the externally partially plated cylinder technique and are extended over large temperature ranges. By using isotopes of both constituents and an indentation marker to designate the original boundary to the plated region, the investigators could determine the velocity of both constituents with respect to the lattice. They next deduced values for the respective Z^*'s, using pure metal values for the correlation coefficients, and report that these Z^* are linear functions of the reciprocal resistivity. These relations lead to values for Z_{el} and (ρ_d/N_d) for each constituent at every composition. Interpretations on the basis of a two-band model lead, in most cases, to opposite signs for Z_{el} for each element and even different signs for the direction of the charge carrier wind. Although the authors comment that the situation can be understood in terms of two-band models, it seems more reasonable that perhaps in these cases the vacancy flux is very influential and of the sort to drive the two components in opposite directions (see Section II). However, one is forced to conclude that solid substitutional alloys for these complex band structures offer a much more complex picture for the understanding of charge carrier wind effect than the interstitial alloys (Section V) for which the broad overview seems to show considerable logical pattern.

X. Electromigration in Thin Films: Problem for Integrated Circuitry

Although the deleterious effects of electromigration and, to a lesser extent, thermomigration had been known for some time in incandescent wire filaments (Johnson, 1938) and dc devices carrying heavy current in space vehicles (O'Boyle, 1965), it came as a rude shock to the computer industry to discover that the elements of integrated circuitry were susceptible to massive failure in the metal stripes that serve as connecting wires in the highly miniaturized circuits. These stripes were in the form of aluminum films a few thousand Angstroms thick and typically a few microns wide and perhaps 100 μm long. The films were deposited on an insulating SiO_2 substrate and this intimate contact kept the foil temperature ~ 100–$200°C$ even though they carried current densities of 10^6 A/cm^2. The aluminum had many important advantages (mechanically tough, corrosion resistant, thermally compatible with the SiO_2, and easily handled by evaporation) but it was prone to void formation and failure after rather short lifetimes. It was immediately apparent from the results of Penney (1964) that the mass transport could not be a bulk process at such low temperatures and as a result the industry mounted an intensive research effort to explore grain boundary and surface electromigration.

The primary techniques in this exploration were resistance measurements of the various section of the stripe and electron optics. The resistivity measurements were sensitive enough to observe the onset of damage well before it was optically apparent. Blech and Meieran (1968) were the first to present a complete story by transmission electron microscopy of the actual failure. They dissolved away the substrate leaving the stripe freestanding and supported mainly by its oxide skin. The story by Ciné film showed bumps and hillocks at the anode side and small voids appearing at the cathode side. The voids appeared usually at the grain boundaries, grew, spread sideways, and finally coalesced to cause failure. Such deterioration was largely reversible and the reversal of the current could heal the voids and deflate the hillocks. In a similar way, Black (1969) viewed the same phenomena by scanning electron microscope. This presentation had two advantages: (1) it was three-dimensional and (2) it was not necessary to dissolve the substrate. The latter advantage meant that the stripe was under actual operating conditions, i.e., heat sunk and free of a large scale thermal gradient. As a result, the voids and hillocks were no longer localized at different ends but occurred quite homogeneously throughout the specimen. Figure 11 shows the growth of voids and one particular hillock in the aluminum film as seen by scanning electron microscope.

Measurements of failure rate as a function of temperature can be interpreted in terms of an activation energy ~ 0.6 eV, a value consistent with grain boundary diffusion. Black reports that large grained specimens showed an energy nearer 0.8 eV and that covering the free surface with a 1 μm layer of SiO_2 could raise the activation energy to about 1.2 eV in the range of the activation energy for bulk diffusion. Not all investigators are so convinced of the importance of the free surface motion. Rosenberg and Berenbaum (1971) have concentrated on the grain boundary aspects. According to them, there are two main causes for a locality to accrue or lose matter. The first is the presence of a grain triple point where three or more grain boundaries come together, since the flow divergence can be substantial at such a line. The second is a change in grain size down the

6. ELECTROMIGRATION IN METALS

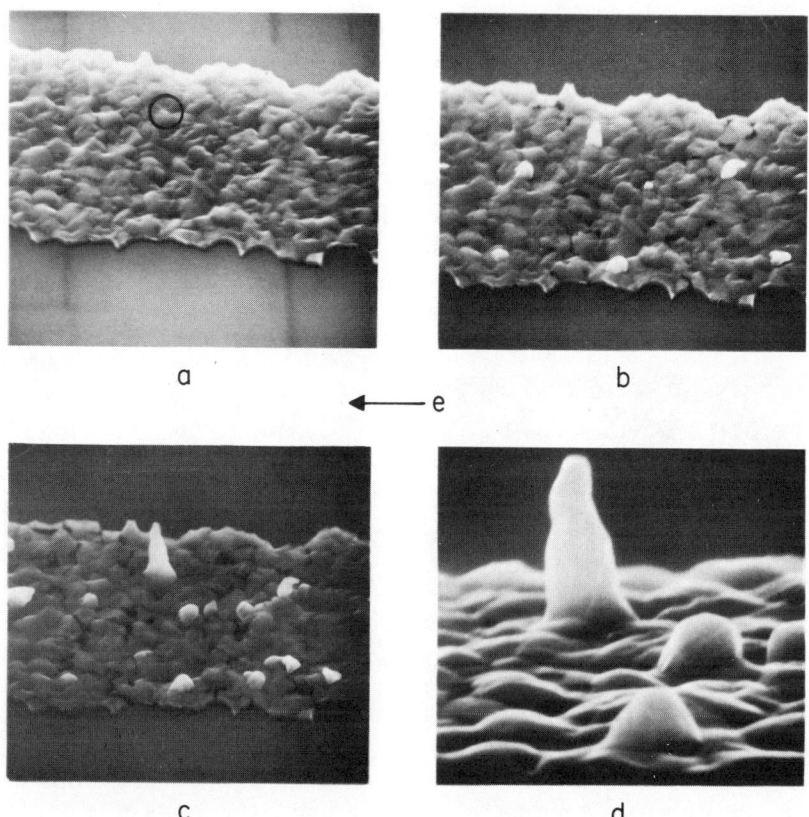

Fig. 11. Development of hillocks and voids in aluminum stripes by scanning electron microscope (Courtesy of Rosenberg and Berenbaum, 1971).

length of the stripe where here again one can expect that the mass transport will be greater in the region of the smaller grain size. The voids in general first appear at the grain boundaries and sometimes are seen to move along them. Occasionally, failure is not so much by the presence of voids as by general thinning of the stripe in the depletion region. In these cases, surface diffusion can be quite important and surface treatment beneficial. While single-crystal films are easily obtained epitaxially in the laboratory, it does not appear feasible to attempt them in production on the polycrystalline SiO_2 substrate.

In this intensive program of investigating electromigration in thin films, aluminum was not the only metal studied. Films of the noble metals, copper, silver, and gold, were all examined as well as chromium. In general, the impressive hillocks seen on aluminum (Fig. 11) were not observed. Instead, the surface of the metal became rippled after powering as can be seen in Fig. 12. It is suggested that the explanation for this difference may lie in the oxide film which forms on the aluminum and inhibits surface diffusion. As a result, the hillock which forms at some small region, perhaps a triple grain point, grows outward from there without spreading.

Several years ago, there was a report by Klotsman *et al.* (1962) that for electromigration at low temperatures ($\sim 500°C$) in silver the direction of motion was reversed and the mass flow was directed toward the cathode. It was presumed that Z^* for motion in the grain boundary was positive and a qualitative theoretical argument was later advanced as an explanation.

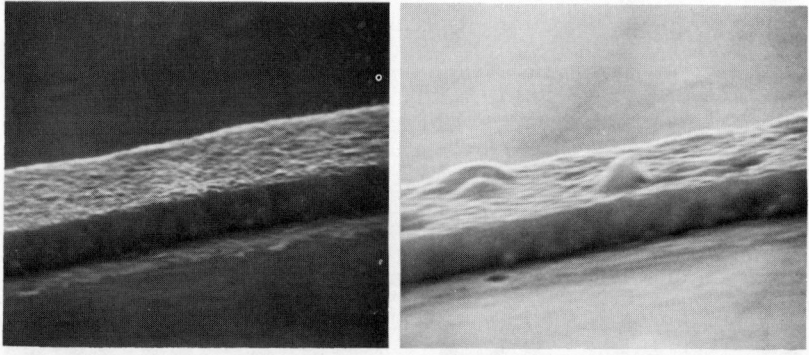

Fig. 12. Ripples of powering silver film by scanning electron microscope. (Courtesy of Rosenberg and Berenbaum, 1971).

In addition, the effect repeated for the motion of silver in other noble metals. Thin film workers who have experimented with the noble metals recently have differed as to the existence of this reversal. The principal evidence for the reversal is the work by Hummel and Breitling (1971) who have used three different techniques to follow electromigration in metals: resistance measurements, tracer studies, and scanning electromicrographs. Their findings show mass motion to the cathode in the noble metals, copper, silver, and gold, but the usual mass motion toward the anode in magnesium, lead, tin, and indium (Breitling and Hummel, 1972). Other observers, however, find little evidence for any reversal.

There have been some efforts to find among other metals a suitable substitute for aluminum in the connecting stripes. Since no other single metal promises quite the advantage of aluminum for this purpose, consideration has been given to composite films. One such is Cr–Ag–Au metallization system (Dong Kang *et al.*, 1969) which shows some aging effects above 300°C but low resistivity and good compatibility with n- and p-type silicon and no failures from electromigration. The Cr film is laid down first for cohesion, the Au film is for external protection, and the intervening Ag film prevents intermetallic formation.

The best answer to the technological problem developed so far appears by long odds to be the judicious use of impurities, such as copper, in aluminum films (Ames et al., 1970; d'Heurle et al., 1971). Impurities generally enhance diffusion in the bulk for most metals and for aluminum in particular; however, it appears that in the grain boundaries the impurities have an inhibiting effect. For the case of copper doping, the optimum results come at about 1 wt % and may correspond to an increase of device lifetime by a factor of 30. A model proposed by d'Heurle et al. (1971) postulates that only a small part of the impurity is needed at any time in the grain boundaries to block the flow of the matrix. However, the impurity itself is subject to electromigration, although with a higher activation energy for motion than the solvent. As in time the "old" impurities are swept out of the grain boundaries, new copper is dissolved from Al_2Cu precipitates into the bulk metal and finds its way to the grain boundary. In the main, the device lifetime is determined by this process and it is only when all the copper in a particular region is exhausted that the aluminum starts to move. This second stage in the process of deterioration then goes quite quickly. While the problem of electromigration damage still imposes some design restrictions on the use of integrated circuitry, the discovery of the beneficial effects of appropriate doping has done a good deal to make the situation less acute.

Acknowledgment

It is a pleasure to thank Professor Alexander Lodding for his critical reading of the manuscript and his thoughtful comments.

References

ADDA, Y., AND PHILIBERT, J. (1966). "La Diffusion dans les Solides," Vol. I, Chapter V, Bibliothèque des Sciences et Techniques Nucléaires, Presses Universitaires de France, Paris.
ALEXANDER, W. B. (1971). Z. Naturforsch. **26a**, 18.
AMES, I., D'HEURLE, F. M., AND HORSTMANN, R. (1970). IBM J. Res. Develop. **14**, 461.
ASCOLI, A., AND POLETTI, G. (1972). Phys. Rev. **B6**, 3681.
BIBBY, M. J., HUTCHINSON, L. C., AND YOUDELIS, W. V. (1966). Can. J. Phys. **44**, 2375.
BLACK, J. R. (1969), IEEE Trans. Electron Devices **16**, 338.
BLECH, I. A., AND MEIERAN, E. S. (1968). Appl. Phys. Lett. **12**, 201.
BOSVIEUX, C., AND FRIEDEL, J. (1962). J. Phys. Chem. Solids **23**, 123.

BREITLING, H. M., AND HUMMEL, R. E. (1972). *J. Phys. Chem. Solids* **33**, 845.
BRYANT, L. (1971). M. S. Thesis, Rensselaer Polytechnic Institute, Troy, New York.
CAMPBELL, D. R., AND HUNTINGTON, H. B. (1969). *Phys. Rev.* **179**, 601.
CARLSON, O. N., SCHMIDT, F. A., AND PEDERSON, D. T. (1966). *J. Less-Common Metals* **10**, 1.
CLAISSE, F., AND KOENIG, H. P. (1956). *Acta Met.* **4**, 650.
D'AMICO, J. F., AND HUNTINGTON, H. B. (1969). *J. Phys. Chem. Solids* **30**, 2607.
DAYAL, P., AND DARKEN, L. S. (1950). *Trans. Met. Soc. AIME* **188**, 1156.
DENBIGH, K. C. (1958). "The Thermodynamics of the Steady State." Methuen, London.
D'HEURLE, F. M., AINSLEE, N. G., GANGULIE, A., AND SHINE, N. C. (1971). Presented at the 5th Vacuum Congress, Boston, Massachusetts.
DOAN, N. V. (1970). *J. Phys. Chem. Solids* **31**, 2079.
DOAN, N. V. (1971). *J. Phys. Chem. Solids* **32**, 2135.
DOAN, N. V. (1972). *J. Phys. Chem. Solids* **33**, 2161.
DOAN, N. V., AND BREBEC, G. (1970). *J. Phys. Chem. Solids* **31**, 475.
DONG KANG, D., BURGESS, R. R., COLEMAN, N. G., AND KEIL, J. G. (1969). *IEEE Trans. Electron Devices* **16**, 356.
DÜBLER, H., AND WEVER, H. (1968). *Phys. Status Solidi* **25**, 109.
DYSON, B. F. (1966). *J. Appl. Phys.* **37**, 2375.
DYSON, B. F., ANTHONY, T. R., AND TURNBULL, D. (1967), *J. Appl. Phys.* **38**, 3408.
EASTMAN, D. E., CASHION, J. K., AND SWITENDECK, A. C. (1971). *Phys. Rev. Lett.* **27**, 35.
EINZIGER, R. E. (1973). Ph.D. Thesis, Rensselaer Polytechnic Institute, Troy, New York.
EINZIGER, R. E., AND HUNTINGTON, H. B. (1974). *J. Phys. Chem. Solids* **35**, 1563.
FEDER, R., AND NOWICK, A. S. (1972). *Phys. Rev.* **5**, 1244.
FEIT, M. D. (1969). Ph.D. Thesis, Rensselaer Polytechnic Institute, Troy, New York.
FIKS, V. B. (1959). *Sov. Phys.—Solid State* (English Trans.) **1**, 14.
FLYNN, C. P., AND STONEHAM, M. (1970). *Phys. Rev.* **B1**, 3966.
FRANTSEVICH, I. N., KALINOVICH, D. F., KOVENSKII, I. I., AND SMOLIN, M. D. (1969). *J. Phys. Chem. Solids* **30**, 947.
FRANTSEVICH, I. N., KALINOVICH, D. F., KOVENSKII, I. I., AND SMOLIN, M. D. (1971). In "Atomic Transport in Solids and Liquids" (A. Lodding and T. Lagenwall, eds.), pp. 100–103. Verlag der Zeitschrift für Naturforschung, Tübingen.
FROHBERG, G. (1971). In "Atomic Transport in Solids and Liquids" (A. Lodding and T. Lagenwall, eds.), p. 19. Verlag der Zeitschrift für Naturforschung, Tübingen.
GERARDIN, M. (1861). *Compt. Rend.* **53**, 727.
GILDER, H. M., AND LAZARUS, D. (1966). *Phys. Rev.* **145**, 507.
GILDER, H. M., AND WALMARK, G. N. (1969). *Phys. Rev.* **182**, 771.
GLINCHUK, M. D. (1959). *Ukr. Fiz. Zh.* **4**, 684.
GRAHAM, D., AND HAINES, E. R., JR. (1970). NASA Tech. Note TND-5905.
GRIMME, D. (1971). In "Atomic Transport in Solids and Liquids" (A. Lodding and T. Lagerwall, eds.), p. 65. Verlag der Zeitschrift für Naturforschung, Tübingen.
GRONE, A. R. (1961). *J. Phys. Chem. Solids* **20**, 88.
GUILMIN, P., TURBAN, L., GERL, M. (1973). *J. Chem. Phys. Solids* **34**, 951.
HAEFFNER, E. (1953). *Nature* **172**, 775.
HEHENKAMP, TH. (1962). *Arch. Eisenhüttenw*, **33**, 501.
HEHENKAMP, TH. (1968). *J. Appl. Phys.* **39**, 3928.
HEHENKAMP, TH. (1970). *Nikrochem. Acta (Wien), Suppl.* **IV**, 147.
HEHENKAMP, TH. (1971). In "Atomic Transport in Solids and Liquids" (A. Lodding and T. Lagerwall, eds.), pp. 68–72. Verlag der Zeitschrift für Naturforschung, Tübingen.
HERING, H., AND WEVER, H. (1967a). *Acta Met.* **15**, 377.
HERING, H., AND WEVER, H. (1967b). *Z. Phys. Chem.* **53**, 1.

6. ELECTROMIGRATION IN METALS

HEROLD, A., MARECHE, J. F., AND RAT, J. C. (1971). *C. R. Acad. Sci.* **273**, 1736.
HEUMANN, TH., AND MEINERS, H. (1966). *Z. Metallk.* **57**, 571.
HEUMANN, TH., MEINERS, H., AND STUER, H. (1970). *Z. Naturforsch.* **25a**, 1883.
HO, P. S. (1966). *J. Phys. Chem. Solids* **27**, 1331.
HO, P. S., AND HUNTINGTON, H. B. (1966). *J. Phys. Chem. Solids* **27**, 1319.
HOFMAN, G. L., AND GUY, A. G. (1972). *J. Phys. Chem. Solids* **33**, 2167.
HUMMEL, R. E., AND BREITLING, R. M. (1971). *Appl. Phys. Lett.* **18**, 3173.
HUNTINGTON, H. B. (1969). *Trans. Met. Soc. AIME*, **243**, 2571.
HUNTINGTON, H. B. (1973). ASM Seminar Volume on "Diffusion." Amer. Soc. Metals, Metals Park, Ohio.
HUNTINGTON, H. B., ALEXANDER, W. B., FEIT, M. D., AND ROUTBORT, J. L. (1971). *In* "Atomic Transport in Solids and Liquids" (A. Lodding and T. Lagerwall, eds.). Verlag der Zeitschrift für Naturforschung, Tübingen.
HUNTINGTON, H. B., AND GRONE, A. R. (1961). *J. Phys. Chem. Solids* **20**, 76.
HUNTINGTON, H. B., AND HO, S. C. (1963). *J. Phys. Soc. Japan Suppl. II* **18**, 202.
JEFFERY, R. N. (1973). *Bull. Amer. Phys. Soc. II* **18**, 428.
JOHNSON, R. P. (1938). *Phys. Rev.* **54**, 459.
JOST, W. (1952). "Diffusion in Solids, Liquids and Gases." Academic Press, New York.
JOST, W. (1960). "Diffusion in Solids, Liquids and Gases," 3rd ptg with Addendum, p. 49, Academic Press, New York.
JOUSSET, J. C., AND HUNTINGTON, H. B. (1969). *Phys. Status Solidi* **31**, 775.
KALENOVICH, D. F., KOVENSKII, I. I., AND SMOLIN, M. D. (1962). *Fiz. Metal. Metalloved.* **13**, 930.
KHAR'KOV, E. A., AND KUZ'MENKO, P. P. (1960). *Ukr. Fiz. Zh.* **5**, 428.
KHOSLA, A. (1973). Ph.D. Thesis, Rensselaer Polytechnic Institute, Troy, New York; *J. Phys. Chem. Solids*, in press.
KIDSON, G. V. (1965). "Diffusion in Body-Centered Cubic Metals," pp. 335–336. Amer. Soc. Metals, Metals Park, Ohio.
KLEMM, A. (1954). *Z. Naturforsch.* **92**, 1031.
KLOTSMAN, S. M., TEMOFEEV, A. M., AND TRAKHTENBERG, I. SH. (1962). *Phys. Metals Metallogr.* **14**, (5) 140.
KNACK, J., AND EICHENAUER, W. (1968). *Z. Naturforschung* **23a**, 1783.
KOVENSKII, I. I. (1963). *Sov. Phys.—Solid State* (English trans.) **5**, 1036.
KUZ'MENKO, P. P. (1962). *Ukr. Fiz. Zh.* **7**, 117.
KUZ'MENKO, P. P., KHAR'KOV, E. I., AND GRINEVICH, G. P. (1960). *Ukr. Fiz. Zh.* **6**, 525.
KUZ'MENKO, P. P., GRINEVICH, G. P., AND DANIL'CHENKO, B. A. (1970). *Fiz. Metal. Metalloved.* **29**, 318.
LANDAUER, R. (1957). *IBM J. Res. Develop.* **1**, 223.
LANDAUER, R., AND WOO, J. W. F. (1974). *Phys. Rev.* **B10**, 1266.
LECLAIRE, A. D., AND LIDIARD, A. B. (1955). *Phil. Mag.* (8), **1**, 518.
LINDE, J. O. (1932). *Ann. Phys.* **15**, 219.
LODDING, A. (1965). *J. Phys. Chem. Solids* **26**, 143.
LODDING, A. (1967). *J. Phys. Chem. Solids*, **28**, 557.
LODDING, A. (1972). Private communication.
LODDING, A., SULLIVAN, G. A., LARSSON, S. R., AND THERNQUIST, P. T. (1972). *Cryst. Lattice Defects* **3**, 29.
MANNING, J. R. (1965). *Phys. Rev.* **A139**, 2027.
MANNING, J. R. (1968). "Diffusion Kinetics for Atoms in Crystals," Van Nostrand-Reinhold, Princeton, New Jersey.
MILLER, J. W. (1969). *Phys. Rev.* **188**, 1074.

MULLEN, J. G. (1961). *Phys. Rev.* **124**, 1723.
MUNDY, J. N. (1971). *Phys. Rev.* **3**, 2431.
NOWICK, A. S., AND FEDER, R. (1972). *Phys. Rev.* **5**, 1238.
O'BOYLE, D. (1965). *J. Appl. Phys.* **36**, 2849.
OKABE, T., AND GUY, A. G. (1970). *Met. Trans.* **1**, 2705.
ORIANI, R. A., AND GONZALES, O. D. (1967). *Trans. Met. Soc. AIME* **239**, 1041.
PATIL, H. R., AND HUNTINGTON, H. B. (1970). *J. Phys. Chem. Solids* **31**, 463.
PENNEY, R. V. (1964). *J. Phys. Chem. Solids* **25**, 335.
PETERSON, D. T., AND SCHMIDT, F. A. (1969). *J. Less-Common Metals*, **18**, 111.
PETERSON, D. T., AND SCHMIDT, F. A. (1972). *J. Less-Common Metals* **29**, 321.
PETERSON, D. T., SCHMIDT, F. A., AND VERHOEVEN, J. D. (1966). *Trans. Met. Soc. AIME* **236**, 1311.
PRIGOGINE, I. (1967). "Introduction to the Thermodynamics of Irreversible Processes," 3rd ed. Wiley (Interscience), New York.
ROSENBERG, R., AND BERENBAUM, L. (1971). *In* "Atomic Transport in Solids and Liquids" (A. Lodding and T. Lagerwall, eds.), p. 113. Verlag der Zeitschrift für Naturforschung, Tübingen.
ROUTBORT, J. L. (1968). *Phys. Rev.* **176**, 796.
RUDMAN, P. S. (1965). Presented at 94th AIME Meeting, Chicago, Illinois.
SCHMIDT, F. A., AND CARLSON, O. N. (1972). *J. Less-Common Metals* **26**, 247.
SCHMIDT, F. A., AND WARNER, J. C. (1967). *J. Less-Common Metals* **13**, 493.
SCHMIDT, F. A., CARLSON, O. N., AND SWANSON, C. E., JR. (1970). *Met. Trans.* **1**, 1371.
SCHWARZ, K. E. (1940). "Elektrolytische Wanderung in flüssigen und festen Metallen" (J. A. Barth, trans., 1945, Leipzig). Edwards, Ann Arbor, Michigan.
SEITH, W. (1955). "Diffusion in Metallen: Platzwechselreaktionen." Springer-Verlag, Berlin and New York.
SEITH, W., AND DAUR, TH. (1938). *Z. Elektrochem.* **44**, 256.
SEITH, W., AND WEVER, H. (1953). *Z. Elektrochem.* **59**, 942.
SKAUPY, F. (1914). *Verh. Deut. Phys. Ges.* **16**, 156.
SORBELLO, R. S. (1970). Ph.D. Thesis, Stanford Univ., California.
SORBELLO, R. S. (1973). *J. Phys. Chem. Solids* **34**, 937.
STONEHAM, A. M. (1972). Rep. T. P. 480, Theoretical Physics Division AERE, Harwell.
STONEHAM, A. M., AND FLYNN, C. P. (1973). *J. Phys. F.* **3**, 503.
SULLIVAN, G. A. (1967a). *J. Phys. Chem. Solids* **28**, 347.
SULLIVAN, G. A. (1967b). *Phys. Rev.* **154**, 605.
SUSSMAN J. A. (1967). *J. Phys. Chem. Solids* **28**, 1643.
SUSSMAN, J. A. (1971). *Ann. Phys.* **6**, 133.
THERNQUIST, P., AND LODDING, A. (1968). *Z. Naturforsch.* **23a**, 627.
THERNQUIST, P., KARRQUIST, CH., AND LODDING, A. (1972). *Phys. Status Solidi* (a) **9**, 171.
VERHOEVEN, J. (1963). *Met. Rev.* **8**, 311.
WARBURTON, W. K. (1972). *Phys. Rev.* **B6**, 2161.
WAGNER, C., AND HELLER, G. (1940). *Z. Phys. Chem.* **B46**, 242.
WEVER, H. (1956). *Z. Elektrochem.* **60**, 1170.
WEVER, H. (1959). "The Physical Chemistry of Metallic Solutions and Intermetallic Compounds," 9th NPL Symp., Paper 2L, H.M's Stationery Office, London.
WEVER, H., AND SEITH, W. (1955). *Z. Elektrochem.* **59**, 942.
WOHLGEMUTH, J. (1973). Ph. D. Thesis, Rensselaer Polytechnic Institute; Troy, New York; *J. Phys. Chem. Solids*, in press.

7

Atom Currents Generated by Vacancy Winds

T. R. ANTHONY

METALLURGY AND CERAMICS LABORATORY
RESEARCH AND DEVELOPMENT CENTER
GENERAL ELECTRIC COMPANY
SCHNECTADY, NEW YORK

List of Symbols	353
I. Introduction	355
II. Theory	355
A. The Phenomenological Diffusion Equations of Irreversible Thermodynamics	355
B. Some Comments about the Phenomenological Coefficients	357
C. Relationship of the Phenomenological Coefficients with the Diffusion Constants	358
D. Relationship of the Phenomenological Coefficients with Coefficients Derived from Atomic Models	359
E. Thermodynamic Forces in Dilute Solutions	361
F. Atom Currents and Vacancy Winds in Dilute Solutions	364
III. Measurement of the Vacancy Wind and the Wind-Generated Solute Current	369
A. Alloy Systems	369
B. The Vacancy Wind	371
C. The Solute Current Generated by the Vacancy Wind	372
IV. Solute Segregation around Vacancy Sinks	374
References	378

List of Symbols

- A   Solvent atoms
- B   Solute atoms
- V   Vacancies
- J_A Solvent current
- J_B Solute current
- J_V Vacancy current (wind)
- X_A Thermodynamic force on solvent A
- X_B Thermodynamic force on solute B
- X_V Thermodynamic force on vacancies V
- L_{ij} Phenomenological coefficients $i = A, B, V; j = A, B, V)$

L_{AA}^f $4Na^2W_0C_V/kT$
C_A Atomic fraction of solvent A
C_B Atomic fraction of solute B
C_V Atomic fraction of vacancies
C_p Atomic fraction of solute–vacancy pairs
D_A Diffusion coefficient of solvent A
D_B Diffusion coefficient of solute B in the alloy
D_B^* Tracer diffusion coefficient of B in pure solvent A
D_V Vacancy diffusion constant in the alloy
D_V^f Diffusion constant of free unpaired vacancies in the alloy
D_p Diffusion coefficient of solute–vacancy pairs in the alloy
μ_A Chemical potential of solvent A in the alloy
μ_B Chemical potential of solute B in the alloy
μ_V Chemical potential of the vacancies in the alloy
μ_A^0 Chemical potential of pure A
μ_B^0 Chemical potential of pure B
γ_A Activity coefficient of the solvent A in the alloy
γ_B Activity coefficient of the solute B in the alloy
∇ Gradient of
W_0 Jump frequency of a vacancy from a position unassociated with any solute to another similar unassociated position
W_1 Jump frequency of a vacancy from a position neighboring a solute atom to a similar associated position via an exchange with a solvent atom
W_2 Jump frequency of a vacancy from a position neighboring a solute atom to a similar associated position via an exchange with a solute atom
W_3 Jump frequency of a vacancy from a position neighboring a solute to a position unassociated with a solute atom
W_4 Jump frequency of a vacancy from a position unassociated with a solute atom to a position neighboring a solute atom
G Gibbs free energy of the solution
$G_0(P, T)$ Gibbs free energy of the pure solvent
g_B Gibbs free energy of solution of solute B in solvent A
g_V Free energy of formation of a vacancy in the solution at a site not associated with a solute atom
Δg_V Free energy of association of a solute–vacancy pair
N_A Number of A atoms
N_B Number of B atoms
N_V Number of vacancies
N_S Total number of lattice sites in the solution
Ω Entropy factor
p Fraction of solute atoms present in solute–vacancy pairs
z Number of nearest neighbors in the solution
K Mass action constant for the solute–vacancy pairing reaction
V_c Volume of a vacancy condensation cavity
S Surface area of a mathematical surface around a vacancy condensation cavity
C_B^0 Initial solute concentration before cavity formation
C_B^f Final solute concentration after cavity formation
U The velocity of the local lattice towards a vacancy sink
J_B' The solute current with respect to a vacancy sink
J_V' The vacancy current with respect to a vacancy sink

I. Introduction

A vacancy is an empty lattice site in a crystalline solid. A vacancy wind, therefore, represents a flow or current of these empty lattice sites through the crystalline solid. Vacancy winds can arise in a number of situations. In some cases, vacancy winds are the direct result of vacancy concentration gradients. These cases include sintering, quenching, and irradiation. In other instances, vacancy winds occur in the absence of vacancy concentration gradients. Kirkendall-type experiments (Jost, 1952) and electromigration experiments (Huntington, 1973) are examples of the latter type of situation.

By its very nature, the flow of vacancies through a lattice requires an equal and opposite flow of atoms. If migrating vacancies could not distinguish between solute and solvent atoms in a solution, then this reverse flow of solute and solvent atoms would be in simple proportion to the respective atomic fractions of solute and solvent in the solution. In real solutions, however, vacancies interchange with solute and solvent atoms with different jump frequencies. The resulting preferential interchange of a migrating vacancy with a particular constituent of the solution will give rise to a reverse atom flow containing a disproportionate number of these favored atoms. The vacancy wind will thus enrich regions from where the wind is coming (vacancy sources) and deplete regions to which the wind is going (vacancy sinks) of these favored atoms. An initially homogeneous substitutional solid solution, consequently, will develop solute and solvent concentration gradients in a vacancy wind.

In this chapter our goal will be to examine and to describe the directions and the magnitudes of solute and solvent atom currents generated by vacancy winds and to investigate the resulting segregation of solute and solvent atoms in a solution subjected to such winds.

II. Theory

A. The Phenomenological Diffusion Equations of Irreversible Thermodynamics

To describe the transport of atoms and vacancies through solids, we will use the phenomenological description of irreversible thermodynamics (deGroot and Mazur, 1962). Consider a dilute alloy containing solvent atoms A, solute atoms B, and vacancies V. Under isothermal conditions, the most general expressions for the currents J_A, J_B, and J_V

of solvent atoms A, solute atoms B, and vacancies V, respectively, are

$$J_A = L_{AA} X_A + L_{AB} X_B + L_{AV} X_V$$
$$J_B = L_{BA} X_A + L_{BB} X_B + L_{BV} X_V \qquad (1)$$
$$J_V = L_{VA} X_A + L_{VB} X_B + L_{VV} X_V$$

where the currents are expressed in the frame of reference of the local crystal lattice of the solution. X_A, X_B, and X_V are the thermodynamic forces on atoms A and B and vacancies V, respectively. In the absence of external forces and under isothermal conditions, these thermodynamic forces are simply the negative of the chemical potential gradients of the constituents of the solution. L_{AA}, L_{AB}, L_{AV}, L_{BA}, L_{BB}, L_{BV}, L_{VA}, L_{VB}, and L_{VV} are the phenomenological coefficients and, as far as irreversible thermodynamics is concerned, are simply ratios of various fluxes and thermodynamic forces measured in experiments. Because vacancies will not, in general, be in equilibrium when a vacancy wind is blowing, it has been necessary to include vacancies explicitly in Eqs. (1) to obtain a complete description of our system for the situations of interest in this chapter.

Equations (1) state that a current of a particular constituent of the ternary A-B-V solution can, in general, be generated not only by its own chemical potential gradient but by the chemical potential gradients of the other two components of the solution. In this regard, the phenomenological coefficients L_{AV}, L_{BV}, and L_{VV} will be of particular interest in this chapter since these coefficients are measures of the ability of a vacancy wind J_V to generate current flows of atoms A and B.

The phenomenological equations in the form of Eqs. (1) contain only two physical factors. First, there is the *equilibrium condition* of no currents ($J_A = J_B = J_V = 0$) when the forces vanish ($X_A = X_B = X_V = 0$). Second, there is the *linear relation* between diffusion fluxes J and thermodynamic forces X that is observed experimentally.

At this point, it is advantageous to introduce three other physical constraints into the phenomenological equations in order to reduce the number of independent phenomenological coefficients and thus the apparent complexity of Eqs. (1). First, in local volumes of the solution away from vacancy sources and sinks, *lattice sites are conserved*. Thus for each vacancy that enters (leaves) a region, an atom must also leave (enter).

$$J_V = -(J_A + J_B) \qquad (2)$$

A second constraint is one of *mechanical equilibrium*. That is, the weighted sum of all the forces acting on the various constituents of the diffusing solid must equal zero,

$$C_A X_A + C_B X_B + C_V X_V = 0 \qquad (3)$$

where C_A, C_B, and C_V are, respectively, the atomic fractions of atoms A and B and vacancies V in the solution.

A third physical constraint of the phenomenological equations is the *microscopic reversibility* of the diffusion process (Denbigh, 1950). Microscopic reversibility implies that the diffusion and intermixing of atoms on an atomic scale occurs simultaneously in both directions in any particular situation so that the observed macroscopic fluxes are only the net differences between larger opposing atomic fluxes. For example, vacancies in a solid containing a vacancy gradient jump and exchange with neighboring atoms on both the uphill and downhill sides of a vacancy gradient. Only the net difference between the uphill and downhill jumps of the vacancies cause a net macroscopic flow of vacancies down the vacancy gradient. This net flow is clearly different from a unidirectional flow situation where all vacancy jumps were restricted to a direction down the vacancy gradient.

By including the three physical constraints of conservation of lattice sites [Eq. (2)], mechanical equilibrium [Eq. (3)], and microscopic reversibility, the number of independent phenomenological coefficients in Eq. (1) can be reduced from nine to three (Hu, 1969). For example, a sufficient condition that $(J_A + J_B) = -J_V$ is that $L_{AA} + L_{BA} + L_{VA} = 0$, $L_{AB} + L_{BB} + L_{VB} = 0$, and $L_{AV} + L_{BV} + L_{VV} = 0$. Microscopic reversibility allows the use of the Onsager reciprocal relations $L_{BA} = L_{AB}$, $L_{VA} = L_{AV}$, $L_{VB} = L_{BV}$ (Onsager, 1931). Equations (1) then become

$$J_A = L_{AA} X_A + L_{AB} X_B - (L_{AA} + L_{AB})X_V$$
$$J_B = L_{AB} X_A + L_{BB} X_B - (L_{BB} + L_{AB})X_V \quad (4)$$
$$J_V = -(J_A + J_B)$$

or, as is it is more commonly written,

$$J_A = L_{AA}(X_A - X_V) + L_{AB}(X_B - X_V)$$
$$J_B = L_{AB}(X_A - X_V) + L_{BB}(X_B - X_V) \quad (5)$$
$$J_V = -(J_A + J_B)$$

In Eqs. (4) and (5), we have decided to use L_{AA}, L_{AB}, and L_{BB} as the independent phenomenological coefficients since these appear most commonly in the diffusion literature. The phenomenological coefficients $L_{AV} = -(L_{AA} + L_{AB})$, $L_{BV} = -(L_{BB} + L_{AB})$, and $L_{VV} = L_{AA} + L_{BB} + 2L_{AB}$) which describe the coupling between the atom currents and the vacancy wind could have been chosen instead.

B. Some Comments about the Phenomenological Coefficients

It is appropriate at this point to make some brief remarks about the phenomenological coefficients. Phenomenalism, it is recalled, is a theory that proposes that knowledge be limited to observable facts and events. In diffusion, such observable facts and events are the measurable *chemical potential gradients* of the various constituents in the solution (the thermodynamic forces X_A, X_B, and X_V) and the *mass transport* of the solution constituents in these chemical potential gradients. Note that both the concentration [by density and lattice parameter determinations (Simmons and Balluffi, 1960)] and the mass transport [by measurement of the motion of inert markers (Jost, 1952)] of vacancies in solids can be directly determined, in principle. The phenomenological coefficients are simply empirical parameters which give the ratios of the various diffusion currents and chemical potential gradients that are observed in experiments. By including other observable facts in our formalism such as the linearity between fluxes and forces, the equilibrium condition for no fluxes, the conservation of lattice sites, and the principle of microscopic reversibility, the number of these empirical parameters that must be measured in an experiment has been reduced from a possibly infinite number to only three. However, it is important to note that the phenomenology of nonequilibrium thermodynamics goes no further and says nothing about the magnitude, concentration dependence, or temperature dependence of the remaining three independent phenomenological coefficients. Because such information is essential in making approximations required for mathematical tractability, we must turn either to experiments or to competing atomic kinetic theories for this information about the phenomenological coefficients.

C. Relationship of the Phenomenological Coefficients with the Diffusion Constants

Fick's first law, $J_i = -D_i \partial(NC)/\partial x$, is a more familiar phenomenological description of diffusion in a solid. Since the phenomenological coefficients of Fick's first law (the diffusion constants D_i) are more commonly tabulated in the literature than are the phenomenological coefficients of irreversible thermodynamics (L_{AA}, L_{BB}, and L_{AB}), the relationship between these two sets of coefficients has been obtained (Howard and Lidiard, 1964).

Consider an isothermal system in which vacancies are everywhere in equilibrium ($X_V = 0$). The currents of atoms A and B are then from Eqs. (4),

$$J_A = L_{AA} X_A + L_{AB} X_B, \quad \text{and} \quad J_B = L_{AB} X_A + L_{BB} X_B \quad (6)$$

The requirement of mechanical equilibrium, Eq. (3), reduces Eqs. (6) to

$$J_A = (L_{AA} - (C_A/C_B) L_{AB}) X_A, \quad \text{and} \quad J_B = [L_{BB} - (C_B/C_A) L_{AB}] X_B \quad (7)$$

The chemical potentials of the constituents A and B are

$$\mu_A = \mu_A^0(P, T) + kT \ln \gamma_A C_A, \quad \text{and} \quad \mu_B = \mu_B^0(P, T) + kT \ln \gamma_B C_B \quad (8)$$

where μ_A^0 and μ_B^0 are the reference chemical potentials of pure A and B, respectively, and γ_A and γ_B are, respectively, the activity coefficients of A and B in the solution. Since $X_A = -\nabla \mu_A$ and $X_B = -\nabla \mu_B$, Eqs. (7) become

$$J_A = -(kT/C_A)[L_{AA} - (C_A/C_B) L_{AB}](1 + \partial \ln \gamma_A / \partial \ln C_A) \nabla C_A$$
$$J_B = -(kT/C_B)[L_{BB} - (C_B/C_A) L_{AB}](1 + \partial \ln \gamma_B / \partial \ln C_B) \nabla C_B \quad (9)$$

A comparison of Eqs. (9) in the notation of irreversible thermodynamics with the phenomenological equations of Fick's first law,

$$J_A = -D_A \nabla(N C_A), \quad \text{and} \quad J_B = -D_B \nabla(N C_B) \quad (10)$$

yields the desired relationship between the phenomenological coefficients and the diffusion coefficients (Howard and Lidiard, 1964).

$$D_A = (kT/N)(L_{AA}/C_A - L_{AB}/C_B)(1 + \partial \ln \gamma_A / \partial \ln C_A)$$
$$D_B = (kT/N)(L_{BB}/C_B - L_{AB}/C_A)(1 + \partial \ln \gamma_B / \partial \ln C_B) \quad (11)$$
$$D_V = (kT/NC_V)(L_{AA} + L_{BB} + 2L_{AB})$$

D_V has been obtained in a manner similar to that used in Eqs. (6)–(10). From Eqs. (11), the magnitude, concentration dependence, and temperature dependence of L_{AA}, L_{BB}, and L_{AB} can be determined if D_A, D_B, and D_V are known. This knowledge allows one to make suitable approximations to solve otherwise mathematically intractable problems and to predict the behavior of moving atoms in a variety of experiments.

D. Relationship of the Phenomenological Coefficients with Coefficients Derived from Atomic Models

An alternative approach to the phenomenological description of diffusion is the modeling of diffusion processes in terms of atomic random-walk theories. These theories construct models in which different jump frequencies are assigned to vacancies jumping between the various classes of lattice sites in the solution. Composition gradients are then introduced into the model and the net flux of atoms and vacancies are calculated in terms of the different vacancy jump frequencies assumed by the model. From these detailed models, the ratios of the various currents to the various concentration gradients can be obtained in terms of the different vacancy jump frequencies and the concentrations of the solution constituents. These ratios derived from an atomic model are equivalent to the phenomenological coefficients L_{AA}, L_{AB}, L_{BB}, etc. These expressions for the phenomenological coefficients are very useful for two reasons. First, they give us explicit information about the magnitude, concentration dependence, and temperature dependence of the phenomenological coefficients. Such information is used repeatedly in later sections of this chapter. Secondly, these expressions allow us to determine the various vacancy jump frequencies in the solution, consistent with the atomic model, if the experimental phenomenological coefficients are known. Such information not only helps in further atomic modeling of the solution but also in estimating such parameters as the interaction energy between solute atoms and vacancies.

Various atomic models have been used to derive expressions for the phenomenological coefficients. We will simply report expressions for L_{AA}, L_{AB}, and L_{BB} derived from one such model (Howard and Lidiard, 1963) and refer the interested reader to the extensive literature on this subject (Manning, 1968). This particular atomic model assumes a dilute face-centered solid solution of B in A in which vacancies V and solute atoms B interact through nearest neighbor forces to form solute–vacancy pairs (Lidiard, 1955). Five different vacancy jump frequencies W_i are used in this model (see Fig. 1). The expressions derived for the three independent phenomenological coefficients L_{AA}, L_{BB}, and L_{AB} from this model are

$$L_{AA} = -\frac{Na^2 C_p}{3kT}\left(-\frac{W_3 W_1 + 40 W_3{}^2 + 14 W_2 W_3 + 4 W_1 W_2}{W_1 + W_2 + 7W_3/2} + \frac{7W_3 W_0}{W_4}\right)$$
$$+ \frac{4N_a{}^2 W_0 (C_V - C_p)}{kT}$$

$$L_{AB} = \frac{Na^2 C_p}{3kT}\frac{W_2(-2W_1 + 3W_3)}{W_1 + W_2 + 7W_3/2} \qquad (12)$$

$$L_{BB} = \frac{Na^2 C_p}{3kT}\frac{W_2(W_1 + 7W_3/2)}{W_1 + W_2 + 7W_3/2}$$

Here N is the number of atoms per unit volume, kT is Boltzmann's constant times the absolute temperature, a is the lattice parameter of the fcc solution, and C_p and C_V are, respectively, the atomic fractions of solute–vacancy pairs and total vacancies in the solution. W_0 is the jump frequency of a vacancy from a position unassociated with any solute to another similar unassociated position; W_1 the jump frequency of a vacancy from a position neighboring a solute atom to a similar associated position via an exchange with a solvent atom; W_2 the jump frequency of a vacancy from a position neighboring a solute atom to a similar associated position via an exchange with a solute atom; W_3 the jump frequency of a vacancy

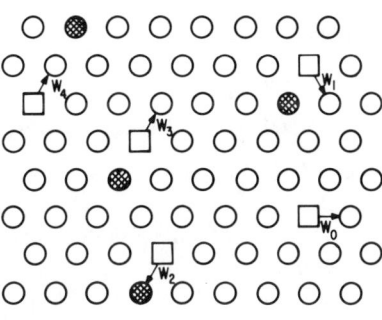

Fig. 1. The jump frequencies W_0, W_1, W_2, W_3, and W_4 of vacancies in various situations in a dilute fcc alloy.

☐ = VACANCY
⬣ = IMPURITY ATOM B
○ = SOLVENT ATOM A

from a position neighboring a solute to a position unassociated with a solute atom; and W_4 the jump frequency of a vacancy from a position unassociated with a solute atom to a position neighboring a solute atom.

E. Thermodynamic Forces in Dilute Solutions

In the absence of temperature gradients and external forces, the thermodynamic forces X_A, X_B, and X_V acting on constituents A, B, and V, respectively, are

$$X_A = -\nabla \mu_A, \qquad X_B = -\nabla \mu_B, \qquad X_V = -\nabla \mu_V \qquad (13)$$

where μ_A, μ_B, and μ_V are, respectively, the chemical potentials of solvent A, solute B, and vacancies V. To determine these three chemical potentials, let us find the Gibbs free energy G of a dilute solution containing solvent A, solute B, and vacancies V. We will require that this solution be sufficiently dilute in solute B and vacancies V so that the solution can be described in terms of isolated vacancies, isolated solute atoms, and solute–vacancy pairs. Although pairs of solute atoms and pairs of vacancies should also be considered, these configurations will be omitted from our description of the solution in the interest of simplicity. Following Howard and Lidiard (1964), consider a solution containing N_A solvent atoms, N_B solute atoms, and N_V vacancies. If p is the fraction of solute atoms in solute–vacancy pairs in the solution, then the number of pairs is $N_B p$, the number of isolated solute atoms is $N_B - N_B p$, and the number of isolated vacancies is $N_V - N_B p$. The Gibbs free energy of this solution is then

$$G = G_0(P, T) + N_B g_B + (N_V - N_B p)g_V + N_B p(g_V + \Delta g_V) - kT\ln \Omega$$
$$= G_0(P, T) + N_B g_B + N_V g_V + N_B p \Delta g_V - kT\ln \Omega \quad (14)$$

where $G_0(P, T)$ is the free energy of the pure solvent lattice, g_B the free energy of solution of the solute atoms in the solvent, g_V the free energy of formation of an isolated vacancy in the pure solvent, $(g_V + \Delta g_V)$ the free energy of formation of a vacancy next to an impurity atom, and $k \ln \Omega$ is the configurational entropy of the solution.

To determine Ω, we must calculate the number of ways of arranging N_A solvent atoms, $N_B(1 - p)$ unpaired solute atoms, $N_V - N_B p$ unpaired vacancies, and $N_B p$ vacancy–solute pairs on a total of $N_S = N_A + N_V + N_B$ lattice sites. This calculation is most simply accomplished in steps by first determining the number of ways Ω_p of setting down the $N_B p$ solute–vacancy pairs on the lattice, then the number of ways Ω_B of setting down $N_B(1 - p)$ unpaired solute atoms on the remaining lattice points, and finally the number of ways Ω_V of setting down $N_V - N_B p$ unpaired vacancies on the rest of the lattice points with the restriction that none of the unpaired vacancies are placed next to any of the unpaired solute atoms (Lidiard, 1960). Since all of the mixing of the solution constituents will have effectively occurred at this point, one would expect that only one distinct way, $\Omega_A = 1$, will be left to place the solvent atoms on the balance of the lattice points.

The total configurational entropy $k \ln \Omega$ of our dilute solution is thus

$$k \ln \Omega = k \ln (\Omega_p \Omega_B \Omega_V \Omega_A) \quad (15)$$

Substituting Eq. (15) into Eq. (14), we arrive at an expression for the Gibbs free energy of the solution in terms of the concentrations of the various constituents of the solution. Because of the excessive length of this expression, we will not write it out in detail here.

Although we will not presuppose that the overall solution is in equilibrium (this will allow the possibility of vacancy supersaturations later), the local solute–vacancy pairing reaction will be assumed to be in balance. This assumption is reasonable since the solute–vacancy pairing reaction which depends only on short range diffusion in small volume elements can be considered to occur instantaneously relative to the long range diffusion required to bring the overall solution in equilibrium. To obtain the equilibrium number of solute–vacancy pairs, the derivative of the Gibbs free energy of the solution with respect to the fractional pair concentration p is set equal to zero,

$$(\partial G/\partial p)_{N_A, N_V, N_B} = 0 \quad (16)$$

which after some rearrangements and approximations becomes (Howard and Lidiard, 1964)

$$[C_p/(C_V - C_p)(C_B - C_p)] = z \exp(-\Delta g_V/kT) = K \tag{17}$$

where $C_p \equiv N_B p/N_S$, $C_V \equiv N_V/N_S$, and $C_B \equiv N_B/N_S$ are the atomic fractions of the various constituents of the solution. Note that $C_V - C_p$ is the fraction of free unpaired vacancies in the solution and $C_B - C_p$ is the fraction of free unpaired solute atoms in the solution. Equation (17) could have also been obtained by utilizing the *law of mass action* for the pairing reaction: unpaired vacancies $(C_V - C_p)$ + unpaired solute $(C_B - C_p) \rightleftarrows$ solute–vacancy pairs (C_p) where K is the mass action constant.

With the pairing reaction in equilibrium, the chemical potentials of the solvent A, solute B, and vacancies V are

$$\mu_A \equiv (\partial G/\partial N_A)_{N_V, N_B, p} = \mu_A{}^0(P, T) - kT(C_V + C_B - C_p) \tag{18}$$

where

$$\mu_A{}^0(P, T) \equiv (\partial G_0(P, T)/\partial N_A)_{N_V, N_B, p}$$

$$\mu_B \equiv (\partial G/\partial N_B)_{N_V, N_A, p} = g_B + kT\ln(C_V - C_p) + kT[zC_V - (z-1)C_p] \tag{19}$$

$$\mu_V \equiv (\partial G/\partial N_V)_{N_A, N_B, p} = g_V + kT\ln(C_B - C_p) + kT[zC_B - (z-1)C_p] \tag{20}$$

where second-order terms in C_B and C_V have been ignored because we have assumed in this dilute solution that $C_B, C_V \ll 1$. Since in simple dilute solutions, the chemical potential μ_X of a component X is $\mu_X = \mu_0 + kT\ln C_X$, we see that Eqs. (18)–(20) differ only from this simple expression by the additional linear terms in C_B, C_p, and C_V in these equations. Also note the complete symmetry between the expressions for μ_V and μ_B as is expected from the fact that both are dilute constituents of the solution and both react to form pairs with each other.

From Eq. (13), the thermodynamic forces that appear in the phenomenological diffusion equations (1), (4), and (5) can be derived from the chemical potentials given in Eqs. (18)–(20).

$$X_A = +kT(\nabla C_B + \nabla C_V - \nabla C_p) \tag{21}$$

$$X_B = -kT[\nabla(C_B - C_p)/(C_B - C_p)] - kT[z\nabla C_V - (z-1)\nabla C_p] \tag{22}$$

$$X_V = -kT[\nabla(C_V - C_p)/(C_V - C_p)] - kT[z\nabla C_B - (z-1)\nabla C_p] \tag{23}$$

F. Atom Currents and Vacancy Winds in Dilute Solutions

The fluxes of A, B, and V generated by these thermodynamic forces can be found from Eqs. (4) and (2) to be (Howard and Lidiard, 1964)

$$J_A = kT(L_{AA} + L_{AB})[\nabla(C_V - C_p)/(C_V - C_p)]$$
$$- kTL_{AB}[\nabla(C_B - C_p)/(C_B - C_p)] + kT[L_{AA}(z+1)$$
$$+ zL_{AB}] \nabla C_B + kT[L_{AA} - zL_{AB}] \nabla C_V - kT[zL_{AA}] \nabla C_p \quad (24)$$

$$J_B = kT(L_{BA} + L_{BB})[\nabla(C_V - C_p)/(C_V - C_p)]$$
$$- kTL_{BB}[\nabla(C_B - C_p)/(C_B - C_p)] + kT[(z+1)L_{AB} + zL_{BB}] \nabla C_B$$
$$+ kT[L_{AB} - zL_{BB}] \nabla C_V - zkTL_{AB} \nabla C_p \quad (25)$$

$$J_V = -(J_A + J_B)$$
$$= -kT(L_{AA} + 2L_{AB} + L_{BB})[\nabla(C_V - C_p)/(C_V - C_p)]$$
$$+ kT(L_{AB} + L_{BB})[\nabla(C_B - C_p)/(C_B - C_p)]$$
$$- kT[L_{AA}(z+1) + (2z+1)L_{AB} + zL_{BB}] \nabla C_B$$
$$- kT[L_{AA} - (z-1)L_{AB} - zL_{BB}] \nabla C_V + kT[zL_{AA} + zL_{AB}] \nabla C_p \quad (26)$$

Equations (24)–(26) are quite cumbersome and some approximations are in order at this point. Because the phenomenological theory does not say anything about the magnitude of the phenomenological coefficients L_{AA}, L_{AB}, and L_{BB}, we turn to the kinetic atomic theories, Eqs. (12), for information about the phenomenological coefficients so that some reasonable approximations can be made. In Table I, the order in terms of concentrations C_B and C_V of the various terms appearing in Eqs. (24)–(26) is given. $L_{AA}^f = 4Na^2W_0 C_V/kT$ is the term in L_{AA} [see Eqs. (12)] proportional to $C_V - C_p$. $L_{AA} - L_{AA}^f$ is then equal to the remaining term in L_{AA} and thus $L_{AA} - L_{AA}^f$ is proportional to C_p.

Since only two of the three fluxes J_A, J_B, and J_V are independent because of the requirement of conservation of lattice sites [Eq. (2)], we will only consider fluxes J_B and J_V from here on since these two fluxes will be the ones directly measured in the experiments described later in this chapter. In making our approximation for the fluxes J_B and J_V, terms of the order of C_B^2, C_V^2 or higher will be dropped and only terms of the order of C_B, C_V, or C_p will be retained. Equations (25) and (26) thus become

$$J_B = kT(L_{BA} + L_{BB})[\nabla(C_V - C_p)/(C_V - C_p)]$$
$$- kTL_{BB}\nabla(C_B - C_p)/(C_B - C_p) \quad (27)$$

$$J_V = -kT(L_{AA} + 2L_{AB} + L_{BB})[\nabla(C_V - C_p)/(C_V - C_p)]$$
$$- kT[\nabla(C_B - C_p)/(C_B - C_p)][L_{AA}^f(1+z)(C_B - C_p) - L_{AB} - L_{BB}] \quad (28)$$

TABLE I

TERMS OCCURRING IN EQUATIONS FOR THE SOLUTE CURRENT, SOLVENT CURRENT, AND VACANCY CURRENT[a]

Term	Order in C_B, C_V	Comments
C_p	$C_B C_V$	See Eq. (21)
L_{AB}	$C_B C_V$	See Eqs. (12) and (21)
L_{AA}^f	C_V	See Eqs. (12)
$L_{AA} - L_{AA}^f$	$C_B C_V$	See Eqs. (12) and (21)
∇C_p	$C_B C_V$	See Eq. (21)
∇C_B	C_B	
∇C_V	C_V	
$\nabla(C_V - C_p)/(C_V - C_p)$	Zeroth order	
$\nabla(C_B - C_p)/(C_B - C_p)$	Zeroth order	

[a] The order of these terms in the atomic fractions of solute C_B and vacancies C_V in the solution are noted. A first-order approximation is made with respect to C_B, C_V, and C_p. Thus terms of the order of C_B, C_V, and C_p are retained while terms of the order of C_B^2, C_V^2, $C_B C_p$, $C_V C_p$, and C_p^2 or higher are dropped.

where one negligible term $-kT\nabla C_p z L_{AA}^f$ has been kept and modified slightly ($z \to z+1$) to make Eq. (28) more symmetric. Before proceeding further, it is helpful to identify the diffusion coefficient D_V^f of *free* vacancies and the diffusion coefficient D_p of pairs in this expression.

To determine D_V^f, consider a homogeneous alloy $\nabla C_B = 0$ in which there is no binding energy between solute atoms and vacancies. In this case, the concentration of pairs C_p is very small ($C_p \approx C_B C_V$) and L_{AB} and L_{BB} can be ignored in comparison to L_{AA}^f. The flux of vacancies then becomes from Eq. (28),

$$J_V^f = -[kTL_{AA}^f/(C_V - C_p)]\nabla(C_V - C_p) \qquad (29)$$

where $C_V - C_p$ is the concentration of free vacancies. Comparing Eq. (29) to Fick's first law, $J_V^f = -D_V^f N \nabla(C_V - C_p)$, we find that the diffusion coefficient D_V^f of free vacancies is

$$D_V^f = kTL_{AA}^f/N(C_V - C_p) \qquad (30)$$

To find the diffusion coefficient D_p of pairs, consider the case of an alloy with a concentration gradient of pairs but no concentration gradient in free vacancies. By Fick's first law, the flux of pairs $J_p = -D_p N \nabla C_p$. We note in the absence of a concentration gradient of free vacancies, $\nabla(C_V - C_p) = 0$, that the flux of solute B from Eq. (27) is $J_B = -kTL_{BB}[\nabla(C_B - C_p)/(C_B - C_p)]$. But since the only solute flowing in the solution under these conditions is that contained in pairs, then $J_B = J_p$

so that $D_p = [kTL_{BB}/N \nabla C_p][\nabla(C_B - C_p)/(C_B - C_p)]$. But from Eq. (21),

$$\nabla C_p/C_p = \nabla(C_B - C_p)/(C_B - C_p)$$

when $\nabla(C_V - C_p) = 0$. Thus the diffusion coefficient D_p of pairs is

$$D_p = kTL_{BB}/NC_p \tag{31}$$

Now, let us now return to calculating the average diffusion coefficient D_V of all vacancies including those in pairs in a homogeneous solution ($\nabla C_B = 0$) where pairs are important. From Eq. (17), it can be shown without approximation that

$$\nabla C_p = [(C_p C_V - C_p^2)/(C_B C_V - C_p^2)] \nabla C_B \\ + [(C_p C_B - C_p^2)/(C_B C_V - C_p^2)] \nabla C_V \tag{32}$$

By eliminating ∇C_p from the general expression for J_V [Eq. (28)] and stipulating a homogeneous alloy ($\nabla C_B = 0$) and an impurity concentration C_B much greater than the vacancy concentration C_V ($C_B \gg C_V, C_p$), we find that (Howard and Lidiard, 1965)

$$J_V = -N\{[(C_V - C_p)/C_V]D_V^f + (C_p/C_V) D_p \\ + [kT/N(C_V - C_p)][(2L_{AB} + L_{AA} - L_{AA}^f)]\}\nabla C_V \tag{33}$$

Since C_V is the total number of vacancies (free vacancies plus vacancies contained in pairs), $(C_V - C_p)/C_V$ is the fraction of free vacancies and C_p/C_V is the fraction of vacancies bound in solute–vacancy pairs in the solution. The vacancy wind J_V in a vacancy concentration gradient ∇C_V is thus the result of the diffusion of free vacancies [first term in Eq. (33)], the diffusion of vacancies contained in pairs [second term in Eq. (33)], and the association–dissociation jumps of vacancies around solute–vacancy pairs [third term in Eq. (33)]. Although experimental evidence indicates that both the second and third terms are probably small relative to the first term in Eq. 33, we will only drop the third term since the second term can be easily handled. Equation (33) then becomes

$$J_V = -N\{[(C_V - C_p)/C_V]D_V^f + (C_p/C_V)D_p\} \nabla C_V \tag{34}$$

where $C_B \gg C_V, C_p$ and $\nabla C_B = 0$.

Let us now consider the solute current J_B generated by the vacancy wind J_V. Returning to Eq. (27) and setting $\nabla C_B \gg C_V, C_p$ and using Eq. (32) to eliminate ∇C_p, we find that

$$J_B = [kT(L_{BA} + L_{BB})/C_V] \nabla C_V \tag{35}$$

Introducing Eq. (31), Eq. (35) becomes

$$J_B = [(L_{BA} + L_{BB})/L_{BB}](NC_p D_p/C_V) \nabla C_V \tag{36}$$

7. ATOM CURRENTS GENERATED BY VACANCY WINDS

where $C_B \gg C_V$, C_p and $\nabla C_B = 0$. This expression for J_B under the conditions stipulated is reasonable since the impurity current is proportional to the diffusion coefficient D_p of solute–vacancy pairs and to the relative fraction C_p/C_V of the vacancies in the solution that are next to a solute atom. (A solute atom cannot move in a substitutional solution unless one of its nearest neighbors is a vacancy.) The factor $(L_{BA} + L_{BB})/L_{BB}$ can be either positive or negative. For tightly bound pairs, we anticipate that this factor will be negative so that vacancies and solute atoms flow in the same direction (migrating as pairs). Conversely, if there is little binding between vacancies and solute atoms, this factor should be positive so that vacancies flowing in one direction produce a reverse current of atoms A *and* B in the opposite direction.

The ratio of the solute current J_B to the vacancy wind J_V is

$$J_B/J_V = -[C_p D_p/(C_V - C_p)][(D_V^f + C_p D_p)][(L_{AB} + L_{BB})/L_{BB}] \quad (37)$$

Equation (37) will not be of much use to us in determining the ratio of the phenomenological coefficients L_{AB}/L_{BB} unless we can express $C_p D_p$ and D_V^f in terms of measurable parameters. D_V^f and $C_p D_p$ are eliminated from Eq. (36) as follows.

Since the concentration of *free* vacancies $C_V - C_p$ in the solution is equal to a first-order approximation to the concentration of vacancies in the pure solvent A, the free vacancy diffusion coefficient D_V^f can be expressed in terms of the self-diffusion coefficient D_A of the pure solvent A,

$$D_A = (C_V - C_p) f D_V^f \quad (38)$$

where f is the correlation factor for vacancy diffusion in the pure solvent. To express $C_p D_p$ in terms of experimentally measurable parameters, consider the diffusion of a small tracer amount of solute B in pure solvent A in an alloy with no vacancy concentration gradients $\nabla C_V = 0$. Then substituting Eq. (32) into Eq. (27) to eliminate ∇C_p, we find

$$J_B = kT\{(L_{BA} + L_{BB})[C_p/(C_B C_V - C_p^2)] - [C_V L_{BB}/(C_B C_V - C_p^2)]\} \nabla C_B \quad (39)$$

Since C_B occurs only in tracer amounts, $C_V \gg C_B$, C_p and Eq. (39) becomes

$$J_B = (kT/C_B)[(L_{BA} + L_{BB})(C_p/C_V) - L_{BB}] \nabla C_B \quad (40)$$

Because $C_p \ll C_V$, we can ignore the first term in Eq. (40). Then comparing Eq. (40) with Fick's first law, $J_B = -D_B^* N \nabla C_B$ where D_B^* is the tracer diffusion coefficient of solute B in pure solvent A, we obtain

$$D_B^* = (kT/N) L_{BB}/C_B \quad (41)$$

The term $C_p D_p$, in terms of parameters measurable in an experiment, is then obtained by combining Eq. (41) with Eq. (31) and eliminating the phenomenological coefficient L_{BB} to give

$$C_p D_p = C_B D_B^* \qquad (42)$$

At this point, it is interesting to return briefly to Eq. (33) and determine the relative proportions of vacancies that diffuse as free vacancies and as members of solute–vacancy pairs. From Eqs. (34), (38), and (42), we have

$$\frac{\text{vacancies diffusing as pairs}}{\text{vacancies diffusing as free vacancies}} = \frac{C_p D_p}{(C_V - C_p)D_V^f} = \frac{fC_B D_B^*}{D_A} \qquad (43)$$

Since f is 0.78 for a fcc lattice and D_B^*/D_A is rarely greater than a factor of 10 for substitutional solutions, then for dilute solutions ($C_B \approx 1\%$) less than 8% of the vacancies diffusing down a vacancy concentration gradient will diffuse as pairs, in general. This probably indicates also that the association–dissociation jump term (third term) of Eq. (33) is small (as we have assumed) in most alloys. Substitution of Eqs. (42) and (38) into Eq. (43) eliminates the unknown terms $C_p D_p$ and D_V^f to give

$$J_B/J_V = [C_B D_B^*/(D_A/f + C_B D_B^*)][(L_{AB} + L_{BB})/L_{BB}] \qquad (44)$$

From Eq. (44), it can be seen that the ratio of the phenomenological coefficients L_{AB}/L_{BB} can be determined from measurements of the self-diffusion constant D_A of the pure solvent A, the tracer diffusion constant of solute B in pure solvent A, and the solute current J_B generated by a vacancy wind $J_V = -D_V \nabla(NC_V)$ in an alloy in which the solute concentration C_B is uniform. (Since $J_V \gg J_B$ in a dilute alloy of B in A, in practice, we need stipulate an *initially* uniform concentration in C_B.)

If we substitute the expressions (12) for the phenomenological coefficients L_{AB} and L_{BB} derived from an atomic model, a more physical understanding of Eq. (44) can be achieved.

$$\frac{J_B}{J_V} = \frac{C_B D_B^*}{D_A/f + C_B D_B^*} \left[\frac{W_1 + (13/2)W_3}{W_1 - (7/2)W_3} \right] \qquad (45)$$

We recall that W_1 is the jump frequency of a vacancy from one nearest neighbor position of a solute atom B to another and W_3 is the jump frequency of a vacancy away from a nearest neighbor position of a solute atom B (thereby breaking apart the solute–vacancy pair) (see Fig. 2). As is expected, when vacancies are tightly bound to solute atoms ($W_3 \ll W_1$) the solute atoms flow in the same direction as the vacancy wind (as solute–vacancy pairs). In contrast, when there is no vacancy–solute interaction and $W_3 \approx W_1$, then the solute current J_B flows in the direction opposite to the

7. ATOM CURRENTS GENERATED BY VACANCY WINDS

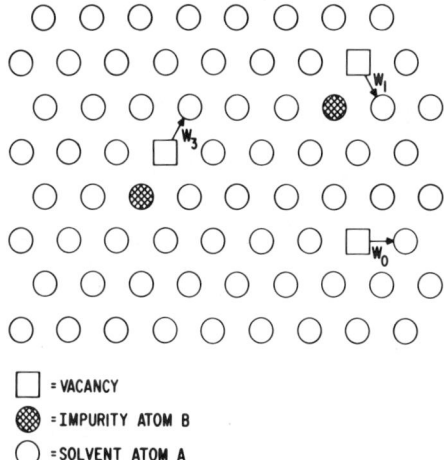

☐ = VACANCY
⬢ = IMPURITY ATOM B
◯ = SOLVENT ATOM A

Fig. 2. The vacancy jump frequencies away from (W_3) and around (W_1) solute atoms in a dilute fcc alloy.

vacancy wind. That is, solute B participates in this case in the normal reverse atom flow that is a consequence of a current of empty lattice sites through a solid.

III. Measurement of the Vacancy Wind and the Wind-Generated Solute Current

A. Alloy Systems

During a high temperature anneal, the equilibrium vacancy concentration in aluminum reaches a relatively large value. Subsequent cooling of the aluminum causes the solution of vacancies to become supersaturated. In electropolished aluminum, a low dislocation density zone 200–300 μm thick exists adjacent to the free surface (Authier et al., 1965). Since the contiguous metal–oxide interface has been shown to be a totally ineffective vacancy sink (Harris and Masters, 1968; Dobson et al., 1968), this deficiency of dislocations implies that except for an occasional intersecting grain boundary, there are relatively few vacancy sinks available to the near surface region. As a consequence, with a suitable cooling rate, the vacancy concentration in the zone neighboring the free surface will reach a critical supersaturation ($\sim 1\%$) at which vacancies begin to precipitate heterogeneously on the metal–oxide interface. The resulting vacancy condensation cavities which contain $\sim 10^{13}$ vacancies each (see Figs. 3 and 4) have been subject of a number of investigations (Dougherty and Davis, 1959; Kasen and Polonis, 1962; Jasperse and Dougherty, 1964).

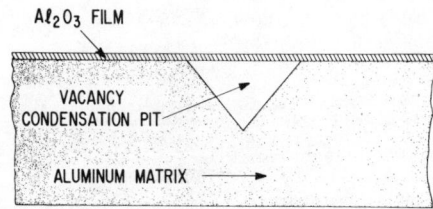

Fig. 3. A schematic diagram of a vacancy condensation cavity on the metal–oxide interface of aluminum formed by cooling from a high temperature anneal.

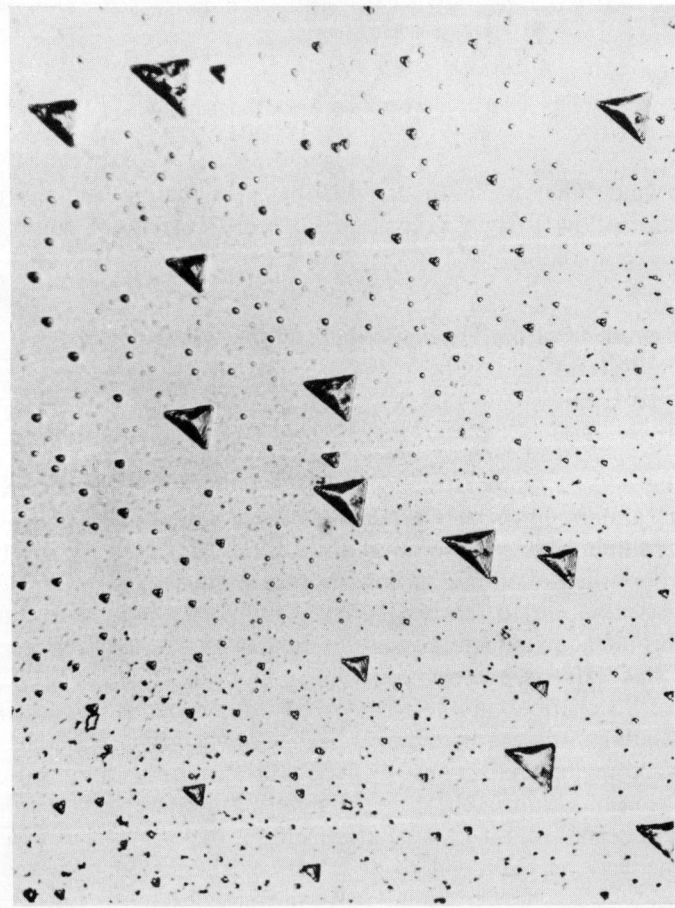

Fig. 4. Photomicrograph of vacancy condensation cavities in an aluminum–zinc alloy.

The appearance of these relatively large accessible vacancy sinks in aluminum and other alloys (Kasen et al., 1966) provides a unique experimental opportunity to study vacancy and vacancy-generated solute currents in metals. It is important that both the solute and vacancy currents be measured, since only their ratio is determined by atomic processes and is characteristic of a particular alloy. As will be shown in Section III.B, both currents can be determined directly in experiments utilizing vacancy condensation cavities. Finally, solute segregation at these condensation cavities is greatly magnified over that which would occur at natural vacancy sinks such as grain boundaries, since the vacancy flow *per lattice site* on the cavity (and thus the vacancy-generated impurity flow) is 10^4 times larger than that impinging on a neighboring grain boundary during the same quench. This magnification results from the concentrating of the excess vacancies at a point sink in the case of a vacancy cavity, as compared to distributing the same number of excess vacancies over a planar sink in the case of a grain boundary. Consequently, experimental detection of currents of vacancies and solute around condensation cavities is much simpler than around other natural vacancy sinks such as grain boundaries or free surfaces.

B. The Vacancy Wind

The total vacancy current that produced a vacancy condensation cavity can be directly determined from the cavity volume V_c. Figure 5 shows a schematic diagram of a vacancy cavity beneath the aluminum oxide film.

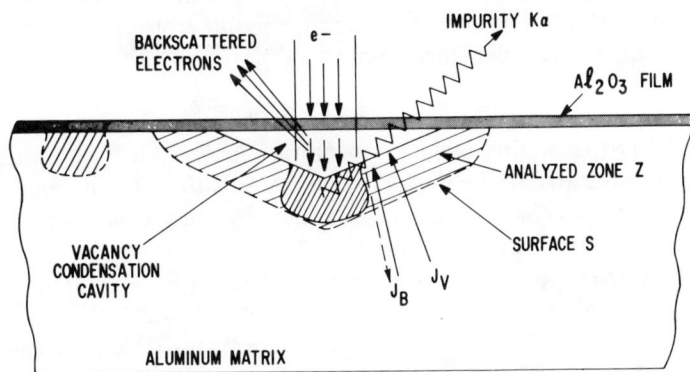

Fig. 5. Schematic diagram of a vacancy condensation cavity on the metal–oxide interface of aluminum, formed by cooling from near the melting point of aluminum. The impinging microprobe beam, the vacancy current, the solute current, zone Z, surface S, and the regions on the free surface and in the vacancy cavity analyzed by the microprobe are shown.

Let S be the area of an imaginary surface surrounding the vacancy cavity. Then one can write

$$\int_{\text{time}} J_V S \, dt = V_c \tag{46}$$

where t is the variable in time, V_c is the volume of the condensation cavity, and J_V is the time-averaged vacancy current. Since the thin Al_2O_3 film over the cavity is transparent in the optical region, both the shape and depth of the cavity can be determined from the displacement of interference fringes viewed in an optical interference microscope. In this manner, the cavity volume was established and the total vacancy flow directly determined (Anthony, 1970a–c).

C. The Solute Current Generated by the Vacancy Wind

After the formation of the vacancy cavity, the solute concentration around the cavity was measured with an electron microprobe. By this means, the solute concentration in zone Z in Fig. 5 was established. For determinations of small solute concentrations in aluminum, the size of this analyzed zone Z is defined by the effective electron penetration distance in aluminum. At the microprobe accelerating voltage of 19 keV used in these experiments (Anthony, 1970a–c), the depth of maximum solute characteristic x-ray production is 0.7 μm in aluminum (Shinoda et al., 1968). To ensure that $95+\%$ of the solute radiation is included, a region double this depth has been used as the analyzed zone Z around the vacancy cavity.

The total solute flow through the surface S induced by the vacancy wind can be computed from the cavity volume V_c, the analyzed zone volume V_z, and the concentration change in the analyzed zone from the original solute level (Anthony, 1970a–c). Thus we have

$$\int_{\text{time}} J_B S \, dt = -C_B{}^0 V_c - [C_B{}^0 - C_B{}^f] V_z \tag{47}$$

where $C_B{}^0$ is the original solute concentration, $C_B{}^f$ the final solute concentration in the analyzed zone following the growth of the vacancy cavity, J_B the time-averaged solute current generated by the vacancy wind forming the cavity, and t the variable of time.

Table II gives the time-averaged experimental ratio of $(J_B/J_V)/C_B$ found for solutes in aluminum subjected to a vacancy wind. Values of this ratio for other alloy systems are included where available. The calculated values for Ag(Zn) and Cu(Zn) are from values of W_3/W_1 available in the literature for these systems.

Also in Table II, values of the ratios of the phenomenological coefficients L_{AB}/L_{BB} and the vacancy jump frequencies W_3/W_1 are given. These ratios are important not only in experiments concerned with vacancy-generated

TABLE II

RATIO OF THE VACANCY WIND AND THE SOLUTE CURRENT GENERATED BY THE VACANCY WIND IN A NUMBER OF ALLOY SYSTEMS[a]

Alloy system solvent (solute)	$(J_B/J_V)/C_B$	Solute redistribution at a vacancy sink[b]	L_{BA}/L_{BB}	W_3/W_1	T/T_m	Ref.
Directly observed						
Al(Zn)	−0.92	E	< −0.79	<0.20	0.92	Anthony (1970c)
Al(Cu)	−1.0	None	−0.40	0.36	0.92	Anthony (1970b)
Al(Ge)	−1.92	D	> −0.77	>0.22	0.92	Anthony (1970a)
Al(Mg)	> −1	E	—	—	0.81 quench from	Cundy et al. (1968)
Calculated						
Ag(Zn)	−1.88	D	—	0.37	0.92	Peterson and Rothman (1969)
Cu(Zn)	−2.80	D	—	0.90	0.92	Rothman and Peterson (1967)
Directly observed						
Cu(Mn)	> −1	E	—	—	0.79	Whittenberger and Dayananda (1970)
Cu(Zn)	< −1	D	—	—	0.79	Whittenberger and Dayananda (1970)
Cu(Ag)	< −1	D	—	—	0.99	Kuczynski et al. (1960)
Ag(Cu)	< −1	D	—	—	0.99	Kuczynski et al. (1960)
Cu(In)	< −1	D	—	—	0.99	Kuczynski et al. (1960)

[a] The currents are expressed with respect to the local crystal lattice. T/T_m is the homologous temperature at which these currents were measured.
[b] D is depletion at a vacancy sink, and E is enrichment at a vacancy sink.

solute currents but also in studies of electromigration and thermal migration. The inequality signs on the L_{AB}/L_{BB} and W_3/W_1 parameters for the Al(Zn) and Al(Ge) systems arise from the fact the Eqs. (44) and (45) do not take into consideration the flow of the solute atoms induced by a developing solute segregation gradient. Thus, Eqs. (44) and (45) tend to overestimate the current generated by a vacancy wind once the initial condition of a uniform solute concentration ($\nabla C_B = 0$) is violated. Consequently, Eqs. (44) and (45) will predict either a greater solute enrichment or a greater solute depletion than actually occurs because of the neglect of the "buffering effect" of the developing solute gradient.

Although the ratio of W_3/W_1 could in principle vary over orders of magnitude (Howard and Manning, 1967), this ratio is nevertheless close to unity in the five different alloy systems so far investigated. This fact indicates that, in these alloys, the solute does not strongly attract a vacancy and cause a large consequential decrease in W_3, nor does the solute greatly disrupt its local environment and produce a resulting large change in W_1.

Consequently, these experiments indicate that the solute–vacancy binding energy in these alloys is small.

A small solute–vacancy binding energy in aluminum is also implied by measurements of solute diffusion in aluminum where the activation energy for solute diffusion and self-diffusion are very nearly equal (Fradin and Rowland, 1969; Peterson and Rothman, 1970; Burke and Ramachandran, 1971). In addition, a small solute–vacancy binding has also been found by simultaneous length and lattice parameter measurements of various aluminum alloys. In these studies, it was found that solute–vacancy binding energies in aluminum are very small [silver (Beaman et al., 1964), copper (King and Burke, 1970)] or zero [magnesium (Beaman et al., 1965)]. Finally, theoretical calculations indicate that solute–vacancy binding energies aluminum are small (Blandin et al., 1963; Edelglass and Ohring, 1969).

In contrast, solute–vacancy binding energies deduced from quenching experiments are large (Doyama, 1966a,b; Doyama and Cotterill, 1966). This inconsistency is most likely related to the possibility of solute clustering in quenching experiments. Thus, for example, some binding energies deduced in quenching experiments may reflect the binding between a vacancy and a cluster of several solute atoms rather than between a vacancy and a single solute atom.

IV. Solute Segregation around Vacancy Sinks

The values of $(J_B/J_V)/C_B$ in Table II are given in terms of currents expressed with respect to the local crystal lattice. However, in order to determine whether solute is flowing toward or away from a vacancy sink such as a grain boundary, we would like to know $(J_B/J_V)/C_B$ with respect to the vacancy sink. Because a vacancy sink is also a lattice plane sink, the local lattice around a vacancy sink collapses inward toward the vacancy sink with a velocity U proportional to the rate of disappearance of lattice sites at the sink. Thus, the fluxes J_B and J_V in the frame of reference of a vacancy sink become

$$J_B' = J_B + N_B U \quad \text{and} \quad J_V' = J_V + N_V U \tag{48}$$

where J_B' and J_V' are the currents of solute atoms B and vacancies V with respect to a vacancy sink. The velocity U of the local lattice toward the vacancy sink is given by the vacancy current J_V disappearing at the vacancy sink divided by the number N of atoms per unit volume.

$$U = J_V/N \tag{49}$$

Substitution of Eq. (49) into Eqs. (48) yields

$$J_B' = J_B + C_B J_V \quad \text{and} \quad J_V' = J_V + C_V J_V \tag{50}$$

The ratio of the solute current to the vacancy wind with respect to a vacancy sink becomes

$$(J_B'/J_V')/C_B = (J_B/J_V)/C_B + 1 \tag{51}$$

where the approximation $1 + C_V \approx 1$ has been made. Thus we see that we need only add 1 to the values of $(J_B/J_V)/C_B$ in Table II to find the values of $(J_B'/J_V')/C_B$ listed in Table III. Positive values of J_B'/J_V' imply that vacancies and solute are both flowing toward a vacancy sink so that solute enrichment then will occur at the sink. Negative values of J_B'/J_V' imply that while vacancies are flowing toward the sink, solute atoms are flowing away, thereby depleting the region around the sink of solute. The actual change ΔC_B in solute concentration at a vacancy sink is

$$\Delta C_B/C_B = \int J_B' \, dt/C_B = [(J_B'/J_V')/C_B] \int J_V' \, dt \tag{52}$$

where the integral is over the time duration of the vacancy flow to the sink.

Because the experiments discussed in this chapter yield only time-averaged values for the flux of solute and vacancies, the fluxes J_B' and J_V' have been treated as time-averaged constants in the derivation of Eq. 52. From Table III, the first factor on the right hand side of Eq. (52) is known. Consequently, we need only know the total vacancy flow to a vacancy sink to estimate the solute concentration change in the region around the sink.

As an example, let us consider the segregation generated by vacancy currents at a grain boundary during the air-cooling of an aluminum alloy from near its melting point (Anthony, 1970c). Figure 6 shows a grain

TABLE III

Ratio of the Vacancy Wind and the Solute Current Generated by the Vacancy Wind in a Number of Alloy Systems[a]

Alloy system solvent (solute)	$\left(\dfrac{J_B'}{J_V'}\right)\bigg/ C_B$	Impurity redistribution at a vacancy sink	References
Al(Zn)	+0.08	E	Anthony (1970c)
Al(Ge)	−0.92	D	Anthony (1970b)
Al(Cu)	0	None	Anthony (1970a)
Al(Mg)	>0	E	Cundy et al (1968)
Ag(Zn)	−0.88	D	Peterson and Rothamn (1969)
Cu(Zn)	−1.80	D	Rothman and Peterson (1967)
Cu(Zn)	<0	D	Whittenberger and Dayananda (1970)
Cu(Mn)	>0	E	Whittenberger and Dayananda (1970)
Cu(Ag)	<0	D	Kuczynski et al. (1960)
Ag(Cu)	<0	D	Kuczynski et al. (1960)
Cu(In)	<0	D	Kuczynski et al. (1960)

[a] The currents are here expressed with respect to a vacancy sink and not the local lattice as in Table II. A positive value of J_B'/J_V' implies a solute current toward a vacancy sink (J_B' and J_V' are in the same direction) while a negative value of J_V'/J_V' implies a solute current away from a vacancy sink (J_B' and J_V' are flow in opposite directions).

TABLE IV

Expected Change of Solute Concentration at a Grain Boundary Relative to the Bulk Concentration

Generation of vacancy current[a]	Alloy system solvent (solute)	$\Delta C_B/C_B \times 100\%$ [b]
(a) Air cool	Al(Zn)	+80% enrichment
	Al(Ge)	−100% depletion
(b) Irradiation	Al(Zn)	+21,600% enrichment
	Al(Ge)	−100% depletion

[a] Generated by vacancy flows to grain boundaries during: (a) air cooling from 630°C; (b) irradiation in a fast neutron reactor for 3 months.

[b] 100% depletion implies that all solute atoms are removed from the grain boundary.

boundary in an aluminum alloy subjected to this treatment. During cooling, vacancy condensation cavities formed at the metal–oxide interface except in regions near the grain boundary. Near the grain boundary, the vacancy concentration was continually depressed because the grain boundary was absorbing vacancies and acting as a vacancy sink. From the width of 2×10^4 atomic planes of the depleted zone in Fig. 6, we estimate that 10 vacancies per atomic site were absorbed by the grain boundary sink. Using Eq. (52) and Table III, we find that $\Delta C_B/C_B$ is considerable in some aluminum alloys (see Table IV).

During irradiation in a fast neutron reactor, both pores and grain boundaries (Brimhall and Mastel, 1969a,b) are acting as vacancy sinks. Thus segregation will occur at pores (Anthony, 1972) and grain boundaries

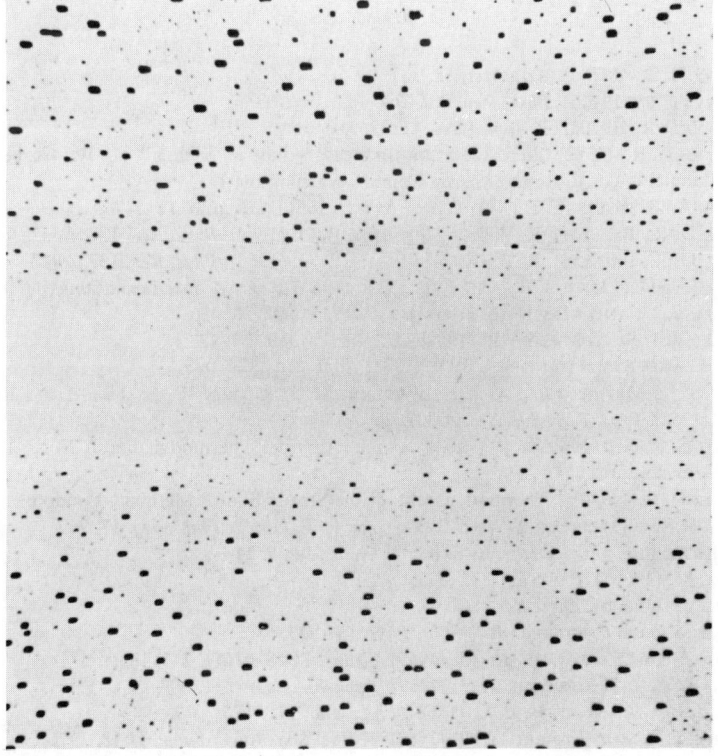

Fig. 6. A 5° tilt boundary around the $\langle 110 \rangle$ direction in an aluminum alloy air-cooled from 620°C. The $\langle 110 \rangle$ direction is perpendicular to the plane of the photograph. Vacancy condensation cavities have formed on the metal–oxide interface of the alloy except near the grain boundary which acted as an efficient vacancy sink during cooling.

(Anthony, 1970c) in reactor materials during irradiation and may make certain alloys susceptible to intergranular corrosion and embrittlement and also cause stresses (Anthony, 1972) to develop around growing pores. Let us consider segregation at a grain boundary. From the width of the zone depleted of pores around a grain boundary in an irradiated alloy (Brimhall and Mastel, 1969a,b), the total vacancy flux impinging on a grain boundary during a three-month irradiation in a 2.5×10^{15} n/cm^2 fast neutron reactor is found to be about 2700 vacancies per boundary lattice site. From the value of the vacancy current, the segregation produced at a grain boundary by this irradiation is calculated in Table IV. As can be seen in Table IV, the greater flow of vacancies to the grain boundary during irradiation causes a much larger amount of solute segregation.

REFERENCES

ANTHONY, T. R. (1970a). *Acta Met.* **18**, 307.
ANTHONY, T. R. (1970b). *Phys. Rev.* **B 2**, 264.
ANTHONY, T. R. (1970c). *J. Appl. Phys.* **41**, 3969.
ANTHONY, T. R. (1972). In "Radiation Induced Voids in Metals" (J. W. Corbett and L. C. Ianniello, eds.), Atomic Energy Comm., Washington, D.C.
AUTHIER, A., RODGERS, C. B., AND LANG, A. R. (1965). *Phil. Mag.* **12**, 547.
BEAMAN, D. R., BALLUFFI, R. W., AND SIMMONS, R. O. (1964). *Phys. Rev.* **134**, A532.
BEAMAN, D. R., BALLUFFI, R. W., AND SIMMONS, R. O. (1965). *Phys. Rev.* **137**, A917.
BLANDIN, A., DÉPLANTÉ, J. L., AND FRIEDEL, J. (1963). *J. Phys. Soc. Japan Suppl. II* **18**, 89.
BRIMHALL, J. L., AND MASTEL, B. (1969a). *J. Nucl. Mater.* **33**, 186.
BRIMHALL, J. L., AND MASTEL, B. (1969b). *J. Nucl. Mater.* **29**, 123.
BURKE, J., AND RAMACHANDRAN, T. R. (1971). *Phil. Mag.*, **24**, 629.
CUNDY, S. L., METHERWELL, A. J. F., WHELAN, M. J., UNWIN, P. N. T., AND NICHOLSON, R. B. (1968). *Proc. Roy. Soc. Ser. A* **307**, 267.
DEGROOT, S. R., AND MAZUR, P. (1962). "Nonequilibrium Thermodynamics." North-Holland Publ., Amsterdam.
DENBIGH, K. G. (1950). "Thermodynamics of the Steady State." Methuen, London.
DOBSON, P. S., KRITZINGER, S., AND SMALLMAN, R. E. (1968). *Phil. Mag.* **17**, 769.
DOUGHERTY, P. E., AND DAVIS, R. S. (1959). *Acta Met.* **7**, 118.
DOYAMA, M. (1966a). *Phys. Lett.* **21**, 395.
DOYAMA, M. (1966b). *Phys. Rev.* **148**, 681.
DOYAMA, M., AND COTTERILL, R. (1966). *Phys. Lett.* **23**, 58.
EDELGLASS, S. M., AND OHRING, M. (1969). *Trans. TMS-AIME* **245**, 186.
FRADIN, F. Y., AND ROWLAND, T. J. (1969). *Appl. Phys. Lett.* **11**, 207.
HARRIS, J. E., AND MASTERS, B. C. (1968). *Phil. Mag.* **17**, 217.
HOWARD, R. E., AND LIDIARD, A. B. (1963). *J. Phy. Soc. Japan Suppl. II* **18**, 197.
HOWARD, R. E., AND LIDIARD, A. B. (1964). *Rep. Progr. Phys.* **27**, 161.
HOWARD, R. E., AND LIDIARD, A. B. (1965). *Phil. Mag.* **11**, 1179.
HOWARD, R. E., AND MANNING, J. R. (1967). *Phys. Rev.* **154**, 561.
HU, S. M. (1969). *Phys. Rev.* **180**, 773.
HUNTINGTON, H. B. (1973). "Advances in Diffusion." Plenum, New York.

JASPERSE, J. R., AND DOUGHERTY, P. E. (1964). *Phil. Mag.* **1**, 635.
JOST, W. (1952). "Diffusion in Solids, Liquids, Gases." Academic Press, New York.
KASEN, M. B., AND POLONIS, D. H. (1962). *Acta Met.* **10**, 821.
KASEN, M. B., TAGGART, R., AND POLONIS, D. H. (1966). *Phil. Mag.* **13**, 453.
KING, A. D., AND BURKE, J. (1970). *Acta Met.* **18**, 205.
KUCZYNSKI, G. C., MATSUMURA, G., AND CULLITY, B. D. (1960). *Acta Met.* **8**, 209.
LIDIARD, A. B. (1955). *Phil. Mag.* **46**, 1218.
LIDIARD, A. B. (1960). *Phil. Mag.* **5**, 1171.
MANNING, J. R. (1968) "Diffusion Kinetics For Atoms in Crystals." Van Nostrand-Reinhold, Princeton, New Jersey.
ONSAGER, L. (1931). *Phys. Rev.* **37**, 407.
PETERSON, N. L., AND ROTHMAN, S. J. (1969). *Bull. Amer. Phys. Soc.* **14**, 389.
PETERSON, N. L., AND ROTHMAN, S. J. (1970). *Phys. Rev.* **B 1**, 3264.
ROTHMAN, S. J., AND PETERSON, N. L. (1967). *Phys. Rev.* **154**, 552.
SHINODA, G., MURATA, K., AND SHIMIZU, R. (1968). *In* "Quantitative Electron Probe Microanalysis" (K. Heinrich, ed.), p. 155, *Nat. Bur. Stand. (U.S.), Spec. Publ.* 298.
SIMMONS, R. O., AND BALLUFFI, R. W. (1960). *Phys. Rev.* **117**, 52.
WHITTENBERGER, J. D., AND DAYANANDA, M. A. (1970). *J. Appl. Phys.* **41**, 840.

8

Diffusion in Alkali Halides

W. J. FREDERICKS

CHEMISTRY DEPARTMENT
OREGON STATE UNIVERSITY
CORVALLIS, OREGON

I. Introduction	381
II. Defects and Their Interactions	382
A. Notation	382
B. Intrinsic Defects	383
C. Extrinsic Defects	384
D. Defect Interactions	385
E. Aggregation and Precipitation	386
F. Long Range Coulomb Interactions	387
III. Theory of Diffusion in Ionic Crystals	388
A. Not Quite Random Walks	388
B. A Nonequilibrium Thermodynamic Approach	392
IV. Experimental Methods	398
A. General	398
B. Significant Factors in Experimental Design	398
V. The Experimental Situation and Numerical Results	408
A. General	408
B. Self-Diffusion	410
C. Diffusion of Monovalent Impurities	424
D. Diffusion of Aliovalent Impurities	424
E. Diffusion of Neutral Species	435
VI. Conclusion	438
References	439

I. Introduction

Our understanding of diffusion processes rests on voluminous work. Many excellent review articles have been written which summarize various aspects of the field; these are listed at the end of this chapter as general references.

The way in which mass transport occurs in solids depends on the defects characteristic of the host–diffusant system. Of the many possible diffusion mechanisms (Manning, 1968) the following three appear to predominate in ionic solids. If the diffusant occupies a substitutional site and migrates by jumping from its initial site to an adjacent vacant site, the diffusion process is called a *vacancy mechanism*. If the diffusant occupies an interstitial position in the crystal and migrates by moving directly from one interstitial site to another, the diffusion is said to occur by an *interstitial mechanism*. If the migration occurs by the interstitial atom moving into a normally occupied site and forcing its original resident into an interstitial position, the process is known as an *interstitialcy mechanism*.

Diffusion by these general mechanisms may occur in any type of crystal. However, in ionic solids the diffusant is restricted to occupy sites in either the cationic or the anionic sublattice when diffusion is by the vacancy or interstitialcy mechanism. Those properties usually considered unique to diffusion in ionic crystals arise from the long range nondirectional binding forces typical of this group of compounds and the strong interactions among defects in such host crystals.

This paper will make an attempt to analyze critically the present status of experimental work in the field. It will be shown that despite all the extensive and careful work, we still lack a detailed understanding of diffusion processes in even such well studied materials as NaCl. In Sections II and III, the necessary theoretical background is briefly outlined. These sections will demonstrate the complexities which are introduced by defect–defect and defect–impurity interactions. Section IV describes in some detail the problem of the design of suitable experiments. This section serves as a necessary background for Section V, in which the current experimental situation is reviewed.

II. Defects and Their Interactions

A. Notation

Only those defects and their interactions required for modeling diffusion processes need to be briefly cataloged here. Because of its convenience, a notation similar to that used by Barr and Lidiard (1970) will be used for thermodynamic expressions. However, when symbols are required to describe defect reactions, a notation related to that developed by Kroger and Vink (1956) will be used. We designate a cation by C or M, an anion by A, a vacancy by V, and an interstitial by I. Superscripts indicate the defect's charge with respect to the lattice or relative position and subscripts denote species or a process. Thus an anion vacancy in an alkali halide is V_A^+. Its concentration expressed as a mole fraction is $x_{V_A^+}$. As this is a cumbersome notation, x_a will be used for $x_{V_A^+}$ and x_c for $x_{V_M^-}$. When referring to thermodynamic parameters for a reaction, the Δ is dropped. For example, the free energy of a reaction between an impurity B and a vacancy V_C should be ΔG_{BV_c}, but will be abbreviated to g_B. The migration enthalpy of a diffusing species on the cation sublattice is given as h_m^c.

B. Intrinsic Defects

In a pure crystal, two types of defects can exist; Schottky defects (equal numbers of anion and cation vacancies) and Frenkel defects (equal numbers of interstitial ions and vacancies in the corresponding sublattice). The concentrations of such defects as a function of temperature can be expressed as ion products. These can be written

$$K_S(T) = x_a x_c = \exp(-g_S/kT) \qquad (1)$$

$$K_{F_C}(T) = x_c x_{I_C} = \exp(-g_{F_C}/kT) \qquad (2)$$

and

$$K_{F_A}(T) = x_a x_{I_A} = \exp(-g_{F_A}/kT) \qquad (3)$$

where K_S is the Schottky product, K_{F_C} is the cationic Frenkel product, K_{F_A} is the anionic Frenkel product, k is Boltzmann's constant, T is the absolute temperature, and g_j is the Gibbs free energy of formation with the subscripts j denoting the specific defect. The free energies can be decomposed into an enthalpy h_j and an entropy s_j of formation by

$$g_j = h_j - T s_j \qquad (4)$$

The enthalpy is the sum of the internal energy of formation u_j and the pressure–volume work,

$$h_j = u_j + P v_j \qquad (5)$$

in which v_j is the volume of formation and P the hydrostatic pressure.

These product relations describe the defect concentrations at thermal equilibrium and are valid whether or not the defect is dominant. As with any ion product, if the concentration of one type of defect is perturbed, the concentration of the other defect must adjust to maintain the product K_j constant. If the crystal contains both Schottky and Frenkel defects, both products K_S and K_F must be satisfied.

Vacancies on two different sublattices bear opposite charges and, thus, are attracted by Coulomb forces. However, when two such vacancies are a nearest neighbor pair, the resulting defect gains additional binding energy through lattice relaxation and it also behaves as a dipole. It is then convenient to consider the pair a distinct defect, a vacancy pair V_p, which is the product of a reaction between the two vacancies. The reaction can be written as

$$V_M^- + V_A^+ = V_M V_A = V_P \qquad (6)$$

The concentration of pairs is given by

$$x_p = x_a x_c z \exp(g_p/kT) = z \exp[-(g_S - g_p)/kT)] \qquad (7)$$

where z is the number of distinct orientations of the pair ($z = 6$ in a NaCl lattice) and g_p is the free energy released when two isolated vacancies form a pair. Unlike the atoms in a chemical reaction, species involving only vacancies are not conserved as are chemical species; their concentration is a function of temperature only.

C. Extrinsic Defects

When an ionic crystal contains an aliovalent impurity (i.e., a cation or anion of differing valence from those of the host material), sufficient vacancies must be present in the host crystal to maintain overall electroneutrality. For example, if a divalent cation is present substitutionally in an alkali halide, a vacant cation site must also be present to compensate for the excess positive charge of the impurity. Again, the aliovalent impurity will attract vacancies bearing the opposite charge and when these two defects in the same sublattice are in their most stable configuration, they are considered a separate defect, an associated impurity–vacancy pair (i–v). Unless the impurity is small compared to the size of the ion for which it substitutes, the most stable configuration is as a nearest neighbor (n.n.). When it is small, then the next-nearest neighbor (n.n.n.) configuration will be more stable (Watkins, 1959a,b). This association can be written as the reaction

$$M_i^+ + V_C^- = M_i V_C \qquad (8)$$

A mass action expression gives the concentration of i–v pairs x_k as

$$x_k/[x_c(c_i - x_k)] = z_k \exp(g_k/kT) \qquad (9)$$

where c_i is the analytical concentration of the impurity i, z_k the number of distinct orientations of the pair ($z_k = 12$ for a n.n. pair and $z_k = 6$ for a n.n.n. pair in a NaCl lattice), and $-g_k$ the free energy of association for the i–v pair.

D. Defect Interactions

If the crystal contains more than a single impurity, the impurities may interact with each other as ions do in solution. In the interpretation of diffusion profiles, these interactions can be significant because, under most circumstances, the ratio of concentration of the various species may vary over a large range along the concentration versus distance curves. In other types of measurements (e.g., ionic conductivity) such interaction effects usually become important only when the various interacting defects or impurities have comparable concentrations.

The simplest such interaction is the common ion effect that occurs when two different defects both react with a third kind of defect (Krause and Fredericks, 1971). Consider a crystal containing two different aliovalent cationic impurities, each of which forms an i–v pair with cationic vacancies. These reactions can be written as

$$M_1^+ + V_C^- = M_1 V_C \tag{10}$$

and

$$M_2^+ + V_C^- = M_2 V_C \tag{11}$$

A mass action expression for either $M_1 V_C$ or $M_2 V_C$ must include the coupling of the two reactions through the common species V_C^-. The equations that describe the equilibrium are

$$x_{k1}/[x_c(x_1 - x_{k1})] = z_1 \exp(g_1/kT) \tag{12}$$

$$x_{k2}/[x_c(x_2 - x_{k2})] = z_2 \exp(g_2/kT) \tag{13}$$

$$x_a x_c = \exp(-g_S/kT) \tag{14}$$

$$x_p = z \exp[-(g_S - g_p)/kT] \tag{15}$$

In many experimental situations, the concentration of one or more of the defects is very small compared to the other and can be neglected.

The interactions between anionic and cationic impurities have been reported to perturb many types of experiments. The effects of the reaction between hydroxide and divalent cations have been observed in ionic conductivity (Fritz et al., 1963), ionic thermocurrent (Cappelletti and Fieschi, 1969), i–v dipole relaxation (Chaney and Fredericks, 1973), and optical properties and diffusion (Allen and Fredericks, 1970a, 1973). Analysis of the vibrational spectra of crystals doped with cyanate or sulfate ions and a wide variety of divalent cations show the formation of anion–cation compounds in alkali halides (Decius et al., 1963; Coker et al., 1961; Decius et al., 1965; Conant and Decius, 1967). While examples of these reactions are widely known, few studies of their properties have been made (Fritz et al.,

1963; Allen and Fredericks, 1973). Various reactions have been proposed for their formation (Stoebe, 1970).

The structure of the cation–anion impurity compound may be complex and involve substitutional and interstitial ions, or may extend over several sites. When the electronic configuration of the impurities differs markedly from the host, the bonding between them may be largely covalent. At present, very little quantitative data on such species are available but, as will be discussed later, they can greatly distort diffusion data.

The spin resonance studies of Mn^{2+} in NaCl and KCl by Watkins (1959a,b) and by Symmons (1970, 1971) provide detailed information on n.n.–n.n.n. i–v pairs. If each pair is considered a distinct species, the mass action expression for the n.n. complex is

$$x_k/[x_c(c_i - x_k - {}^nx_k)] = z_k e^{g_k/kT} \tag{16}$$

and the ratio ${}^nx_k$ to x_k is given by

$$ {}^nx_k/x_k = ({}^nz_k/z_k) \exp[({}^ng_k - g_k)kT] \tag{17}$$

where the superscript n refers to the n.n.n. pair. When the impurity is small compared to the ion for which it is substituted, the n.n.n. pair is the more stable of the two. Watkins (1959a) reports $x_k/{}^nx_k$ as 7.5 and 0.65 for Mn^{2+} in NaCl and KCl at room temperature, respectively. From dielectric relaxation measurements, Dreyfus (1961) found Mg^{2+} forms predominantly n.n.n. i–v pairs in NaCl.

E. Aggregation and Precipitation

When an impurity is introduced into a crystal at high temperatures, its concentration may exceed the solubility limit at a lower temperature. If the crystal is held at the lower temperature for a sufficiently long time, or if the crystal is cooled slowly enough, the impurity may precipitate as a second phase, or it may form an aggregate species of i–v dipoles. The formation of clusters of divalent cation–cation vacancy dipoles into trimers occurs with many such impurities before they precipitate (Cook and Dryden, 1962; Dryden and Harvey, 1969; Chiba et al., 1963; Chiba and Sakamoto, 1965; Symmons and Kemp, 1966). Crawford (1969) has summarized the available results. Cook and Dryden (1962) report that in alkali halides quenched from 350 to 400°C all the impurity ions are initially present as dipoles (2–5% dissociated) so that aggregate species should not be important during the usual diffusion experiment, but may form after the crystal has been removed from the diffusion furnace.

Some divalent ions precipitate as an ordered phase (Suzuki, 1961). The stability of these phases has not been extensively investigated. Toman (1962) found this phase for Cd^{2+} in NaCl to be stable only to 250–300°C and not present during diffusion above this temperature.

F. LONG RANGE COULOMB INTERACTIONS

Mass action expressions written in terms of concentrations, because of their inherent simplicity, are convenient and useful in discussing defect reactions. However, these expressions imply that the reacting defects are randomly distributed which, because of long range Coulomb forces, is not the case at moderate impurity concentrations. For example, around a vacancy in the cation sublattice, the probability of finding a vacancy in the same sublattice is much less than that of finding one in the anion sublattice. The similarity to aqueous ionic solutions is obvious and Lidiard (1957) applied Debye–Hückel theory to a model based on neutral n.n. pairs and otherwise unassociated defects. In the context of ionic solution theory, this is equivalent to using an effective concentration a_j which is equal to $\gamma_j x_j$, where γ_j is an activity coefficient. Lidiard obtained the expression

$$\gamma_j = \exp[-q^2\kappa/2\varepsilon kT(1 + \kappa R)] \tag{18}$$

where κ is given by

$$\kappa^2 = 4\pi \sum x_j q_j^2 / v\varepsilon kT \tag{19}$$

Here v is the molecular volume, ε the dielectric constant, x_j the mole fraction of j, q_j the defect's effective charge, and R the distance of closest approach between unpaired defects. Lidiard took R to be equal to the n.n.n. distance and estimated that the Debye–Hückel correction reduces the enthalpy of formation of the Schottky defect h_S by slightly less than 0.1 eV for typical alkali halides at high temperature.

In Debye–Hückel approximation, the defect "sees" the average potential of all other charged defects in their most probable configuration. As used by Lidiard, the defects are divided into noninteracting close pairs and interacting unassociated defects. A more rigorous treatment using cluster formalism was made by Allnatt and Cohen (1964a,b) and, reviewed by Allnatt (1967), it provides a completely general treatment of interacting defects. This leads to the defect contribution expressed as a viral expansion in the defect concentration. The slow convergence and awkward form make it difficult to use in analysis of experimental data. However, as a first approximation, it reproduces the Lidiard–Debye–Hückel form. Even the latter approach to defect interactions has not been widely used in the analysis of experimental results because of the presence of x_j in the Debye–Hückel screening constant [Eq. (19)].

III. Theory of Diffusion in Ionic Crystals

A. Not Quite Random Walks

The theory of matter transport in solids is based on models in which the atoms migrate by jumping from one position to another. If these jumps comprise a random walk in a simple cubic crystal, the diffusion coefficient D is related to the displacement distance λ and jump rate Γ by the expression

$$D = \tfrac{1}{6}\Gamma\lambda^2 \qquad (20)$$

However, if the diffusion occurs by a process which destroys the local symmetry around the diffusing atom, its movement will no longer be random. For example, immediately after a tracer atom has moved by a vacancy mechanism the vacancy is still adjacent to the tracer atom. The tracer is more likely to return to its original position than to proceed by random diffusion (Bardeen and Herring, 1951). The tracer atom then follows a path in which each jump is correlated with the preceding jump and only a fraction $f\Gamma$ of the tracer atom's jumps lead to random diffusion, and Eq. (20) becomes

$$D = \tfrac{1}{6}f\Gamma\lambda^2 \qquad (21)$$

The correlation factor f depends on the coordination number of the tracer site and on the diffusion mechanism.

For self-diffusion, only a single jump frequency is required and a unique value of f can be obtained for each lattice and mechanism. In all cases, f is less than but of the order of unity. [For compilations of f, see Manning (1968) and LeClaire (1970); for vacancy-pair mechanism, see Nelson and Friauf (1970).]

In impurity diffusion, f is a function of the various jump frequencies (Fig. 1) which enter into the diffusion process. Those ions in a region near the impurity will have their jump frequencies perturbed by it. The jump frequency of an unperturbed solvent ion into a vacancy is called w_0, the impurity–vacancy exchange frequency is w_2, the jump frequency of a vacancy from one n.n. site of the impurity to an equivalent n.n. site is w_1, and the jump frequency of a vacancy to a more distant site from the impurity (i.e. dissociation of the vacancy–impurity, n.n. → n.n.n.) is w_3.

If it is assumed that once the dissociative jump occurs, the dissociated vacancy returns to a n.n. site with only random probability (i.e., they are "lost"), the correlation factor is given by (LeClaire and Lidiard, 1956)

$$f = (2w_1 + 7w_3)/(2w_1 + 2w_2 + 7w_3) \qquad (22)$$

This degree of approximation is consistent with the assumptions inherent in a simple mass action treatment of defect equilibrium.

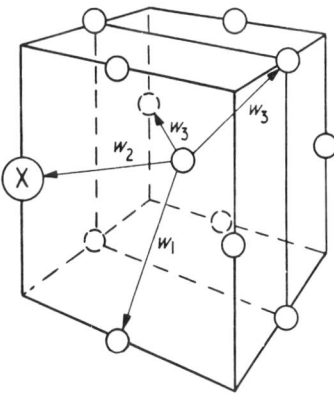

Fig. 1. A section of a fcc lattice. × impurity ion, ○ nearest neighbor. The central site is vacant and the site marked × is occupied by a solute atom; all other sites are occupied by solvent atoms. The rates of movement of the vacancy to the various sites are as indicated. It is assumed that all jumps which carry the vacancy away from the solute atom occur with the same frequency w_3 even though they are not crystallographically equivalent. The w_4 jump is the inverse of the w_3 jump (Howard and Lidiard, 1964). (© Institute of Physics.)

However, as a result of the long range of Coulomb forces, it is very unlikely that this assumption will be valid for aliovalent impurities and not even possibly for homovalent impurities. Manning (1962b, 1964) has considered this problem in detail and obtained

$$f = [2w_1 + 7w_3 F(w_4/w_0)]/[2w_1 + 2w_2 + 7w_3 F(w_4/w_0)] \quad (23)$$

where w_4 is the jump frequency of a vacancy from a n.n.n. to a n.n. position to an impurity and differs from w_0 and w_3. The quantity F is a function of w_4/w_0 and is shown in Fig. 2. The jump rates are not all independent. At equilibrium, $x_k w_3 = {}^n x_k w_4$ and Eq. (17) then gives the ratio w_3/w_4. As noted in Section II, divalent impurities which are small compared to the host ion for which they substitute tend to form more stable n.n.n. pairs than n.n. pairs making $w_3 > w_4$.

Using Eq. (23) is consistent with using Eqs. (1), (7), (16), and (17) to describe the defect equilibria. A correlation factor consistent with the Debye–Hückel approximation should include more distant vacancy–impurity interactions. Manning (1964, 1965b) has considered the case of impurity–vacancy interaction at second neighbors.

The effect of a dilute solute on solvent diffusion is important in self-diffusion in slightly impure host crystals. It is probably the most complex case to which diffusion kinetics have been applied. Even when the impurity is restricted to sites that are nearest neighbor to either the vacancy or the solvent tracer, the number of tracer–vacancy jump frequencies introduced into the diffusion process is large. For example, in a fcc lattice, there are 18 such sites which introduce 13 distinct jump frequencies for the tracer. A general solution to this problem, based on partial correlation factor calculations, was given by Howard and Manning (1967). Compaan and

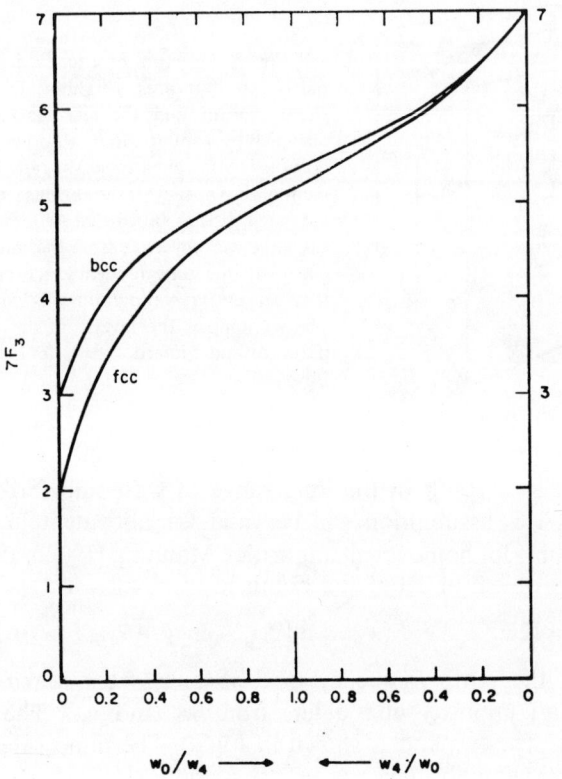

Fig. 2. Values of $7F_3$ for the bcc and fcc lattices (Manning, 1964).

Haven (1956) and Howard (1966) have considered this problem in the more restricted case of tight binding (i.e., the tracer always jumps to a vacancy bound to an impurity). Lidiard (1960) discussed this problem in a weak binding approximation. Howard's and Manning's results reduce to Lidiard's when their average partial correlation factors are equal to the correlation factor in the pure material. There is insufficient space here to discuss the details of this problem. The reader is referred to LeClaire's (1970) review for a summary of its current status.

An interesting consequence of the inequality of the various jump frequencies that occur in f for diffusion of tracer ions in dilute solutions or for impurity diffusion is that f itself must be temperature dependent (Manning, 1958). This is illustrated in Fig. 3 which shows possible values of f calculated from Symmons' (1970) values of w_i for the NaCl : Mn^{2+}

8. DIFFUSION IN ALKALI HALIDES

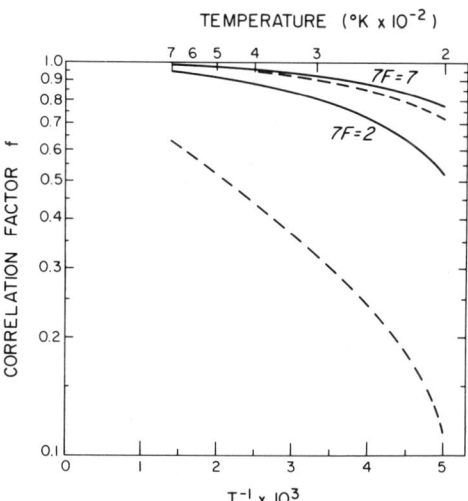

Fig. 3. The correlation factor f calculated for Symmons' (1970) values of w_i assuming they are valid at high temperature. The three upper curves are calculated from Eq. (23) with $7F(w_4/w_0) = 7$, 5.15, and 2; i.e., $w_4/w_3 = 0$, 1, and ∞, respectively. The lower curve was calculated from Eq. (22) assuming $w_3 = 0$.

system. The temperature dependence of f is given by

$$C = k \, \partial \ln f / \partial T^{-1} \tag{24}$$

Most experimental measurements are presented in the form of an Arrhenius equation

$$D_s = D_0 \exp(-U/kT) \tag{25}$$

where U is an activation energy and D_0 contains the temperature independent constants. Since the solute diffusion coefficient is also given by

$$D_s = a^2 f x_k w_2 \tag{26}$$

U must contain the temperature dependent parts of x_k, w_2, and f. The activation energy U is then

$$U = h + h_m - C \tag{27}$$

where h is the enthalpy of formation of a vacancy at a n.n. site, and h_m the enthalpy of migration. The quantity C is negligible only when $f \to 1$. Thus, the correlation factor through the temperature dependence of C can cause curvature of the Arrhenius plot. Calculations by LeClaire (1962) suggest this curvature will be small and negative.

The correlated walk approach to diffusion in ionic crystals can be especially useful when combined with conductivity measurements in establishing the diffusion mechanism for self-diffusion (Friauf, 1957) or else when using isotope effects‡ (Barr and LeClaire, 1964). However, when applied to impurities or self-diffusion in dilute solution, the various jump frequencies required to evaluate f are generally unavailable. They can be calculated when both dielectric and anelastic relaxation data are available if an independent measure of dipole concentration can be made (Symmons, 1970).

While this theory was developed primarily for solute-enhanced diffusion in alloys, a similar situation is found in self-diffusion in NaCl-type lattices containing aliovalent impurities. Unfortunately, there is not yet a treatment for the frequently encountered case of the simultaneous diffusion of a tracer solvent ion and an aliovalent impurity.

B. A Nonequilibrium Thermodynamic Approach

A major advance in applying thermodynamics to diffusion in ionic crystals came with the recognition that for a vacancy mechanism it is the tracer ion–vacancy pair that diffuses and the development of this approach by Lidiard (1957). Later Howard and Lidiard (1964) generalized these ideas into an elegant and comprehensive theory discussing the phenomena associated with matter transport in solids. We give only the details necessary to illustrate the assumptions inherent in the results and to compare them with the simpler theory (Lidiard, 1957) frequently used to describe experimental results.

Pure, strongly ionic crystals as normally grown are stoichiometric, electrolytic conductors with the anions and cations confined to separate sublattices and, with the exception of any external or internal surfaces, are electrically neutral everywhere. Thus the thermodynamic cross terms between anions and cations may be neglected and in this approximation, it is satisfactory to treat the processes occurring on different sublattices separately. Some coupling of the diffusion fluxes may arise through the electrical diffusion potentials (Nernst potential). Howard and Lidiard represent this by including the corresponding electric fields in the forces used in the phenomenological equations. The kind of coupling which arises from vacancy interactions of the type producing Debye–Hückel screening is omitted.

For isothermal diffusion on one sublattice containing two different

‡ See also Peterson, Chapter 3.

8. DIFFUSION IN ALKALI HALIDES

species, A and B, the phenomenological equations of irreversible thermodynamics can be written as

$$J_A = -L_{AA}(\nabla\mu_A - \nabla\mu_V) - L_{AB}(\nabla\mu_B - \nabla\mu_V) + (q_A L_{AA} + q_B L_{AB})E \quad (28a)$$

$$J_B = -L_{BA}(\nabla\mu_A - \nabla\mu_V) - L_{BB}(\nabla\mu_B - \nabla\mu_V) + (q_A L_{BA} + q_B L_{BB})E \quad (28b)$$

where J_A and J_B are the flux of species A and B, respectively, referred to the local crystal lattice which is in mechanical equilibrium. The L's are the phenomenological coefficients and, as written, obey the Onsager reciprocal relations. The thermodynamic forces are written conventionally in terms of the gradient in the isothermal chemical potential μ_i and the total electric field E. Here q_A and q_B are the charges on ions A and B, respectively.

In the pure region of the crystal, the field can only be the external field E_{ext} and the conductivity is Λ_0, while in the region of the crystal which contains some species B, the field is $E = E_{ext} + E_{int}$ and the conductivity is Λ. In purely ionic conductors, the internal field must be such that it ensures the uniformity of the total electric current. One obtains

$$E_{int} = \{(q_A L_{AA} + q_B L_{BA})(\nabla\mu_A - \nabla\mu_V) + (q_A L_{AB} + q_B L_{BB})(\nabla\mu_B - \nabla\mu_V) \\ - (\Lambda - \Lambda_0)E_{ext}\}\Lambda^{-1} \quad (29)$$

Substitution of $E = E_{int} + E_{ext}$ into (28), with E_{int} given by (29), on rearrangement gives the flux of B ions as

$$J_B = \frac{(L_{AA}L_{BB} - L_{AB}L_{BA})[q_A q_B(\nabla\mu_A - \nabla\mu_V) - q_A^2(\nabla\mu_B - \nabla\mu_V)]}{q_A^2 L_{AA} + q_A q_B(L_{AB} + L_{BA}) + q_B^2 L_{BB}} \\ + (q_A L_{BA} + q_B L_{BB})(\Lambda_0/\Lambda)E_{ext} \quad (30)$$

The flux J_B in the absence of an external field is just the diffusion flow. When diffusion is by a vacancy mechanism, the diffusion of B is given by

$$J_B = -ND_B \nabla x_k \quad (31)$$

Now it is necessary to depart from a general ionic solid limited only in that it be an ionic conductor and introduce specific defect models to obtain $\nabla\mu_A$, $\nabla\mu_B$, and $\nabla\mu_V$. Howard and Lidiard consider diffusion in a simple association model. The diffusion coefficients were evaluated for two cases, $q_A = q_B$ and $2q_A = q_B$. To simplify these expressions, we write the Schottky product [Eq. (1)] with $x_a = x_V' - c_B$ and $x_c = x_V' - x_k$. Thus, x_V' is the equilibrium vacancy concentration in a pure crystal (i.e., $x_V' = K_S^{1/2}$).

When $q_A = q_B$, there should be little pairing between impurities and vacancies, i.e., $x_k \ll c_B$, because there is no electrostatic binding and

$$D_B = [(L_{AA}L_{BB} - L_{AB}L_{BA})kT]/[Nc_B(L_{AA} + L_{AB} + L_{BA} + L_{BB})] \quad (32)$$

When $L_{AA} \gg L_{AB}$ and L_{BB}, this reduces to

$$D_B = kTL_{BB}/Nc_B \quad (33)$$

When $2q_A = q_B$ the general expression for D_B is

$$D_B = \frac{2kT(L_{AA}L_{BB} - L_{AB}L_{BA})}{N(L_{AA} + 2L_{AB} + 2L_{BA} + 4L_{BB})} \frac{(x_V' - x_k)}{(2x_V'c_B - x_k^2 - c_B^2)} \quad (34)$$

Two limiting cases are of special interest. When $c_B \ll K_S^{1/2}$, the low impurity concentration limit with $L_{AA} \gg L_{AB}$ and L_{BB}, Eq. (34) simplifies to Eq. (33). In the other limit $(c_B \gg K_S^{1/2})$, assuming the same relationships among the phenomenological coefficients, it reduces to

$$D_B = 2kTL_{BB}/Nc_B(1 + p) \quad (35)$$

where p is the fraction of impurity ions associated with vacancies

$$p = x_k/c_B$$

To evaluate the L_{ij} coefficients, it is necessary to consider the specific lattice in which the particular diffusion process occurs. Howard and Lidiard discussed the NaCl structure in which each sublattice is fcc and for which both the J_A and J_B, derived from a kinetic treatment of pair diffusion and a thermodynamic approach, are available (Lidiard, 1955).

Substitution of Howard and Lidiard's (1963) expression for L_{BB} in Eq. (33) gives

$$D_B = \tfrac{1}{3}a^2 w_2 \, fp \quad (36)$$

when

$$L_{AA} \gg L_{AB}, L_{BB}, \qquad q_A = q_B, \qquad x_k \ll c_B$$

or

$$L_{AA} \gg L_{AB}, L_{BB}, \qquad 2q_A = q_B, \qquad c_B \ll K_S^{1/2}$$

Under similar conditions, Eq. (34) reduces to either

$$D_B = \tfrac{2}{3}a^2 w_2 \, f\{p(x_V' - pc_B)/[2x_V' - c_B(p^2 + 1)]\} \quad (37)$$

when

$$L_{AA} \gg L_{AB}, L_{BB}, \qquad 2q_A = q_B$$

or

$$D_B = \tfrac{2}{3}a^2 w_2 \, f[p/(1 + p)] \quad (38)$$

when
$$L_{AA} \gg L_{AB}, L_{BB}, \quad 2q_A = q_B, \quad c_B \gg K_S^{1/2}$$

The condition $c_B \gg K_S^{1/2}$ is often satisfied for diffusion of moderately soluble divalent ions in monovalent crystals at normal experimental temperatures. When the attraction between the impurity and vacancy is small or in the limit of low impurity concentration, the assumption that $L_{AA} \gg L_{AB}, L_{BB}$ is reasonable, but in the limit of $c_B \gg K_S^{1/2}$ this assumption is not necessarily valid, although it is frequently assumed.

Figure 4 illustrates the striking concentration dependence of D_B as calculated from (38). The strong dependence of D_B on c_B at low concentrations of impurity is expected if the flux of B ions is interpreted as a flux of B-ion–vacancy pairs. This appears reasonable as the impurity can only diffuse when a vacancy is adjacent to it. The flux of B ions is then proportional to the gradient ∇x_k until $p \to 1$ and D_B becomes concentration independent at its saturation value D_s. However, from Eq. (30), it is evident that the electrical diffusion potential affects the form of Eq. (34) and the relations derived from it. Therefore, to neglect this potential will lead to wrong results. Experimental measurements, which will be discussed in Section V, confirm the concentration dependence predicted in Eq. (38).

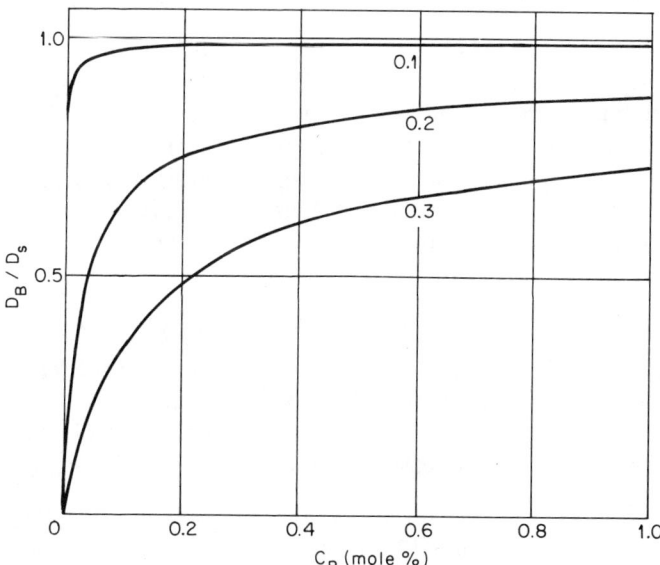

Fig. 4. Theoretical concentration dependence of the diffusion coefficient D_B relative to its saturation value D_s of divalent ions in NaCl-type crystals for three different temperatures. Numbers on the curves are values of kT in units of the association free energy Δg (Howard and Lidiard, 1964). (© Institute of Physics.)

Howard and Lidiard (1964) also considered the case by which the host crystal contains a uniform concentration of B into which a radioactive isotope B* diffuses. Using similar arguments for three species, A, B, and B*, they found

$$D_{B^*} = (a^2 w_2 \, f/3)\{p + [1 - (c_{B0}/c_B)][pc_B(1-p)^2]/[2x_{V'} - c_B(p^2+1)]\} \quad (39)$$

where c_B is the total concentration of B and B* and c_{B0} is the concentration of B before diffusion of B*. When $c_{B0} \ll c_B$, Eq. (39) reduces to Eq. (37) and $D_{B^*} = D_B$. If the crystal is heavily doped, i.e., $c_B = c_{B0}$, throughout the diffusion, then D_{B^*} is given by Eq. (36) and D_{B^*} is concentration dependent through the factor p. In the intermediate region, Eq. (39) describes D_{B^*}. Comparison of (36) and (38) shows that for diffusion of a tracer into a doped crystal, D_{B^*} will vary more slowly with c_B than in a pure crystal.

As yet, the diffusion potential has not been studied directly in ionic solids. If it is omitted from the calculation, but the pair diffusion model and the assumption of local equilibrium are retained, D_B is given by (Lidiard, 1957)

$$D_B' = (a^2 w_2 \, f/3)\{1 - [1 + 48c_B \exp(\Delta g/kT)]^{-1/2}\} \quad (40)$$

This equation exhibits a concentration dependence exactly like Eq. (38) (Fig. 4). When $p \to 1$, Eqs. (40) and (38) approach the same limit

$$D_B' = D_B = a^2 w_2 \, f/3 \equiv D_s \quad (41)$$

As yet, this thermodynamic approach has not been extensively applied to other ionic lattices‡ or to other mechanisms in them. However, it has been applied to other diffusion problems in ionic crystals, the most important of these being the generalized Nernst–Einstein relation.

The relation of the mobility of a charge carrier to its diffusion coefficient, the Nernst–Einstein ratio, is usually given as

$$\lambda_B/D_B = q_B/kT \quad (42)$$

where λ_B is the mobility of the charge carrier B. This relation was approximately verified for monovalent ions in NaCl, but λ_B/D_B was much less than q_B/kT for divalent ions in the same host. The Nernst–Einstein relation for an ionic conductor obtained by a thermodynamic treatment (Howard and Lidiard, 1964) is

$$\lambda_B/D_B = |q_A L_{BA} + q_B L_{BB}|/kT L_{BB} \quad (43)$$

when $x_k \ll c_B$ for the case $q_A = q_B$ or when $c_B \ll K_S^{1/2}$ and $2q_A = q_B$ such

---

‡ A discussion of mass transport in fluorite lattices has recently appeared (Lidiard, 1974).

that D_B is given by Eq. (33). If $L_{BA} = 0$, the usual form, Eq. (42), of the Nernst–Einstein equation results. In solids, the off-diagonal term does not usually vanish even though there is a single mechanism for mass and charge transport.

When B is an isotope A* of A, as in self-diffusion experiments, McCombie and Lidiard (1956) have shown that for a vacancy mechanism

$$\lambda_{A*}/D_{A*} = \{f^{-1}\}|q_A|/kT \qquad (44)$$

and for a direct interstitialcy mechanism

$$\lambda_{A*}/D_{A*} = \{f^{-1}\}2|q_A|/kT \qquad (45)$$

are exact expressions for the Nernst–Einstein ratio. Experimentally, λ is obtained from a measurement of the ionic conductivity. When ionic conductivity and diffusion measurements are made on the same material, a value of f can be calculated. Since f depends on the kinetics of the diffusion process, these measurements provide a method of determining the diffusion mechanism. Manning (1962a,b, 1965) has developed explicit equations for λ/D for both vacancy and interstitialcy mechanisms in many lattice types.

When two or more types of mobile defects are present, they contribute additively to the conductivity, to the diffusion of charge and to the diffusion of tracer. In such a case, one cannot obtain f directly from measurements of D and λ (LeClaire, 1970). LeClaire suggests using the term "Haven ratio" to represent

$$Dq/kT\lambda = DNq^2/kT\sigma \equiv H_r, \qquad (46)$$

in order to distinguish this experimental ratio from the true correlation factor. (Here $\sigma = N\lambda q$ is the conductivity and N is the number of carriers per unit volume.)

Simple isothermal diffusion studies do not provide a means of estimating the magnitude of the Nernst potential [i.e., the first two terms of Eq. (29)]. The inclusion of this potential gives the general solution for D_B [Eq. (34)]. Only under the conditions used to derive Eq. (38) do the general thermodynamic results agree with Lidiard's (1957) solution [see Eqs. (40) and (41)]. Yet many experiments are interpreted with equations derived by Lidiard's (1957) method and it is not always established that the required restrictions have been imposed on the experiment. It would be useful to know when the Nernst potential can be neglected and the simpler derivation is a good approximation. Two experiments offer promise of providing such an estimate. One is the Soret effect; the other is measurement of the homogeneous thermoelectric power. As yet, neither measurement has been made with sufficient accuracy to permit this, but recent work offers great promise.

The theory of the Soret effect has been developed in detail (Wirtz, 1943; Howard, 1957; Allnatt and Chadwick, 1966). The thermoelectric power of ionic crystals has been discussed by Howard and Lidiard (1964).

At present, only simple association theory has been used in the derivation of matter transport equations. The Lidiard–Debye–Hückel theory attempts to allow for nonideality caused by Coulomb interactions among unassociated defects (Lidiard, 1957). With the assumptions given in Section II, the effect is to alter the meaning of K. An alternate formulation expresses the chemical potential in the form (Allnatt and Cohen, 1964a,b) of a cluster expansion.

Recently, simple approximations have been developed which represent the exact cluster integral to $\sim 1\%$ accuracy in the concentrations used in most impurity experiments in strongly ionic solids (Allnatt et al., 1972; Allnatt and Loftus, 1973).

IV. Experimental Methods

A. General

Measurement of a diffusion coefficient in an ionic solid should be a relatively simple experiment. For a controlled set of experimental conditions, a measurement of the penetration of the diffusant into the crystal, or of the change of total diffusant concentration with time, and the solution of Fick's second law applicable to the initial and boundary conditions established for the experiment will suffice. However, the measurements are tedious, generally cannot be automated, and unfortunately are subject to many perturbations which may sometimes escape detection.

Many details on techniques are given by Tomizuka (1959) in his review of some general methods used in diffusion studies. Solutions of Fick's laws for many initial and boundary conditions have been given by Crank (1956).

The general sequence of operations occurring in such diffusion experiments have been described many times (cf. Keneshea and Fredericks, 1963, or Krause and Fredericks, 1971). Here only those items will be discussed which may introduce errors in the experiment.

B. Significant Factors in Experimental Design

1. Pure Host Crystals

Any impurity can perturb the profile. The most serious are aliovalent ions or ions on the sublattice opposite that of the diffusant that form bonds with the diffusant which differ markedly from those the normal ions in that sublattice form with the diffusant. The more the impurities resemble the host lattice ions, the less likely they will perturb the profile.

The major source of diffusant–impurity interaction problems arises from unsuspected contamination in the host crystals. Such effects may not be as small in self-diffusion measurements (Feit et al., 1973) as is suggested by introducing a small error in the diffusion coefficient through f (see Section III.A). If the diffusant is a different ionic species but homovalent with the ions in the corresponding sublattice, a perturbing interaction between random impurities and the diffusant can occur. The magnitude of the effect depends on the difference in binding energy of the diffusant–impurity compound and the host–impurity compound. If the diffusant and impurity are aliovalent, the interactions are Coulomb and long range. The only such interaction specifically studied by diffusion involves a homovalent anionic impurity OH^- effusing from KCl while an aliovalent cationic impurity Hg^{2+} diffuses into the crystal. The $Hg^{2+}-OH^--KCl$ system forms a complex set of impurity compounds. The interrelations among them and the conditions under which they form are given by Allen and Fredericks (1970a). As the Hg^{2+} diffuses into the crystal it encounters OH^- effusing from the crystal. These two ions form a strongly bound compound which appears in comparison with Hg^{2+} or OH^- to be immobile. This produces a concentration versus distance curve with an unusual hump of tracer ^{203}Hg well within the crystal where the mercury and hydroxide compound concentrates. Figure 5 illustrates one of these unusual profiles (Allen and Fredericks, 1973). The lower curve shows the profile of Hg^{2+} diffused into pure KCl under identical conditions. As the diffusion continues for longer times, the hump moves into the crystal and diminishes in amplitude. If the diffusion anneal is extended to very long times, the mercury–hydroxide hump becomes a knee. Note that the amount of Hg^{2+} in the crystal is greatly increased over that found in a pure crystal (Fig. 5) and that at penetration distances on the decreasing side of the hump, the Hg^{2+} concentration falls below the normal profile. The effect in this case is dramatic because of the very low solubility of the tracer ion. However, when the tracer is present in much greater concentrations than the other impurity, the interaction is difficult to detect and leads to underestimates of D and overestimates of activation energies.

When unsuspected aliovalent impurities are present on the same sublattice as the diffusant, they may interact through the common ion effect of the vacancies with which they both form complexes. To obtain reliable diffusion parameters when these interactions are strong, the random impurity concentrations must be several orders of magnitude less than the lowest concentration of the diffusant required to satisfy the boundary condition $(dc/dx)_{c\to 0} = 0$. The sensitivity and precision of the analytical method

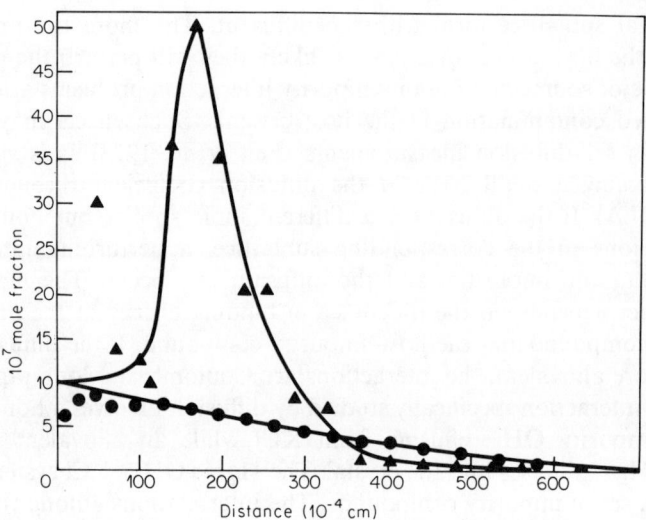

Fig. 5. Comparison of mercury diffusion in purified and hydroxide doped KCl. Source temperature was 180°C. Crystal temperature was 473°C. The diffusion time was 7.58×10^4 sec. ● purified KCl, ▲ KCl : OH^- with an initial OH^- concentration of 2×10^{-6} mole fraction. Diffusion atmosphere $\frac{1}{3}$ atm Cl_2 with no preanneal (Allen and Fredericks, 1973).

usually sets the lower limit on the measurement of diffusant concentration. At present, there is no quantitative method of estimating the magnitude of these interactions. Although many such impurity–impurity compounds are known, no association energies have been measured.

With the exception of some alkali halides (Rosenberger, 1972), very few ionic crystals have been prepared with the purity required for the study of aliovalent ion diffusion. While many ionic crystals are available commercially, none can be relied upon to be sufficiently pure for such studies. The safest procedure is to analyze each boule from which samples are to be cut for diffusion studies. Unfortunately simple assay analyses are generally not sufficiently accurate or specific enough to establish the quality of the crystal and rather difficult and expensive trace analyses should be performed. When doped crystals are stored for long periods of time, they may become contaminated with OH^- (cf. Chaney and Fredericks, 1973). The problem of obtaining sufficiently good ionic host crystals is a serious limitation on making accurate aliovalent ion studies.

2. *Pure Diffusant*

Most current diffusion studies use radioactive tracers. The supplier specifies the radioactive purity and unfortunately this is not related to the chemical purity.

8. DIFFUSION IN ALKALI HALIDES

Obviously, short lived isotopes may require separation of the decay product from the isotope before use as a source. For example, $^{203}_{80}$Hg decays to $^{203}_{81}$Tl with a 49-day half-life. A fresh $^{203}_{80}$Hg source will be 25% Tl after 20 days, a moderate time from the beginning of an experiment until the counting is complete. Use of the same batch of isotope for a diffusion experiment from a deposited surface layer would in fact be a study of the simultaneous diffusion of mercury and thallium. In this particular case, assuming TlCl and $HgCl_2$ form an ideal solution, a vapor phase diffusion experiment should be valid since the heats of vaporization (Brewer, 1950) vary by almost a factor of 2. (The corresponding vapor pressures are $P_{HgCl_2} = 7.5 \times 10^{-3}$ and $P_{TlCl} = 5.9 \times 10^{-11}$ at 369°K over the 20-day-old solution.)

Most isotopes and carriers are only available as aqueous solutions or are made into solutions to convert the isotopes to the salt required or to dilute the isotope with carrier. Many salts of interest crystallize as hydrates. If the hydrate is strongly heated to drive off the water, some salts form a hydroxychloride which can cause impurity–hydroxide compounds to form during the diffusion (Feitknecht, 1953). These may distort the diffusion profile. Each salt requires individual treatment. With halides, the treatment may vary from drying under a stream of warm argon to dissolving the corresponding halogen or hydrogen halide in the molten salt. A suitable drying procedure should be established prior to the experiment, but a post facto indication of H_2O or OH^- in the source salt is visible etching of the Vycor or quartz capsule around the region where the source was deposited.

3. Suitable Method for the Measurement of the Penetration of the Diffusant

Obviously there are many techniques available, but the choice of the best type of experiment is governed by the system being investigated. If the diffusion coefficient is concentration dependent, the concentration versus distance function must be measured. If the diffusion coefficient is constant it is usually sufficient to measure the total concentration change of the diffusant as a function of time at a fixed temperature. If the diffusant forms a compound or eutectic with the host, the presumed boundary conditions of the experiment may not be met in diffusion from a surface deposit. In desorption experiments, reflection of the effusing ion can occur at the surface. The amount of reflection depends on the chemical potential of the desorbed ion in the atmosphere surrounding the crystal.

An ideal experiment could use any method appropriate to the system. The most reliable method for obtaining concentration versus distance curves is sectioning the host crystal following the diffusion anneal. This well-known method need not be discussed here (see Mapother *et al.*, 1950; Shirn *et al.*,

1953; Keneshea and Fredericks, 1963). Probably because sectioning is tedious and sometimes inconvenient, other methods of obtaining the diffusion coefficient have been developed. They are generally useful only when it can be demonstrated that the diffusion coefficient is not concentration dependent. Some examples of the use of methods other than sectioning will be mentioned here.

Ikeda (1964) has followed the diffusion of Cd^{2+} in NaCl by measuring the penetration microscopically after producing colloidal coloration by electron bombardment. After the sample had been annealed for a known time at constant temperature and the penetration depth measured, the total amount of Cd^{2+} in the sample was determined spectrophotometrically. Using a method due to Wagner (1950), Ikeda estimated D_s and with Lidiard's theory (1957) obtained the free energy of association. Comparison of his data with tracer diffusion measurements suggested the method was not valid (Allen et al., 1967b).

Reisfeld et al. (1965a) have measured the diffusion of Tl^+ into KCl by following the absorbance of the 247 nm band of KCl:Tl as a function of distance from the edge of the crystal. This method suffers from several difficulties such as the need for collimation of the incident light beam, precise orientation of the crystal to the light beam, and the fact that the simple form of Beer's law does not apply when a concentration gradient exists perpendicular to the light beam. This last effect is small when the absorbance is small.

Another type of such measurement utilizes the variation of some parameter M that can be related to the total concentration of the diffusant as a function time at constant temperature (Crank, 1956). This type of experiment was used to obtain D for OH^- in KCl by desorption from single crystals using optical absorption at 206 nm as a measure of the OH^- concentration (Allen and Fredericks, 1970b). Glasner and Reisfeld (1961) mixed a powder of pure diffusant with a fine powder of the host crystal and used the optical absorption characteristic of the diffusant in that host as a measure of concentration. Care must be taken to avoid contamination in preparing the powders. In fact, Pringsheim (1949) reports that even gentle grinding will introduce enough Tl^+ to KCl to produce a phosphor.

A somewhat related method (cf. Patterson et al., 1956) measures the desorption of a radioactive isotope from a crystal by following the rate at which it appears in the atmosphere surrounding the crystal. Lagerwall (1965) has shown that for large diffusion distances the shape of the sample influences the results. The isotope exchange method is very useful in measuring small diffusion coefficients and has been widely applied to anion diffusion measurements in alkali halides.

Diffusion parameters can be obtained from many other types of measurements which are not primarily diffusion experiments. The more important of these are the various relaxation studies that involve the reorientation of impurity–vacancy dipoles. These have recently been reviewed by Nowick (1972) and need not be discussed here. However, it will be noted that most such measurements are perturbed by unsuspected impurities, but not in quite the same way as in diffusion. For instance, if cation–vacancy studies are being made on a crystal which contains OH^- ions, an additional relaxation time may be introduced (Chaney and Fredericks, 1973).

4. Boundary Condition Problems

Even after a pure host crystal and pure diffusant have been obtained and a suitable diffusion technique devised, events can occur during the anneal that perturb the results. These generally occur by some circumstance that changes the boundary conditions of the experiment. Three such problems are considered here as illustrations.

In impurity diffusion, if the diffusant forms a eutectic or compound with the host, specified source conditions are not maintained during anneal. When surface deposited sources are used, this problem cannot be avoided if the impurity and host form such eutectics or compounds. In vapor source experiments, techniques have been developed to avoid this problem (Krause and Fredericks, 1971, 1973).

In desorption experiments, some diffusion from the surface back into the host crystal (reflection) can occur. If the amount of reflection is known, this is not a problem (Crank, 1956). More often it is unknown, and it is safest to remove the diffusant as it arrives at the surface by reaction with the surrounding atmosphere. Obviously, the atmosphere and reaction products should not diffuse at a rate comparable to the substance being studied.‡ Reflection can cause appreciable variation in diffusing parameters (Allen and Fredericks, 1970b).

When the anneal is terminated in a vapor source experiment the capsule cools first, the diffusant vapor condenses on it, and diffusant desorbs from the crystal. This produces a profile that has abnormally low diffusant concentration near the surface. Effective quenching reduces this effect, but it cannot be completely eliminated from high temperature anneals. It is then necessary to show by varying the quenching rate that the parameters derived from the deeper portion of the profile are unperturbed by this desorption.

‡ *Note added in proof:* Allen and Fredericks assumed that Cl_2 reacted with the effusing OH^- as it reached the surface of the crystal. However, Ikeda (1973) has shown the OH^- profile is unusual in the presence of Cl_2 and that an additional optical absorption band at 240 nm (O_2^-) occurs at an interface between a region free of OH^- and the OH^--doped region. This observation suggests a reaction responsible for the decrease in the 204 nm (OH^-) band occurs deep within the crystal.

5. Mathematical Analysis

The preceding part of this section demonstrates the necessity of careful analysis of diffusion data. Ideally, the diffusion profile itself is calculated from a model that considers all factors contributing to the diffusion. If a simplified model is used, it should be justified and proven acceptable by showing that the profile behaves in a fashion that fits the model.

When there are no interactions of the diffusant with its environment, the diffusion coefficient is constant and the analysis of the diffusion data is relatively simple. Many methods and the required solutions of Fick's laws for the experimental boundary conditions are known (cf. Crank, 1956; Jost, 1960). A migration energy can be obtained from measurement of D at several temperatures and, if the mechanism is known, the correlation factor can be obtained for the particular lattice under study. This case requires no discussion here.

If the diffusion coefficient is concentration dependent, the analysis of data is somewhat more complex. Frequently the method of Matano (1933), or some modification of it, is used to obtain the concentration dependent diffusion coefficient $D(c)$. The diffusion parameters are usually calculated by a least-squares fitting of equations such as (37) with p obtained from Eq. (9) or Eqs. (1) and (9) together. After $D(c)$ at a particular T has been obtained, the free energy of association and the value of $D(c)$ as $p \to 1$ can be obtained fitting an equation giving the dependence of D on the defect concentration (see Section II). The value of $D(c)$ as $p \to 1$ is often called the "saturation diffusion coefficient" D_s and should be distinguished from $D_{c \to c_s}$ or D_{c_∞} as the latter may be quite different because of higher order clusters or precipitation along the diffusion profile.

The values of D_s at various T are then fitted to

$$D_s = D_0 \exp(-U/kT) \tag{47}$$

where D_0 is a temperature independent constant and U the migration energy for the diffusant–vacancy pair.

F. Bénière et al. (1972) considered the case of diffusion from a thin layer of divalent ion deposited on the surface of an alkali halide. Extending Lidiard's 1957 derivation [Eq. (40)] to include K_S, they calculated the distribution of the divalent ion $c(x, t)$ by a finite difference method. Their analysis fits data for surface deposited diffusant experiments extremely well (Fig. 6). In particular, it fits the slight curvature near the surface evident in the plot. The curvature alters the entire penetration plot and thus the practice of using the apparently linear region deeper in the crystal will give erroneous values of D. If such a practice is used, D must be shown to be independent of the time of the diffusion anneal.

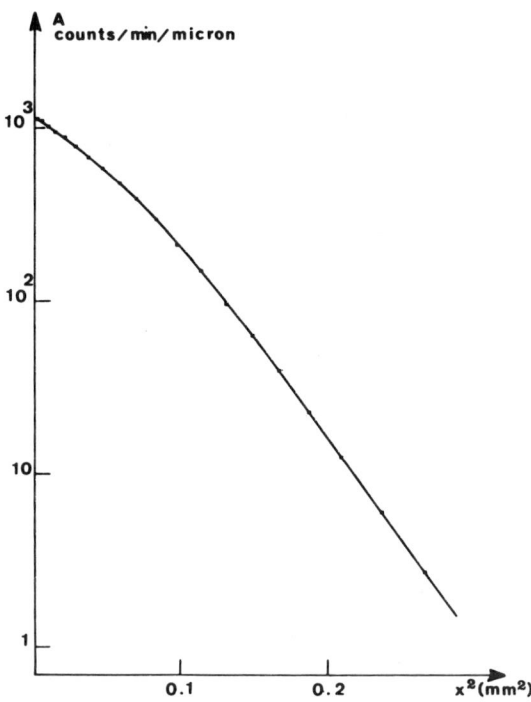

Fig. 6. Experimental distribution of ^{45}Ca plotted as $\log(A)$ against x^2 after diffusion in NaCl ($\tau = 644.5°C$, $t = 236{,}000$ sec). The solid curve is the calculated diffusion profile (F. Bénière et al., 1972).

The origin of the curvature in the penetration plot observed by F. Bénière et al. is not the only source of curvature which has been reported. Even in self-diffusion experiments, curved penetration plots have been observed (Rothman et al., 1972; Feit et al., 1973). In their study of "carrier-free" ^{22}Na diffusion from a surface-deposited source on NaCl, Feit et al. discovered that the tracer was contaminated with Mg^{2+}. A model was developed for the simultaneous diffusion of a divalent cation and monovalent tracer into a monovalent host. Feit et al. assume the local tracer diffusivity is proportional to the local vacancy concentration and is a linear function of the local impurity concentration. This is equivalent to assuming very little association of the impurity (i.e., $p \ll 1$). Their analysis gave excellent agreement (Fig. 7) with complex profiles which experimentally differed only in the diffusion time. This work has shown that disruption in the local vacancy concentration may cause curvature of the diffusion profile. The analysis is rather complex, but appears to handle a wide range of experimental conditions.

When the source contains more than one aliovalent diffusant, the penetration plots can be deceptive. Krause and Fredericks (1971) have studied the simultaneous diffusion of two divalent ions into a monovalent host. When both impurities diffuse by a vacancy mechanism, the diffusants are coupled through a common ion effect. Equations (12)–(15) express the equilibrium conditions of the reactions involved. The vacancy-pair contribution, Eq. (15), can be neglected when the impurity concentrations are large enough to suppress the formation of anion vacancies. The concentrations

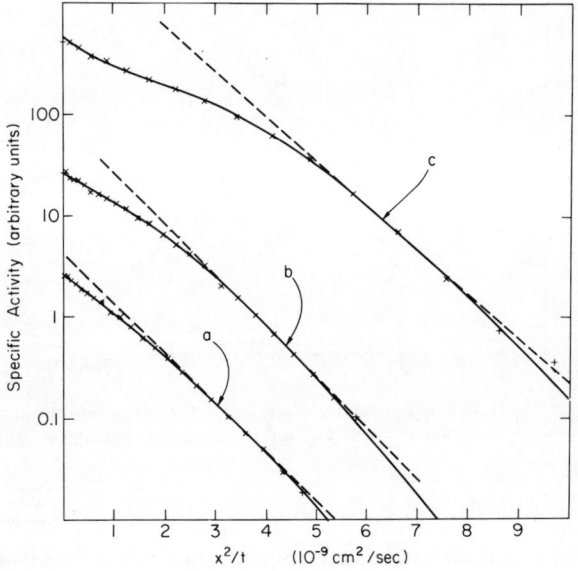

Fig. 7. Penetration profiles of ^{22}Na into NaCl taken at the same temperature ($T = 616°C$) but with differing diffusion times: (a) longest time, (b) intermediate time, (c) shortest time. (—) best fit calculated by the Feit–Mitchell–Lazarus method; (---) limiting slopes (Feit *et al.*, 1973).

and free energies of association of the i–v complexes can be such that the penetration plots of each diffusant appear almost normal (Fig. 8), but neither curve alone will give correct values of D_s or of the free energy of association. If the diffusant or host crystal contains an unknown aliovalent ion, this effect can cause errors which may be difficult to detect. Under some conditions, one of the diffusants can form i–v complexes at the expense of the other diffusant and diffuse rapidly. Eventually, the more slowly

8. DIFFUSION IN ALKALI HALIDES

diffusing ion becomes so low in concentration that it no longer provides a source of vacancies for the faster ion and the concentration of the faster ion increases in that region of the host crystal. Such profiles can even show both maximum and minimum. This has been observed (Krause and

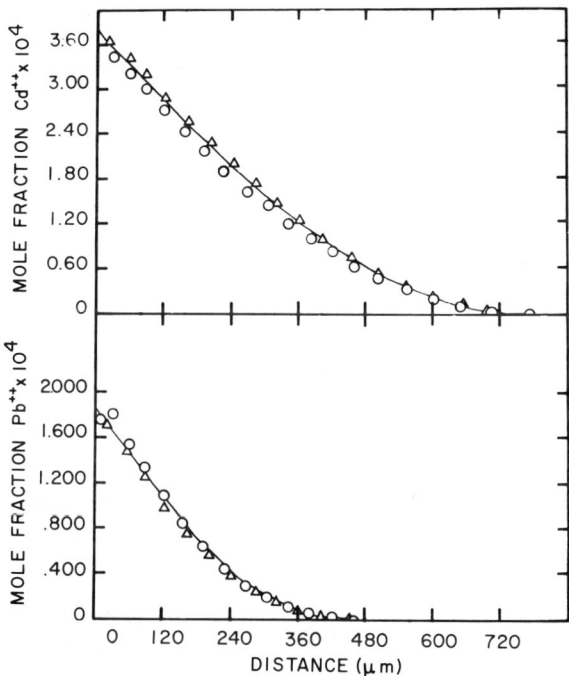

Fig. 8. Penetration profiles for the diffusion of cadmium and lead ions in NaCl at 460°C. ○ crystal A, △ crystal B. Curves are profiles calculated using common ion model (Krause and Fredericks, 1971).

Fredericks, 1973) in the KCl : Cd^{2+} : Pb^{2+} system (Fig. 9). When the host crystal contains a small but relatively uniform distribution of the ion contributing vacancies to the diffusant, the experimental profiles can give diffusion parameters quite different from the true values.

Actual experimental work often does not show the care which the above discussion has shown to be required. The major problem appears to be quality of host crystals and diffusant. Sometimes unknown interactions of the diffusant are undetected or neglected. In general, diffusion experiments on ionic crystals are subject to rather interesting and often difficult problems.

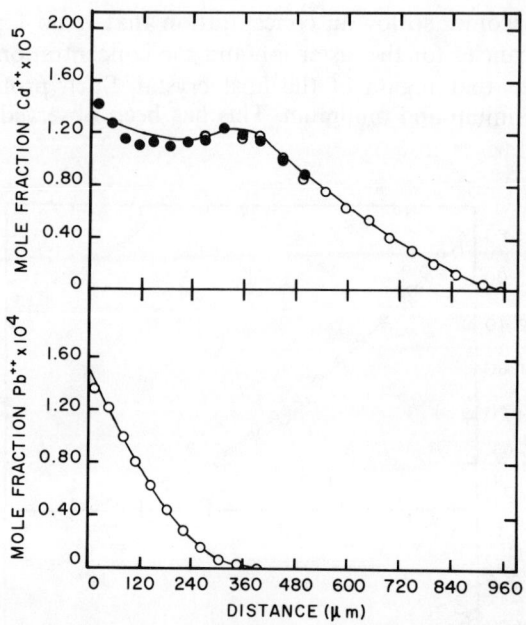

Fig. 9. Penetration profiles for the diffusion of cadmium and lead ions in KCl at 455°C. ○ crystal A. Curves are profiles calculated using common ion model (Krause and Fredericks, 1973).

V. The Experimental Situation and Numerical Results

A. General

In this section, a catalog of various diffusion parameters obtained experimentally will be provided for several different alkali halides. These results will be grouped according to diffusion-host relationships as was done in the preceding discussions. Not all early experimental results will be given since they may be found in other works (Friauf, 1963, 1972; Süptitz and Teltow, 1967; F. Bénière, 1972). No attempt has been made to present a critical evaluation of the tabular data given; in the opinion of this author, there are few systems on which a sufficient number of independent diffusion experiments have been made to justify it. Furthermore, in many cases it is impossible to estimate the extent and kind of interactions that may perturb the diffusant–host system because of inadequate specification of the quality of the host crystal and the diffusant. Even when all the species present are known, it is generally impossible to separate the effects of the various interactions because few of the impurity reactions involved are understood to the extent that the thermodynamic parameters required are available.

Since many otherwise excellent experiments can be questioned on this point, an illustration of the problems introduced by neglecting the points of Section IV.B will be given for a hypothetical experiment on the diffusion of ^{22}Na in NaCl. The host crystal is obtained commercially as "pure" NaCl (which means the manufacturer did not intentionally add anything to it and the crystal is suitable for the use to which it is usually put, i.e., an infrared window). A measurement of the ionic conductivity as a function of temperature is used to estimate the divalent ion content from the knee of the $\log \sigma$ vs T^{-1} plot. The value so obtained is either the divalent ion concentration present or, if the crystal contained OH^- (as most commercial alkali halides do, especially crystals grown in air), this measurement gives the uncompensated divalent ion concentration. (See Fritz et al., 1963.)‡

The tracer used is obtained from a supplier of radioactive chemicals as "carrier-free" ^{22}Na with a minimum radiopurity of 99.9%. (In this case, the purity only specifies the cleanness of the radioactive emission and carrier-free indicates the only isotope of Na present should be ^{22}Na.) Nothing is promised about the total chemical purity of the isotope solution. In this case, the most likely impurity is Mg^{2+} as ^{22}Na is most often produced by the reaction

$$^{24}_{12}Mg + ^{2}_{1}H \longrightarrow ^{4}_{2}He + ^{22}_{11}Na$$

With most such mixtures, quantitative separation can be accomplished in several ways. For radioactive mixtures, ion exchange is convenient because of the relatively simple equipment required and the small amount of chemical manipulation involved (Strelow et al., 1968; Kraus and Nelson, 1958).

Thus from the specifications reported here, which are more precise than is typical in diffusion papers, it is impossible to define the diffusion experiment with any certainty. It could be ^{22}Na in NaCl, or ^{22}Na, ^{24}Mg in NaCl, or ^{22}Na in NaCl : Mg^{2+} : OH^-, or any other combination of these species. Since these ambiguities arise from the nature of the system under study, they are present in all types of measurements which are not completely specified chemically. With this note of caution, we will proceed to specific host crystals.

‡ This ambiguity can be reduced by measuring the uv absorption band (185 nm in NaCl), in a thick sample of the host crystal and, with greater difficulty, further reduced by estimating total divalent ion content from direct analysis (Fredericks et al., 1966; reviewed by Rosenberger, 1972).

B. SELF-DIFFUSION

Since Chemla's (1952, 1953, 1954) early work on alkali halides, they have been the most popular ionic host crystals for diffusion experiments. Possibly their popularity is due to the easy availability of large single crystals, either natural or synthetic, and their simple structure. Unfortunately, their quality is quite variable, as it is with all ionic crystals.

The case in which host–diffusant interactions should be least important is self-diffusion. Recently, both F. Bénière (1972) and Fuller (1972) have reviewed this subject. Since there is neither a charge nor a size difference between the diffusing tracer and the host ion for which it substitutes, there is no preferred association between the vacancy and the tracer and a simple kinetic expression is sufficient to give D for a vacancy mechanism. For the diffusion of cations, one obtains

$$D^c = 4a^2 w_c \, f x_c \exp(-g_m{}^c/kT) \tag{48}$$

where the c refers to cation vacancies and m to migration, respectively. In a pure crystal x_c is given by the Schottky product and Eq. (48) is frequently written as

$$D^c = 4a^2 f w_c \exp\{[s_m{}^c + (s_S/2)]/k\} \exp\{[h_m{}^c + (h_S/2)]/kT\} \tag{49}$$

where Eq. (4) has been used to write the free energies as entropies and enthalpies. This simple expression is applicable when the only source of cation vacancies are Schottky defects and the only diffusion process contributing to mass transport is the vacancy mechanism. An analogous expression can be written for anion self-diffusion under similar restrictions. If diffusion occurs only by this single-vacancy mechanism, then comparison of the ionic conductivity with the diffusion coefficient through the Haven ratio, Eq. (46), confirms the mechanism. Such measurements of diffusion in "pure" and divalent cation-doped crystals show the diffusion coefficient to be greater than expected from the conductivity. Thus a defect which contributes an additional component to diffusion, but not to the conductivity, is required to account for these results. The presence of a neutral vacancy pair will account for the observed behavior (Laurance, 1960; Barr *et al.*, 1965; Fuller, 1966). Doping with divalent cations surpresses the single-anion vacancy concentration and permits the vacancy–pair contribution $D_p{}^c$ to be evaluated and thus the single-vacancy component D^c obtained by subtraction

$$D^c_{\text{expt}} = D^c + D_p{}^c \tag{50}$$

The vacancy–pair contribution to the diffusion can be written in the form of Eqs. (48) and (49) by using Eq. (7) to obtain x_p. The resulting expression for cation diffusion is

$$D_p{}^c = (4a^2 f_{pr} w_c/3) \exp - (g_m{}^c/kT) x_p \tag{51}$$

For an NaCl lattice, $z = 6$ in Eq. (7) and substitution in Eq. (51) using (4) to separate g_m, g_S, and g_p gives

$$D_p^c = 8a^2 f_{pr} w_c \exp[(s_S/2 + s_p + s_m^c)/k] \exp[(-h_S/2 + h_p + h_m^c)/kT] \quad (52)$$

Similar equations can be written for anion self-diffusion.

As long as the only source of vacancies is the formation of Schottky defects, the equations given above describe self-diffusion in ionic crystals of the NaCl type. However, in any real crystal, as the diffusion temperature is lowered, the number of vacancies arising from extrinsic sources eventually exceeds those generated thermally by formation of Schottky defects. This region is easily recognized by the large decrease in slope of the Arrhenius plot at lower temperatures. In this temperature region, the diffusion coefficient is a function of the impurities present in the crystal, its dislocation content and grain boundaries. For anion diffusion at low temperatures, the latter two imperfections dominate (Barr et al., 1960; Laurent and Bénard, 1958; Dawson and Barr, 1967a; Fuller, 1966). Cation diffusion in the extrinsic region proceeds by both the single-vacancy mechanism and by neutral impurity–vacancy pairs. In principle, an expression similar to (48) can be developed for this region if x_c is obtained from the proper mass action expressions and terms included for the neutral impurity–vacancy-pair component. When $c_B \gg K_S^{1/2}$, and at temperatures where the association between impurities and vacancies is small, the experimentally observed activation energy approaches h_m; however, in most cases, the data obtained in the extrinsic region arise from rather complex interactions between unknown impurity systems and, as such, are difficult to interpret.

Table I contains results for self-diffusion measurements in alkali halides. In the intrinsic region, data from various diffusion experiments should agree within experimental error and the details of the models should converge to a consistent combination of diffusion processes.

Recently, several self-diffusion experiments in NaCl have been reported (F. Bénière et al., 1970;‡ Downing and Friauf, 1970; Nelson and Friauf, 1970; Rothman et al., 1972). Apparently, these four studies represent measurements on essentially similar systems. Thus, a comparison of the results of these experiments, as given in Table II, provides a unique opportunity to assess the current state of self-diffusion in alkali halides. There are minor but, as we shall see, not insignificant differences among the various experiments. All used host crystals from the same source (Harshaw Chemical Company) except Friauf and his co-workers who used crystals from Harshaw as well as crystals grown in air from a quartz crucible. Sodium chloride grown in air incorporates OH⁻ (Fredericks et al., 1966)

‡ See F. Bénière (1970) for more details.

TABLE I Self-Diffusion in Alkali Halides

Crystal	T (°C)	$D_0{}^a$ (cm² sec⁻¹)	U^a (eV)	Remarks	Reference
Cations					
LiF	800–650	9	1.90		Naumov and Ptashnik (1968)
	830–730	5.6	3.0		Matzke (1971)
	730–580	0.8	2.0		Matzke (1971)
NaCl	797–587	76.9	2.04	Isotope effect	Rothman et al. (1972)
	726–600	33.1	1.975	σ, $D(\mathrm{Na})$, D_p	F. Bénière et al. (1970)
	790–640	790	2.23	$D(\mathrm{Na})$, $D(\mathrm{Cl})$, D_p, electric field	Nelson and Friauf (1970)
	790–570	0.5	1.61		Laurent and Bénard (1957)
	720–550	3.1	1.80		Mapother et al. (1950)
	695–585	3.2	1.78	σ, $D(\mathrm{Na})$	Downing and Friauf (1970)
	–550	390	1.86		Harrison (1961)
	570–450	$3.5(10^{-6})$	0.72	27 ppm Sr^{2+}	F. Bénière et al. (1970)
	550–	$2(10^{-6})$	0.87		Guccione et al. (1959)
	550–275	$3.5(10^{-6})$	0.84		Riesfeld and Fredericks (1970)
NaBr	680–500	0.67	1.53		Mapother et al. (1950)
NaI			1.60	(intrinsic), h_S 2.27	Hoodless et al. (1971)
	560–400			$h_m(\mathrm{Na})$ 0.47	
	627–112			0.45 by NMR	
KF	820–600	2	1.78		Laurent and Bénard (1957)
KCl	738–590	137	2.15		M. Bénière et al. (1970)
	675–526	$1.84(10^{-5})$	0.79	61 ppm Sr^{2+}	
KBr	720–460	$1(10^{-2})$	1.26		Laurent and Bénard (1957)
KI	680–460	$1(10^{-5})$	0.64		Laurent and Bénard (1957)
RbCl	697–568	33.3	1.99		Arai and Mullen (1966)
CsCl	610–490	0.1	1.39		Laurent and Bénard (1957)
	460–360	1.10	1.35		Hoodless and Turner (1972b)
CsI	500–294	80	1.64		Laskar et al. (1969)
		17.6	1.54		Klotsman et al. (1972)
Anions					
NaCl	750–500	60.7	2.14		F. Bénière et al. (1968)
	740–520	56	2.12		Laurance (1960)
	700–300		2.16		Barr et al. (1965)
NaBr	750–650	1.0	1.70		O. Dobrovinskaya et al. (1967)
KCl	750–560	61	2.12		Fuller (1966)
	730–530	10	2.00		F. Bénière et al. (1969b)
	740–500	178	2.25		M. Bénière et al. (1970)
KBr	700–	$3(10^{-4})$	2.61		Dawson and Barr (1967a)
	–400	$3(10^{-3})$	1.49		
KI	680–460	$1.2(10^{-3})$	1.12		Laurent and Bénard (1957)
RbCl	697–568	33.3	1.99		Arai and Mullen (1966)
	700–600	$3.37(10^{-5})$	1.35		Kakaishi and Sensui (1969)
				Vacancy-pair component 50%	Hoodless and Turner (1972a)
CsCl	610–490	0.7	1.56		Laurent and Bénard (1957)
	460–280	1.51	1.27		Harvey and Hoodless (1967)
	460–360	1.64	1.29	Pure and Ba^{2+} doped	Hoodless and Turner (1972b)
CsI	560–405	0.39	1.27		Klotsman et al. (1967a)

[a] Here and in all following tables these represent the constant terms in $D = D_0 \exp(-U/kT)$.

8. DIFFUSION IN ALKALI HALIDES

and estimates of divalent ion content of such crystals from conductivity measurements are unreliable (e.g., Section II.; Fritz et al., 1963). Rothman et al. estimated the OH⁻ concentration from the uv adsorption band as 10 ppm and gave the "conductivity knee" as $\sim 500°C$, almost the same value (490°C) given by Nelson and Friauf. F. Bénière et al. used ^{22}Na prepared from ^{24}Mg; Friauf and co-workers did not specify the source of their tracer; Rothman et al. reported 300 μg/ml of nonradioactive Ca^{2+} and, in one experiment, Pb^{2+} was reported. Rothman et al. specially purified the ^{22}Na after discovery of the Ca^{2+} and reported the ^{24}Na they used to be free of harmful impurities. All diffusion measurements were from a surface layer and the profiles were obtained by the sectioning technique.

From diffusion measurements alone, it is not possible to deduce the mechanism of diffusion. Some combination of measurements must be made to separate the various components. F. Bénière et al. measured the diffusion of both ^{22}Na and ^{36}Cl, the conductivity and the transport number. Nelson and Friauf measured the conductivity and the diffusion of ^{22}Na and ^{36}Cl under the influence of an electric field, while Downing and Friauf measured conductivity and Na diffusion in NaCl. Rothman et al., in the best specified self-diffusion experiment yet reported, used the isotope effect of ^{22}Na and ^{24}Na to set limits on the pair component contribution to Na self-diffusion.

F. Bénière et al. used expressions equivalent to (49) with $f = 1$ for both anion and cation diffusion, but for the vacancy-pair diffusion coefficient used

$$D_p = D_{0p} \exp(-h_p/kT) \tag{53}$$

This rather than Eq. (52) gave the total anion and cation diffusion as

$$D_{(Cl^-)} = D(Cl^-) + D_p \tag{54}$$

and

$$D_{(Na^+)} = D_V(Na^+) + D_p \tag{55}$$

As the neutral pairs do not contribute to the ionic current, the conductivity was given as

$$\sigma = \sigma_{Na^+} + \sigma_{Cl^-} = (Ne^2/kT)4a^2w_0 \exp[-(h_S - Ts_S/2kT)]$$
$$\times \{\exp(s_m^c/k)\exp(-h_m^c/kT) + \exp(s_m^a/k)\exp(-h_m^a/k)\} \tag{56}$$

The transport numbers t, obtained from a three-crystal Tubandt (1932) type experiment, were used to get the ratio

$$(D_{(Na^+)} - D_p)/(D_{(Cl^-)} - D_p) = \sigma_{Na^+}/\sigma_{Cl^-} = t_+/t_- \tag{57}$$

and the Nernst–Einstein relation was given as

$$\sigma = (Ne^2/kT)(D_{(Na^+)} + D_{(Cl^-)} - 2D_p) \tag{58}$$

TABLE IIa

RECENT EXPERIMENTS ON INTRINSIC DIFFUSION IN SODIUM CHLORIDE

Experiment	Diffusant	D_0 (cm^2 sec^{-1})	U (eV)	ΔT (°C)	t_+ [at T (°C)]	$\dfrac{D_p}{D(\mathrm{Na})}$ [at T (°C)]
(1) F. Bénière et al.						
(1970)	$D(\mathrm{Na})$	33.16	1.975	606–721	0.81[580]	0.11[mp]
(1968)	$D(\mathrm{Cl})$	60.7	2.14		0.32[600]	
(1970)	$D_V(\mathrm{Cl})$	36.2	2.11		0.84[650]	
(1970)	$D_p(\mathrm{Na, Cl})$	5200	2.63	606–721		
Conductivity				600–720		
(1970)	$\sigma T = 2.006 \times 10^9 \exp(-1.998/kT)$					
(2) Downing and Friauf						
(1970)	$D(\mathrm{Na})$	3.2	1.78	560–720		0.31[690]
(3) Nelson and Friauf						
(1970)	$D_V(\mathrm{Na})$	118[b]	2.10	640–790[a]		0.40[mp]
	$D_p(\mathrm{Na})$	1130	2.35	640–790		
	$D(\mathrm{Cl})$	(two expt. points not calculated)		690, 770		
Conductivity				550–800		
Best fit	$\sigma T = 2.0 \times 10^9 \exp(-1.98/kT)$				0.66[550–800]	$D_p(\mathrm{Na})/D_p(\mathrm{Cl}) \sim 5.6$
	$+ 1.5 \times 10^{17} \exp(-3.77/kT)$					
σ_{Na} fixed by	$\sigma T = 3.6 \times 10^9 \exp(-2.06/kT)$					~ 0.179
M_V (No)	$+ 2.53 \times 10^9 \exp(-2.08/kT)$					
(4) Rothman et al.						
(1971)	$D(\mathrm{Na})$	77	2.04		587.796	0.30–0.45[mp]
						$D_p(\mathrm{Na})/D_p(\mathrm{Cl}) \sim 5$

[a] Electric field 600–700.
[b] Corrected value, private communication.

TABLE IIb
Derived Transport Parameters in NaCl[a]

Source	h_S	$s_S(10^3)$	h_m^c	$s_m^c(10^3)$	h_m^a	$s_m^a(10^3)$
Allnatt et al. (1971)	$2.4-2.5^b$	$0.7-0.9^b$	0.65 ± 0.01^b	0.15 ± 0.1^b	$(1.75)^c$	$(1.07)^c$
F. Bénière et al. (1970)	2.5	0.96	0.72	0.178	0.86	0.195
	$h_m^c + h_s/2$	$s_m^c + s_s/2$			$h_m^a + h_s/2$	$s_m^a + s_s/2$
Nelson and Friauf (1970)	2.06	$3.83(10^{-3})$		2.08		$3.66(10^{-3})$

[a] All units eV or eV/degree.
[b] Maximum interval within which parameter lies with 95% probability.
[c] A typical value. Authors conclude that no useful results can be obtained about "anion" parameters from their analysis.

Thus, from measurements of $D_{(Na^+)}$, $D_{(Cl^-)}$, σ, and t_+, the enthalpies and entropies of the various transport processes can be evaluated. Note, however (as Rothman et al. point out) the assumption that $D_p(Na^+) = D_p(Cl^-)$ in Eq. (53).‡

Nelson and Friauf expressed the diffusion coefficient as the sum of two components, as in Eqs. (54) and (55), except that they specifically noted that the pair component of cation and anion diffusion is not necessarily equal and must be individually evaluated as $D_p(Na^+)$ and $D_p(Cl^-)$. They used the Nernst–Einstein relation in a more conventional way. Writing this relation for each ion and including the correlation factor f for an fcc lattice (0.78146), they obtained

$$D_V(Na^+) = f(kT/Ne^2)\sigma_{Na^+} \qquad (59)$$

and

$$D_V(Cl^-) = f(kT/Ne^2)\sigma_{Cl^-} \qquad (60)$$

where σ_{Na^+} and σ_{Cl^-} are the contributions of Na^+ and Cl^-, respectively, to the total conductivity.

Nelson and Friauf measured the total conductivity, D_{Na^+}, D_{Cl^-}, and diffusion of Na^+ in an applied electric field. With these measurements, they could obtain values for all components of intrinsic ionic transport in NaCl. The unique measurement in their work was Na^+ diffusion in an external field. This transport experiment consisted of placing a thin layer of salt containing the radioactive tracers between two flat crystal slabs which were thick compared to the diffusion distance. A steady electric field was applied perpendicular to the interface containing the tracer. The diffusion profile was then expected to drift in the field by an amount proportional to the mobility [Eq. (30)], but the field was not expected to influence the spreading (cf. Howard and Lidiard, 1964). Figure 10 illustrates the types of profiles obtained. The discontinuity of the profile at the interface between the crystals is obvious. Nelson and Friauf's results appear in Table II.

Rothman et al. used the isotope effect to set limits on the pair contribution to Na^+ diffusion. (The isotope effect in diffusion is discussed in Chapter 3.) The observed isotope effect E for two mechanisms operating simultaneously is

$$E = E_1[D_1/(D_1 + D_2)] + E_2[D_2/(D_1 + D_2)] \qquad (61)$$

where subscripts 1 and 2 denote single vacancies and vacancy pairs, respectively. If E_1 and E_2 are known, then the vacancy-pair contribution $D_2/(D_1 + D_2)$, can be calculated from E. The E_i for a specific mechanism i

‡ *Note added in proof:* In later work on KCl, Chemla and F. Bénière (1973) note that the assumption that $D_p(Na^+) = D_p(Cl^-)$ is not valid.

8. DIFFUSION IN ALKALI HALIDES

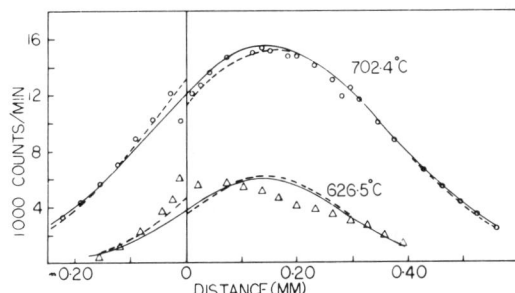

Fig. 10. Diffusion profiles for Na in NaCl with a dc electric field, ○, △ actual points obtained by sectioning for two different runs, (—) least-squares fit without a barrier, (- - -) fit with a barrier. The initial interface is at $x = 0$ and the field is applied along the positive axis (Nelson and Friauf, 1970).

is $E_i = f_i \Delta K_i$, where ΔK_i is the factor that accounts for the fact that the preexponential in the Arrhenius expression for the atomic jump frequency is not exactly proportional to $m^{-1/2}$. Rothman et al. take f_1 as 0.782 and note $1.0 \geq \Delta K_1 \geq 0.8$ (Brown et al., 1971; Barr and LeClaire, 1964), and thus know E_1 to 10%. To estimate E_2 requires a value for f_2, which must have limits between 0 and 0.782. From these data, Rothman et al. estimate possible range in the magnitude of the pair component to the diffusion of Na^+ in NaCl; their conclusions are summarized in Table IIa.

Allnatt et al. (1971) have carefully considered their results of conductivity measurements on NaCl grown in a vacuum and have given values of the migration and formation parameters derivable from conductivity measurements. Their values are also listed in Table IIb.

There is poor agreement between the various experimenters; although Nelson and Friauf's "best fit" conductivity expression and the single-exponential fit of F. Bénière et al. agree, the actual values of σT at 720°C calculated from the two expressions differ by 23%.

Nelson and Friauf gave preference to their second σT (Table IIa) expression which has a more reasonable enthalpy parameter $(h_m^a + h_S/2)$ if the second term represents the anion contribution to σT. Their first expression would give an anion contribution of 5.7% to σT at 720°C while the second gave 35%. The preferred expression requires t_+ to be 0.66 while the value experimentally measured by F. Bénière et al. is about 0.83, increasing slightly as T increases.

F. Bénière et al. and Rothman et al. agree on the values of $D_{(Na)}$ at various temperatures. Rothman et al. used curved profiles to obtain $D_{(Na)}$, doing so on the basis of an earlier argument by Rothman et al. (1966).

However the recent work by Feit et al. (1973) strongly questions the validity of such an argument. (See Section IV.)

Nelson and Friauf's electric field measurements have been criticized (Rothman et al., 1972) on the quality of the contact between the two crystals. The assumption of a uniform field in the diffusion region becomes questionable if the tracer contained a divalent impurity. Two recent experiments (Fiet et al., 1973; Rothman et al., 1972) have shown that such effects may be common with ^{22}Na tracers.

In summary, the results of these experiments alter the accepted parameter values of defects in NaCl. The formation enthalpy for a Schottky defect now appears to be ~ 2.4–2.5 eV, considerably larger than the earlier value, 2.12 eV, of Dreyfus and Nowick (1962), but in good agreement with recent calculations by Boswarva (1972). The cation vacancy migration enthalpy appears to be 0.65–0.72 eV and that of the anion vacancy is ~ 0.86 eV. The calculated values for cation migration are 0.70–0.92 eV and for anion migration 0.85–1.00 eV (Barr and Lidiard, 1970).

With the exception of the isotope effect, similar studies combining diffusion measurement with other studies have been also made on KCl. Fuller et al. (1968) used conductivity and anion diffusion data (Fuller, 1966) to compare simple association theory and Lidiard–Debye–Hückel theory. Both Harshaw KCl- and Sr^{2+}-doped crystals were used. The basic treatment was to write σT and $D_{(Cl^-)}$ equations to describe conductivity and diffusion over a large temperature range, then introduce defect equilibrium quantities into these equations. The resulting set contains eleven adjustable parameters. Least squares fitting of exponential functions can be difficult when the values of the exponent do not differ much (Barr and Lidiard, 1970). The eleven-parameter fit was made from ~ 350 to 760°C. The average rms deviation obtained in the least-squares analysis of the conductivity using the simple association theory was 3.0%; with the Lidiard–Debye–Hückel, it was 2.4%. However, neither theory enables this five-defect model to describe precisely the conductivity data. Fuller et al. found nonrandom deviations between the calculated and experimental conductivity results. It should be noted that their deviation plots for various samples show the experimental conductivity to be 10% greater than that calculated from the best fit parameters in the region of 750°C.

Another set of experiments in which several measurements were combined in an effort to understand the complex diffusion mechanism in KCl was the work of M. Bénière et al. (1970)‡. The self-diffusion and conductivity of

‡ *Note added in proof:* Chemla and F. Bénière (1973) have reported new results for detailed and careful transport studies in KCl. Their main results are: $h_S = 2.54$ eV; $s_S = 2.19 \times 10^{-4}$; $w_c = 4.72 \times 10^{13} \exp(-0.731 \text{ eV}/kT) \text{ cm}^2 \text{ sec}^{-1}$; $w_a = 1.05 \times 10^{14} \exp(-0.85 \text{ eV}/kT) \text{ cm}^2 \text{ sec}^{-1}$; $D_p(K^+) = 5480 \exp(-2.65 \text{ eV}/kT) \text{ cm}^2 \text{ sec}^{-1}$; $D_p(Cl^-) = 133 \exp(-2.39 \text{ eV}/kT) \text{ cm} \text{ sec}^{-1}$.

^{42}K and ^{36}Cl were measured as before (M. Bénière et al., 1970). In addition, the experiments included both anion and cation diffusion in Sr^{2+}-doped crystals. In the pure crystals, no curvature was evident in the penetration plots. The source and purity of crystals and tracers was unspecified, but lack of curvature in the low penetration region may indicate a relatively pure system. Their analysis was similar to that used for NaCl and they again obtained $f = 1$.

In Table III, we have compiled the results obtained from the various combined diffusion and conductivity measurements. They appear to be in relatively good agreement. However, as Jacobs and Pantelis (1971) have shown, they are not sufficiently accurate to provide a choice from among several models proposed to account for the abnormally high conductivity of KCl in the region near its melting point. Using their own conductivity data measured on relatively pure crystals, Jacobs and Pantelis calculate diffusion coefficients for anion and cation diffusion using various models to represent the conduction process (Debye–Hückel, added Frenkel defects, an added trivacancy component, and an excess conductance model). The models were fitted to the σT data and the diffusion coefficients calculated from the Nernst–Einstein relation with no pair contribution correction in $D_V(K^+)$, while the $D_V(Cl^-)$ was corrected for pair diffusion using Fuller's (1966) parameters. Acceptable values of $D_V(K^+)$ resulted from several models, but the anion diffusion coefficients were not acceptable. Use of Fuller and co-workers (1968) pair diffusion parameters improved the correlation between the experimental diffusion coefficients and the calculated coefficients using the Debye–Hückel parameters. However, the improvement was not sufficient to permit a clear choice between this model and those which introduce additional current carriers. The values given in the lower portion of Table III are from these and other conductivity experiments. In their paper, Jacobs and Pantelis thoroughly discuss the problems and limitations of these values. They note that better values of diffusion coefficients $D_V(K^+)$ and $D_V(Cl^-)$ and of the pair component are required if these processes are to be understood.

One other mechanism has been identified as operating in alkali halides. At low temperatures, anion diffusion occurs by an extrinsic process down dislocations and grain boundaries (Barr et al., 1960; Laurent and Bénard, 1958; Dawson and Barr, 1967a,b; Barr and Dawson, 1969; Fuller, 1966). Here we will only discuss the results of Dawson and Barr.

Bromide diffusion coefficients were measured in KBr and KBr doped with $CaBr_2$ and with $SrBr_2$ by both sectioning and the isotope-exchange techniques, and the two methods are shown to give equivalent results. The sensitive isotope-exchange technique was then used to measure $D_{(Br^-)}$ down to 338°C $(D_{(Br^-)} = 1.32 \times 10^{-14}$ cm^2/sec). Diffusion along dislocations becomes important in anion diffusion in alkali halides at temperatures

TABLE III

DEFECT PARAMETERS IN POTASSIUM CHLORIDE

Reference	h_S (eV)	s_S (10^3 eV/°K)	h_m^a (eV)	s_m^c (10^3 eV/°K)	h_m^c (eV)	s_m^a	D_{0p} (cm²/sec)	U_p (eV)	Expt.
Fuller (1966)	2.31	0.651					8560	2.654	$D(Cl)$
Fuller et al. (1968)[a]									
Simple assoc. model	2.59	0.800	0.79	0.246	0.95	0.361	5430	2.59	$D(Cl)$, σT
Lidiard–Debye–Hückel model									
M. Bénière et al. (1970)	2.49	0.658	0.76	0.220	0.82	0.253	5150	2.62	
Barr and Lidiard (1970)	2.64	0.939	0.79	0.245	0.89	0.240	987	2.49	$D(K)$, $D(Cl)$, σT
Theory	1.75–2.26		0.77–0.96		0.80–1.21	0.213		2.65	
Beaumont and Jacobs (1966)	2.26	0.462	0.705	0.163	1.04	0.541			
Chandra and Rolfe (1970)	2.59	0.828	0.73	0.232	0.99	0.356			
Jacobs and Pantelis (1971)[a]									
added Frenkel defects model	2.236	0.244	0.675	0.228	0.978	0.573	$h_F^c = 3.614$		$s_F^c = 0.094$
excess cond. model	2.222	0.231	0.676	0.226	0.943	0.543			
Debye–Hückel model	2.302	0.358	0.665	0.226	1.303	0.884			

[a] See original for error limits.

below 0.75 of the melting temperature. Assuming the dislocation diffusion coefficient D_d could be expressed in the usual Arrhenius form, the anion diffusion coefficient consisted of three components; the single vacancy, the vacancy pair, and the dislocation portions. Doping the crystals with divalent cations was assumed to reduce the single-vacancy component to a negligible level. The diffusion coefficient for doped crystals is then fitted to an expression which is the sum of two exponentials in $D_p(\text{Br}^-)$ and $D_d(\text{Br}^-)$. The pure crystal diffusion coefficient would require the single-vacancy component $D_V(\text{Br}^-)$ to be added to the other two. With this technique, Dawson and Barr were able to obtain each component (Table IV).

TABLE IV

ANION DIFFUSION IN KBR

h_S (eV)	$h_m{}^a$	$h_p{}^a$	h_d
2.54	0.83	2.60	1.3

In the last few years some progress has been made toward understanding self-diffusion in alkali halides. Many of the mechanisms involved are well established, but some high temperature conductivity data indicate that an additional current carrier may exist.

A number of questions remain. If Feit et al. (1973) are correct in their analysis of the effect of an aliovalent impurity on self-diffusion, why do Rothman et al. (1972) obtain the same diffusion coefficients as F. Bénière et al. (1970)? If F. Bénière and co-workers, like Rothman et al., had divalent impurities present, their apparent diffusion coefficient should have depended on the time of anneal (Feit et al., 1973) and it seems unlikely that the two results would agree. Rothman et al. find the vacancy-pair component to be between 30 and 45% of the total cation diffusion in sodium chloride at its melting point in rough agreement with Nelson and Friauf (1970). But Nelson and Friauf's conductivity parameters show poor agreement with those of Allnatt et al. (1971) (Table IIb). The identity of the unidentified species which transports charge (at high temperature) but not tracer still appears unknown (Nelson and Friauf, 1970; Allnatt and Pantelis, 1968a; Allnatt et al., 1971; Jacobs and Pantelis, 1971). NaCl and KCl are the only alkali halides for which much data on transport properties are available. It appears that many more experiments free of disturbing side effects remain to be performed on these two salts and work has only begun on the remaining ones.

TABLE V
Diffusion Parameters for Monovalent Impurities in Alkali Halides

Crystal	Diffusant	ΔT (°C)	D_0 (cm² sec⁻¹)	U (eV)	Remarks	Reference
Cations						
NaF	Li⁺	830–730	5.6	3.0		Matzke (1971)
	K	730–580	0.8	2.0		Matzke (1971)
NaCl	K	650–400	—	0.5		Geguzin et al. (1966)
	Rb⁺	787–599	205	2.11		Arai and Mullen (1966)
		746–600	28.5	1.98		F. Bénière et al. (1969b)
	Cs⁺	698–596	1.62	2.00		F. Bénière et al. (1969b)
	Cu⁺	650–350	33.8	1.43		Haneda et al. (1968)
	Ag⁺	726–576	380	2.00		Chemla (1954)
	Tl⁺	620–443	100	1.8		Geguzin et al. (1965)
NaI		635–400	5×10^{-3}	1.17		Schmidt and Staube (1968)
KCl	Li⁺	730–500	20	1.52		Hanson (1968, 1970)
	Na⁺	750–570	2.2	1.75		Arnikar and Chemla (1956)
	Rb⁺	763–607	26.8	2.04		Arai and Mullen (1966)
	Cs⁺	750–570	0.7	1.75		Arnikar and Chemla (1956)
	Tl⁺	727–457	1.34×10^{-3}	1.30	Calc. h_S as 2.12 eV	Tierman and Wuensch (1971)
		457–230	5.81×10^{-11}	0.24	Electron microprobe meas. profile	Tierman and Wuensch (1971)
		730–550	2	1.70		E. Dobrovinskaya and Podorshanskaya (1966)
		550–225	2×10^{-8}	0.43		E. Dobrovinskaya and Podorshanskaya

8. DIFFUSION IN ALKALI HALIDES

Host	Ion	Temp. range	D_0	Q	Notes	Reference
KBr	Cu^+	400–270	1.04×10^{-3}	0.88	Optical abs. profile	Reisfeld et al. (1965a)
	Ag^+	200–150	7.14	1.09	Powder optical abs.	Glasner and Reisfeld (1961)
	Tl^+	210–176	1.29×10^3	1.36	Powders "pure" and doped	Glasner et al. (1967)
		176–160	1.95×10^{-3}	0.836		Glasner et al. (1967)
		650–350	10.6	1.23		Haneda et al. (1968)
		650–200	—	0.42		Geguzin et al. (1966)
		500–350	50	2.01		Illingworth (1963)
KI	Tl^+	207–142	0.69–1.45	1.02	Powder-optical abs.	Glasner et al. (1961)
RbCl	Tl^+	205–141	9.1×10^{-2}	0.99	TlCl in KBr powders	Glasner et al. (1961)
	Na^+	650–400	8×10^{-3}	1.17		Geguzin et al. (1965)
	Na^+	707–500	1.03×10^3	2.06		Peterson et al. (1969)
CsI	Na^+	593–368	78.3	1.52		Klotsman et al. (1967b)
		—	—	—	Showed Na^+ does not diffuse by grain boundaries	Klotsman et al. (1970)
	Rb^+	—	9×10^{-2}	1.09		Klotsman et al. (1972)
	Tl^+	570–440	27.3	1.50		E. Dobrovinskaya and Podorshanskaya (1966)
	In^+	560–394	9	1.5		Birman et al. (1968)
Anions						
NaCl	Br^-	748–612	2.8×10^{-4}	2.66		Chemla (1954)
	I^-	650–500	20	1.94		O. Dobrovinskaya et al. (1967)
KCl	I^-	736–592	80	2.29		Chemla (1954)
	I^-	700–530	500	2.24		Beaumont and Cabane (1961)
	I^-	650–500	50	2.0		Beaumont and Cabane (1961)
	OH^-	515–397	$3(10^5)$	2.0	Cl_2 atmosphere	Allen and Fredericks (1970b)
	OH^-	515–397	$1.2(10^3)$	2.00	Vacuum of 10^{-6} Torr	
KI	Cl^-	650–500	1.5×10^{-3}	1.13		Beaumont and Cabane (1961)

C. Diffusion of Monovalent Impurities

A complex of mechanisms similar to those found in self-diffusion should be present in the diffusion of monovalent impurities in alkali halides although they have not been identified. The study of monovalent impurity diffusion has been rather neglected in the last few years and drift (Chemla, 1956) and isotope-effect measurements have not been pursued. Table V lists most of the results available. Even a casual inspection reveals the scattered nature of the D_0 and U values obtained in these experiments. Frequently, when the temperature range is sufficiently large, the diffusion coefficient exhibits two regions, one at high temperatures with values of D_0 and U larger than those measured at low temperatures. This behavior is like the intrinsic and extrinsic regions found in self-diffusion but is better termed the "Schottky or Frenkel region" than "intrinsic impurity region." Only for a few systems have independent, duplicate measurements been made and those show a wide variation in the parameters obtained. For example, compare Geguzin *et al.* (1965) and Schmidt and Staube (1968) on Tl^+ in NaI, or Tierman and Wuensch (1971) and E. Dobrovinskaya and Podorshanskaya (1966) on Tl^+ in KCl. There is no reason to prefer one set of results over the other.

At present, the diffusion of monovalent impurities in alkali halides appears to be a fruitful area for diffusion studies on well defined and carefully controlled systems.

D. Diffusion of Aliovalent Impurities

The diffusion of divalent impurities in the alkali halides can provide an almost ideal system to test Howard and Lidiard's (1964) theoretical development of the dependence of the diffusion coefficient on concentration [Eqs. (36) and (40)], if sufficient care is used to assure that the actual experimental conditions satisfy those imposed by their model. Unfortunately, in addition to the desired divalent ion–vacancy interactions, a wide variety of undesired interactions can occur. As discussed in general in Section II and illustrated with specific examples in Section IV, the two most troublesome of these are impurity cation–impurity anion compound formation and common ion interactions. When the radius of the diffusant is small, n.n.–n.n.n. equilibria must be included in the model.

One of the techniques followed is to choose a temperature range where $c_B \simeq 10 K_S^{1/2}$, use especially purified crystals, diffuse under a controlled atmosphere, and obtain a profile by microtome sectioning. This profile is then analyzed by one of the methods given in Section IV to obtain $D_B(c)$, then D_s and g_B are calculated by least-squares fitting to the appropriate

Howard and Lidiard expression for $D(c)$. In all cases investigated to date, the ratio $D(c)/D_s$ plotted against the concentration c_B exhibits the behavior predicted by the Howard and Lidiard theory (see Section III, Fig. 4) as illustrated by Fig. 11.

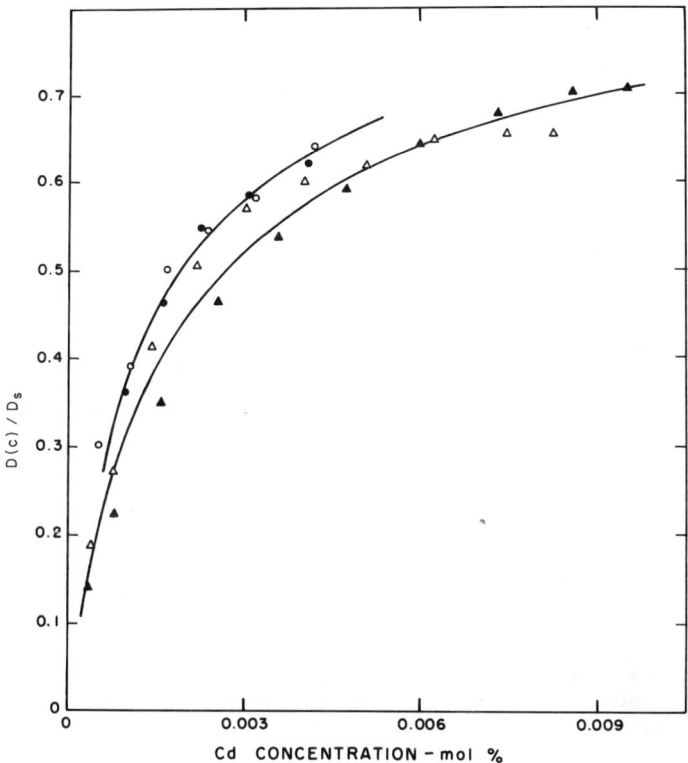

Fig. 11. Diffusion of Cd^{2+} in KCl as a function of Cd concentration at 471°C (●,○) and 505°C (▲,△) compared with theoretically calculated curves (—) (Keneshea and Fredericks, 1965b).

If the diffusion conditions are adjusted to allow $p \to 1$, then a plot of $D(c)$ vs c will reach a constant value of $D(c)$ at large c. Figure 12 shows an example of a system approaching this condition. When the diffusion coefficient becomes independent of the concentration, $D(c) = D_s$. Values of D_s evaluated in this way agree closely with those obtained by the least-squares fitting of the profile (Mannion et al., 1968; Allen et al., (1967a,b).

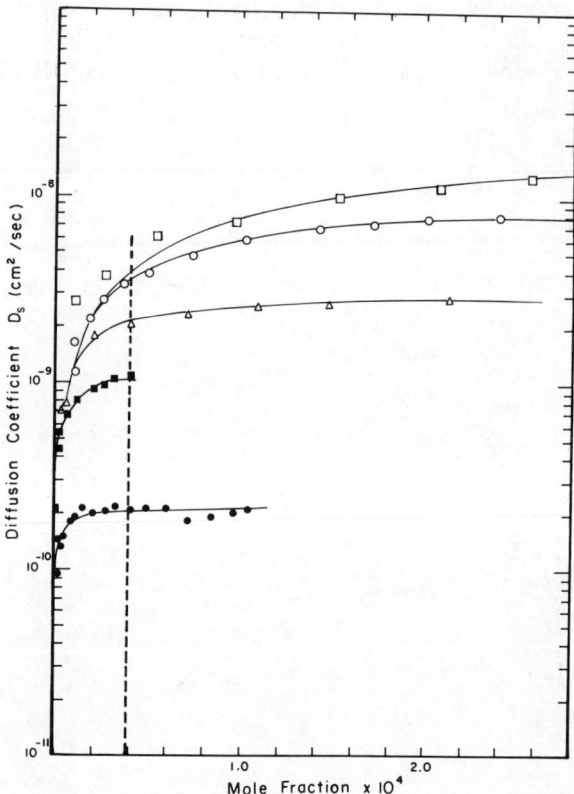

Fig. 12. Diffusion coefficient of Cd^{2+} as a function of Cd^{2+} concentration in NaCl at various temperatures. (---) at constant concentration provides the values of D_s used in the dashed curve of Fig. 13 (Allen *et al.*, 1967).

Arrhenius plots of the values of D_s give the migration enthalpy of the divalent ion as $p \to 1$ [i.e., $U = h_m(M^{2+})$]. Over the temperature ranges in which this technique has been applied, if D_s is correctly evaluated, the Arrhenius plots are linear. However, if the values of D_s actually correspond to a condition in which $p < 1$, the $\log D$ vs T^{-1} plot will exhibit some downward curvature. An example of such a case is shown in Fig. 13. The dashed line is drawn for values of D corresponding to $c = 2 \times 10^{-5}$ and the solid line is drawn for the correctly evaluated D_s. An alternate method of obtaining D_s is to diffuse the divalent tracer into a crystal doped with enough of the same divalent ion to cause $p \to 1$. Slifkin and Brébec (1968) and Rothman *et al.* (1966) have used this technique. Howard and Lidiard's treatment shows Eq. (39) to be applicable to this experiment in general, but

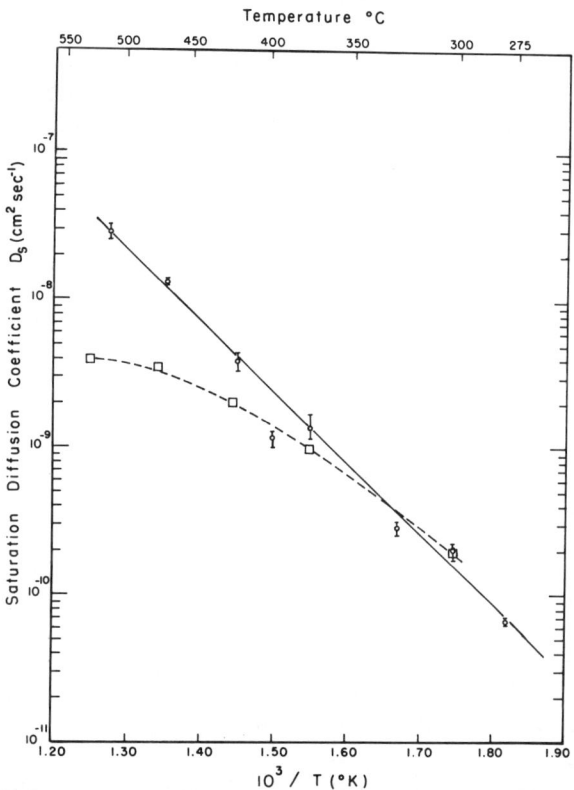

Fig. 13. Log D_s vs T^{-1} from Cd^{2+} diffusion in NaCl. (---) constructed from constant concentration data of Fig. 12 and illustrates curvature caused by incorrect determination of D_s (Allen et al., 1967).

when $p = 1$, a simpler relation can be used to obtain D_s. If the amount of diffusant is very small (i.e., $c_B \ll K_S^{1/2}$) and the host pure, the experimental conditions approximate those of Eq. (35). If an independently measured value of h_S is known, it is possible to obtain g_B using the value of h_m from the high concentration experiment. The initial concentration of impurity present in the crystal must be sufficiently high to cause $p \rightarrow 1$ (see Fig. 12), but not so high that a precipitate is present during the diffusion. Obtaining a uniform concentration of dopant, even at high temperatures, may be difficult (Allnatt and Chadwick, 1967). At the required doping levels, the chemical potential of the impurity in the crystal can exceed its value in an external vapor, so usually the experiment is from a deposited surface layer. Anomalous increases in impurity diffusion coefficients at high diffusant concentrations have been reported by Keneshea and Fredericks (1963) and Allen et al. (1967b). An example of this effect is shown in Fig. 14. Its cause has not been specifically determined, but it is frequently observed under conditions in which the surface concentration of the diffusant becomes high and the temperature is above a eutectic point for an impurity–host solution.

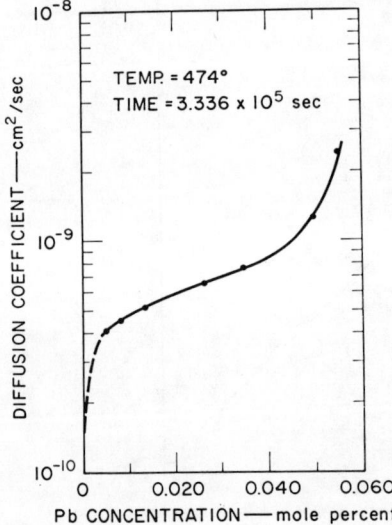

Fig. 14. Diffusion of Pb^{2+} ions in KCl doped with 0.003 mole % $PbCl_2$ showing anomalous increase in $D(c)$ (Keneshea and Fredericks, 1963).

The conditions necessary for the low concentration experiment can be met by adjustment of time and temperature or by use of a remote vapor source (Allen and Fredericks, 1973). The problems of very low concentration experiments usually arise from control of extraneous impurities. The host and diffusant must be extremely clean. Experiments under these conditions are especially subject to OH^- reaction with the diffusant (Allen and Fredericks, 1973) and to common ion effects (Krause and Fredericks, 1971, 1973).

The free energy of association of an impurity with a vacancy can be obtained by using the equation [cf. Eqs. (36)–(40)] that fits the experimental conditions to analyze diffusion data as a function of concentration (Keneshea and Fredericks, 1963; Allen and Fredericks, 1973). This technique has limitations similar to those described above. Unsuspected anion–cation reactions or common ion effects can markedly affect the profile and thus both g_B and D_s. A rather sensitive test for such interactions is to extract the entropy from g_B by plotting g_B vs T and obtaining a constant entropy s_B over the entire temperature range. Results of such experiments are in Table VI. In

8. DIFFUSION IN ALKALI HALIDES

many studies, g_B was found to show marked curvature as a function of temperature. By using highly purified crystals and simultaneously calculating $D(c)(Cd^{2+})$ and $D(c)(Pb^{2+})$ from expressions which account for the common interactions, g_i's were obtained which exhibited no curvature over the entire temperature range (Fig. 15).

When the divalent diffusant is small in size compared to the ion for which it substitutes, there is ample evidence (see Section II) that n.n.n. impurity–vacancy complexes are more stable than n.n. complexes. Next-nearest neighbor effects, however, are usually neglected in the analysis of experimental diffusion data; when the diffusant is a small divalent ion, they should not be.

Very few experiments involving diffusion of trivalent ions in alkali halides have been reported. In general, such ions are not very soluble because of the energy needed to form the two cation vacancies required for electroneutrality. The rare earth Ce^{3+} has been studied using the sectioning technique with the tracer initially a surface deposit. The profiles showed considerable scatter near the surface but were linear in $\ln D$ vs x^2 in the interior (Keneshea and Fredericks, 1965c). The lack of concentration dependence at the low concentration found in the crystal was taken as evidence that the second vacancy remains associated with the Ce^{3+} ion. Reisfeld and Honigbaum (1968a) investigated diffusion of $BiCl_3$ from a surface layer into KCl using a spectrophotometric method (cf. Section IV). They found a low temperature region with an activation energy for migration U of 0.63 eV and a high temperature region with a U of 0.97 eV. It was suggested on the basis of the magnitudes of these migration energies, that the low temperature region was due to diffusion of Bi^{3+} by an interstitial mechanism and the high temperature region due to diffusion of Bi^{2+} by a vacancy mechanism.

In Table VI, much of the data available on the diffusion of aliovalent ions in alkali halides is listed. An attempt was made to classify the limiting conditions under which the diffusion parameters were determined. Three cases are listed: low concentration, $c \to 0$; high concentration, $c \to \infty$, and $p \to 1$ which corresponds to D_s as used in this article. When the measurement did not appear to fit these simple limits, no designation was made. The distinction between $D_{p \to 1} = D_s$ and $D_{c \to \infty}$ was made to distinguish between the experimental techniques. In $D_{p \to 1}$ the concentration of the diffusant varies from $c \to 0$ to near saturation, and plots of $D(c)$ vs c readily expose high concentration effects (Fig. 14) while, in the $c \to \infty$ case, the concentration of a chemical species indistinguishable from the diffusant tracer is high over the entire profile. Thus, anomalies due to high concentration effects may be undetected. If not otherwise specified, the results listed are for radioactive tracer–sectioning experiments from a deposited layer source

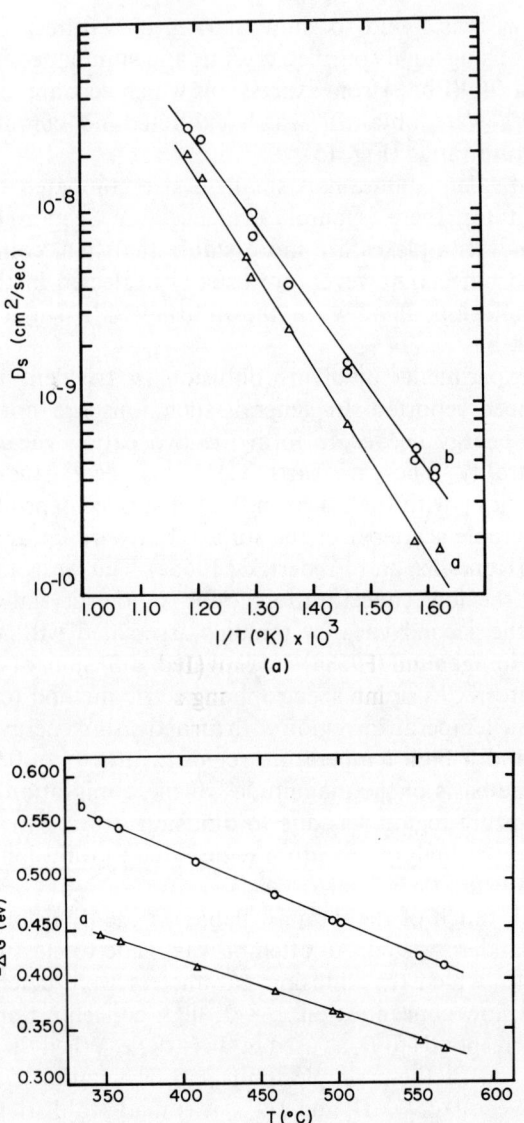

Fig. 15. (a) Log D_s vs $1/T$ from simultaneous diffusion of Pb^{2+} and Cd^{2+} in NaCl. The Pb^{2+} results are fit by line a and those of Cd^{2+} by line b. (b) Gibbs free energies of association of impurity–vacancy complexes in NaCl as a function of temperature. The Pb^{2+} results are fit by line a and those of Cd^{2+} by line b. Note the linearity shown by these data (Krause and Fredericks, 1971).

TABLE VI

DIFFUSION OF ALIOVALENT IONS IN ALKALI HALIDES

Diffusant	D_i^a	ΔT (°C)	D_0 (cm² sec⁻¹)	U (eV)	g_k	Remarks	Reference
NaCl							
Ca^{2+}		750–430	6.0×10^{-4}	0.90		Drift expt.	Banasevich et al. (1960)
	$c \to 0$	795–680	0.13	1.55	$h_k = -0.52$ to -0.57^b	$h_m(Ca) + h_k + h_s/2$	Slifkin and Brébec (1968)
	$c \to \infty$	500–360	2.35×10^{-4}	0.37		$h_m(Ca)$	Slifkin and Brébec (1968)
		737–619	9.4×10^{-4}	1.15		$h_k = -0.57$ from conductivity	F. Bénière et al. (1969a)
	$c \to \infty$	650–350	—	0.96	$-0.67 + (4.9 \times 10^{-4})T$	Doped at 0.2 mole % $CaCl_2$ Arrhenius plot bent down at high temperature.	Murin (1961)
Sr^{2+}		750–600	4.13×10^{-2}	1.36		Simultaneous with Co^{2+}, surface source, not conc. dependent, bent profiles	Allnatt and Pantelis (1968b)
	$c \to \infty$	730–500	7.6×10^{-3}	1.25		Not treated as conc. dependent	F. Bénière et al. (1969a)
		795–117	1.7×10^{-3}	1.31			F. Bénière (1970)
			1.1×10^{-3}	0.91		Drift expt.	Chemla (1956)
Cd^{2+}		655–530	3.9×10^{-3}	0.26		Atomic abs. spectroscopy	Dominguez and Munoz (1970)

continued

TABLE VI—continued

Diffusant	D_i^a	ΔT (°C)	D_0 (cm² sec⁻¹)	U (eV)	g_k	Remarks	Reference
		530–350	$1.2c$	0.64		c is m.f. CdCl$_2$, Colloid formation.	Ikeda (1964)
	$p \to 1$	500–275	2.06×10^{-2}	0.92	$-1.085 + (8.95 \times 10^{-4})T$	Vapor source	Allen et al. (1967b)
	$p \to 1$	569–347	3.57×10^{-3}	0.857	$-0.972 + (6.65 \times 10^{-4})T$	Vapor source, simultaneous Pb and Cd in purified NaCl.	Krause and Fredericks (1971)
Mn²⁺	$c \to \infty$	700–350	3.48×10^{-3}	0.95		Epr analysis, conc. dependent, large source	Stewart and Reed (1965)
		750–450	2×10^{-5}	0.66	$-0.7 + (1.9 \times 10^{-4})T$	Drift expt. also epr analysis	Lure et al. (1963)
		700–400	1.05×10^{-4}	0.675			Riveros et al. (1971)
Zn²⁺	$c \to 0$	800–590	2×10^{-2}	1.03	$h_k = -0.78$–0.83^b	$h_m(\text{Zn}) + h_k + h_{IS}/2$	Rothman et al. (1966)
	$c \to \infty$	780–550	1.5×10^{-4}	0.51		$h_m(Z)$	Rothman et al. (1966)
		720–540	4.0×10^{-2}	1.06			Dominguez and Munoz (1970)
Pb²⁺	$p \to 1$	553–348	1.75×10^{-2}	0.98	$-0.780 + (5.31 \times 10^{-4})T$	Purified cryst., vapor source	Mannion et al. (1968)
	$p \to 1$	528–321	1.5×10^{-2}	0.98_8	$-0.632 + (2.60 \times 10^{-4})T$	Vapor source	Allen et al. (1967a)
	$p \to 1$	569–347	1.40×10^{-2}	0.982	$-0.775 + (5.29 \times 10^{-4})T$	Purified cryst., vapor source, simultaneous Cd²⁺ and Pb²⁺	Krause and Fredericks (1971)
Co²⁺	$c \to \infty$		2.8×10^{-3}	–1.06		Simultaneous Co²⁺ and Sr²⁺, surface source, no conc. dependence, bent profile	Allnatt and Pantelis (1968b)

8. DIFFUSION IN ALKALI HALIDES

Ion		T range	Value	Ratio	Notes	Reference
Ni^{2+}	$c \to \infty$	760–610	8×10^{-3}	1.1		From surface, not conc. dependent, discuss possibility of interstitial Co	Iida and Tomono (1964)
	$c \to \infty$	750–620	2×10^{-2}	1.3		From surface conc. independent	Iida and Tomono (1964)
Hg^{2+}		650–200	—	0.26			Geguzin et al. (1966)
	$p \to 1$	549–449	8.18×10^{-5}	0.57	$g_k = 0.80$–0.75	Purified KCl and KCl:OH, g_k obtained by fitting Eq. (37); profiles scattered at surface	Allen and Fredericks (1973)
KCl							
Cd^{2+}	$p \to 1$	500–350	4.68×10^{-5}	0.54	$g_k = 0.52$–0.53	Vapor source	Keneshea and Fredericks (1965b)
Pb^{2+}	$p \to 1$	474–229	1.02×10^{-3}	1.01	$g_k = 0.45$ (474–433) and 0.58 (373)	"Pure" and Pb^{2+}-doped crystals	Keneshea and Fredericks (1963)
	$p \to 1$	500–373	0.11	1.18	$g_k \sim 0.41$–0.48	Above data plus new new data from Harshaw crystals	Keneshea and Fredericks (1964)
		465–275	1.82×10^{-2}	1.11	—	Powder samples, spectrophotometric	Reisfeld et al. (1965)
Eu^{2+}	$c \to \infty$	550–418	6.45×10^{-2}	1.28		Spectrophotometric, EuCl$_3$ surface to EuCl$_2$ interior, conc. independent profiles, assume complete association	Reisfeld and Honigbaum (1968b)
Co^{2+}		650–200		0.20			Geguzin et al. (1966)

continued

TABLE VI—continued

Diffusant	D_i^a	ΔT (°C)	D_0 (cm² sec⁻¹)	U (eV)	g_k	Remarks	Reference
Bi^{2+}		674–400	5.6×10^{-3}	0.97	$g_k \approx 0.57; s_k \sim 0$	Spectrophotometric method	Reisfeld and Honigbaum (1968a)
Bi^{3+}		400–275	1.7×10^{-3}	0.63		Spectrophotometric, authors suggested interstitial diffusion of Bi^{3+}	Reisfeld and Honigbaum (1968a)
Ce^{3+}		700–500	1.1×10^{-3}	1.03		Surface source, conc. independent, assumed $Ce(Cl)_3$ did not completely dissociate	Keneshea and Fredericks (1965c)
KBr							
Pb^{2+}	$c \to \infty$	400–275	1.5×10^{-3}	0.91		Powders, spectrophotometric	Reisfeld and Glasner (1965)

[a] Estimate of conditions of diffusion parameters.
[b] Estimated here using the values of Allnatt et al. (1971) of h_S from Table II.

using commercial crystals or crystals with their quality minimally specified.

Examination of Table VI reveals little consistency among various experiments on specific impurities. Some measurements agree well while others are difficult to reconcile. With few exceptions, there is not an obvious improvement with time. If one adopts a conservative point of view, then all parameters derived from profiles which exhibit kinks or bends which are not predicted by model calculations should be regarded as artifacts.

Well-behaved profiles which give plots of $D(c)/D_s$ vs c which saturate at unity as c becomes large are not sufficient to prove the impurity system studied is the one proposed by the investigator. All the equations given here which describe the concentration dependence of an aliovalent impurity will behave in similar manner if the preexponential term in the $D(c)$ equation is assumed to be $D_s = (a^2 w_2 f)/3$. Their differences lie in the initial slope or in the temperature dependence of such plots. Even when these have been determined, the free energy of association of the supposed reaction should be shown to be linear over the temperature range studied as in Fig. 15b.

E. Diffusion of Neutral Species

We deal here with diffusion of water molecules and inert gases in alkali halides. The former is of interest in work on color centers, while the latter is important in connection with problems of reactor materials. Because of this interest, the results in these fields are summarized here.

Gründig and Rühenbeck (1965) developed a method of studying the diffusion of H_2O into potassium halides by observing the bleaching of the color of crystals to which K_2O or K had been added.

Gründig and Rühenbeck (1972) have identified the reactions that occur as the bleaching proceeds into the crystal as

$$2H_2O + 2K \longrightarrow 2KOH + H_2 \quad (62a)$$

$$\tfrac{1}{2}H_2 + K \longrightarrow KH \quad (62b)$$

and

$$H_2O + KH \longrightarrow KOH + H_2 \quad (62c)$$

for K-doped crystals and

$$H_2O + K_2O \longrightarrow 2KOH \quad (62d)$$

for the K_2O-doped crystals.

Rühenbeck (1967) found the solubility of H_2O divided into two regions. At lower temperatures, the solubility decreased as the temperature increased, while at higher temperatures, the opposite behavior was observed. The temperature and the range over which this transition occurs depends on the purity of the crystal. With pure zone refined crystals, the transition is relatively well defined, but when Ca^{2+} was added to $KCl:K_2O$, a temperature independent region was found extending from well below to well above that found for pure crystals. The diffusion coefficient of K^+ in KCl is about one-fifth that of H_2O. The results of these experiments are given in Table VII. It was suggested that at low temperatures water diffuses interstitially.

Other neutral diffusants of current interest are inert gas atoms. These are formed as decay products of isotopes of the ions of the host lattice

TABLE VII

SOLUBILITY AND DIFFUSION OF H_2O AND ITS PRODUCTS IN K_2O- AND K-DOPED CRYSTALS

		Solubility		
Crystal	pC_0 (cm^{-1})^c	W (eV)	Temp. (°C)^a	Reference
KCl	8.9×10^9	-0.20	$T < \sim 350$	Rühenbeck (1967)
	9.1×10^{16}	0.67	$T > \sim 350$	
KBr	6.2×10^{10}	-0.15	$T < \sim 325$	
	7.6×10^{16}	0.53	$T > \sim 325$	
KI	1.5×10^{13}	-0.06	$T < \sim 290$	
	1.8×10^{16}	0.33	$T > \sim 290$	

		Diffusion			
Crystal	Diffusant	ΔT (°C)	D_0 (cm^2 sec^{-1})	U (eV)	Reference
KCl	H_2O	700–180	62	0.80	Rühenbeck (1967)
	H_2	420–180	630	0.97	Gründig and Rühenbeck (1972)
	O^{-2}	—	1.5×10^{-3}	1.25	Stasiw (1935)^b
KBr	H_2O	600–180	7.7	0.69	Rühenbeck (1967)
	H_2	400–180	50	0.81	Gründig and Rühenbeck (1972)
KI	H_2O	600–180	0.48	0.56	Rühenbeck (1967)
	H_2	420–180	5.6	0.64	Gründig and Rühenbeck (1972)

^{a} For doped zone refined crystals.
^{b} Given by Rühenbeck (1967). For OH^- diffusion, see Section V and Table V.
^{c} p(Torr)C_0(cm^{-3} Torr^{-1}).

TABLE VIII DIFFUSION OF INERT GASES IN ALKALI HALIDES

Crystal	Diffusant	ΔT (°C)	D_0 (cm² sec⁻¹)	U (eV)	Remarks	Reference
KCl	Ar	$T > 400$	7.9×10^{-4}	0.38	Dose 10^{13}–10^{15} n cm⁻²	Schmeling (1965)
	Kr	500–250	1×10^6	2.1	Ion implanted	Pronko and Kelly (1972)
		250–175	8×10^{-5}	~1.1	^{85}Kr	
KBr	Ar	300–20	1×10^5	1.5	Ion bombardment	Matzke (1967)
KI	Xe	300–20	10	1.4		
	Xe	500–150	1.49	1.03	Dope with precursor ^{133}I → ^{133}Xe	Mears and Elleman (1971)
RbF	Kr	727–298	2.5	1.38	Low dose ~10^{15} n cm⁻². Only one diffusion region. $D_{mp} = 4.5 \times 10^{-6}$ cm² sec⁻¹	Felix (1968)
RbCl	Kr	mp–308	7.5×10^{-3}	0.56	$D_{mp} = 6.3 \times 10^{-6}$ cm² sec⁻¹	Felix (1968)
		308–197	1.4×10^{10}	2.04		Felix (1968)
		mp–150	—	1.62		Kalbitzer (1962)
		700–360	7.9×10^{-4}	0.44	Dose 10^{19} n cm²	Bannasch and Schmeling (1965)
		360–250	—	2.60		Bannasch and Schmeling (1965)
RbBr	Kr	mp–322	5×10^{-4}	0.30	$D_{mp} = 1 \times 10^{-5}$ cm² sec⁻¹	Felix (1968)
		322–192	4×10^7	1.68		Felix (1968)
		mp–150	—	1.32		Kalbitzer (1962)
RbI	Kr	mp–337	1.6×10^{-3}	0.31	$D_{mp} = 2 \times 10^{-5}$ cm² sec⁻¹	Felix (1968)
		337–181	1.3×10^6	1.41		
		mp–150		1.34		Kalbitzer (1962)
CsF	Xe	500–150	8.2×10^{-2}	0.93	^{133}I → ^{133}Xe	Mears and Elleman (1971)
CsCl	Xe	600–450	≤1×10^2	≤2.0	$D_{mp} \leq 1 \times 10^{-8}$ cm² sec⁻¹	Felix (1967)
	Xe	470–350	1×10^{-1}	0.90	CsCl structure $D_{mp} = 1.4 \times 10^{-6}$ cm² sec⁻¹	Felix and Meier (1969)
	Xe	650–470	1×10^{-1}	0.87	NaCl structure $D_{mp} = 1.8 \times 10^{-6}$ cm² sec⁻¹	Felix and Meier (1969)
CsBr	Xe	580–260	1×10^2	1.45	$D_{mp} = 8 \times 10^{-7}$ cm² sec⁻¹	Felix (1967)
CsI	Xe	500–150	0.57	1.01		Elleman et al. (1969)
		to mp	3×10^{-1}	0.98	$D_{mp} = 9.7 \times 10^{-7}$ cm² sec⁻¹	Felix and Meier (1969)
		500–150	0.57	1.00		Mears and Elleman (1971)
		550–200	2.0	1.14	$D_{mp} = 8 \times 10^{-7}$ cm² sec⁻¹	Felix (1967)

when it is exposed to radiation. Norgett and Lidiard (1968) have calculated the energies for various kinds of diffusion processes for inert gases in alkali halides. These calculations suggested a model in which diffusion occurs interstitially with the diffusant being trapped in vacancies or vacancy pairs. Atoms which are large compared to the host ions require vacancy pairs or clusters as traps while smaller atoms can be trapped in single vacancies. In the latter case, the addition of divalent impurities to the host should markedly slow the desorption rate. Qualitatively, this model leads to three regions of desorption. (1) At low temperatures, the traps are mostly filled and the observed enthalpy h_1 approaches the migration enthalpy h_m. (2) At slightly higher temperatures, the traps are partially empty and diffusion occurs by migration with trapping and the observed enthalpy h_2 is greater by the enthalpy of trapping h_T. (3) When the temperature is sufficiently high, the traps are mostly empty and h_3 approaches h_m again. The trap concentration is a complex function of temperature and depends on the actual impurity system present in the crystal. Most data which have been reported can be explained on the basis of this model. Gaus (1965) also has developed a macroscopic theory based on diffusion with trapping.

The diffusion coefficient is obtained by doping the crystal with inert gas by neutron irradiation, by the addition of a precursor isotope such as ^{133}I, or by ion implantation. Table VIII provides a selection of D_0 and U values for various hosts and inert gas diffusants, as well as providing references to work using the various doping techniques.

Schmeling (1967) observed an S-shaped Arrhenius plot for Ar diffusing in KCl. Felix (1968) reported extensive studies on the diffusion of Kr in rubidium halides produced by low dose neutron irradiation. He observed the high and intermediate temperature regions. His results can be explained qualitatively by Norgett and Lidiard's model. For RbF, the Arrhenius plot was linear, indicating that Kr was too large to be trapped, while in the chloride, bromide and iodide trapping occurred. As the size of the anion increased, the enthalpy h_1 followed the expected order RbF > RbCl > RbBr ≈ RbI. If the principal source of traps arises from Schottky defects, then the "knees" separating region 3 from region 2 should occur at temperatures in decreasing order RbI > RbBr > RbCl as they do. (As g_S is not known for the rubidium halides it was assumed that Barr and Lidiard's empirical expression $h_S = 2.14 \times 10^{-3}\ T_{melt}$ is valid.)

VI. Conclusion

Usually, in a review that concerns ionic solids, an extensive tabulation of various energies of formation, migration, etc., which have been calculated theoretically, are compared with those obtained experimentally. This has

8. DIFFUSION IN ALKALI HALIDES

not been done here, partly because many already exist (see list of general references), but primarily because we are only now beginning to appreciate the chemical complexity of these systems. Furthermore, there are relatively few cases of independent measurements yielding parameters which agree within the experimental errors. When different experimental methods are used to measure the same parameter, different values are obtained. These variations appear to arise most often because the impurity–host system is not that proposed by the investigator.

However, the macroscopic theories, both thermodynamic and kinetic, are powerful tools and, when applied to appropriately controlled systems, yield accurate explanations and parameters. The problem is now with experimentalists to perform a proper measurement on a known system.

ACKNOWLEDGMENTS

I want to express my thanks to Professor D. Lazarus for keeping me informed about his work, to Dr. John R. Manning for his comments and help on Section III.A, and to Dr. A. B. Lidiard who provided many useful comments which were of great assistance in preparing this chapter.

GENERAL REFERENCES

A. Diffusion

ADDA, Y., AND PHILIBERT, J. (1966). "La Diffusion dans les Solides." Presses Univ. de Frances, Paris.
BÉNIÈRE, F. (1972). Diffusion in Ionic Crystals, *in* "Physics of Electrolytes" (J. Hladik, ed.), p. 203. Academic Press, New York.
FULLER, R. G. (1972). Ionic Conductivity (Including Self-Diffusion), *in* "Point Defects in Solids" (J. H. Crawford, Jr., and L. M. Slifkin, eds.), Vol. I, p. 103. Plenum, New York.
HOWARD, R. E., AND LIDIARD, A. B. (1964). *Rep. Progr. Phys.* **27**, 161.
LECLAIRE, A. D. (1970). Correlation Effects in Diffusion in Solids, *in* "Physical Chemistry" (H. Eyring, D. Henderson, and W. Jost, eds.), Vol. X. Academic Press, New York.
MANNING, J. R. (1968). "Diffusion Kinetics for Atoms in Crystals." Van Nostrand-Reinhold, Princeton, New Jersey.
NOWICK, A. S. (1972). Defect Mobilities in Ionic Crystals Containing Aliovalent Ions, *in* "Point Defects in Solids" (J. H. Crawford, Jr., and L. M. Slifkin, eds.), Vol. I, p. 151. Plenum, New York.
SHEWMON, P. G. (1963). "Diffusion in Solids." McGraw-Hill, New York.

B. Properties of Ionic Solids

ALLNATT, A. R. (1967). Statistical Mechanics of Point-Defect Interactions in Solids, *in* "Advances in Chemical Physics" (I. Prigogine, ed.), Vol. XI, p. 1. Wiley (Interscience), New York.

BARR, L. W., AND LIDIARD, A. B. (1970). Defects in Ionic Crystals, in "Physical Chemistry" (H. Eyring, D. Henderson, and W. Jost, eds.), Vol. X. Academic Press, New York.
FRANKLIN, A. D. (1972). Statistical Thermodynamics of Point Defects in Crystals, in "Point Defects in Solids" (J. H. Crawford, Jr., and L. M. Slifkin, eds.), Vol. I, p. 1. Plenum, New York.
KRÖGER, F. A. (1964). "Chemistry of Imperfect Crystals." North-Holland Publ., Amsterdam.
KRÖGER, F. A., AND VINK, H. J. (1956). Solid State Phys. 3, 307.
LIDIARD, A. B. (1957). Ionic Conductivity, in "Handbuch der Physik" (S. Flügge, ed.), Vol. XX, p. 246. Springer-Verlag, Berlin.
SÜPTITZ, P., AND TELTOW, J. (1967). Phys. Status Solidi 23, 9.
TOSI, M. P. (1964). Solid State Phys. 16, 1.
VAN BUEREN, H. G. (1961). "Imperfections in Crystals," 2nd ed. North-Holland Publ., Amsterdam.

SPECIFIC REFERENCES

ALLEN, C. A., AND FREDERICKS, W. J. (1970a). Phys. Status Solidi 3a, 143.
ALLEN, C. A., AND FREDERICKS, W. J. (1970b). J. Solid State Chem. 1, 205.
ALLEN, C. A., AND FREDERICKS, W. J. (1973). Phys. Status Solidi 55b, 615.
ALLEN, C. A., IRELAND, D. T., AND FREDERICKS, W. J. (1967a). J. Chem. Phys. 46, 2000.
ALLEN, C. A., IRELAND, D. T., AND FREDERICKS, W. J. (1967b). J. Chem. Phys. 47, 3068.
ALLNATT, A. R., AND CHADWICK, A. V. (1966). Trans. Faraday Soc. 62, 1726.
ALLNATT, A. R., AND CHADWICK, A. V. (1967). Trans. Faraday Soc. 63, 1929.
ALLNATT, A. R., AND COHEN, M. H. (1964a). J. Phys. Chem. 40, 1860.
ALLNATT, A. R., AND COHEN, M. H. (1964b). J. Phys. Chem. 40, 1871.
ALLNATT, A. R., AND LOFTUS, E. (1973). J. Phys. (Paris) 34 (Suppl. 11–12), C9-283.
ALLNATT, A. R., AND PANTELIS, P. (1968a). Solid State Commun. 6, 309.
ALLNATT, A. R., AND PANTELIS, P. (1968b). Trans. Faraday Soc. 64, 2100.
ALLNATT, A. R., PANTELIS, P., AND SIME, S. J. (1971). Proc. Phys. Soc. London (Solid State Phys.) 4, 1778.
ALLNATT, A. R., LOFTUS, E., AND ROWLEY, L. A. (1972). Cryst. Lattice Defects 3, 77.
ARAI, G., AND MULLEN, J. G. (1966). Phys. Rev. 143, 663.
ARNIKAR, H. J., AND CHEMLA, M. (1956). C.R. Acad. Sci. Paris 252, 2132.
BANASEVICH, S. N., LURE, B. G., AND MURIN, A. N. (1960). Sov. Phys.—Solid State 2, 72.
BANNASCH, W., AND SCHMELING, P. (1965). J. Phys. Chem. Solids 26, 1999.
BARDEEN, J., AND HERRING, C. (1951). In "Atom Movements," p. 18. Amer. Soc. Metals, Metals Park, Ohio.
BARR, L. W., AND DAWSON, D. K. (1969). Rep. AERE-R 6324.
BARR, L. W., AND LECLAIRE, A. D. (1964). Proc. Brit. Ceram. Soc. 1, 109.
BARR, L. W., HOODLESS, I. M., MORRISON, J. A., AND RUDHAM, R. (1960). Trans. Faraday Soc. 56, 697.
BARR, L. W., MORRISON, J. A., AND SCHROEDER, P. A. (1965). J. Appl. Phys. 36, 624.
BEAUMONT, J. C., AND CABANE, J. (1961). C.R. Acad. Sci. Paris 252, 113, 266.
BEAUMONT, J. H., AND JACOBS, P. W. M. (1966). J. Chem. Phys. 45, 1496.
BÉNIÈRE, F. (1970). Thesis, Univ. of Paris, Orsay.
BÉNIÈRE, F., BÉNIÈRE, M., AND CHEMLA, M. (1968). C.R. Acad. Sci. Paris 267, 633.
BÉNIÈRE, F., BÉNIÈRE, M., AND CHEMLA, M. (1969a). C.R. Acad. Sci. Paris 268, 1461.

BÉNIÈRE, F., BÉNIÈRE, M., AND CHEMLA, M. (1969b). *J. Chem. Phys.* **66**, 898.
BÉNIÈRE, F., BÉNIÈRE, M., AND CHEMLA, M. (1970). *J. Phys. Chem. Solids* **31**, 1205.
BÉNIÈRE, F., BÉNIÈRE, M., AND CHEMLA, M. (1972). *J. Chem. Phys.* **56**, 549.
BÉNIÈRE, M., BÉNIÈRE, F., AND CHEMLA, M. (1970). *J. Chem. Phys.* **67**, 1312.
BIRMAN, B. I., ZAKHARIN, YA. A., AND PODORSHANSKAYA, N. M. (1968). *Monokrist. Stsintill. Org. Lyuminofory* **3**, 8.
BOSWARVA, I. M. (1972). *Proc. Phys. Soc. London (Solid State Phys.)* **5**, L5.
BREWER, L. (1950). The Fusion and Vaporization of the Halides, in "The Chemistry and Metallurgy of Miscellaneous Materials" (L. L. Quill, ed), p. 193. McGraw-Hill, New York.
BROWN, R. C., WORSTER, J., MARCH, N. H., PERRIN, R. C., AND BULLOUGH, R. (1971). *Phil. Mag.* **23**, 555.
CAPPELLETTI, R., AND FIESCHI, R. (1969). *Cryst. Lattice Defects* **1**, 69.
CHANDRA, S., AND ROLFE, J. (1970). *Can. J. Phys.* **48**, 412.
CHANEY, R. E., AND FREDERICKS, W. J. (1973). *J. Solid State Chem.* **6**, 240.
CHEMLA, M. (1952). *C.R. Acad. Sci. Paris* **234**, 2601.
CHEMLA, M. (1953). *C.R. Acad. Sci. Paris* **236**, 484.
CHEMLA, M. (1954). Thesis. Univ. of Paris, Paris.
CHEMLA, M. (1956). *Ann. Phys. (Paris)* **1**, 959.
CHEMLA, M., AND BÉNIÈRE, F. (1973). *J. Phys. (Paris)* **34** (Suppl. 11-12), C9-11.
CHIBA, Y., AND SAKAMOTO, M. (1965). *J. Phys. Soc. Jap.* **20**, 1284.
CHIBA, Y., UEKI, K., AND SAKAMOTO, M. (1963). *J. Phys. Soc. Jap.* **18**, 1092.
COKER, E. H., DECIUS, J. C., AND SCOTT, A. B. (1961). *J. Chem. Phys.* **35**, 745.
COMPAAN, K., AND HAVEN, Y. (1956). *Trans. Faraday Soc.* **52**, 786.
CONANT, D. R., AND DECIUS, J. C. (1967). *Spectrochim. Acta A* **23**, 2931.
COOK, J. S., AND DRYDEN, J. S. (1962). *Proc. Phys. Soc. London* **80**, 479.
CRANK, J. (1956). "The Mathematics of Diffusion." Oxford Univ. (Clarendon Press), London and New York.
CRAWFORD, J. H., JR. (1969). *J. Phys. Chem. Solids* **31**, 399.
DAWSON, D. K., AND BARR, L. W. (1967a). *Proc. Brit. Ceram. Soc.* **9**, 171.
DAWSON, D. K., AND BARR, L. W. (1967b). *Phys. Rev. Lett.* **19**, 846.
DECIUS, J. C., COKER, E. H., AND BRENNA, G. L. (1963). *Spectrochim. Acta* **19**, 1281.
DECIUS, J. C., JACOBSON, J. L., SHERMAN, W. F., AND WILKINSON, G. R. (1965). *J. Chem. Phys.* **43**, 2180.
DOBROVINSKAYA, E. R., AND PODORSHANSKAYA, N. M. (1966). *Ukr. Fiz. Zh.* **11**, 227.
DOBROVINSKAYA, O. R., SOLUNSKII, V. I., AND SHAKHOVA, A. G. (1967). *Ukr. Fiz. Zh.* **12**, 868.
DOMINGUEZ, H. A., AND MUNOZ, P. Z. (1970). *Rev. Mex. Fis.* **19**, 375.
DOWNING, H. L., AND FRIAUF, R. J. (1970). *J. Phys. Chem. Solids* **31**, 845.
DREYFUS, R. W. (1961). *Phys. Rev.* **121**, 1675.
DREYFUS, R. W., AND NOWICK, A. S. (1962). *J. Appl. Phys.* **33**, 473.
DRYDEN, J. S., AND HARVEY, G. G. (1969). *Proc. Phys. Soc. London* **2**, 603.
ELLEMAN, T. S., FOX, C. H., AND MEARS, L. D. (1969). *J. Nucl. Mater.* **30**, 89.
FEIT, M. D., MITCHELL, J. L., AND LAZARUS, D. (1973). *Phys. Rev.* **B8**, 1715.
FEITKNECHT, W. (1953). *Fortschr. Chem. Forsch.* **2**, 670.
FELIX, F. W. (1967). *Z. Naturforsch.* **22a**, 2075.
FELIX, F. W. (1968). *Phys. Status Solidi* **27**, 529.
FELIX, F. W., AND MEIER, K. (1969). *Phys. Status Solidi* **32**, K 139.
FREDERICKS, W. J., SCHUERMAN, L. W., AND LEWIS, L. C. (1966). "An Investigation of Crystal Growth Processes." Final Reps. AF-AFOSR 217-63 and AF-AFOSR-217-66, Corvallis, Oregon.
FRIAUF, R. J. (1957). *Phys. Rev.* **105**, 843.

FRIAUF, R. J. (1963). Properties of Ionic Crystals, in "American Institute of Physics Handbook" (D. E. Gray, ed.), 2nd ed., p. 9-63. McGraw-Hill, New York.
FRIAUF, R. J. (1972). Properties of Ionic Crystals, in "American Institute of Physics Handbook" (D. E. Gray, ed.), 3rd ed., p. 9-74. McGraw-Hill, New York.
FRITZ, B., LÜTY, F., AND ANGER, J. (1963). Z. Phys. **174**, 240.
FULLER, R. G. (1966). Phys. Rev. **142**, 524.
FULLER, R. G., MARQUARDT, C. L., REILLY, M. H., AND WELLS, J. C., JR. (1968). Phys. Rev. **176**, 1036.
GAUS, H. (1965). Z. Naturforsch. **20a**, 1298.
GEGUZIN, YA. E., DOBROVINSKAYA, E. R., AND PODORSHANSKAYA, N. M. (1965). Zh. Prikl. Spektrosk. **2**, 552.
GEGUZIN, YA. E., DOBROVINSKAYA, E. R., LEV, I. E., AND MOZHAROV, M. V. (1966). Fiz. Tverd. Tela **8**, 3248.
GLASNER, A., AND REISFELD, R. (1961). J. Phys. Chem. Solids **18**, 345.
GLASNER, A., REJOAN, A., AND REISFELD, R. (1961). J. Phys. Chem. Solids **19**, 331.
GLASNER, A., REISFELD, R., AND LINENBERG, A. (1967). Phys. Status Solidi **24**, 695.
GRÜNDIG, H., AND RÜHENBECK, C. (1965). Z. Phys. **183**, 274.
GRÜNDIG, H., AND RÜHENBECK, C. (1972). Z. Phys. **249**, 269.
GUCCIONE, R., TOSSI, M. P., AND ASDENTE, J. (1959). J. Phys. Chem. Solids **10**, 162.
HANEDA, K., IKEDA, T., AND YOSHIDA, S. (1968). J. Phys. Soc. Jap. **25**, 643.
HANSON, R. C. (1968). Bull. Amer. Phys. Soc., Ser. II, **13**, 902.
HANSON, R. C. (1970). Phys. Status Solidi (a) **1**, 109.
HARRISON, L. G. (1961). Trans. Faraday Soc. **57**, 1191.
HARVEY, P. J., AND HOODLESS, I. M. (1967). Phil. Mag. **16**, 3408.
HOODLESS, I. M., AND TURNER, R. G. (1972a). Phys. Status Solidi (a) **11**, K 55.
HOODLESS, I. M., AND TURNER, R. G. (1972b). J. Phys. Chem. Solids **33**, 1915.
HOODLESS, I. M., STRANGE, J. H., AND WYLDE, L. E. (1971). Proc. Phys. Soc. London (Solid State Phys.) **4**, 2727.
HOWARD, R. E. (1957). J. Chem. Phys. **27**, 1377.
HOWARD, R. E. (1966). Phys. Rev. **144**, 650.
HOWARD, R. E., AND LIDIARD, A. B. (1963). J. Phys. Soc. Jap. (Suppl. II) **18**, 197.
HOWARD, R. E., AND MANNING, J. R. (1967). Phys. Rev. **154**, 561.
IIDA, Y., AND TOMONO, Y. (1964). J. Phys. Soc. Jap. **19**, 1264.
IKEDA, T. (1964). J. Phys. Soc. Jap. **19**, 858.
IKEDA, T. (1973). Jap. J. Appl. Phys. **12**, 1810.
ILLINGWORTH, R. (1963). J. Phys. Chem. Solids **24**, 129.
JACOBS, P. W. M., AND PANTELIS, P. (1971). Phys. Rev. B **4**, 3757.
JOST, W. (1960). "Diffusion in Solids, Liquids and Gases," p. 32. Academic Press, New York.
KAKAISHI, T., AND SENSUI, Y. (1969). Trans. Faraday Soc. **65**, 131.
KALBITZER, S. (1962). Z. Naturforsch. **17a**, 1071.
KENESHEA, F. J., AND FREDERICKS, W. J. (1963). J. Chem. Phys. **38**, 1952.
KENESHEA, F. J., AND FREDERICKS, W. J. (1964). J. Chem. Phys. **41**, 3271.
KENESHEA, F. J., AND FREDERICKS, W. J. (1965a). J. Chem. Phys. **43**, 2925.
KENESHEA, F. J., AND FREDERICKS, W. J. (1965b). J. Phys. Chem. Solids **26**, 501.
KENESHEA, F. J., AND FREDERICKS, W. J. (1965c). J. Phys. Chem. Solids **26**, 1787.
KLOTSMAN, S. M., POLIKARPOVA, I. P., TIMOFEEV, A. N., AND TRAKHTENBERG, I. S. (1967a). Fiz. Tverd. Tela **9**, 2487.
KLOTSMAN, S. M., POLIKAPROVA, I. P., AND TIMOFEEV, A. N. (1967b). Fiz. Tverd. Tela **11**, 2710.
KLOTSMAN, S. M., POLIKARPOVA, I. P., AND TIMOFEEV, A. N. (1970). Fiz. Tverd. Tela **12**, 3364.
KLOTSMAN, S. M., POLIKARPOVA, I. P., AND TIMOFEEV, A. N. (1972). Phys. Status Solidi (b) **49**, 423.

KRAUS, K. A., AND NELSON, F. (1958). ASTM Spec. Publ. No. 195, 27.
KRAUSE, J. L., AND FREDERICKS, W. J. (1971). *J. Phys. Chem. Solids* **32**, 2673.
KRAUSE, J. L., AND FREDERICKS, W. J. (1973). *J. Phys. (Paris)* **34** (Suppl. 11–12), C9-13.
LAGERWALL, T. (1965). *Trans. Chalmers Univ. Techn., Gothenburg* No. 307.
LASKAR, A. L., BATRA, A. P., AND SLIFKIN, L. (1969). *J. Phys. Chem. Solids* **30**, 1173.
LAURANCE, N. (1960). *Phys. Rev.* **120**, 57.
LAURENT, J. F., AND BÉNARD, J. (1957). *J. Phys. Chem. Solids* **3**, 7.
LAURENT, J. F., AND BÉNARD, J. (1958). *J. Phys. Chem. Solids* **7**, 218.
LECLAIRE, A. D. (1962). *Phil. Mag.* **7**, 141.
LECLAIRE, A. D., AND LIDIARD, A. B. (1956). *Phil Mag.* **1**, 518.
LIDIARD, A. B. (1955). *Phil. Mag.* **46**, 1218.
LIDIARD, A. B. (1960). *Phil. Mag.* **5**, 1171.
LIDIARD, A. B. (1974). In "The Physics of Fluorite Compounds" (W. Hayes, ed.). Oxford Univ. Press, London and New York.
LURE, B. G., MURIN, A. N., AND BRIGEVICH, R. F. (1963). *Sov. Phys.—Solid State* **4**, 1432.
MANNING, J. R. (1958). *Phys. Rev. Lett.* **1**, 365.
MANNING, J. R. (1962a). *Phys. Rev.* **125**, 103.
MANNING, J. R. (1962b). *Phys. Rev.* **128**, 2169.
MANNING, J. R. (1964). *Phys. Rev.* **A 136**, 1758.
MANNING, J. R. (1965). *Phys. Rev.* **A 139**, 2027.
MANNION, W. A., ALLEN, C. A., AND FREDERICKS, W. J. (1968). *J. Chem. Phys.* **48**, 1537.
MAPOTHER, D., CROOKS, H. N., AND MAURER, R. J. (1950). *J. Chem. Phys.* **18**, 1231.
MATANO, C. (1933). *Jap. J. Phys.* **8**, 109.
MATZKE, HJ. (1967). *Z. Naturforsch.* **22a**, 507.
MATZKE, HJ. (1971). *J. Phys. Chem. Solids* **32**, 437.
MCCOMBIE, C. W., AND LIDIARD, A. B. (1956). *Phys. Rev.* **101**, 1210.
MEARS, L. D., AND ELLEMAN, T. S. (1971). *Phys. Status Solidi (a)* **7**, 509.
MURIN, A. N. (1961).
NAUMOV, A. N., AND PTASHNIK, V. B. (1968). *Fiz. Tverd. Tela* **10**, 3710.
NELSON, V. C., AND FRIAUF, R. J. (1970). *J. Phys. Chem. Solids* **31**, 825.
NORGETT, M. J., AND LIDIARD, A. B. (1968). *Phil. Mag.* **18**, 1193.
PATTERSON, D., MORRISON, J. A., AND ROSE, G. S. (1956). *Phil. Mag.* **1**, 393.
PETERSON, N. L., AND ROTHMAN, S. J. (1969). *Phys. Rev.* **177**, 1329.
PRINGSHEIM, P. (1949). "Fluorescence and Phosphorescence." Wiley (Interscience), New York.
PRONKO, P. P., AND KELLY, R. R. (1972). *J. Phys. Chem. Solids* **33**, 1761.
REISFELD, R., AND FREDERICKS, W. J. (1970). *Israel J. Chem.* **8**, 959.
REISFELD, R., AND GLASNER, A. (1965). *J. Chem. Phys.* **42**, 2983.
REISFELD, R., AND HONIGBAUM, A. (1968a). *J. Chem. Phys.* **48**, 5565.
REISFELD, R., AND HONIGBAUM, A. (1968b). *Israel J. Chem.* **6**, 53.
REISFELD, R., GLASNER, A., AND HONIGBAUM, A. (1965a). *J. Chem. Phys.* **42**, 1892.
REISFELD, R., GLASNER, A., AND HONIGBAUM, A. (1965b). *J. Chem. Phys.* **43**, 2923.
RIVEROS, H. G., MUNOZ, E. P., AND RUIZ, C. M. (1971). *Rev. Mex. Fis.* **20**, Suppl. FA101-8.
ROSENBERGER, F. (1972). Purification of Alkali Halides, in "Ultrapurity: Methods and Techniques" (M. Zief and R. M. Speights, eds.). Dekker, New York.
ROTHMAN, S. J., BARR, L. W., ROWE, A. H., AND SELWOOD, P. G. (1966). *Phil. Mag.* **14**, 501.
ROTHMAN, S. J., PETERSON, N. L., LASKAR, A. L., AND ROBINSON, L. C. (1972). *J. Phys. Chem. Solids* **33**, 1061.
RÜHENBECK, C. (1967). *Z. Phys.* **207**, 446.
SCHMELING, P. (1965). *Phys. Status Solidi* **11**, 175.
SCHMELING, P. (1967). *J. Phys. Chem. Solids* **28**, 1185.
SCHMIDT, K., AND STAUBE, H. (1968). *Z. Phys. Chem. N.F.* **60**, 90.

SHIRN, G. A., WAJDA, E. S., AND HUNTINGTON, H. B. (1953). *Acta Met.* **1**, 513.
SLIFKIN, L., AND BRÉBEC, G. (1968). Rep. CEA-DM/1750.
STASIW, O. (1935). *Z. Tech. Phys.* **16**, 343.
STEWART, W. H., AND REED, C. A. (1965). *J. Chem. Phys.* **43**, 2890.
STRELOW, F. W. E., LIEBENBERG, C. J., AND VON S. TOERIEN, F. (1968). *Anal. Chim. Acta* **43**, 465.
STOEBE, T. G. (1970). *J. Phys. Chem. Solids* **31**, 1291.
SUZUKI, K. (1961). *J. Phys. Soc. Jap.* **16**, 67.
SYMMONS, H. F. (1970). *Proc. Phys. Soc. London (Solid State Phys.)* **3**, 1846.
SYMMONS, H. F. (1971). *Proc. Phys. Soc. London (Solid State Phys.)* **4**, 1945.
SYMMONS, H. F., AND KEMP, R. C. (1966). *Brit. J. Appl. Phys.* **17**, 607.
TIERMAN, R. J., AND WUENSCH, B. J. (1971). *J. Chem. Phys.* **55**, 4996.
TOMAN, K. (1962). *Czech. J. Phys.* **12**, 542.
TOMIZUKA, C. T. (1959). Diffusion, *in* "Methods of Experimental Physics" (L. Marton, ed.), Vol. VI, p. 364. Solid State Physics, Pt. A (K. Lark-Horovitz and V. A. Johnson, eds.), Academic Press, New York.
TUBANDT, C. (1932). "Handbuch der Experimentalphysik," Ist ed., Vol. XII, p. 383. Akad. Verlagsgesellschaft, Leipzig.
WAGNER, C. (1950). *J. Chem. Phys.* **18**, 1227.
WATKINS, G. D. (1959a). *Phys. Rev.* **113**, 79.
WATKINS, G. D. (1959b). *Phys. Rev.* **113**, 91.
WIRTZ, K. (1943). *Physik. Z.* **44**, 221.

9

Very Rapid Ionic Transport in Solids

ROBERT A. HUGGINS

CENTER FOR MATERIALS RESEARCH
STANFORD UNIVERSITY
STANFORD, CALIFORNIA

I. Introduction	445
II. Special Characteristics of Fast Ionic Conductors	446
A. Historical Remarks	446
B. Characteristic Properties	447
C. Parameters in the Simple Hopping Model	450
III. Materials Which Exhibit Fast Ionic Motion	452
A. Introduction	452
B. Cation Conductors	452
C. Anion Conductors	471
IV. Theoretical Approaches to Fast Ion Conduction	475
V. Outlook for the Future	482
References	483

I. Introduction

This chapter will be focused upon diffusion and ionic motion in several groups of materials called fast ionic conductors in which mass transport occurs with unusual rapidity, sometimes at surprisingly low temperatures. As will be seen, several characteristics of these materials are substantially different from those found in the simple metals and ionic conductors with relatively close packed crystal structures that have received the bulk of our attention to date. Indeed, these differences can be so large that it may be questioned whether the traditional conceptual framework and kinetic models, developed over a period of many years and discussed in some detail in other chapters of this book, are appropriate for materials of this type.

After a brief discussion of the unusual properties of such fast ionic conductors, the special structural features and the resultant transport behavior of presently known groups of such materials will be discussed. It will be seen that one of their important characteristics is that their crystallographic structures are not close packed. Instead they contain ion-sized tunnels or passageways comprised of connected void polyhedra within which the ions on one of the sublattices reside and move. Typically, there are more sites than ions in these tunnels and, in at least some cases, the occupation of the available sites is random rather than ordered. The openness of these structures sometimes produces a tendency for structural instability at lower temperatures, resulting in phase transformations to more closely packed atomic arrangements.

Because ions upon only one of the sublattices typically reside within the crystallographic tunnels, ionic transport is quite selective in such materials. Fast cation conductors are generally not fast anion conductors, and vice versa. Because of this, these two groups will be described separately, with each section organized according to important features of the crystal structure.

Following this review of current information, various attempts to provide a theoretical understanding of the basis for the unusual properties of fast ionic conductors will be discussed. Particularly instructive is a recent calculation of potential energy profiles within a crystallographic tunnel. This model explains, for the first time, the experimental observation of unusually low values of both σ_0 and activation energy in such structures, and why ions of intermediate size can have the greatest mobility. It also predicts a variation in transport path and site preference within tunnels for ions of different sizes and shows that small ions will not always tend to reside in the centers of void polyhedra.

II. Special Characteristics of Fast Ionic Conductors

A. Historical Remarks

Electrical conductivity due to the motion of ions in solids has been recognized and has attracted scientific attention for a long time (Warburg, 1884; Warburg and Tegetmeier, 1888; Haber and Tolloczko, 1904; Bruni and Scarpa, 1913; Tubandt and Lorenz, 1914; Tubandt and Eggert, 1920; Tubandt and Reinhold, 1923; Tubandt, 1932). In addition to the measurement of conductivity values, much of this early work was directed toward the determination of the fraction of the electric current carried by the various species. This is commonly expressed in terms of transference numbers, where the transference number of any specie is the ratio of the partial conductivity of that specie to the total conductivity.

Measurements of ionic conductivity and diffusion on a large number of materials, predominantly the halides, led to the establishment of the presently standard theoretical models both for the defect structures and for mass and charge transport in materials with mobile ions. Experimental results showed that materials with Frenkel disorder, and thus interstitial species, have substantially greater values of ionic conductivity than those with Schottky disorder, where the predominant defects are vacancies.

During the course of this early work, it also became recognized that a number of simple ionic compounds undergo phase transformations at elevated temperatures. In some cases, the ionic conductivity goes through a corresponding abrupt transition, with much higher values in the high temperature phase, where the temperature dependence (the activation energy) also is reduced to unusually low values. All of the materials which were recognized at that time as having high temperature phases with unusually fast ionic conductivity were silver or copper compounds.

It has been shown recently that this type of "high temperature" behavior, characterized by especially rapid ionic motion, can persist to ambient temperatures and below in some materials. Furthermore, fast ionic conduction has now been found in several families of compounds containing alkali ions, and it has become recognized that this is a general phenomenon, not just characteristic of silver and copper salts.

These two factors, as well as greatly increased understanding in the general field of solid state electrochemistry, have caused a rebirth of interest in ionic conductors in recent years. This has been accentuated by the recognition of a number of important potential practical applications of these fast ionic conductors.

B. CHARACTERISTIC PROPERTIES

The highest value of ionic conductivity yet reported for a solid at room temperature is found in rubidium silver iodide ($RbAg_4I_5$). Measurements (Owens and Argue, 1967, 1970; Raleigh, 1970) indicate a value of about 0.3 $(ohm\text{-}cm)^{-1}$ at 300°K. This is not only within an order of magnitude of typical values for liquid electrolytes commonly used in battery systems, but is about 14 orders of magnitude greater than the ionic conductivity of sodium chloride, a typical alkali halide, at the same temperature (Dreyfus and Nowick, 1962). While the difference between the ionic conductivities of these two solids is considerably reduced at elevated temperatures because their temperature dependences are vastly different, the great disparity between the conductivity values of materials such as $RbAg_4I_5$ and the alkali halides, traditionally considered model ionic conductors, indicates that there is an appreciable substantive difference in the transport phenomena in these two groups of materials.

In order to illustrate how extremely rapid ionic transport in some fast ionic conductors can be, it is interesting to utilize a simple model (Rickert, 1973) to calculate the maximum value that the diffusion coefficient might have in any solid with fixed lattice positions. If we make the assumption that the ions move at thermal velocity v directly from one lattice site to the next without oscillating at each site, we can calculate the jump frequency from $\nu = v/d$. Assuming a jump distance of 1 Å, and picking a temperature of 300°C, ν is found to be 3.4×10^{12} sec^{-1}. Inserting this value into the relation $D = \alpha d^2 \nu$ and setting $\alpha = \frac{1}{6}$ for a cubic structure, it is found that $D = 5.6 \times 10^{-5}$ cm^2/sec.

Experimental conductivity data for AgI, a fast ion conductor, can be compared with this limiting value of the diffusion coefficient by use of the Nernst–Einstein relation, as explained later. The maximum theoretical value of the conductivity thus obtained is 2.8 (ohm-cm)$^{-1}$, whereas experimental measurements (Kvist and Josefson, 1968) have shown that AgI actually has a conductivity of 1.97 (ohm-cm)$^{-1}$ at 300°C.

Thus we see that, in this material, the magnitude of the ionic conductivity is very close to the value that would be predicted by this simple limiting model which assumes that *all* of the silver ions are free to move and that *every* jump attempt is successful. These experimental conductivity data also lead to a value of diffusion coefficient for the silver ions inside solid AgI that is comparable to those typically found in liquids.

As an even more extreme example of the unusually rapid mass transport that can sometimes be found in fast ionic conductors, recent measurements (Chu, *et al.*, 1973) have shown that the chemical diffusion coefficient for silver in α-Ag_2S is 0.47 cm^2/sec at 200°C. This is a number of orders of magnitude greater than even the largest values found for diffusion coefficients in typical metals and close packed ionic materials. In this case, there are two factors of importance. Not only are the silver ions relatively mobile in this phase of silver sulfide, but this material is a mixed conductor, that is, both the electrons and ions contribute to the electric charge transport. The interaction of the fluxes of these two species in the diffusion process results in a large internal electrical field and great enhancement of the rate of ionic motion and thus of the chemical diffusion coefficient. This behavior, sometimes expressed in terms of a thermodynamic enhancement factor (Darken, 1948; Wagner, 1951), can lead to values of diffusion coefficient several orders of magnitude greater than the limit derived from this simple limiting case theory. Therefore, it becomes quite obvious that

9. VERY RAPID IONIC TRANSPORT IN SOLIDS

a simple single-jump model cannot be realistic for such cases. There must be some sort of cooperative motion or multiple-site jumps taking place.

Other important differences between fast ionic conductors and more conventional materials can be seen in Fig. 1, where the variation with temperature of the ionic conductivity of four materials is shown: KCl (an alkali halide with Schottky disorder), AgCl (which has Frenkel disorder), AgI, and sodium beta alumina (with the nominal formula $MAl_{11}O_{17}$), the last two being examples of fast ionic conductors. The AgI undergoes a phase transformation at 146°C, and the low temperature β phase is not in the fast ionic conductor class. On the other hand, beta alumina appears to be stable in the high conductivity form to well below room temperature.

From Fig. 1, we can see that there are three principal features of the experimental data relating to fast ionic conductors that differentiate them from normal ionic conductors and deserve special attention. As already mentioned, the absolute magnitude of the ionic conductivity can be unusually high.

The second unusual experimental characteristic of fast ionic conductors is the small temperature dependence of the conductivity. In some cases,

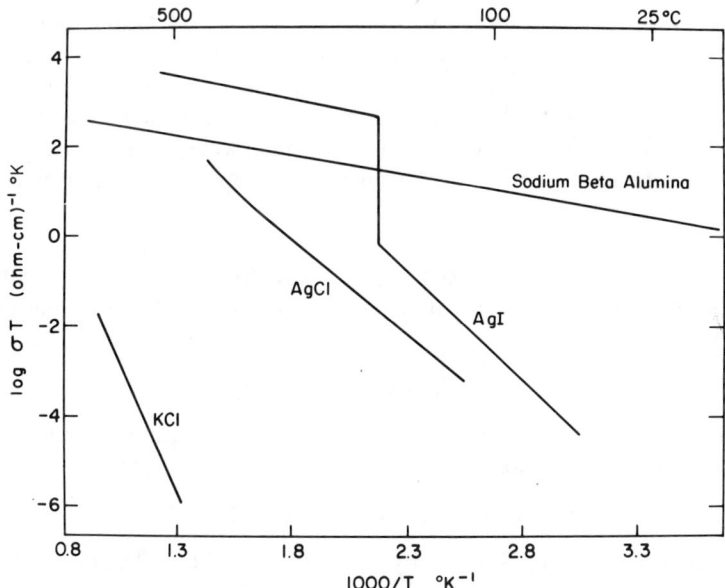

Fig. 1. Temperature dependence of the ionic conductivity of several materials, illustrating the large variation between different groups.

the activation energy can be as low as 2–5 kcal/mole. Since the mobile carrier concentration is very large as well as temperature independent, this is interpreted as a low enthalpy of motion.

The third unique feature is the very small value of σ_0 found in these materials, of the order of 10^3 (ohm-cm)$^{-1}$ °K. (σ_0/T is the preexponential of the conductivity.) As will be shown shortly, this implies either an especially low value of attempt frequency or a very small value of the entropy of migration.

These novel characteristics are, of course, interrelated, rather than independent. However, the important point is that they clearly indicate that ionic motion in fast ionic conductors is vastly different from that found in the groups of materials with which the materials science community has been primarily concerned over the last few decades.

C. Parameters in the Simple Hopping Model

If we follow the normal random-walk formulation for defect transport by isolated jumps, we can express the diffusion coefficient of a particle species by

$$D = \alpha d^2 v \tag{1}$$

where α is a geometric factor which is the reciprocal of the number of possible jump directions from a given site. The value of α is $\frac{1}{6}$ for a close packed cubic lattice, and $\frac{1}{2}$ if jumps are constrained to only one dimension. In this expression, d is the distance moved during a single jump and v the average jump frequency. It should be mentioned that this equation would have to include the correlation factor if we were dealing with the diffusion of radiotracers.

We can write the average jump frequency in terms of the fraction of the species that is free to move, β, the attempt frequency v_0, and the probability of success $\exp(-\Delta G/RT)$, where ΔG is the free energy of migration,

$$v = \beta v_0 \exp(-\Delta G/RT) \tag{2}$$

If we divide the free energy into its entropy and enthalpy components, we get the following expression for the diffusion coefficient:

$$D = \alpha d^2 \beta v_0 \exp(\Delta S/R) \exp(-\Delta H/RT) \tag{3}$$

We can go one step further by consideration of the factor β. When transport occurs by the motion of defects, the value of β is obviously the defect concentration. In conventional high purity ionic conductors where the defect concentration is dominated primarily by the probability of the formation of intrinsic defect pairs, a further temperature dependence is introduced, for

$$\beta = \exp(-\Delta G_F/RT) \qquad (4)$$

where ΔG_F is the free energy of formation of the intrinsic defect pair. In such a case, β can be quite small at lower temperatures, severely constraining the magnitude of the diffusion coefficient. The slope of a plot of $\ln D$ vs T^{-1} is then given by $-(\Delta H + \Delta H_F)/R$, where ΔH_F is the enthalpy of defect pair formation. In fast ionic conductors, however, essentially all of the ions on one of the sublattices can be considered mobile. As a result, β is both very large and effectively independent of temperature. When this is the case, the slope of the $\ln D$ vs T^{-1} plot becomes only $-\Delta H/R$.

If we can assume that charge and mass are transported by the same mobile species, the ionic conductivity is related to the particle diffusion coefficient by the Nernst–Einstein relation

$$\sigma = DCz^2F^2/RT \qquad (5)$$

where C is the concentration of the mobile species, z their charge number, and F the Faraday constant.

In accordance with the Nernst–Einstein relation and the temperature dependence of the diffusion coefficient in Eq. (3), one expects a linear relationship between the logarithm of the product σT and T^{-1}, as shown in Fig. 1.

This relationship can be expressed as

$$\sigma = (\sigma_0/T) \exp(-\Delta H/RT) \qquad (6)$$

where σ_0 is the value of the extrapolated intercept at $T^{-1} = 0$ on such a plot. By combining Eqs. (3) and (5), it follows that

$$\sigma_0 = C\beta d^2 v_0 \alpha(z^2F^2/R) \exp(\Delta S/R) \qquad (7)$$

The interesting thing to note is that the fast ion conductors clearly have a *much lower* value of σ_0 than do the normal ionic conductors. This difference can be as large as 5 orders of magnitude. Which terms in Eq. (7) could be responsible for this? The values of C, d, α, and z do not vary appreciably from one material to the next, so the only possibilities are β, v_0, and ΔS. However, β is expected to be both temperature independent and *much larger* in fast ion conductors than those in which the defect concentration is determined by intrinsic thermally activated defect-pair formation, so this is obviously not the dominant effect.

Values of ΔS for the migration of species in fast ion conductors are not presently known, but it is reasonable to assume that they may be unusually small, as motion of the fast ions may entail relatively insignificant changes in the lattice frequencies of the surrounding crystal structure. The attempt frequency v_0 may also be unusually low in such materials if the mobile ions sit in relatively shallow potential wells.

From these simple considerations, it seems clear that any useful theoretical approaches to the understanding of the transport process in fast ionic conductors must address themselves not just to the observations of very low activation energy, but also to the existence of unusually low values of either ΔS or v_0.

III. Materials Which Exhibit Fast Ionic Motion

A. Introduction

Fast ion conductors are all characterized by crystallographic structures which, if modeled by the use of hard spheres of appropriate radii to represent the constituent ions, can be seen to contain tunnels or pathways of connected space, with lateral dimensions comparable to ionic diameters. Because of this geometric feature of their crystal structures, the transport or long range motion of ions from point to point can occur in such materials without involving significant ionic displacements and strain energy effects, such as provide the primary contribution to the activation energy for motion in close packed metallic systems.

For purposes of this discussion, fast ion conductors will be divided into two categories, those in which ionic conductivity is predominantly due to cationic transport, and those which are primarily anion conductors. In each case, emphasis will be given to the important crystallographic features characteristic of these types of materials.

B. Cation Conductors

1. *Materials with Cubic Structures*

The high temperature (α phase) structure of silver iodide (AgI) has been known for some time to have an unusually high value of ionic conductivity. In fact, it was shown by Tubandt and Lorenz (1914) to have such a high value of ionic conductivity in the solid state just below the melting point that there is a drop of more than 20% upon melting. These results have been confirmed much more recently by Kvist and Josefson (1968), and are illustrated in Fig. 2. X-ray diffraction experiments showed (Strock, 1934, 1936; Krug and Sieg, 1952; Hoshino, 1957) that the iodide ions exist in a bcc lattice. The silver ion positions are hard to determine from the diffraction information, and the experimental data have been interpreted

9. VERY RAPID IONIC TRANSPORT IN SOLIDS

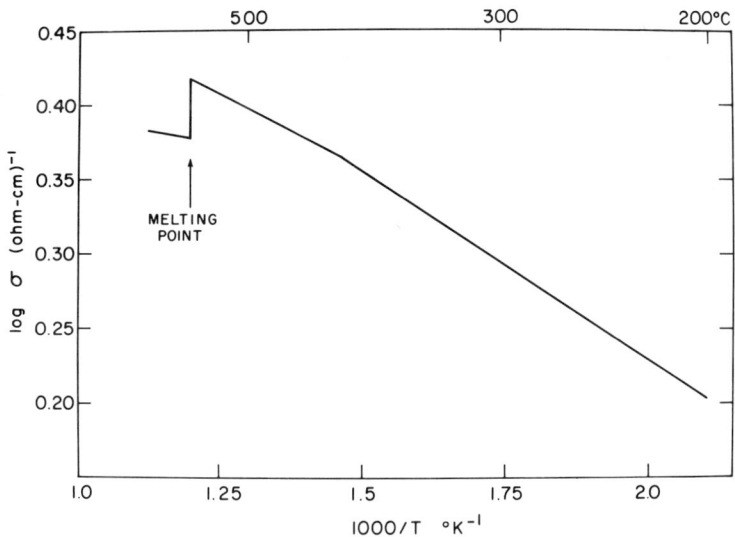

Fig. 2. Temperature dependence of the ionic conductivity of α-AgI showing the drop at the melting point.

in terms of a distribution among sites with three-fold and four-fold coordination in this structure. These sites can be viewed (Flygare and Huggins, 1973) as special positions within the tunnels which run in the cube edge directions between the iodide ions. There are many more such possible sites for the silver ions than the number of ions present, and the large Debye–Waller factors obtained from the diffraction data imply unusually large mean displacements (from 0.2 Å at 150°C to 0.5 Å at 450°C) of the atoms from the assumed crystallographic positions. Such materials are sometimes described as having a relatively rigid anion skeleton containing silver ions in a liquid state. The question of cation site occupancy and mobility in this structure will be discussed in more detail in Section IV. There are several materials which have this same crystal structure, and these are listed in Table I.

Two compounds having the general formula Ag_3SX, where X can be either iodine or bromine, have been studied by several investigators (Reuter and Hardel, 1960, 1961, 1965, 1966a,b; Takahashi and Yamamoto, 1964, 1965, 1966; Reuter *et al.* 1967). The high temperature α form of Ag_3SI has the same structure as alpha silver iodide except that half of the iodine ions are randomly replaced by sulfur ions. The silver ions are distributed in the tunnels, and this material also has an unusually high value of silver ion conductivity. In the case of the β (low temperature) structure of Ag_3SI and in Ag_3SBr, the anions are ordered upon the bcc lattice, forming an

TABLE I

PHASES WITH CRYSTAL STRUCTURES IN WHICH THE ANIONS ARE ARRANGED IN A BCC LATTICE

		Structural information		
Phase	Stability range (°C)	Lattice constant (Å)	Temp. (°C)	Reference
α-AgI	146–555	5.058	146	Strock (1934)
α-Ag$_2$S	179–825	4.88	250	Rahlfs (1935)
α-Ag$_2$Se	133–880	4.98	250	Rahlfs (1935)
α-CuBr	472–490	4.59	485	Krug and Sieg (1952) Hoshino (1952)
(Na, Li)$_2$SO$_4$[a]	521–600	5.77	556	Førland and Krogh-Moe (1958)
(Ag, Li)$_2$SO$_4$[a]	417–574	5.76	545	Øye (1963)
α-Ag$_3$SI	245–>400			
β-Ag$_3$SI	20–235			
Ag$_3$SBr	0–300			

[a] Data are given for equal molar compositions. These phases are stable over a wide range of cation ratios.

antiperovskite structure. This ordering appears to reduce the mobility of the silver ions within the tunnels somewhat. It has also been shown that β-Ag$_3$SI and Ag$_3$SBr exhibit a continuous solid solution range.

Data on the ionic conductivity of these materials as a function of temperature are presented in Fig. 3. There is a small electronic contribution to the conductivity in addition to the much larger ionic conductivity in Ag$_3$SI. The data that have been reported for this partial electronic conductivity have not been consistent among different investigators, and the value clearly depends upon the thermodynamic conditions, particularly the sulfur activity, imposed during sample preparation and measurement. It seems that the greater the sulfur activity and the lower the iodine activity, the lower the electronic conductivity. Lower silver activity also seems to reduce the electronic component of the total conductivity. However, it must be recognized that these are ternary phases, and so two compositional variables must be specified as well as the total pressure and temperature in order to define the thermodynamic conditions.

The low temperature β phase of silver iodide has the zinc blende structure, in which every silver ion is surrounded by a tetrahedron of iodine ions, and vice versa. This may also be described as interpenetrating fcc cation and anion lattices. As this structure is close packed, the silver ion conductivity is much lower than in the high temperature α phase. In

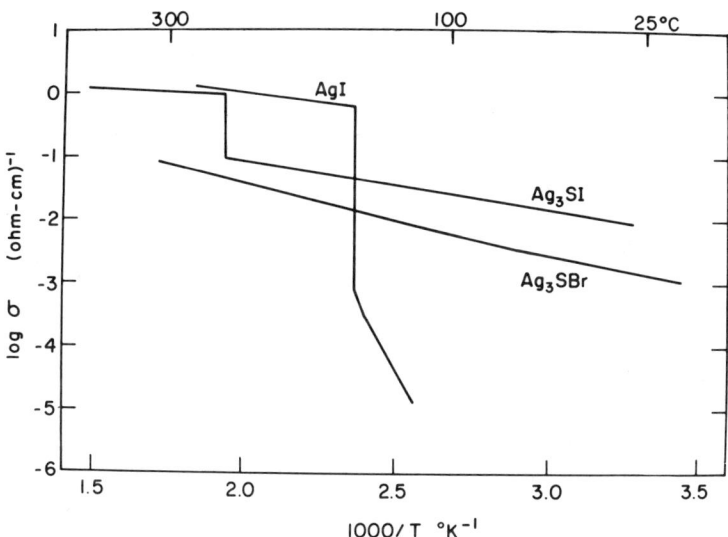

Fig. 3. Ionic conductivity of several materials with a bcc anion lattice.

addition, it has a much greater temperature dependence in the β phase, as is characteristic of transport in close packed lattices.

The high temperature α phase of the compound Ag_2HgI_4 has a structure (Ketelaar, 1934a,b, 1938; Suchow and Pond, 1953; Hoshino, 1955) which is a modification of the zinc blende arrangement in which two of the silver ions are replaced by one Hg^{2+} in accordance with charge balance considerations. As a result, there is a very large concentration of empty sites. The geometry of this structure can be described in terms of the partial occupation by cations of the tetrahedral interstices in a fcc anion lattice. Motion of the silver ions among the tetrahedral sites is very easy and it is found that the ionic conductivity is very high in this structure.

There is also a copper analog with partial cationic occupation of the tetrahedral sites within a fcc anion lattice Cu_2HgI_4, and conductivity data have been reported (Heyne, 1970) for a solid solution of the silver and copper phases, $(Ag,Cu)HgI_4$. It has also been found (Rahlfs, 1936; Krug and Sieg, 1952; Miyake et al., 1952) that the α-Ag_2Te, α-CuI, α-Cu_2S, and α-Cu_2Se phases consist of face-centered anion arrangements, with the cations distributed in various ways among the interstitial sites. Lithium sulfate also has a face-centered cubic anion arrangement at high temperatures, and its ionic conductivity has been investigated by Kvist (1965). It has also been inferred from conductivity measurements (Kvist and Lunden, 1966) that lithium tungstate probably forms a cubic or pseudocubic high temperature

TABLE II

PHASES WITH CRYSTAL STRUCTURES IN WHICH THE ANIONS ARE ARRANGED IN A FCC LATTICE

Phase	Stability range (°C)	Phase	Stability range (°C)
α-Ag$_2$HgI$_4$	50–93	α-Cu$_2$S	>91
α-Cu$_2$HgI$_4$	67–90	α-Cu$_2$Se	>110
α-(Ag, Cu)HgI$_4$		α-Li$_2$SO$_4$	574–860
α-Ag$_2$Te	>150	α-Li$_2$WO$_4$	684–738
α-CuI	407–600		

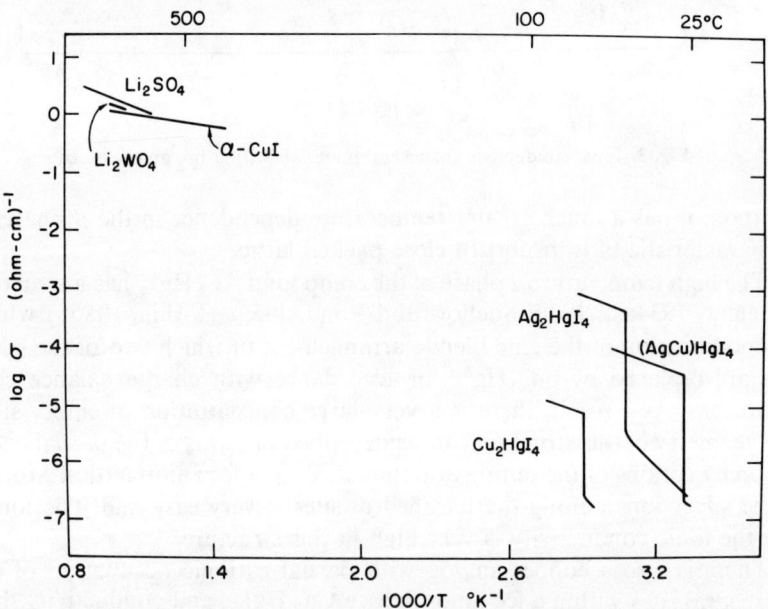

Fig. 4. Ionic conductivity of several materials with fcc arrangement of anions.

phase. These phases, along with information about the temperatures at which they are stable are listed in Table II, and ionic conductivity data are presented in Fig. 4.

Another group of cubic materials having related structures with dilute site occupancy in interstitial tunnels is based upon the alkali silver iodides whose nominal formula is MAg$_4$I$_5$, where M is Rb, K, or NH$_4$. The unusually high values of silver ion conductivity in these materials were discovered independently by Bradley and Greene (1966, 1967a), and Owens and Argue (1967, 1970).

This group of ionic conductors has the highest value of ionic conductivity at room temperature presently known. The primary member of this group is $RbAg_4I_5$ whose crystal structure has been found (Bradley and Greene, 1967b; Geller, 1967) to be cubic with four formula units per unit cell. The four rubidium ions are surrounded by distorted iodine octahedra. This provides 56 tetrahedral interstitial positions which might be occupied by the 16 silver ions present in the structure. The distribution of the silver ions among the three slightly different types of tetrahedral positions has been discussed by Wiedersich and Geller (1970).

Although the related cubic α-silver iodide structure is stable only above 146°C, the presence of the large M ions evidently prevents the structure from collapsing until a considerably lower temperature is reached. In the case of KAg_4I_5, the cubic high temperature structure transforms to one of lower symmetry at $-136°C$, with an accompanying decrease of a factor of 250 in the ionic conductivity. This structural change is deferred to $-139°C$ in samples with the nominal formula $(K_{0.75}, Rb_{0.25})Ag_4I_5$. In the case of $RbAg_4I_5$ there are two low temperature modifications. A phase transformation occurs at $-155°C$ that is associated with a conductivity drop of close to 2 orders of magnitude. Another structural change has been found to occur in the vicinity of $-65°C$, but evidently does not cause a discontinuity in the ionic conductivity. In this latter case, thermodynamic measurements (Johnston and Lindberg, 1968) indicate that this is a second-order transition. It is interesting that there have also been recent reports (Perrot and Fletcher, 1968, 1969) of a structural change evidently involving a discontinuous change in the order parameter in the high temperature body-centered alpha silver iodide phase at about 410°C. As in the case of $RbAg_4I_5$, it has been found (Josefson et al., 1971) that there is an accompanying change in slope, but no discontinuity, in the ionic transport properties.

In the case of the potassium member of this group, there is a narrow temperature range (257–332°C) in which the analogous copper compound KCu_4I_5 is stable.

A number of related silver ion-conducting materials of the MAg_4I_5 type have been described recently, in which the M ion is replaced by an organic group. These developments were discussed in a review by Owens (1971). (See also the paper by Geller, 1973.)

Another direction of current interest involves phases in which some of the iodine ions are replaced by other large anions. Examples include $Ag_4HgI_2Se_2$ (Takahashi et al., 1967), $Ag_7I_4PO_4$ which is stable up to 79°C (Takahashi et al., 1972), $Ag_{19}I_{15}P_2O_7$ which is stable to 147°C

(Takahashi et al., 1967, 1972), and $Ag_6I_4WO_4$ which decomposes at 293°C (Takahashi et al., 1973). These are essentially pure silver ion conductors and have conductivity values that are not quite as high as those found for the MAg_4I_5 phases. Data concerning the conductivity of a group of the silver iodide-related materials at 25°C are presented in Table III.

TABLE III

VALUES OF IONIC CONDUCTIVITY AT 25°C FOR A NUMBER OF FAST ION CONDUCTORS

Material	Conductivity at 25°C $(\Omega\text{-cm})^{-1}$	Material	Conductivity at 25°C $(\Omega\text{-cm})^{-1}$
β-Ag_3SI	1×10^{-2}	$Ag_4HgI_2Se_2$	2×10^{-2}
Ag_3SBr	2×10^{-3}	$Ag_7I_4PO_4$	2×10^{-2}
$RbAg_4I_5$	3×10^{-1}	$Ag_{19}I_{15}P_2O_7$	9×10^{-2}
KAg_4I_5	2×10^{-1}	$Ag_6I_4WO_4$	5×10^{-2}
$NH_4Ag_4I_5$	2×10^{-1}		

2. *Materials with Hexagonal Structures*

Both Ag_2SO_4 and K_2SO_4 have been found to have hexagonal structures. In the case of K_2SO_4, the hexagonal structure is stable from 584°C to the melting point at 1068°C (Majumdar and Roy, 1965; Schroeder and Kvist, 1969).

3. *Materials with Unidirectional Tunnel Structures*

Many materials are now known in which the structure is comprised of cation-centered octahedra (or BX_6 groups) which share corners or edges. These BX_6 groups are often arranged in such a manner that there are parallel open tunnels of ionic size running through the structure between them. In a number of cases, these tunnels may be either partially or fully occupied by cations, typically monovalent or divalent species. It has been found that these tunnel-resident cations are quite mobile in some cases (e.g., Singer et al., 1973) and it is reasonable to expect that further experimental evidence for fast ionic transport in materials with such tunnel structures will appear in the near future, as this is an area of considerable current interest.

Two of the more well known of these tunnel structures are the tetragonal I and hexagonal tungsten bronze structures, which are found in a number of materials having the general formula M_xWO_3, where M might be any of a number of cations, usually monovalent, and x is between zero and unity. The existence of these structures and of their stability ranges depends, of course, upon the identity of the cation M. These structures are illustrated in Fig. 5.

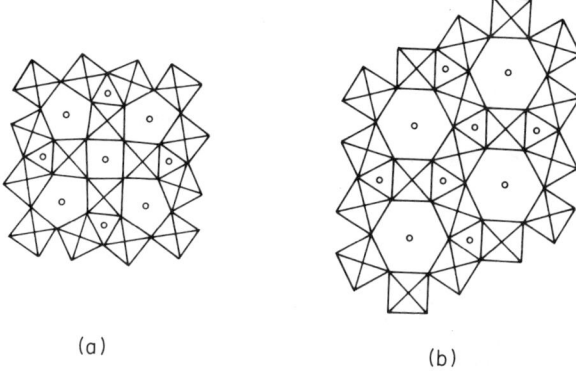

(a)   (b)

Fig. 5. Schematic drawings of portions of the (a) tetragonal I and (b) hexagonal tungsten bronze structures. The diagonally crossed squares are cation-centered anion octahedra, and the tunnel locations are indicated by the small circles. Note that there are 3-sided, 4-sided, and 5-sided tunnels in the tetragonal structure, and 3-sided and 6-sided tunnels in the hexagonal structure.

There is presently little information about diffusion of species in these tunnel structures. However, evidence of relatively rapid transport was was presented by the demonstration (Whittingham and Huggins, 1971b, 1972) that such materials can be used as reversible electrodes in the ac measurement of ionic conductivity of materials containing several different monovalent cations. This requires that both ionic and electronic species are mobile. Also, it has been found (Remeika *et al.*, 1967; King, 1972) that the superconducting properties of the hexagonal rubidium tungsten bronze can be markedly altered by acid leaching at 100°C, which evidently removes about half of the rubidium ions from the tunnels. Specific heat measurements (King, 1972) have implied that potassium ions have a greater mobility than rubidium ions in this structure. Furthermore, observations (Clark, 1970; Murphy, 1971; Clark *et al.*, 1972) of the narrowing of the hydrogen nuclear magnetic resonance line in samples of the hexagonal ammonium tungsten bronze $(NH_4)_x WO_3$ have indicated motion of the hydrogen within the structure. It has been found that the temperature dependence of the line width depends sensitively upon the value of x, which is a measure of the filling of the channels. By use of both pulse and steady state NMR techniques, it has been found that two different types of ionic motion occur. One involves rotation of the ammonium ions at sites within the channels, and the other some type of translational motion, probably from site to site along the channel. An additional complication is that neutral ammonia molecules may also be resident within the channels, so that translation of hydrogen ions may be due to jumps from ammonium ions to nearby ammonia molecules as well as to the translation of ammonium ions or ammonia molecules through the tunnels.

There have been as yet only a few other reports of ionic transport measurements in materials with noncubic tungsten bronze-type structures. Electrical conductance measurements (Whittingham and Huggins, 1973) on a tantalum-substituted potassium tungsten bronze, $K_x Ta_x W_{1-x} O_3$, showed a marked frequency dependence at lower temperatures, indicating an ionic component to the total conduction process. Dielectric loss measurements (Layden, 1968) have also shown ionic motion in the channels of $BaTa_2O_6$, which has a related structure.

In addition to the tungsten bronze group, there are a number of other structures with linear tunnels; some of these also contain double or triple tunnels, as illustrated in Fig. 6.

The tetragonal rutile (TiO_2) structure has small tunnels which run parallel to the c axis, and it has been found (Johnson, 1964) that lithium diffuses in a very anisotropic manner in TiO_2, evidently moving through these tunnels with relative ease.

Vanadium also forms a series of mixed-conducting oxide bronzes containing tunnels. The β phase, $M_x V_2 O_5$, can be described in terms of linear $(V_2O_5)_n$ chains, with alkali metal ions in the tunnels between the chains. The γ phase, $M_{1+x} V_3 O_8$, consists of ribbons of VO_6 octahedra joined into sheets by chains of VO_5 trigonal bipyramids. The channels between the oxygen atoms of adjacent ribbons contain both octahedral and tetrahedral sites in which the alkali metal M atoms can reside. These structures are included in Fig. 6. Pulsed nuclear magnetic resonance studies of lithium transport in the beta lithium vanadium bronze phase were made by Gendell et al. (1962). Halstead et al. (1973) also recently reported the results of a study of lithium diffusion in the gamma lithium vanadium bronze by evaluation of the temperature dependence of the steady state NMR line width.

Experimental observations indicating high ionic conductivity at low temperatures make materials with the tetragonal hollandite structure of particular interest. Dryden and Wadsley (1958) prepared samples with the formula $Ba_x(Mg_x Ti_{8-x})O_{16}$ with this structure, and interpreted anisotropic dielectric loss results in terms of barium ion motion through the unidirec-

9. VERY RAPID IONIC TRANSPORT IN SOLIDS

tional tunnels. Their experiments on the temperature and frequency dependence of the loss led to a value of 3.9 kcal/mole for the activation energy for barium ion transport. Recently, Singer, *et al.* (1972) reported dielectric loss, capacitance, and dc polarization measurements of the potassium analog $K_{2x}(Mg_xTi_{8-x})O_{16}$, as well as the related $K_x(Al_xTi_{8-x})O_{16}$.

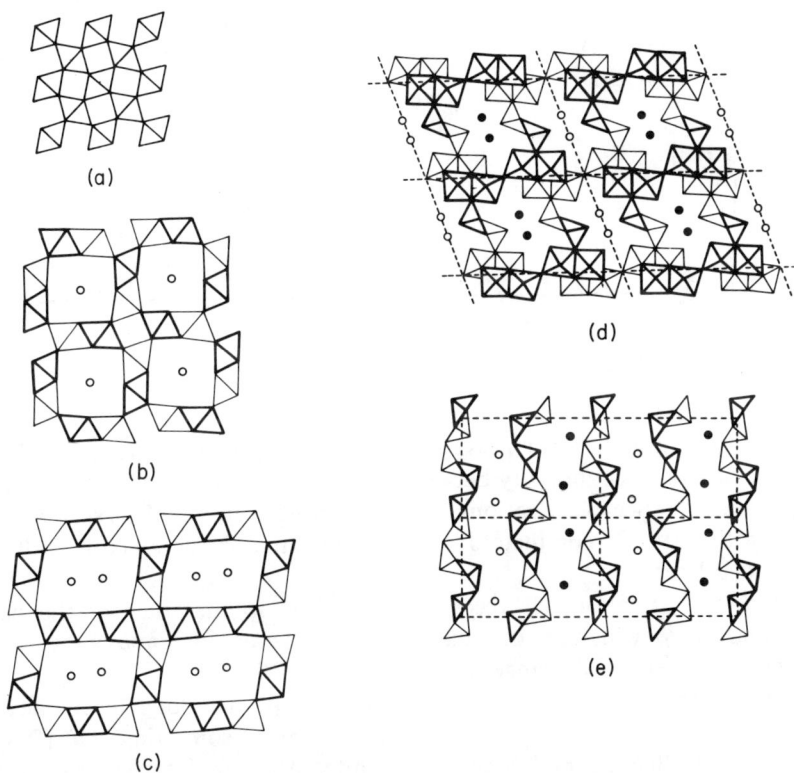

Fig. 6. Schematic drawings of several additional structures containing tunnels; (a) rutile, (b) hollandite, (c) psilomelane, and (d) beta and (e) gamma vanadium bronzes. Light and dark lines indicate different elevations in these projections.

Singer *et al.* (1972, 1973) also made dielectric loss measurements on the potassium tungstate phase, $K_2W_4O_{13}$, and found an activation energy (for the transport of the potassium ions?) of about 3.9 kcal/mole. Direct measurements of the ionic conductivity from 25 to 750°C (Nagel and Huggins, 1973) have confirmed Singer's results on the potassium hollandite. Results on these tunnel structure materials are presented in Table IV.

TABLE IV
Transport Data on Some Tunnel Structure Materials

Material	Activation energy (kcal/mole)	Reference
$Ba_x(Mg_xTi_{8-x})O_{16}$	3.9	Dryden and Wadsley (1958)
	5.3	Singer et al. (1973)
$K_{2x}(Mg_xTi_{8-x})O_{16}$	5.1	Singer et al. (1973)
$K_x(Al_xTi_{8-x})O_{16}$	5.6	Singer et al. (1973)
$K_2W_4O_{13}$	3.9	Singer et al. (1973)
$BaTa_2O_6$	17.8	Layden (1968)
$(NH_4)_{0.33}WO_3$	1.8	Clark et al. (1972)
$(NH_4)_{0.25}WO_3$	3.3	Murphy (1971)
$Li_{0.35}V_2O_5$	1.6	Gendell et al. (1962)
$Li_{1.1}V_3O_8$	7.2	Halstead et al. (1973)
Li in TiO_2	7.6	Johnson (1964)

4. Materials with Layer Structures

An increasing number of materials are being identified which have layer-type crystallographic structures in which certain species can have unusually high values of ionic mobility between the layer planes. In some cases these highly mobile species are cations, in others they are anions. This section will deal with cationic conductors of this type; anionic conductors will be described later.

Greatest attention has recently been given to materials of the beta alumina family, which are unusually good conductors for some of the alkali metal ions over a wide range of temperatures.

Although materials of the beta alumina type had long been known, and the basic crystallographic structure determined some time ago (Bragg et al., 1931; Beevers and Ross, 1937), intense recent interest arose as a result of the important work at the Ford Motor Company (Yao and Kummer, 1967; Radzilowski et al., 1969; Kummer, 1972). Researchers not only called attention to the unusual physical properties of these materials, but also (Weber and Kummer, 1967) described how it is possible to use sodium beta alumina as a solid electrolyte with liquid sodium and sodium sulfide phases in the electrode compartments to produce a new type of high energy density battery. The use of liquid electrodes and a solid electrolyte is just the inverse of traditional battery designs and can lead to some very distinct advantages.

There are two major structural types in this group. One of these is beta alumina itself, which has the nominal formula $MAl_{11}O_{17}$, or more properly,

$M_{1+x}Al_{11-x/3}O_{17}$. The other is generally called β'', and has the nominal formula MAl_5O_8. In both cases, M can be any of a long list of monovalent cations, even Ga^+ or NO^+. The beta alumina structures are composed of blocks (generally four oxygen layers thick) of Al_2O_3 with the gamma alumina (spinel) structure connected by bridging layers containing only M^+ ions and oxide ions. These bridging layers contain a hexagonal array of parallel tunnels in which the M^+ ions reside. The site occupation in the tunnels is quite dilute; all the sites would be filled if x were unity, but x values are typically 0.1–0.3. In contrast to normal gamma alumina, where aluminum ions reside on only two-thirds of the cation sites of the spinel structure, all these sites are occupied in the spinel-type blocks of beta alumina. This structure is illustrated schematically in Fig. 7.

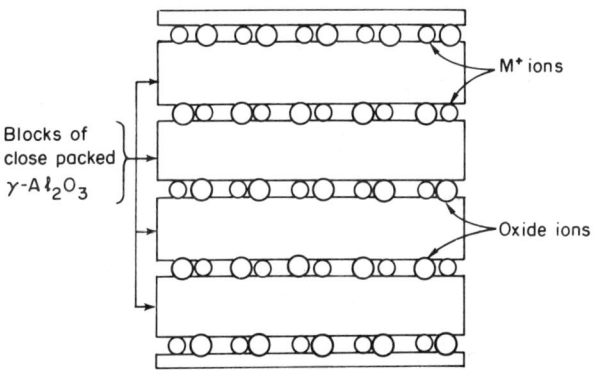

Fig. 7. Schematic representation of the structure of beta alumina showing γ blocks separated by bridging layers containing M^+ and oxide ions.

The primary difference between the β phase and the β'' phase in this family lies in the manner in which the gamma alumina blocks, in which the oxygens have a close packed cubic arrangement, are stacked upon one another. In the β structure, the bridging layers which contain the M^+ ions are mirror planes, so that the overall structure has hexagonal symmetry, with lattice parameters of about $a_0 = 5.59$ and $c_0 = 22.53$ Å in the sodium case; the distance between bridging layers is therefore about 11.27 Å. In the case of the β'' structure (Yamaguchi and Suzuki, 1968; Thery and Briancon, 1964; Bettman and Peters, 1969; Whittingham et al., 1969) the repeat distance is also about the same, but the c_0 value is approximately 33.9 Å and the symmetry is rhombohedral. The spinel blocks in β'' are related to each other by rotation about a threefold screw axis normal to the bridging layers, whereas in β case, the screw axis is twofold. Recent observations utilizing electron microscopy (LeCars et al., 1973) have indicated that the β and β'' phases can exhibit coherent intergrowth.

In addition to β and β'', there have been reports (Bettman and Terner, 1971; Weber and Venero, 1970) of additional related structures, designated as β''' and β''''. The structure of β''' is reported to be similar to that of beta alumina, except that the γ blocks contain six layers of oxygen atoms, and the bridging layers are thus 15.9 Å apart. The β'''' structure is similar to β'', but again has six layers of oxygens in the γ blocks.

Yamaguchi and Suzuki (1968) reported another phase intermediate between β and β'', which they labeled β'. However, it evidently differs from the β structure only with respect to the positions deduced for the sodium ions in the bridging layer. As a result, it has become generally assumed that their results related to a stoichiometric variation of the normal β phase. It has been found (Weber and Venero, 1970) that the presence of MgO or Li_2O stabilizes the β'' structure, and much of the work done on this phase has involved samples containing modest amounts of one or the other of these constituents. This structure contains some 50% more M^+ ions in the bridging layers than does the β phase; presumably the magnesium ions replace some of the aluminum ions in the γ blocks, most probably along the midplane.

The arrangement of the ions within the bridging layer in the β structure is indicated in Fig. 8. It can be seen that the oxygen ions occupy one-fourth of the positions that would be filled in a close packed plane. The M^+ ions reside in the hexagonally disposed tunnels that permeate this oxygen array. The normal M^+ ion positions are octahedrally surrounded by oxygens above and below the bridging layer. An alternative and evidently higher energy type of site is also present, in which the M^+ ions lie between pairs of oxygen ions. It was mentioned earlier that samples of beta alumina typically contain excess M^+ ions—that is, more M^+ ions than can be accommodated by filing only the preferred sites in the bridging layer. X-ray diffraction results by Roth (1972) on silver beta alumina indicated that the excess silver ions occupy these alternative sites. On the other hand, work by Peters et al. (1971) on sodium beta alumina appeared to show that excess sodium is accommodated in a different way. In this case, the M^+ ions occupy octahedrally coordinated sites directly in line between pairs of oxygen ions in the bridging layer. Obviously, if an M^+ ion were in such a position it would have to dislodge one of the nearby ions from a normal position to a similar nearby in-line site.

Thus, the addition of one extra M^+ ion can be accommodated either by its residing upon an alternative site or by the occupation of a pair of in-line sites at the expense of one ion upon a normal site.

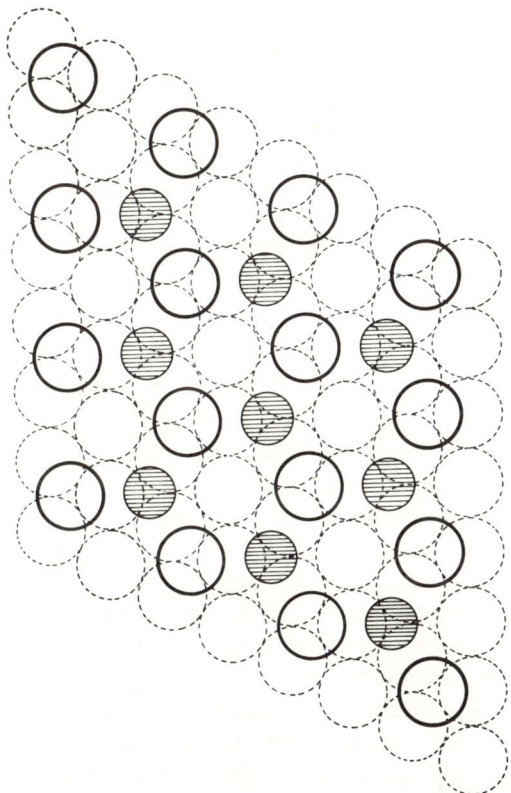

Fig. 8. Arrangement of ions in the bridging layer of beta alumina, assuming ideal stoichiometry. Actual materials have excess cations. Oxide ions in adjacent close packed plane of γ-block structure are shown by dashed circles, oxide ions in bridging layer by heavy circles, and mobile cations by smaller cross-hatched circles.

However, this situation is complicated even more by the suggestion (Roth, 1972; van Gool and Bottelberghs, 1973) that deviations from stoichiometry might be accomplished by the presence of a domain-type structure in the bridging layers, with the filling of excess sites at the domain walls. It is fair to say that the distribution of cations within the bridging layer in these materials has not yet been fully resolved.

The mobile ion in the bridging layer can be exchanged with a number of other cations by immersion in appropriate molten salts (Yao and Kummer, 1967) or by electrochemical pumping (Whittingham et al., 1969). There is also evidence that water intercalates into these structures (Kummer, 1972; Saalfeld et al., 1968), although it is not clear whether the water is

TABLE V

Self-Diffusion Data Obtained on Cations in the β-Alumina Structure by Use of Radiotracer Techniques in the Temperature Range 200–400°C[a]

Ion	D_0 (cm^2/sec)	ΔH (kcal/mole)
Na$^+$	2.4×10^{-4}	3.81
Ag$^+$	1.65×10^{-4}	4.05
K$^+$	0.78×10^{-4}	5.36
Rb$^+$	0.34×10^{-4}	7.18
Li$^+$	14.5×10^{-4}	8.71
Tl$^+$	0.65×10^{-4}	8.22

[a] Kummer (1972).

present as neutral H_2O molecules or as H_3O^+ ions. In any case, its presence impedes the transport of the other cations present.

Information on the self-diffusion of several different ions in the beta alumina structure was presented by Yao and Kummer (1967) and Radzilowski et al. (1969) by the use of radiotracer techniques. The results of these measurements are presented in Table V.

The dielectric loss method was also used by Radzilowski et al. (1969) to evaluate ionic motion in polycrystalline beta alumina containing various different M^+ ions. In some cases, they found quite good agreement between these results (which relate to ionic motion at relatively low temperatures) and data obtained from the radiotracer experiments, whereas in others, there was a considerable discrepancy in the apparent activation energy.

Direct measurements of the ionic conductivity of members of the beta alumina family using ion-blocking electrodes and alternating current techniques have been reported by a number of authors. However, the results have typically shown a frequency dependence and comparison between the results of experiments by different investigators has been disappointing. A major step forward was the introduction (Whittingham and Huggins, 1971b) of the use of mixed conductors as ionically reversible electrodes. These are electronic conductors in which the M^+ ion is also quite mobile. By using this technique, frequency independent values of the ionic conductivity could be obtained over a wide range of temperature. Using properly selected tungsten bronze phases as electrodes, reliable data have been obtained for the ionic conductivity of single crystals of beta alumina containing silver (Whittingham and Huggins, 1971a), sodium (Whittingham and Huggins, 1971b), and potassium, thallium, and lithium (Whittingham and Huggins, 1972) from over 800°C to well below ambient temperature.

The results of these measurements of the conductivity are linear over an unusually wide temperature range. As an example, the data for sodium β alumina from 820 to 150°C, are shown in Fig. 9.

Ideally, the information acquired by the use of these three methods, radiotracer self-diffusion, dielectric loss, and ionic conductivity, should be consistent, as they all measure ionic transport. The remarkably good agreement found in this case is illustrated in Fig. 10, in which the ionic conductivity calculated from the radiotracer and dielectric loss data measured by the Ford group (Yao and Kummer, 1967; Radzilowski et al., 1969) are plotted along with the directly measured single-crystal conductivity values obtained at Stanford (Whittingham and Huggins, 1971b) for sodium beta alumina. The values extracted from the radiotracer diffusion data by use of the Nernst–Einstein relation are expected to be somewhat lower than the true conductivity values because of correlation effects. From this difference, one can evaluate the Haven ratio H_R where $H_R \equiv D_T/D_\sigma$. Because the M^+ ions move much more rapidly than the other species present, the value of H_R is the same as the correlation factor in this case. This allows conclusions to be drawn concerning the detailed jump mechanism involved in the transport process.

Similar data for beta alumina containing silver and potassium in the bridging layer are shown in Figs. 11 and 12. In the silver case, the conductivity data were not extended below ambient temperature. It can also be seen that although the slopes of the radiotracer and dielectric loss data are quite different in the case of potassium beta alumina, their overall magnitudes are reasonably consistent, and they are tied together quite well by the direct conductivity measurements. Similar comparisons of results obtained by the use of these three methods in the cases of thallium and lithium beta alumina can be found in the paper by Whittingham and Huggins (1972).

The ionic conductivity data obtained by the reversible electrode single-crystal method for these five different M^+ ions in the beta alumina structure are collected in Table VI, assuming an Arrhenius relation of the form $\sigma = (\sigma_0/T) \exp(-\Delta H/RT)$. Also shown are the experimental values of the Haven ratio, where they are available. In the case of silver, sodium, and thallium beta alumina, they were found to be very close to the theoretical value of 0.6 calculated (Whittingham and Huggins, 1971a) for an interstitialcy (indirect interstitial) mechanism in the beta alumina structure.

The lithium-containing material is evidently somewhat different from the others. The directly measured conductivity data were found to fall upon two connecting straight lines, meeting at about 180°C. The conductivity and radiotracer data essentially coincide in the high temperature region, giving comparable values of the activation energy. However, the Haven ratio

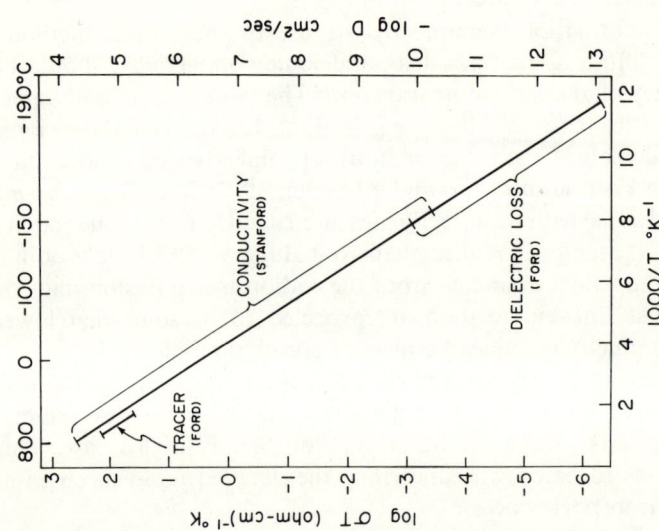

Fig. 10. Temperature dependence of the ionic conductivity and self-diffusion coefficient of sodium beta alumina measured by the use of three different techniques (after Whittingham and Huggins, 1972). Ford data from Yao and Kummer (1967) and Radzilowski *et al.* (1969). Stanford data from Whittingham and Huggins (1971b).

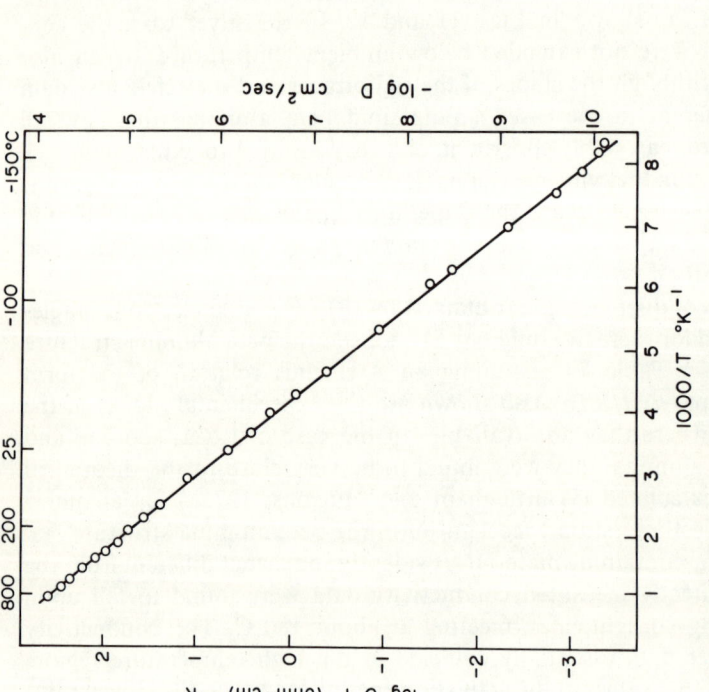

Fig. 9. Temperature dependence of the ionic conductivity of sodium beta alumina (from Whittingham and Huggins, 1971b).

9. VERY RAPID IONIC TRANSPORT IN SOLIDS

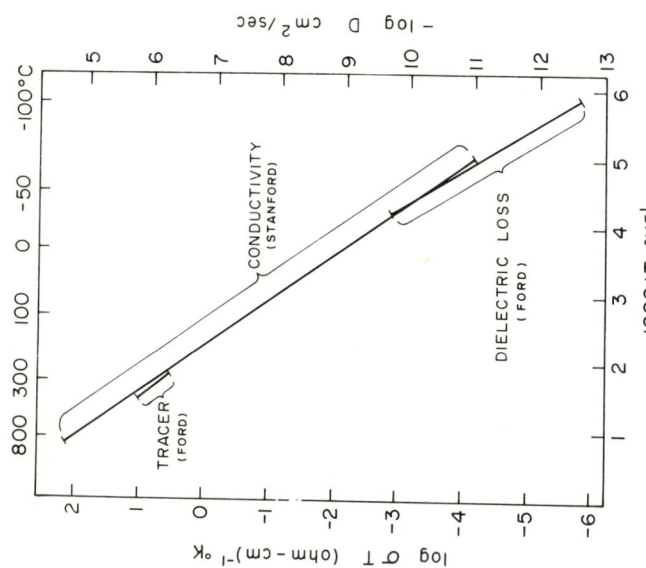

Fig. 12. Temperature dependence of the ionic conductivity and self-diffusion coefficient of potassium beta alumina (after Whittingham and Huggins, 1972).

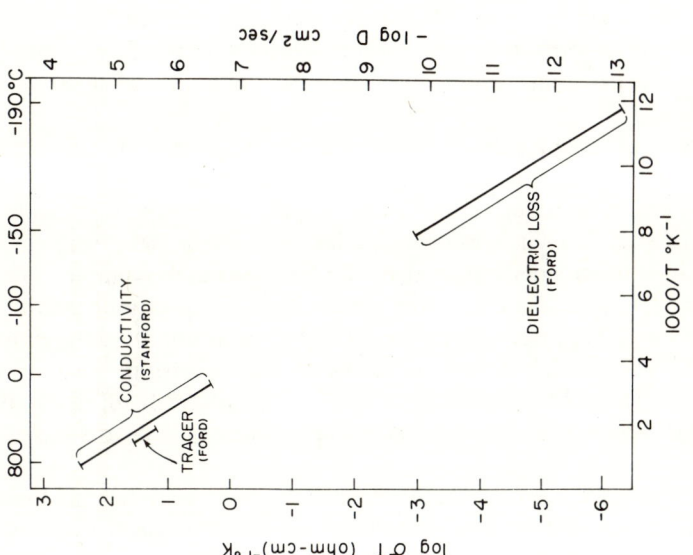

Fig. 11. Temperature dependence of the ionic conductivity and self-diffusion coefficient of silver beta alumina (after Whittingham and Huggins, 1972). In this case, the Stanford data are from Whittingham and Huggins (1971a).

TABLE VI

Ionic Conductivity of Single-Crystal β Alumina Containing Different Mobile Ions

Ion	Temperature range (°C)	σ_0 ((ohm-cm)$^{-1}$ °K)	Activation energy (kcal/mole)	Conductivity at 25°C (ohm-cm)$^{-1}$	Correlation factor (D_T/D_σ)
Ag	25 → 800	1.6×10^3	3.98	6.7×10^{-3}	0.61
Na	$-150 \to 820$	2.4×10^3	3.79	1.4×10^{-2}	0.61
K	$-70 \to 820$	1.5×10^3	6.79	6.5×10^{-5}	—
Tl	$-20 \to 800$	6.8×10^2	8.2	2.2×10^{-6}	0.58
Li	180 → 800	9.7×10^3	8.56	—	1.0
Li	$-100 \to 180$	5.4×10^1	4.3	1.3×10^{-4}	—

is evidently unity in this region, implying that the very small lithium ion is moving primarily by a direct interstitial mechanism in this temperature range. Because of its lower temperature dependence, another process (perhaps the normal interstitialcy mechanism) evidently dominates the conductivity at lower temperatures. A further complication in lithium-substituted beta alumina is the observation that annealing at temperatures above 800°C causes some structural change—perhaps diffusion of the lithium into the γ blocks—that results in a decrease of the low temperature conductivity.

As a result of these transport measurements, it appears that there is an optimum ionic radius for achieving the maximum mobility in the beta alumina structure, with both smaller and larger ions slower than those of intermediate size. The importance of this size factor was very nicely confirmed by experiments reported by Radzilowski and Kummer (1971) which showed a linear change in the ionic conductivity of ion-substituted beta alumina with hydrostatic pressure. At higher pressures, the conductivity of lithium β increased, whereas it decreased for potassium β, and was essentially unchanged in the sodium case. A theoretical approach that produces this same type of ionic size dependence will be discussed in Section IV.

One further aspect of the beta aluminas should also be mentioned. Because of the interest in their potential use as solid electrolytes, it is important to know the electronic transference number, the fraction of the total conductivity carried by electronic (as distinct from ionic) species. This has been measured in the one case of single-crystal silver beta alumina

(Whittingham and Huggins, 1971a) using the Wagner asymmetric polarization method (Wagner, 1956) from 555 to 790°C and over a wide range of oxygen partial pressure. The value of the electronic transference number was found to be about 3×10^{-5} at 750° and 10^{-6} at 560°C. It was also found that below ~ 750°C, changes in the ambient oxygen partial pressure did not appreciably influence the electronic conductivity. Presumably, the structure is effectively "frozen" below that temperature. If one can extrapolate these data to lower temperatures, it is obvious that the beta aluminas should have extremely low values of electronic conductivity at the lower temperatures at which its ionic conductivity makes it a very attractive solid electrolyte.

It is generally considered that the ionic conductivity of the β'' phase is higher than that found for the β materials, although well defined and reproducible data for different ions in this structure are not yet available. Some values for sodium β'' samples can be found in the papers by Kummer (1972) and Whittingham and Huggins (1972).

Although most attention has been given to the beta alumina family, there are many other materials with layer structures, and we should expect information about some of these to become available in the near future.

C. Anion Conductors

1. *Materials with Cubic Structures*

There are a number of materials which have the fluorite (CaF_2) structure in which the ionic conductivity, which occurs by the transport of anions, can have very high values at elevated temperatures. These fall into two general classes, oxides and fluorides. It was shown quite a long time ago that CaO-doped ZrO_2 is a useful solid electrolyte in which the moving species is the oxide ion (Baur and Preis, 1937). Two decades later a pair of papers were published by Kiukkola and Wagner (1957) which called attention to the unusual properties of this material (as well as several cation conductors) and to a number of potential applications of such solid electrolytes for a variety of purposes. In the same year, Ure (1957) studied the ionic conductivity of both pure and doped CaF_2, which has the same structure. His results indicated that this material can be employed as a useful fluoride ion-conducting solid electrolyte. Since that time, there have been a number of studies of both these and other oxide and fluoride materials, and they have been used as solid electrolytes or electrochemical transducers in an increasing number of applications.

The fluorite structure can be represented as a fcc arrangement of the cations, with the anions residing in the tetrahedral interstices. Ionic transport takes place by anions moving among these interstitial positions

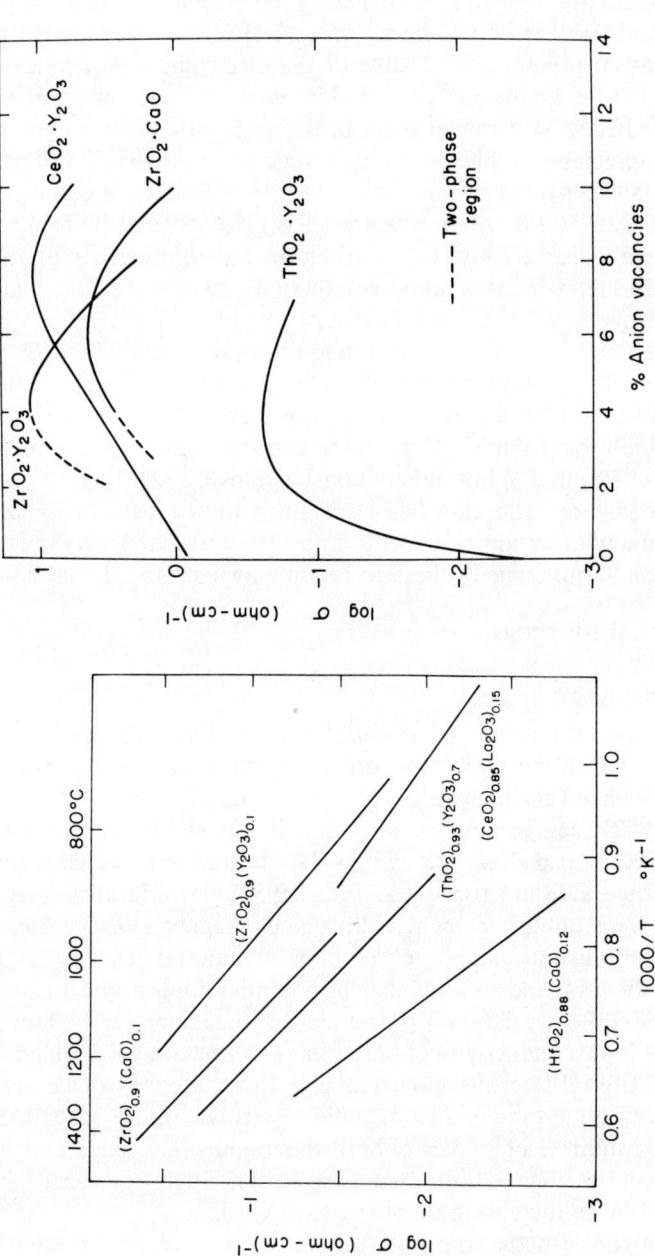

Fig. 13. Ionic conductivity of several oxide ion-conducting materials with the fluorite structure (after Kvist, 1972).

Fig. 14. Influence of anion vacancy concentration upon ionic conductivity in several oxide ion-conducting materials the fluorite structure (after Steele, 1972).

by a vacancy diffusion mechanism, and it is well known that doping in such a way as to increase the vacancy concentration substantially enhances the ionic conductivity.

Although the magnitude of the conductivity can be very high in some of these materials, its temperature dependence is very large. As a result, the conductivity does not attain appreciable values until rather high temperatures are reached, and such materials are probably not properly classified as fast ion conductors. They will, therefore, be discussed only briefly here.

The oxide ion conductors which have attracted the greatest interest are those based upon ZrO_2, ThO_2, and CeO_2. These are typically doped with 5–15 mole % of another oxide (e.g., CaO or Y_2O_3) to increase the anion vacancy concentration and thereby enhance ionic conductivity. However, ionic transport in these materials is not actually so straightforward as implied by this simple model because of the high concentrations of vacancies which result from such large numbers of aliovalent cations. There have been reports of time dependent changes in the transport properties at temperatures around 800–1000°C which may relate to ordering phenomena, phase transformations, or the presence of extended defects, etc. There have been several quite exhaustive reviews and discussions of such matters (Carter and Roth, 1968; Etsell and Flengas, 1970; Kvist, 1972; Steele, 1972), and so they will not be reiterated here.

The temperature dependence of the ionic conductivity of several oxide ion-conducting materials with the fluorite structure is shown in Fig. 13. The influence of the concentration of anion vacancies introduced by doping with oxides containing aliovalent cations upon the conductivity at 1000°C in some cases is presented in Fig. 14. It is seen that the conductivity goes through a maximum, rather than continuing to increase at large vacancy concentrations.

2. Materials with Layer Structures

There are also a number of anion conductors having layer structures, and in some cases they have the characteristics of fast ion conductors. The primary examples are the materials with the tysonite structure (Bauman and Porto, 1967) such as LaF_3. In this structure, the lanthanum and one of the fluorines form hexagonal layers similar to those formed by boron and nitrogen in hexagonal boron nitride. The other fluorine atoms lie in planes between these hexagonal layers, and evidently are highly mobile.

The tysonite structure is formed by a number of the fluorides of the lighter lanthanide group elements, as well as some of the actinides. Several ternary compounds with the general formula AB_2F_8, where A is an alkaline earth element and B a lanthanide or yttrium, can also have this structure (Garashina and Sobolev, 1971).

Nuclear magnetic resonance experiments (Goldman and Shen, 1966; Lee, 1969) showed that the fluorine ions in the fluorine-only layer are very mobile at low temperatures. At 200–250°C, interchange between the fluorines in the two types of positions begins to occur, and all the anions then contribute to the conductivity.

Ionic conductivity measurements on tysonite structure fluorides (Sher et al., 1966; Fielder, 1969; Nagel, 1972; Nagel and O'Keeffe, 1973) have been somewhat contradictory, although all indicate very high values extending down to ambient temperatures. Experimental and interpretational problems arose due to an apparent frequency dependence in certain temperature ranges, aging effects, and the possibility of oxygen contamination.

The trifluorides of yttrium and some of the heavier lanthanides have an orthorhombic structure which transforms to the tysonite type of structure at high temperatures (Thoma and Brunton, 1966). Ionic conductivity measurements on these materials have also been reported recently (O'Keeffe, 1973b), and typical fast ion conductivity is found at temperatures above the phase transition.

Data for the ionic conductivity of a group of fluoride ion conductors are shown in Fig. 15.

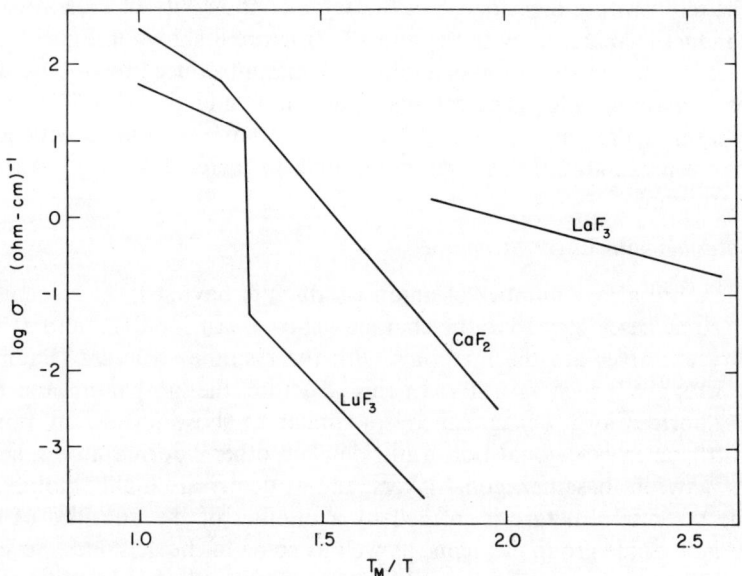

Fig. 15. Ionic conductivity data for several fluoride ion conductors plotted on a homologous temperature scale where T_M is the melting point (after O'Keeffe, 1973b).

IV. Theoretical Approaches to Fast Ion Conduction

It was recognized more than three decades ago by Ketelaar (1938) that a few substances were then known which exhibited unusually high values of ionic conductivity at relatively low temperatures and in which the conductivity is much less temperature dependent than is the case with normal ionic conductors. He also recognized that these fast ionic conductors are relatively insensitive to the presence of either physical defects or chemical impurities. Based upon his earlier determination (Ketelaar, 1934b) of the crystal structure of the high temperature modification of Ag_2HgI_4, as well as the work of Strock (1934, 1936) on AgI, he concluded that an important structural characteristic contributing to unusually good ionic conductivity is the presence of a larger number of available lattice positions than ions to populate them, with the ions randomly distributed among such sites.

Ketelaar (1938) also pointed out that the α–β-phase transformation in Ag_2HgI_4 is due to ordering of the cations upon the available sites as the temperature is decreased. By analysis of the results of an experimental determination of the specific heat and the heat of transformation (Ketelaar, 1935), he was able to show that the transformation in Ag_2HgI_4 is of the second-order type first described by Bragg and Williams (1934, 1935a,b).

This is followed (at a higher temperature) by a heterogeneous transformation. He also pointed out that the temperature dependence of the energy content indicates a stronger than linear dependence upon the degree of order, implying a considerable amount of cooperation. Evidently the first few disordered atoms expand and distort the lattice, considerably reducing the energy necessary for further disorder. The degree of order in the low temperature β phase at the heterogeneous transformation point at which the isothermal structural change occurs was found to be 0.90. In the α phase, in which the ionic conductivity is much higher, the cations are completely disordered.

More recently, Wiedersich and Geller (1970) again emphasized that (cation-conducting) fast ion conductors have cation-disordered structures, and pointed out that one can describe their structures in terms of networks of passageways (tunnels) resulting from the face-sharing of anion polyhedra. Both the mobile cations and a large concentration of vacant sites exist in these tunnels, and the order–disorder reaction involves their distribution. Wiedersich and Geller also pointed out that many materials of this type which can be construed as showing large amounts of ionic disorder have unusually large values of specific heat. In some cases, the heat capacity at constant pressure C_p decreases with increasing temperature over a wide temperature range.

In the bcc anion structures such as alpha silver iodide, the tunnels or passageways can be described in terms of the face-sharing of tetrahedra. The Ag_3SX materials also can be viewed as having tunnels comprised of face-sharing tetrahedra through which the silver ions move. In the case of the MAg_4I_5 materials, the tunnels are formed by the alternation of two slightly different sets of tetrahedra which share faces. In structures based upon a fcc anion arrangement, the tunnel structure consists of the sharing of faces between tetrahedra and octahedra alternately.

In addition to the importance of dilute site occupation within crystallographic tunnels, which results in a very large fraction of the ions being mobile, it is generally recognized that the number of mobile ions is essentially independent of temperature rather than being controlled by a Boltzmann factor containing the formation energy of the transporting defects, as is the case in more normal ionic conductors. This results, of course, in a smaller overall temperature dependence in the diffusion coefficient and ionic conductivity.

Ketelaar (1938) introduced the concept that the primary determinant of the activation energy for motion in such cases is the "squeezing" of the cations between the nearby anions (through the faces of the shared polyhedra). Wiedersich and Geller (1970) estimated that less squeezing is necessary in materials with the bcc anion structure than in the case of the Ag_3SX materials, in which the anion arrangement is of the antiperovskite type. In the case of the fcc materials such as Ag_2HgI_4, the cations tend to reside in tetrahedral interstices which are separated by normally unoccupied octahedral interstices. They predicted that even greater squeezing would be necessary for motion in such materials. Because of the difference in energy between ions in these tetrahedral and octahedral sites, one can expect a contribution to the temperature dependence of the conductivity in such materials comparable to a defect formation energy term, if an appreciable population of these secondary (octahedral) sites is not already caused by the presence of an aliovalent species or by stoichiometric variations.

A further observation that may contribute to the understanding of the unusual character of fast ion conductors was recently made by Armstrong *et al.* (1973), who again stressed the importance of dilute site occupation and a relatively small energy difference between ordered and disordered distributions of the mobile ions upon these sites. They also pointed out that both monovalent silver and copper ions which are present in a number of fast ion conductors are unusual in that they are stable in four-coordinated (tetrahedral) as well as existing in three-coordinated and two-coordinated configurations in some solids. The other monovalent cations typically have coordination numbers of six or eight in their simple salts. The stability of these low coordination numbers indicates a partial covalent bonding.

Armstrong et al. (1973) also noted that fast ionic conduction typically occurs in structures containing highly polarizable anions (e.g., I, S, Se, Te), which have some $d\pi$-bonding ability.

Another interesting possibility was suggested by Kvist and Bengtzelius (1973) for the case of materials containing large nonspherical anions. These authors reported experimental data on the diffusion of various ions in cubic Li_2SO_4. Not only are a number of cations very mobile, but the same is true of some of the smaller anions as well. In addition, the presence of smaller anions, such as Cl^- evidently enhances the cation transport. They postulated that coupling between the rotational motion of the sulfate anions, which act as cogwheels, and the motion of the other ions could play an important role in these materials, and that the presence of other anions influences the rotational properties of the nonspherical anions.

A theoretical approach to transport in fast ionic conductors was presented by Rice and Roth (1972) based upon a free ion-like model for the translational state in which it was assumed that an ion moves from one localized state (position) to another. The density of states for such excitation was assumed to be continuous for energies greater than a critical value, which defined an "energy gap," and to vanish below that value. From this type of model, they computed the lifetime, translational velocity, and mean free path of such "excited" ions. However, this model has been questioned with regard to several points, including both the assumed density of states function and the assumed velocity of the excited ions. It was recently pointed out by Haas (1973) that all of the substantial conclusions that resulted from the Rice and Roth model can also be derived from the conventional harmonic oscillator hopping type of model. In addition, the concept that transport occurs by the motion of a small number of ions in "free ion" states over appreciable distances, with the others remaining fixed in place, is difficult to reconcile with crystallographic considerations. Since this phenomenological approach also does not provide useful guidance with regard to structural criteria, which apparently are of great importance in fast ion conductors, it will not be discussed further here.

There have been several recent attempts to extend the traditional hopping model of diffusion to cases in which the defect concentration is large and interactions cannot be neglected (O'Keeffe, 1970, 1973a). One of the more detailed approaches to this problem has been the development of the "path probability method" by Kikuchi (1951, 1966) and its application to cation diffusion and ionic conductivity (Sato and Kikuchi, 1971; Kikuchi and Sato, 1971; Kikuchi, 1973).

The path probability method results in a modification of the random-walk approach by the introduction of the influence of interdefect interactions upon the apparent correlation factor when the defect concentration becomes large. The tracer diffusion coefficient for vacancy diffusion in a material with mobile cations is expressed as

$$D = d^2 v_0 \exp(-U/RT) V W f \tag{8}$$

where V is the vacancy availability factor, which is proportional to the vacancy concentration, W is a function of the cation–cation repulsive interaction, and f is the normal "geometric" correlation factor. According to this model, both the overall apparent activation energy and the magnitude of the diffusion coefficient are predicted to depend strongly upon the temperature and the extrinsically controlled vacancy concentration.

It appears that this prediction of a temperature dependent activation energy is not borne out by the available experimental results in a number of cases. In addition, this approach leads to a correlation factor of 1.0 in the beta alumina structure, which is also not in accord with experiments. Furthermore, the predicted strong influence of composition upon the activation energy cannot presently be evaluated because of the lack of appropriate experimental data on fast ion conductors.

Another important question that has arisen in connection with considerations of transport mechanism in fast ion conductors is the matter of possible cooperative motion. It is now recognized (Anderson and Hyde, 1967; Hyde and Bursill, 1970) that variations in stoichiometry are accommodated in many materials by the presence of organized or extended defects, one class of which are called "crystallographic shear planes." Changes in composition occur by the formation and migration of such structural features. Likewise, it has been proposed that in materials such as β alumina the excess cation concentration might be present in domain walls separating regions within the structure that are crystallographically similar but have a different distribution of cations among the available sites (Roth, 1972; van Gool, 1973a,b; van Gool and Bottelberghs, 1973).

To the present time, there is no firm experimental evidence for the existence of domain structures in fast ionic conductors. If cation rich or cation poor regions were to be present, there must be a reduction in some other contribution to the total free energy (e.g., strain energy) to more than balance the expected increase in energy due to cation–cation interactions and ordering. In the absence of such a compensating energy term, one would expect the free energy to be minimized by a distribution of the excess cations such that they are as mutually far apart as possible.

Taking a different approach, van Gool and Piken (1969) calculated the Madelung electrostatic energy for sodium ions in both the normal position and the alternate position in the beta alumina structure, and found that this contribution to the total energy is very nearly the same in both cases. Although their calculation did not take into account repulsion and polarization contributions, it did indicate that there is not a large preference between these two types of sites within the tunnels in the bridging layer of this important fast ion conductor.

More recently, initial steps have been taken (Flygare and Huggins, 1973) to develop an explicit atomistic model for the transport of ions through crystallographic tunnels. This theoretical approach used the general method initiated by Born and Mayer (1932), in which the total interaction energy between an ion i and the surrounding j lattice ions is expressed as the sum of three types of terms, electrostatic point charge (Coulomb or Madelung) interactions, dipolar polarization interactions, and overlap repulsion between closed shell ions. That is,

$$E_T = E_C + E_P + E_R \tag{9}$$

where

$$E_C = e^2 \sum_j q_i q_j / r \tag{10}$$

$$E_P = -(e^2/2) \sum_j \alpha_j q_i / r^4 \tag{11}$$

$$E_R = b \sum_j \exp[(r_i + r_j - r_{ij})/\rho] \tag{12}$$

and q_i and q_j are the charges, α_j the polarizability, r_i and r_j repulsion radii, b and ρ constants, and r_{ij} the distance between the test cation and the jth lattice ion.

With the help of a digital computer, the total interaction energy between a lone test cation at any arbitrary position within the crystal structure and the rest of the lattice could be calculated, assuming that all the other ions remain fixed in position.

This type of calculation was carried out for the case of the α-AgI structure and the total energy found as a function of the cation position within the tunnels which run through the bcc anion lattice, assuming a number of different values of the cation radius. In addition to a series of points along the centerline of the tunnel, a three-dimensional raster of off-center positions was also investigated.

By use of this method, the minimum energy path through the tunnel was found to be strongly influenced by the value of the cation radius. In the case of small cations, there is a symmetrical pair of preferred paths which deviate toward the nearest neighbor anions along the tunnel wall, due to the influence of the polarization energy term. For large cations just the opposite is true, and a pair of equivalent paths are preferred which deviate from the centerline away from the nearby anions, due to the relatively greater importance of the repulsion term in the total energy.

In addition to showing the influence of ionic size upon the location of the minimum energy path, these calculations gave information about the variation in potential energy during translation along this preferred transport path for cations of different radii. These results are illustrated for a series of different cationic sizes in Fig. 16. It is seen that there is, as expected, a periodic variation in energy along the transport path. However, the phase is reversed between relatively small and comparatively large cations. The amplitudes

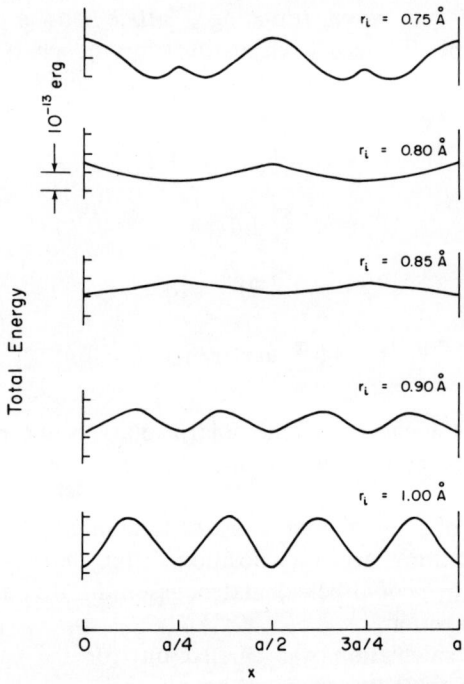

Fig. 16. Variation of potential energy along preferred transport path within unit cell of α-AgI structure for cations of various sizes (after Flygare and Huggins, 1973).

also increase as the size becomes more extreme in either direction, as both small and large cations encounter substantial energy barriers due to the dominance of one or another of the terms in the total energy. In the case of cations of intermediate size, these factors tend to balance each other, and the energy varies less as the ion moves through the tunnel.

If one interprets the difference between maxima and minima in energy along the preferred transport path as an activation energy for motion, it is clear why ions of intermediate size should have the greatest mobility in materials with structures containing such tunnels. The influence of cationic size upon the activation energy determined in this way is shown in Fig. 17.

Although several important simplifications were made to facilitate this calculation, including the assumptions of no cation–cation interaction, negligible cationic polarization, and a rigid anion lattice, the results are remarkably reasonable. For example, the theory predicts that cations with repulsion radii of approximately the proper value should be most mobile in this structure (silver is between 0.8 and 0.95 Å), and the calculated

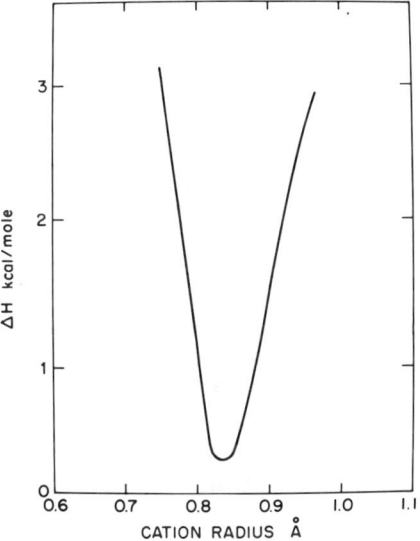

Fig. 17. Calculated influence of cation radius upon activation energy for migration in the α-AgI structure (after Flygare and Huggins, 1973).

activation energy values are quite close to those found from experiments (1.5 kcal/mole for α-AgI).

These results, although admittedly only approximate, do indicate that the potential wells in which the cations will tend to reside are broad and shallow, leading to the expectation of an unusually low vibrational frequency, as inferred from the especially small values of σ_0 found experimentally for fast ion conductors, mentioned in Section I. This is also in accord with the observations of unusually large values of the Debye–Waller factor, which is a measure of the mean thermal displacement of ions from their lowest energy sites, in the α-AgI structure (Hoshino, 1957). Large Debye–Waller factors have also been found in other fast ion conductors

(Krug and Sieg, 1952; Hoshino, 1955), as well as in some tunnel structure materials that may well have high cationic mobility (Mumme and Wadsley, 1967; Mumme and Reid, 1968).

One other aspect of the results of this calculation is the indication of a shift in preferred sites within the tunnels in this structure for cations of different radii. It appears that the small cations will prefer threefold coordinated positions, while the larger ones would have lower values of energy at fourfold coordinated positions.

V. Outlook for the Future

Although ionic conductivity in solids has been studied for a great many years, and some fast ionic conductors were recognized almost six decades ago, this area is suddenly attracting a large amount of attention in many quarters.

The primary reason for this is the current recognition of the broad range of possible applications for fast ionic conductors in both science and technology. Such materials have been employed as solid electrochemical transducers in a wide variety of scientific investigations in the last decade or so for the determination of thermodynamic and kinetic parameters. Until recently, however, these studies were restricted to systems involving oxygen ions at high temperatures or silver ions at ambient temperatures. It is now obvious that similar behavior can be found in a number of other systems. Particularly important was the discovery of fast ionic conductors with substantial thermodynamic stability within which the alkali ions are mobile. This has greatly increased the potential range of application of solid electrolytes.

Technological developments are already being pursued along a number of lines. These include several new types of high energy density battery and fuel cell systems, novel thermoelectric devices with very high efficiency, sensors for species in liquids and gases, ionic pumps, timers, memory elements, and ion-transparent membranes.

Along with this interest in applications, greatly increased scientific attention is being given this area, with the recognition that there is much yet to be learned about the fundamental nature of transport processes in materials with relatively open structures and unusually mobile ions. The apparent generality of these phenomena and the accumulation of increased knowledge of the structure–property relationships is also leading to an intensive search for additional materials which exhibit large values of ionic conductivity and which might be of practical importance.

This is clearly a field currently characterized by great excitement and rapid progress. Exploratory fervor is rampant, and there is much to be done.

Acknowledgments

Work in this area within the Solid State Electrochemistry Group at Stanford has been supported by the Office of Naval Research, the Environmental Protection Agency, the National Science Foundation, and the Advanced Research Projects Agency (through Stanford's Center for Materials Research).

References

ANDERSON, J. S., AND HYDE, B. G. (1967). *J. Phys. Chem. Solids*, **28**, 1393.
ARMSTRONG, R. D., BULMER, R. S., AND DICKENSON, T. (1973). *In* "Fast Ion Transport in Solids" (W. van Gool, ed.), p. 269. North-Holland Publ., Amsterdam; *J. Solid State Chem.* **8**, 219.
BAUER, E., AND PREIS, H. (1937). *Z. Elektrochem.* **43**, 727.
BAUMAN, R. P., AND PORTO, S. P. S. (1967). *Phys. Rev.* **161**, 842.
BEEVERS, C. A., AND ROSS, M. A. S. (1937). *Z. Kristallogr., Kristallgeometrie, Kristallphys., Kristallchem.* **97**, 59.
BETTMAN, M., AND PETERS, C. R. (1969). *J. Phys. Chem.* **73**, 1774.
BETTMAN, M., AND TERNER, L. (1971). *J. Inorg. Chem.* **10**, 1442.
BORN, M., AND MAYER, J. E. (1932). *Z. Phys.* **75**, 1.
BRADLEY, J. N., and GREENE, P. D. (1966). *Trans. Faraday Soc.* **62**, 2069.
BRADLEY, J. N., AND GREENE, P. D. (1967a). *Trans. Faraday Soc.* **63**, 424.
BRADLEY, J. N., AND GREENE, P. D. (1967b). *Trans. Faraday Soc.* **63**, 2516.
BRAGG, W. L., AND WILLIAMS, E. J. (1934). *Proc. Roy. Soc. A* **145**, 699.
BRAGG, W. L., AND WILLIAMS, E. J. (1935a). *Proc. Roy. Soc. A* **151**, 540.
BRAGG, W. L., AND WILLIAMS, E. J. (1935b). *Proc. Roy. Soc. A* **152**, 231.
BRAGG, W. L., GOTTFRIED, C., AND WEST, J. (1931). *Z. Kristallogr., Kristallgeometrie, Kristallphys., Kristallchem.* **77**, 255.
BRUNI, E., AND SCARPA, O. (1913). *Rend. Reale Accad. Naz. Lincei* **22**, 438.
CARTER, R. E., AND ROTH, W. L. (1968). *In* "Electromotive Force Measurements in High-Temperature Systems" (C. B. Alcock, ed.), p. 125. Inst. Mining and Metallurgy, London.
CHU, W. F., RICKERT, H., AND WEPPNER, W. (1973). *In* "Fast Ion Transport in Solids" (W. van Gool, ed.), p. 181. North-Holland Publ., Amsterdam.
CLARK, L. D. (1970). Ph.D. Dissertation, Stanford Univ., California.
CLARK, L. D., WHITTINGHAM, M. S., AND HUGGINS, R. A. (1972). *J. Solid State Chem.* **5**, 487.
DARKEN, L. S. (1948). *Trans. AIME* **175**, 184.
DREYFUS, R. W., AND NOWICK, A. S. (1962). *Phys. Rev.* **126**, 1367.
DRYDEN, J. S., AND WADSLEY, A. D. (1958). *Trans. Faraday Soc.* **54**, 1574.
ETSELL, T. H., AND FLENGAS, S. N. (1970). *Chem. Rev.* **70**, 339.
FIELDER, W. L. (1969). *NASA Tech. Note* D-5505.
FLYGARE, W. H., AND HUGGINS, R. A. (1973). *J. Phys. Chem. Solids* **34**, 1199.
FØRLAND, T., AND KROGH-MOE, J. (1958). *Acta Crystallogr.* **11**, 224.
GARASHINA, L. S., AND SOBOLEV, B. P. (1971). *Sov. Phys.-Crystallogr.* **16**, 254.
GELLER, S. (1967). *Science* **157**, 310.
GELLER, S. (1973). *In* "Fast Ion Transport in Solids" (W. van Gool, ed.), p. 607. North-Holland Publ., Amsterdam.

GENDELL, J., COTTS, R. M., AND SIENKO, M. J. (1962). *J. Chem. Phys.* **37**, 220.
GOLDMAN, M., AND SHEN, L. (1966). *Phys. Rev.* **144**, 321.
HAAS, C. W. (1973). *J. Solid State Chem.* **7**, 155.
HABER, F., AND TOLLOCZKO, A. (1904). *Z. Anorg. Chem.* **41**, 407.
HALSTEAD, T. K., BENESH, W. U., GULLIVER, R. D., II, AND HUGGINS, R. A. (1973). *J. Chem. Phys.* **58**, 3530.
HEYNE, L. (1970). *Electrochim. Acta* **15**, 1251.
HOSHINO, S. (1952). *J. Phys. Soc. Jap.* **7**, 560.
HOSHINO, S. (1955). *J. Phys. Soc. Jap.* **10**, 197.
HOSHINO, S. (1957). *J. Phys. Soc. Jap.* **12**, 315.
HYDE, B. G. AND BURSILL, L. A. (1970). *In* "Chemistry of Extended Defects in Non-Metallic Solids" (L. Eyring and M. O'Keeffe, eds.), p. 347. North-Holland Publ., Amsterdam.
JOHNSON, O. W. (1964). *Phys. Rev.* A **136**, 284.
JOHNSTON, W. V., AND LINDBERG, G. W. (1968). Abstract No. 97, 156th Nat. Meet. Amer. Chem. Soc., Atlantic City, New Jersey.
JOSEFSON, A-M., KVIST, A., AND TÄRNEBERG, R. (1971). *In* "Atomic Transport in Solids and Liquids" (A. Lodding and T. Lagerwall, eds.), p. 291. Verlag der Z. Naturforsch., Tübingen.
KETELAAR, J. A. A. (1934a). *Z. Phys. Chem.* B **26**, 327.
KETELAAR, J. A. A. (1934b). *Z. Kristallogr. Kristallgeometrie, Kristallphys., Kristallchem.* A **87**, 436.
KETELAAR, J. A. A. (1935). *Z. Phys. Chem.* B **30**, 53.
KETELAAR, J. A. A. (1938). *Trans. Faraday Soc.* **34**, 874.
KIKUCHI, R. (1951). *Phys. Rev.* **81**, 988.
KIKUCHI, R. (1966). *Progr. Theor. Physics (Kyoto) Suppl.* **35**, 1.
KIKUCHI, R. (1973). *In* "Fast Ion Transport in Solids" (W. van Gool, ed.), p. 249. North-Holland Publ., Amsterdam.
KIKUCHI, R., AND SATO, H. (1971). *J. Chem. Phys.* **55**, 702.
KING, C. N. (1972). Ph.D. Dissertation, Stanford Univ., California.
KIUKKOLA, K., AND WAGNER, C. (1957). *J. Electrochem. Soc.* **104**, 308, 379.
KRUG, J., AND SIEG, L. (1952). *Z. Naturforsch.* **7a**, 369.
KUMMER, J. T. (1972). *Progr. Solid State Chem.* **7**, 141.
KVIST, A. (1972). *In* "Physics of Electrolytes" (J. Hladik, ed.), Vol. 1, p. 319. Academic Press, New York.
KVIST, A., AND BENGTZELIUS, A. (1973). *In* "Fast Ion Transport in Solids" (W. van Gool, ed.), p. 193. North-Holland Publ., Amsterdam.
KVIST, A., AND JOSEFSON, A-M. (1968). *Z. Naturforsch.* **23a**, 625.
KVIST, A., AND LUNDEN, A. (1965). *Z. Naturforsch.* **20a**, 235.
KVIST, A., AND LUNDEN, A. (1966). *Z. Naturforsch.* **21a**, 1509.
LAYDEN, G. K. (1968). *Mater. Res. Bull.* **3**, 349.
LECARS, Y., THERY, J., AND COLLONGUES, R. (1973). Presented at 24th Meet. ISE, Eindhoven, Netherlands.
LEE, K. (1969). *Solid State Commun.* **7**, 367.
MAJUMDAR, A. J., AND ROY, R. (1965). *J. Phys. Chem.* **69**, 1684.
MIYAKE, S., HOSHINO, S., AND TAKENAKA, T. (1952). *J. Phys. Soc. Jap.* **7**, 19.
MUMME, W. G., AND REID, A. F. (1968). *Acta Crystallogr.* **24**, 625.
MUMME, W. G., AND WADSLEY, A. D. (1967). *Acta Crystallogr.* **23**, 754.
MURPHY, D. J. (1971). Ph.D. Thesis, Oxford Univ.
NAGEL, L. E. (1972). Ph.D. Dissertation, Arizona State Univ., Tempe.
NAGEL, L. E., AND HUGGINS, R. A. (1973). Unpublished results.

NAGEL, L. E., AND O'KEEFFE, M. (1973). *In* "Fast Ion Transport in Solids" (W. van Gool, ed.), p. 165. North-Holland Publ., Amsterdam.
O'KEEFFE, M. (1970). *In* "Chemistry of Extended Defects in Non-Metallic Solids" (L. Eyring and M. O'Keeffe, eds.), p. 609. North-Holland Publ., Amsterdam.
O'KEEFFE, M. (1973a) *In* "Fast Ion Transport in Solids" (W. van Gool, ed.), p. 233. North-Holland Publ., Amsterdam.
O'KEEFFE, M. (1973b). *Science* **180**, 1276.
OWENS, B. B. (1971). *Advan. Electrochem. Electrochem. Eng.* **8**, 1.
OWENS, B. B., AND ARGUE, G. R. (1967). *Science* **157**, 308.
OWENS, B. B., AND ARGUE, G. R. (1970). *J. Electrochem. Soc.* **117**, 898.
ØYE, H. A. (1963). Ph.D. Thesis, Tech. Univ. of Norway, Trondheim.
PERROTT, C. M., AND FLETCHER, N. H. (1968). *J. Chem. Phys.* **48**, 2143, 2681.
PERROTT, C. M., AND FLETCHER, N. H. (1969). *J. Chem. Phys.* **50**, 2770.
PETERS, C., BETTMAN, M., MOORE, J., AND GLICK, M. (1971). *Acta Crystallogr. B* **27**, 1826.
RADZILOWSKI, R. H., AND KUMMER, J. T. (1971). *J. Electrochem. Soc.* **118**, 714.
RADZILOWSKI, R. H., YAO, Y. F., AND KUMMER, J. T. (1969). *J. Appl. Phys.* **40**, 4716.
RAHLFS, P. (1935). *Z. Phys. Chem.* **31**, 157.
RALEIGH, D. O. (1970). *J. Appl. Phys.* **41**, 1876.
RAPP, R. A., AND SHORES, D. A. (1970). *In* "Physicochemical Measurements in Metals Research" (R. A. Rapp, ed.), Part 2, p. 123. Wiley (Interscience), New York.
REMEIKA, J. P., GEBALLE, T. H., MATTHIAS, B. T., COOPER, A. S., HULL, G. W., AND KELLY, E. M. (1967). *Phys. Lett. A* **24**, 565.
REUTER, B., AND HARDEL, K. (1960). *Angew. Chem.* **72**, 138.
REUTER, B., AND HARDEL, K. (1961). *Naturwissenschaften* **48**, 161.
REUTER, B., AND HARDEL, K. (1965). *Z. Anorg. Allg. Chem.* **340**, 158, 168.
REUTER, B., AND HARDEL, K. (1966a). *Z. Electrochem.* **70**, 82.
REUTER, B., AND HARDEL, K. (1966b). *Ber. Bunsenges. Phys. Chem.* **70**, 82.
REUTER, B., PICKARDT, J., AND HARDEL, K. (1967). *Z. Phys. Chem.* **56**, 309.
RICE, M. J., AND ROTH, W. L. (1972). *J. Solid State Chem.* **4**, 294.
RICKERT, H. (1973). *In* "Fast Ion Transport in Solids" (W. van Gool, ed.), p. 3. North-Holland Publ., Amsterdam.
ROTH, W. L. (1972). *J. Solid State Chem.* **4**, 60.
SAALFELD, H., MATTHIES, H., AND DATTA, S. K. (1968). *Ber. Deut. Keram. Gesell.* **45**, 212.
SATO, H., AND KIKUCHI, R. (1971). *J. Chem. Phys.* **55**, 677.
SCHROEDER, K., AND KVIST, A. (1969). *Z. Naturforsch.* **24a**, 844.
SHER, A., SOLOMON, R., LEE, K., AND MULLER, M. W. (1966). *Phys. Rev.* **144**, 593.
SINGER, J., KAUTZ, H. E., FIELDER, W. L., AND FORDYCE, J. S. (1972). Presented at Spring Meet., Electrochem. Soc., Houston, Texas.
SINGER, J., KAUTZ, H. E., FIELDER, W. L., AND FORDYCE, J. S. (1973). *In* "Fast Ion Transport in Solids" (W. van Gool, ed.), p. 653. North-Holland Publ., Amsterdam.
STEELE, B. C. H. (1972). *In* "Solid State Chemistry" (L. E. J. Roberts, ed.), p. 117. Butterworths, London.
STROCK, L. W. (1934). *Z. Phys. Chem. B* **25**, 441.
STROCK, L. W. (1936). *Z. Phys. Chem. B* **32**, 132.
SUCHOW, L., AND POND, G. R. (1953). *J. Amer. Chem. Soc.* **75**, 5242.
TAKAHASHI, T., AND YAMAMOTO, O. (1964). *J. Electrochem. Soc. Jap.* **32**, 174.
TAKAHASHI, T., AND YAMAMOTO, O. (1965). *J. Electrochem. Soc. Jap.* **33**, 191.
TAKAHASHI, T., AND YAMAMOTO, O. (1966). *Electrochim. Acta* **11**, 779, 911.
TAKAHASHI, T., YAMAMOTO, O., AND KUWABARA, K. (1967). *Denki Kagaku* **35**, 264.

TAKAHASHI, T., IKEDA, S., AND YAMAMOTO, O. (1972). *J. Electrochem. Soc.* **119**, 477.
TAKAHASHI, T., IKEDA, S., AND YAMAMOTO, O. (1973). *J. Electrochem. Soc.* **120**, 647.
THERY, J., AND BRIANCON, D. (1964). *Rev. Hautes Temp. Refract.* **1**, 221.
THOMA, R. E., AND BRUNTON, G. D. (1966). *Inorg. Chem.* **5**, 1937.
TUBANDT, C. (1932). *In* "Handbuch der Experimentalphysik," Vol. XII, p. 383.
TUBANDT, C., AND EGGERT, S. (1920). *Z. Anorg. Allg. Chem.* **110**, 196.
TUBANDT, C., AND LORENZ, E. (1914). *Z. Phys. Chem.* **87**, 513.
TUBANDT, C., AND REINHOLD, H. (1923). *Z. Elektrochem.* **29**, 313.
URE, R. W. (1957). *J. Chem. Phys.* **26**, 1363.
VAN GOOL, W. (1973). *In* "Fast Ion Transport in Solids" (W. van Gool, ed.), p. 201. North-Holland Publ., Amsterdam.
VAN GOOL, W., AND BOTTELBERGHS, P. H. (1973). *J. Solid State Chem.* **7**, 59.
VAN GOOL, W., AND PIKEN, A. G. (1969). *J. Mater. Sci.* **4**, 105.
WAGNER, C. (1951). *In* "Atom Movements." Amer. Soc. Metals, Metals Park, Ohio.
WAGNER, C. (1956). *Z. Elektrochem.* **60**, 4.
WARBURG, E. (1884). *Wiedemann. Ann. Phys.* **21**, 622.
WARBURG, E., AND TEGETMEIER, F. (1888). *Wiedemann. Ann. Phys.* **32**, 455.
WEBER, N., AND KUMMER, J. T. (1967). *Proc. Annu. Power Sources Conf.* **21**, 37.
WEBER, N., AND VENERO, A. (1970). Presented at Annu. Meet., Amer. Ceram. Soc., Philadelphia, Pennsylvania.
WHITTINGHAM, M. S., AND HUGGINS, R. A. (1971a). *J. Electrochem. Soc.* **118**, 1.
WHITTINGHAM, M. S. AND HUGGINS, R. A. (1971b). *J. Chem. Phys.* **54**, 414.
WHITTINGHAM, M. S., AND HUGGINS, R. A. (1972). *In* "Solid State Chemistry" (R. S. Roth and S. J. Schneider, eds.), p. 139. Nat. Bur. Stand. U.S. Spec. Publ. 364, Washington, D.C.
WHITTINGHAM, M. S., AND HUGGINS, R. A. (1973). *In* "Fast Ion Transport in Solids" (W. van Gool, ed.), p. 645. North-Holland Publ., Amsterdam.
WHITTINGHAM, M. S., HELLIWELL, R. W., AND HUGGINS, R. A. (1969). *U.S. Govt. Res. Develop. Rep.* **69**, 158.
WIEDERSICH, H., AND GELLER, S. (1970). *In* "The Chemistry of Extended Defects in Non-Metallic Solids" (L. Eyring and M. O'Keeffe, eds.), p. 629. North-Holland Publ., Amsterdam.
YAMAGUCHI, G., AND SUZUKI, K. (1968). *Bull. Chem. Soc. Jap.* **41**, 93.
YAO, Y. F. Y., AND KUMMER, J. T. (1967). *J. Inorg. Nucl. Chem.* **29**, 2453.

Index

A

Actinides
  fast diffusion in, 193-194
Activation volume, 123, 125
  for noble metals in lead, 180-184
Alkali halides, 149-153, 155-157, 381-439
  aliovalent impurities in, 424-435
  H_2O in, 435
  inert gas atoms in, 436
Alkali metals, *see also* Sodium
  electromigration in, 329, 334-336
  fast diffusion in, 191-193
Alloys, *see also* various metals
  vacancy wind effects in, 369-378
Aluminum
  condensation cavities in, 369-372
  Cu in, 373, 376
  electromigration in, 329, 338-339, 345-347
  Ge in, 373, 376
  H in, 292
  Mg in, 373, 376
  solute segregation in, 375-378
  vacancy wind in, 373, 376
  Zn in, 373, 376
Anharmonic effects, 3, 10-14, 24-26, 58-69
Arrhenius relation, 1
  deviations from, 51, 264-266

B

Barium
  H in, 292
Beta alumina, 449, 462-471

C

Cobalt
  H in, 292
Cobalt oxide
  self-diffusion, 158
Common ion effect, 385, 406
Copper
  Ag in, 373, 376
  H in, 60, 261, 292
  impurity diffusion in, 141
  In in, 373, 376
  jump frequency in, 63
  localized modes, 46
  Mn in, 373, 376
  quantum effects in, 63, 66-69
  self-diffusion, 126-128
  vacancy preexponential, 52-56
  vacancy saddle-point configuration, 53
  vacancy wind in, 373, 376
  Zn in, 373, 376
Copper-palladium alloys
  H in, 268, 269

Copper-zinc alloys, 147-148
Correlation effects, 311-312
Correlation factor, 58, 116, 209, 343,
    388-392, 478, see also Isotope effect
  impurity diffusion, 139-143, 146, 147
  for interstitials, 176
  numerical values, 118
  partial, 136

D

DNA
  tunneling in, 60
Defect interactions
  coulomb, 387
Dielectric loss
  in beta alumina, 466-469
Dielectric relaxation, 386, 403
Diffusion, see specific headings
Dislocations
  diffusion down, 343
Divalent metals
  electromigration in, 329, 336-338
Dynamical theory, 145

E

Electrolytes, solid, see Fast ionic conductors
Electromigration, 303-349, 355
  boundary and surface, 346
  dilute alloy, 312, 330
  and fast diffusion, 344
  interstitial, 309-310, 325-328
  self-, 311-312, 328-330
  techniques, 314-320
  in thin films, 345-349
  vacancy flow, 310-313
Electron channeling, 198
Electron spin resonance
  in alkali halides, 386
Electron wind, 321-324
Enhancement factor, 178-179, 186-188,
    209, 217
Extended defects, 478

F

Fast diffusion
  in metals, 143-145, 171-226

Fast ionic conductors, 445-482
  anion type, 471-474
  cation type, 452-471
  theory, 475-482
$Fe_{1-x}O$
  self-diffusion, 159
Frenkel defect, 383

G

Germanium
  Cu in, 174
  interstitials in, 47
  localized modes, 45
  self-diffusion, 137
Gold
  H in, 292
Grain boundaries
  solute segregation to, 375-378
Grain boundary diffusion
  correlation effects, 163-167
  in silver, 164-166

H

Hafnium
  H in, 292
Haven ratio, 397, 467
Hillocks
  in electromigration, 346
Hydrogen in metals, 231-302
  dependence on alloying, 266, 268, 269
  deviations from Arrhenius relation, 264-266
  experimental methods, 240-246
  influence of structure, 261-264, 268
  influence of traps, 267
  isotope effects, 260-264, 270
  phase diagrams and solubility, 233-240
  tabulated data, 272-294

I

Impurity diffusion, 395, 399-402
  correlation factor, 139-143, 146-147
  dilute alloys, 138-143
  in NiO, 160-163
Impurity-vacancy pair, 384-385, 406, 411
  free energy of association, 428

INDEX
489

Indium, 182, 190
  noble metals in, 225
Internal friction, 202-205, 221
  Gorsky effect, 243, 253, 257
  Snoek effect, 242, 255
Interstitial defects, 196-205
Interstitial mechanism, 144, 211-225, 382
  diplon, 211-212, 214-215, 221
Interstitial-vacancy pairs, 216-219
  in semiconductors, 176
Interstitialcy machanism, 153
Interstitials
  comparison of, 270
Ionic conductivity, 149, 396-397, 409, 410, 413, 416-417
Ionic crystals
  diffusion theory, 388-398
  impurity diffusion, 388-391
Iron
  C in, 2
  H in, 250-252, 265, 268, 281-284
  localized modes, 46
  self-diffusion, 131, 135
Irradiation
  solute segregation during, 377
Irreversible thermodynamics, 355-369, 392-398
  and electromigration, 307-313
  phenomenological coefficients, 355-361
  thermodynamic forces, 361-369
Isotope effect, 56-58, 116-167, 215
  anharmonic effects, 57, 111
  concentrated alloys, 146
  and ΔK, 58, 95, 111
  dilute alloys, 138-146
  H in metals, 260-264, 270
  molecular dynamics calculation, 107-111
Isotope exchange method, 402

J

Jump frequency, 5-14, 16, 31
  anharmonic effects, 10-14
  harmonic approximation, 6-10, 87

K

Kirkendall effect, 355

L

Lanthanides
  fast diffusion in, 193-194
Lanthanum
  H in, 292
Lattice vibrations, 14-50, 69
  anharmonic theory, 24-26
  calculations, 44-48
  correlation functions, 41
  defect modes, 18-24, 26-50
  Green's function, 33-36, 40
  harmonic theory, 17-24
  impurity adjacent to a vacancy, 36-39, 46
  inhomogeneity matrix, 33-36
  localized modes, 23, 27-33, 45, 46, 47
  molecular defects, 42-44
Layer structures
  and ionic conduction, 462-471, 473-474
Lead
  cadmium in, 225
  group IIB elements in, 180-181, 184-187
  mercury in, 225
  noble metals in, 177-205, 221-224
Local modes, 270

M

Many-body effects (ΔK), 121, 139
Mass dependence, see Isotope effect
Molecular dynamics calculations, 74-79, 88-104
Molecular statics calculations, 75-77, 84-88, 101
Molybdenum
  H in, 292
Monte Carlo calculations, 75-77, 79-84, 89-105
  defect concentrations, 81-84
  free energy differences, 81-84, 104
Mössbauer effect, 199

N

Nernst-Einstein relation, 310, 396, 413, 416, 451
Neutron scattering, 245
Nickel
  H in, 60, 248, 261, 265, 278-280
Nickel oxide

impurity diffusion, 160-163
self-diffusion, 158
Niobium
  H in, 252-255, 258-260, 261, 265, 267, 285-287
Nitrogen
  traps for H, 267
Noble metals
  electromigration in, 328-334, 348
Nonstoichiometric defects, 157-159
Nuclear magnetic resonance, 245

O

Onsager relations, 308-309, 393
Optical absorption, 402
Ordered alloys, 147
Oxide-ion conductors, 471-473

P

Palladium
  H in, 247, 261-265, 272-277, 325-326
  self-diffusion, 131
Partial molar volume, 198
Platinum
  H in, 293
Point defects, *see also* Impurity diffusion, Interstitials, Vacancy
  phonon theory of, 3
Potassium bromide, 419-421
Potassium chloride, 418-420, 425, 428, 433-434
Precipitation, 386
  kinetics, 200

Q

Quantum effects, 4, 58-69, 121, 146, 270
  quantum statistics, 4, 58-60
  tunneling, 4, 59
  zero-point energy, 60

R

Radial distribution functions, 197-198
Random walk theory, 117

Resistivity measurements, 201
Resonance modes, 145
Rubidium chloride
  sodium diffusion, 155-157

S

Scandium
  H in, 293
Schottky defects, 383
  formation enthalpy, 418
Self-diffusion
  alkali halides, 410-423
  dilute alloys, 140-142
  pure metals, 124-138
    divacancy contribution, 125-131, 134
Semiconductors
  fast diffusion in, 174-177
Silicon
  interstitials in, 47
  localized modes, 45, 47
  self-diffusion, 137
Silver
  Cu in, 373, 376
  H in, 292
  impurity diffusion in, 141
  self-diffusion, 128-131
  vacancy wind in, 373, 376
  Zn in, 373, 376
Silver halides, 149, 153-155
  interstitialcy diffusion, 153-155
Silver palladium
  H in, 266
Silver salts, 448-449, 452-455
Sodium
  localized modes, 46
  self-diffusion, 132-135
Sodium chloride, 151-153, 411-418, 426-427, 430-433
  color centers in, 46
  localized modes, 45, 46
Soret effect, 32, 397

T

Tantalum
  H in, 255-257, 261, 265, 268, 288-290
Thallium, 183, 190
  noble metals in, 225

INDEX

Theory of diffusion, 1-70
  anharmonic effects, 3, 10-14, 24-26, 58-69
  computer calculations, 73-112
  jump frequency, 5-14
  lattice vibrations, 14-50, 69
  preexponential, 49-56
  quantum effects, 4, 58-69
  thermodynamic description, 3
Thermodynamics, *see* Irreversible thermodynamics
Thermoelectric power
  of ionic crystals, 398
Thermomigration, 334, 340, 341
Thin films
  electromigration in, 345-349
Thorium
  H in, 293
Tin, 189
  alloys, 197
  group IIB elements in, 181-183
Titanium
  H in, 268, 293
Tracer contamination, 401, 405
Tracer diffusion, 396
Transference (transport) number, 413, 447
  electronic, 470
Transition metal oxides, 157
Transition metals
  electromigration in, 329, 340-341
Tungsten
  H in, 293
Tunnel structures, 453, 456, 458-461, 476, 479

U

Uranium
  H in, 293

V

Vacancy
  concentrations, 99
  formation and motion energies, 107
  jump calculations, 96-103, 105-107
  persistence of jumps, 95
Vacancy jump time, 91, 94
Vacancy mechanism, 207-210, 382
Vacancy pair, 384, 410
  jump frequency, 105
  in NaCl, 151-155
Vacancy sinks
  solute segregation around, 374-378
Vacancy wind, 353-378
  definition, 355
  measurement of, 369-374
  solute currents generated by, 369-378
Vanadium
  H in, 257, 261, 265, 290
Voids
  in electromigration, 346
  solute segregation at, 377

Z

Zinc
  self-diffusion, 137
Zirconium
  electromigration in, 341-343
  H in, 293